Coloring of Plastics

Second Edition

COLORING OF PLASTICS

Fundamentals
Second Edition

Edited by
ROBERT A. CHARVAT
Charvat and Associates, Inc.
Cleveland, Ohio

A JOHN WILEY & SONS, INC., PUBLICATION

Copyright © 2004 by John Wiley & Sons, Inc. All rights reserved.

Published by John Wiley & Sons, Inc., Hoboken, New Jersey.

Published simultaneously in Canada.

No part of this publication may be reproduced, stored in a retrieval system, or transmitted in any form or by any means, electronic, mechanical, photocopying, recording, scanning, or otherwise, except as permitted under Section 107 or 108 of the 1976 United States Copyright Act, without either the prior written permission of the Publisher, or authorization through payment of the appropriate per-copy fee to the Copyright Clearance Center, Inc., 222 Rosewood Drive, Danvers, MA 01923, 978-750-8400, fax 978-750-4470, or on the web at www.copyright.com. Requests to the Publisher for permission should be addressed to the Permissions Department, John Wiley & Sons, Inc., 111 River Street, Hoboken, NJ 07030, (201) 748-6011, fax (201) 748-6008, e-mail: permreq@wiley.com.

Limit of Liability/Disclaimer of Warranty: While the publisher and author have used their best efforts in preparing this book, they make no representations or warranties with respect to the accuracy or completeness of the contents of this book and specifically disclaim any implied warranties of merchantability or fitness for a particular purpose. No warranty may be created or extended by sales representatives or written sales materials. The advice and strategies contained herein may not be suitable for your situation. You should consult with a professional where appropriate. Neither the publisher nor author shall be liable for any loss of profit or any other commercial damages, including but not limited to special, incidental, consequential, or other damages.

For general information on our other products and services please contact our Customer Care Department within the U.S. at 877-762-2974, outside the U.S. at 317-572-3993 or fax 317-572-4002.

Wiley also publishes its books in a variety of electronic formats. Some content that appears in print, however, may not be available in electronic format.

Library of Congress Cataloging-in-Publication Data:

Coloring of plastics / edited by Robert A. Charvat.– 2nd ed.
   v. cm.
Includes index.
Contents: v. 1. Fundamentals
  ISBN 0-471-13906-8 (v. 1 : acid-free paper)
  1. Plastics—Coloring. I. Charvat, Robert A., 1930–
TP1170.C64 2003
668.4′1—dc21   2003005321

Printed in the United States of America.

10 9 8 7 6 5 4 3 2 1

Contents

Preface   vii

Acknowledgments   ix

Contributors   x

1. **Introduction**   1
   Robert A. Charvat

2. **Color as a Science**   4
   William R. Mathew and Cathy J. Hanlin

3. **Colorimetry, Color Specifications, and Production Tolerances by Visual and Instrumental Means**   23
   Danny C. Rich

4. **Advanced Color Formulation Technology**   48
   Robert Olmsted

5. **Visual Color Matching of Plastic Materials**   61
   Frances A. Leiby

6. **ISO 9000 and Other ISO Standards, Guides, and Test Procedures**   79
   Charles T. Bradshaw

7. **Introduction to Colorants**   85
   Robert A. Charvat

8. **Organic Colorants**   100
   Peter A. Lewis

9. **Inorganic Colored Pigments**   127
   George Rangos

10. **Titanium Dioxide Pigments**   146
    Dwight A. Holtzen and Austin H. Reid, Jr.

11. **Carbon Black Pigments**   159
    Scott A. Brewer

12.	**Soluble Dyes** Craig Weadon and Walter Martin	175
13.	**Photochromic and Thermochromic Colorants** John J. Luthern	185
14.	**Metallic Pigments** Bernhard Klein and Hans-Henning Bunge	202
15.	**Pearlescent Pigments/Flakes** Gerhard Pfaff and Joachim Weitzel	226
16.	**Fluorescence** Christopher Newbacher and Apparao Jatla	242
17.	**Introduction to Colorant Selection and Application Technology** Dennis I. Meade	258
18.	**Color Compounding** Scott D. Russell	268
19.	**Dry Color Concentrate Manufacture** Joseph M. Cameron	276
20.	**Liquid Color Concentrates** Richard L. Abrams	287
21.	**Use of Color Concentrates** Scott D. Russell and Ralph A. Helfer	301
22.	**Product Testing** Ronald M. Harris	320
23.	**Effect of Additives on Coloring Plastics** Bruce M. Mulholland	341
24.	**Recycling** Jack W. Blakeman	353
25.	**The Envrinoment and Government Regulations** Patrick Surgeon	358
26.	**Use of Statistics in the Coloring of Plastics** Thomas V. Edwards	380
	Index	403

Preface

This publication represents the work of many people working in the color and appearance industry. Each one is regarded as most knowledgeable in the field they have covered in their chapters. Their contributions to the basic science of coloring plastics and other polymeric materials will serve industry for years to come.

Coloring of materials must start with an understanding of the basics of light, how light interacts with objects, and, finally, how humans react to and respond to the visible light reaching the eye and then interpreted by the brain. It should not be a surprise to anyone that no two people will see and interpret a colored object exactly the same. It is these variations and others that make the industry of coloring of plastics so interesting. The subjects in this publication have been designed to cover the most important technical and scientific issues involved in the coloring of plastic materials. This publication will be a valuable resource to those looking for information on the many aspects of plastics coloring.

This book covers our understanding of color as a science. It provides the foundation for many additional technological subjects. Measurement information along with matching, visually and instrumentally, will give the reader an understanding of the issues involved. Color specifications and a look at how statistical analysis can improve consistency, not only of colored polymer production runs but also of the colorants used to match the color, are addressed. The basic families of colorants are explained to give the reader an understanding of their properties. Basic information on the techniques usually employed to incorporate colorants into polymers as compounds or concentrates is presented. Environmental issues as well as issues of reuse of discarded materials are covered. An all-important issue, the potential interaction between colorants and other additives, is described to make the reader aware of potential problems with his or her projects. A diligent reader of this volume will come away with an enhanced appreciation of the technology, issues, potential problems, and considerations a colorist must consider if a plastics-coloring project is to succeed.

Volume 2 will cover polymers, by giving an overview of the major polymer issues. The differences between a commodity and an engineering (exotic) polymer will be discussed. The impact of these differences will influence the reader when evaluating a strategy for a coloring of plastic design. The other chapters in Volume 2 will describe the major polymer families. This description will cover such things as properties, advantages disadvantages, and typical markets and applications for the polymer families. This will be followed by general requirements for colorants in these polymers. Next, basic information will be given on methods normally used to incorporate colorants into these polymer. Special issues concerning the coloring of the polymers, including any particular problem or pitfalls that should be avoided are described. This should give the reader or researcher vital fundamental information that will help to avoid major upsets in a coloring project. An important additional piece of data given in each polymer chapter is a coloring matrix. This matrix lists the more important colorant chemical family types and their expected performance in the polymer under discussion. This fundamental information will supply information that will serve as a basic guideline to the reader, thus providing data which will help avoid a catastrophic failure involving the colorants used to color that specific polymer. This should be particularly useful to anyone approaching a coloring project using a polymer and/or colorants new to the reader. Basic attributes such as, but not limited to, colorant heat stability, physical properties, dispersability and any special issues connected to the colorant/polymer combination are covered.

Finally, one should not overlook the important issue that coatings and inks are polymers. It is quite possible the technical and process information contained here will also apply directly or indirectly to these associated polymeric materials. Therefore, this publication should have serious application capability to the coatings, ink and other related industries.

Robert A. Charvat

Acknowledgments

This publication is dedicated to the authors and the many hours they committed to their work on each chapter. At the personal level, I thank the authors for responding cordially to my many calls and reacting professionally to my frustrations in completing this work. Also, a special thanks to my wife, Nancy, for her valuable assistance with the mechanics of good writing.

Robert A. Charvat

Contributors

Richard L. Abrams
Ferro Corporation
7500 East Pleasant Valley Road
Independence, OH 44131

Joachim Weitzel
EMD Chemicals Inc.
Merck KGaA, Darmstadt Germany
7 Skyline Drive
Hawthorne, New York 10532

Jack W. Blackeman
505 Sackett Street
Maumee, OH 43537

Charles T. Bradshaw
Quality Measurement Corporation
428 Spruce Place
Concord, NC 28026-1023

Scott A. Brewer
Cabot Corporation
157 Concord Road
Billerica, MA 01821

Hans-Henning Bunge
Eckhart Americas, L.P.
32644 Luke Road
Avon Luke, OH 44012-1646

Joseph M. Cameron
RTP Company
580 East Front Street
Winona, Minnesota 55987

Robert A. Charvat
Charvat and Associates, Inc.
374 Bradley Road
Cleveland, OH 44140-1149

Thomas V. Edwards
489 Windsor Drive
Elyria, OH 44035

Cathy J. Hanlin
Americhem, Inc.
225 Broadway East
Cuyahoga Falls, OH 44222-0375

Ronald M. Harris
Ferro Corporation
7500 East Pleasant Valley Road
Independence, OH 44131

Ralph A. Helfer
Polyone
2900 Shawnee Industrial Way
Suwanee, GA 30024

Dwight A. Holtzen
E. I. du Pont de Nemours & Company
Chestnut Run Plaza
Wilmington, DE 19880-0709

Apparao Jatla
DAYGLO Color Corporation
4515 St. Clair Ave
Cleveland, OH 44103

* Deceased

Bernhard Klein
ECKART-Wercke
Kaiserstrasse 30
Fürth, D907763
Germany

Francis A. Leiby*

Peter A. Lewis*

John J. Luthern
University of Akron
2174 Thomas Road
Hubbard, OH 44225-9785

Walter Martin
Rainbow Quest Color Consulting
504 Leatherwood Court
Virgine Beach, VA 23462

William R. Matthew
Americhem, Inc.
225 Broadway East
Cuyahoga Falls, OH 44122-0375

Dennis I. Meade
Accel Corporation
38620 Chester Road
Avon, OH 44011

Bruce M. Mulholland
Ticona
8040 Dixie Highway
Florence, KY 41042

Christopher Newbacher
DAYGLO Color Corporation
4515 St. Clair Avenue
Cleveland, OH 44103

Robert Olmsted
GretagMacbeth
617 Little Britain Road
New Windsor, NY 12553

Gerhard Pfaff
Merck KGaA, Pigments Division
Frankfurter Strasse 250
Darmstadt, Germany

George Rangos
Ferro Corporation
West Wylie Avenue
Washington, PA 15301

Austin H. Reid, Jr.
E. I. du Pont de Nemours & Company
Chestnut Run Plaza Bldg. 709
Wilmington, DE 19880-0709

Danny C. Rich, Ph.D
Sun Chemical Corporation
631 Central Avenue
Carlstadt, NJ 07072

Scott O. Russell
Polyone
33587 Walker Road
Avon Luke, OH 44012

Patrick Surgeon
323 Danielson Way
Chandler, AZ 85225

Craig Weadon
Polysolve, Inc.
Crown Point, IN 46307

CHAPTER 1

Introduction

Robert A. Charvat

Charvat and Associates, Inc.
374 Bradley Road
Cleveland, Ohio 44140-1149

The goal of this textbook is to expose the reader to the many aspects of coloring of plastics. To accomplish this objective, the technologist must understand colorants for what they are and what they are not. He or she must understand the performance of colorants not only during the processing and manufacturing steps but also during the life cycle of the final product. Today it is important to consider the issues of recycling the product after its useful life has come to an end. This publication will not make the reader a world-recognized expert. However, the color technologist will find useful information within these pages. The information will improve his or her capabilities, as the knowledge of many technical experts are contained here. This book should be a resource center, or a starting point, for anyone beginning a coloring of plastics project where they are proceeding into unfamiliar territory. This book should also provide support to the accomplished colorist who desires to fill in that one area or areas where his or her background is not as strong or complete as it might be.

The creation of this publication started with the Color and Appearance Division (CAD) of the Society of Plastics Engineers. In 1979, a small volume was published titled *Coloring of Plastics*. This volume, with numerous authors, was edited by Thomas G. Webber under the sponsorship of the Color and Appearance Division. This was the first publication dedicated totally to the coloring of plastic materials. Thomas Webber's book was the first and only volume truly focusing on the subject. This book was the primer for the coloring of plastics industry. However, many years later the book was hopelessly outdated. The need for such a volume now is as great,

---

Coloring of Plastics, Fundamentals, 2nd edition. Edited by Robert A. Charvat
ISBN 0-471-13906-8   Copyright © 2004 by John Wiley and Sons, Inc.

if not greater, than ever. The CAD Board of Directors has promoted the preparation of a new, comprehensive, and up-to-date book for some time. The CAD Board of Directors approved the development of a new *Coloring of Plastics* book. This publication hopes to meet the challenge presented by the CAD Board of Directors. The board also selected the Chairman of its Education Committee Robert A. Charvat as editor. This started the long road of the development of a new *Coloring of Plastics* book under the sponsorship of the CAD.

The organization of the new book needed to include a number of issues and subjects not covered or not known at the time of the Webber book. The technology of colorants is light years ahead of where it was those many years ago. The ability to measure color and colored materials has made significant advances. The ability of computers to match colors accurately and quickly is standard procedure today but was very difficult, if not impossible, years ago. Our understanding of how we see color is significantly better now than years ago. The number of polymers available today is tremendously larger than at the time of the *Coloring of Plastics* book by Webber. The ability of new polymers to perform in demanding applications requires colorants also have the ability to meet these same demanding requirements. Keeping up with the introduction of these new polymers and their new applications is a problem not faced years ago.

All the above is prolog to this publication. This new publication is divided into two major parts. Volume I deals with the many technology issues that impact the coloring of plastics today. Volume II will cover the major polymer families and deliver *basic* information on the coloring of those polymers.

Volume I covers our understanding of color as a science. It provides the foundation for the many additional technological subjects. Measurement information along with matching, visually and instrumentally, will give the reader an understanding of the issues involved. Color specifications and a look at how statistical analysis can improve the consistency not only of colored polymer production runs but also of the colorants used to match the color are addressed. The basic families of colorants are explained to give the reader an understanding of the properties of these families. Basic information on the techniques usually employed to incorporate colorants into polymers as compounds or concentrates is discussed. Environmental issues as well as reuse of discarded materials are covered. An all-important issue, the potential interaction between colorants and other additives, is described to make the reader aware of potential problems with his or her projects. A diligent reader of Volume I will come away with an enhanced appreciation of the technology, issues, potential problems, and considerations a colorist must take into account if the coloring of a plastics project is to succeed.

Volume II will consider polymers, by giving an overview of the major polymer issues. The differences between a commodity and an engineering (exotic) polymer will be discussed. The impact of these differences will influence the reader when evaluating a strategy for the coloring of a plastic design. Volume II also describes the major polymer families. This description will cover such things as properties, advantages, disadvantages, and typical markets and applications for these polymer families. This followed by general requirements for colorants in these polymers. Basic information will be given on methods normally used to incorporate colorants into the polymers under discussion. Special issues concerning the coloring of polymers, including any particular problems or pitfalls that should be avoided, are described. This should give the reader or researcher vital information that will help to avoid major upsets in a coloring project. Each polymer chapter has a coloring matrix, which lists the more important colorant chemical family types and their expected performance in the polymer under discussion. This will supply information that will serve as a basic guideline to the reader, providing data that may help avoid a catastrophic failure involving the colorants used to color a specific polymer. This should be particularly useful in a coloring project using a polymer and/or colorant new to the reader. Basic attributes such as, but not limited to, colorant heat stability, physical properties, dispersability, and any special issues connected to the colorant/polymer combination are covered.

Finally, it is important to note that coatings and inks are polymers. The technical and process information contained here may also apply directly or indirectly to these polymeric materials. Therefore, this publication should have serious application capability to the coating, ink, and related industries.

CHAPTER 2

Color as a Science

William R. Mathew and Cathy J. Hanlin

Americhem, Inc.
225 Broadway East
Cuyahoga Falls, Ohio 44122-0375

2.1. INTRODUCTION

In our everyday life we are surrounded by the sights and sounds of our world. For the most part we take for granted the complexity of these sights and sounds and how our lives are impacted by our surroundings, that is, until we decide to take music lessons of some sort or we try to match the paint we bought three years ago to paint our family room. Then these complexities take on a new reality for us. For those of us in industry who have to deal with color problems every day, it is extremely frustrating how much the "science of coloring plastics" is trivialized. The result of this trivialization is often a crises situation where "color" was considered a nonissue until parts are rejected for color: "They don't look the same as they did in my office" or "All we did was change to a different polymer. That should not impact the color, should it?" or "You mean it makes a difference if I use the 2° observer instead of the 10° observer in my color difference calculations?" We could list hundreds of quotes to make our point, but it is not our intent to "preach" to the reader. We only wish to establish that the coloring of plastics is an elaborate, multivariable puzzle that requires a sound scientific approach and a clear understanding of all the components that make up this intricate world of color. It is the intent of this chapter to establish for the reader an accurate perception of the complexity of color science as well as a means to deal with this science in a logical manner. We will do this by helping the reader to gain a clear understanding of the variables associated with color science and how they interact. It is not the intent to re-create a text in detail

---

Coloring of Plastics, Fundamentals, 2nd edition. Edited by Robert A. Charvat
ISBN 0-471-13906-8   Copyright © 2004 by John Wiley and Sons, Inc.

of the science of color, for there are already several good books on the subject and we will reference these as we go along. Now, let us begin to explore this complex world of coloring plastic materials.

2.2. THE TRIAD

As we see color from strictly a physiological point of view we must consider a special "triad." This triad consists of a source of energy (the *light source*), an object that is illuminated by the energy (the *object*), and a detector (the *Observer*). The observer could be a human observer or alternatively a photosensitive detector attached to a computer. In the case of the human observer the eye is the detector, with the brain as the perceiver of the information sent to it from the eye. This combination of the eye and the brain creates the unique situation of *interpretation* of the physical information. This means that color exists in the mind of the viewer. This adds another dimension to color, the psychological dimension. We will cover this in more detail in another section. It is key to remember as we go through the discussion of the physical aspects of color that this human interpretative quality is, for us, the most important aspect. This triad is in a constant state of interaction. When it comes to the physical color stimulus, the three components of the triad do not act independent of each other. This means that as one progresses through this chapter, you keep in mind this constant interaction between the light source, the object, and the observer.

2.3. THE LIGHT SOURCE

Even though its the same color, it looks different. Why? Many conditions affect the way color appears. One of these is the light source. For instance, an apple may appear a bright delicious red under the sunlight at the farmer's market, but somehow does not look as good under the fluorescent lights at home. Except in rare situations, we do not see color without light. Furthermore, the light we see depends upon the characteristics of the light source under which an object is seen.

The light source is that which illuminates the object we are viewing. This light source emits energy in the electromagnetic spectrum. Light is the segment of electromagnetic radiation that also includes X rays, ultraviolet and infrared radiation, radio and TV waves, and cosmic radiation. The human eye can respond to electromagnetic radiation between 380 and 780 nm as light (colorimeter range is 400–700 nm). The part of this continuum that we are interested in is known as the visible spectrum. This is the portion of the electromagnetic to which our eyes respond. The visible spectrum, like all other portions of the entire spectrum, are divided into small segments that are described either by their frequency (cycles per second) or by the wavelength of one cycle. For the visible portion of the spectrum wavelength is most used. These wavelengths are expressed in units called nanometers ($1 nm = 10^{-9} m$). Light with a short wavelength appears blue or violet. As the wavelength increases, the color appears to change through green, yellow, orange, and red. Radiation combining all the wavelengths of the visible spectrum in about equal amounts is

6   Color as a Science

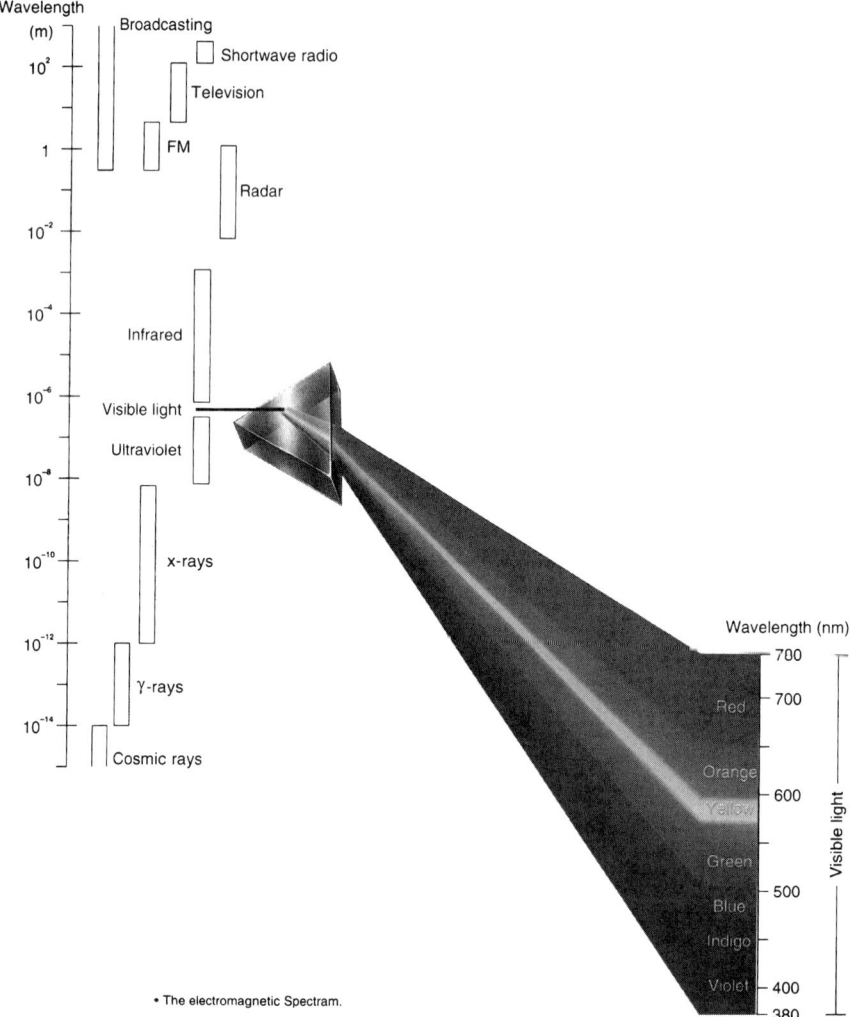

Figure 2.1. The electromagnetic spectrum and the portion that we see as visible light (Minolta, 1993).

perceived as white. This is best described with an illustration (see Fig. 2.1) (Minolta, 1993).

Radiant energy also affects the perception of color. More important to the perception of color than the radiant energy is the spectral distribution of the radiation from the light source. Everyone has experienced a colored object looking different in the sunlight, under cloudy skies, under fluorescent light, or under normal electric light bulbs. It is therefore essential to select a light source before defining color tolerances, whether these are for use in visual assessment or in colorimetry. Even northern light varies greatly depending on the time of day or season for reproducible result to be obtained, so as a general rule, an artificial light source is used.

Light sources can be described by their spectral power or energy distribution. This simply plots the amount of light (relative power) as a function of wavelength. Figures 2.1–2.5 are examples of standard light sources (Billmeyer and Saltzman, 1981).

For artificial light sources, two values are important: the color temperature in degrees Kelvin and the relative spectral distribution of the light source $S(\lambda)$, where λ shows that the radiant energy is dependent on wavelength.

2.3.1. Blackbodies

An important group of light sources are called *blackbodies*. We will not take the time here to discuss in detail the science of these sources, because they are well described in other texts (e.g., Billmeyer and Saltzman, 1981; Judd, 1964). The color temperature of a lamp is based on the radiation emitted by a blackbody. The important point to note here is that these sources are based on absolute temperature in degrees Kelvin (K = degrees Celsius + 273) of standard blackbodies such as tungsten-filaments. Thus, a 6500 K (D65) source is a blackbody heated to 6500 K. Figure 2.2 shows two such sources. The main principle to gain from this section on the light source is; each source has its own unique spectral power distribution and therefore will interact with the other two components of the triad according to this uniqueness. In other words, an object illuminated with a 6500 K source will appear different from the same object illuminated with a 2854 K source.

The total amount of energy from the source is also an important factor in the perception of color. If the energy is too low, we are not able to see the full color results. If the energy is too high, we are "blinded by the light." The ASTM standard D 1729-89 addresses the issue of level of illumination under section 5.1.2, Photometric Conditions; this is an excellent practice to adopt for color laboratories as it focuses on the visual evaluation of color in a controlled environment (ASTM D 1979-89).

The CIE (Commission Internationale de l'Eclairage) has selected a number of light sources from the wide range available with different spectral distribution and color temperature. The use of these standard illuminants is recommended and those used most commonly are listed here:

- Standard illuminant D65 (6500 K) represents average daylight.
- Standard illuminant A (2856 K) represents evening lighting in rooms (e.g., normal light bulb).
- Illuminant F2 (4230 K) refers to a cool white fluorescent (CWF) lamp.
- Illuminant F11 (4000 K) refers to a triple-band lamp (TL).

D65 is considered to be a better representation of natural daylight than illuminant C, which has a lower proportion of ultraviolet (UV) radiation than D65 (Huff, 1994).

Color assessment should be carried out using the illuminants specified and in a color assessment booth, such as is available from a number of manufacturers. The booth is usually equipped with at least three lamps to produce the various illuminants. A UV lamp is normally included as well. The inside of the color booth is painted neutral gray in order to prevent inaccuracies produced by reflection from the walls. *External light must be blocked out.*

8   Color as a Science

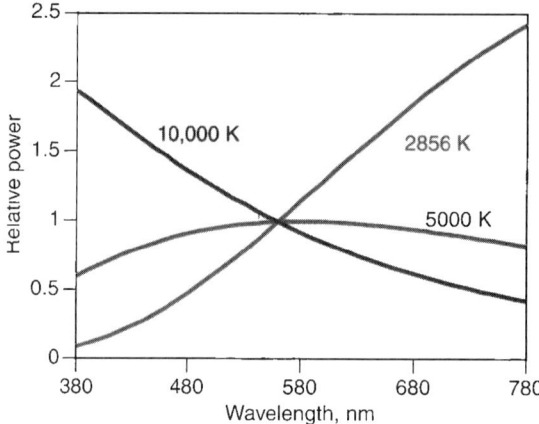

Figure 2.2. Spectral power distribution of blackbodies with color temperatures of 2854 K (source A) and 6500 K (Pivovonski, 1963; Billmeyer and Saltzman, 1981).

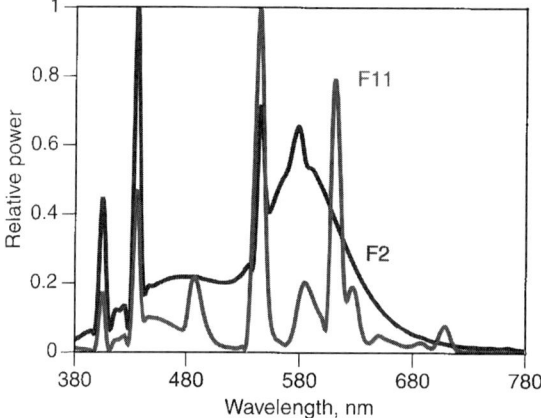

Figure 2.3. Spectral power distribution of cool white fluorescent lamp (IES, 1981; Billmeyer and Saltzman, 1981).

The objects that we often consider sources of light (the sun, light bulbs of many types, etc.) are seen by our eyes as white or almost white. Through a series of experiments using a prism, Newton discovered that this white light actually consists of all of the visible wavelengths described above (Newton, 1730). This is true when the source is said to be polychromatic, meaning it contains all wavelengths of light from 400 to 700 nm. It is important to note that some sources that "appear" to be white may not contain all of the wavelengths of the visible spectrum. Some of the mixed halide lamps such as sodium and mercury are such sources. These present problems when viewing objects because they are not polychromatic. Later in the chapter we will discuss why this is true.

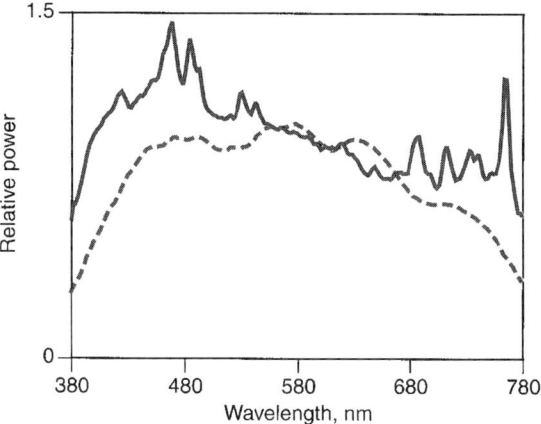

Figure 2.4. Spectral power distribution of a typical line source, a mercury arc lamp (IES, 1981; Billmeyer and Saltzman, 1981).

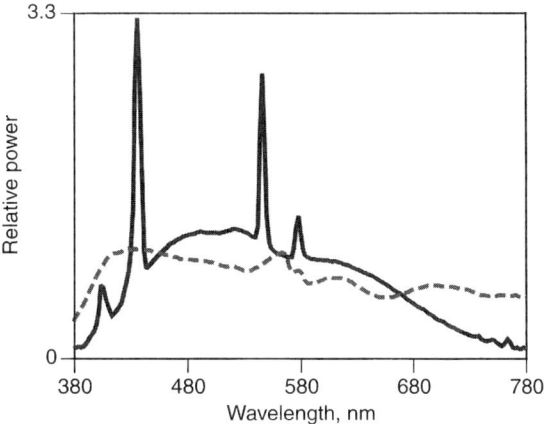

Figure 2.5. Spectral power distribution for some standard light sources used in describing color (Wyszecki, 1967, 1970; Billmeyer and Saltzman, 1981).

2.4. THE OBJECT

The object is the second part of our triad. Here we will discuss how materials interact with the energy from the light source. In this book we consider objects made of polymeric materials. Generally, polymers are colorless or at best weakly colored, and the aim always seems to be to "cover up" the undesirable color of the polymer in favor of the more desirable color selected by the designers. This requires the addition of colorants (pigments and dyes) to the polymer. The subject of colorants will be covered in depth in another chapter. However, because colorants become part of the object, we will first discuss some aspects of how they interact with light.

2.4.1. Refractive Index and Transmission

Looking at the model in Figure 2.3, note that the first few micrometers of the surface of our object is polymer rich. This fact becomes very important as we consider how our object interacts with light at the surface of the model. When energy strikes our model, one or more things occur. First the energy can be *transmitted* through the object. The energy passes through the object and is essentially unchanged; however, a small amount (about 4% for refractive indices of 1.5) is scattered at the surface of the object (Fig. 2.4). This scattering is due to a property of the material referred to as the *refractive index*. This property is why we must consider what happens at the surface of our model. Two materials (two different polymers) having different refractive indices will scatter light differently. This means that the two materials would have different appearances even though they could be colored with exactly the same colorants. Because scattering at the surface is wavelength dependent, which is why a prism can break white light into its components (Newton, 1730), if our two materials above are viewed at different angles, their appearance will change in relationship to the angle at which they are viewed. This is often called "flop." This property has a pronounced effect on both visual and instrumental color assessment, especially in darker, more chromatic colors. This subject is covered in detail by Donald and Mathew (SPE, 1989), who give a detailed discussion of the impact of the refractive index on color measurement. The important principle to remember here is that the refractive index is a property of the material (polymer) and light will interact at the surface differently if we have two different materials possessing different refractive indices.

2.4.2. Absorption

The second way a material can interact with light is by *absorption*—the conversion of visible light to longer wavelengths (heat) or the energy of the visible light being used in a photochemical process called polymer or colorant degradation. If the material absorbs only part of the light, it will appear colored. For example, if the object absorbs all but the green portion of visible light, the object will appear green. This of course assumes the light source is polychromatic and contains the green portion of the spectrum. If the light source would happen not to contain this green portion of the spectrum, then our object would appear to be black or gray. This is why it is important to understand the makeup (spectral energy distribution) of the light source. If the object absorbs all of the visible spectrum of a polychromatic source, then the object will appear black.

There are two fundamental laws to consider when we discuss absorption: Beer's law and Lambert's law. Lambert's law states that equal thickness of materials cause equal amounts of absorption. Beer's law states that equal amounts of absorbing material cause equal absorption. Figure 2.5 illustrates these laws (Billmeyer and Saltzman, 1981). Both laws will only work in the absence of scattering and are not applicable in opaque or translucent materials. They are very useful in transparent materials, where the scattering due to the colorants and the polymers themselves (low refractive indices) is very low.

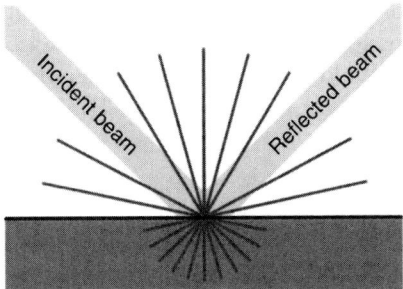

Combination of diffuse and specular reflection due to scattering from beneath, plus reflection from, a smooth surface.

Figure 2.6. Illustration of relationship between the particle of a material and its scattering ability Billmeyer and Saltzman, 1981).

2.4.3. Scattering

The last and most complicated interaction that our object has with light is *scattering*. For the most part scattering in plastic materials is caused by the inclusion of particulate matter into the plastic. This particulate matter we call pigments and fillers. Pigments, like all other forms of matter, have the properties of refractive index, transmission, absorption, and scattering. When we put pigments into a plastic material of a given refractive index, we are introducing materials of a different refractive index, different absorption, and different scattering (see Fig. 2.3). This means the color of our object depends on the accumulative amount and kind of scattering and absorption that take place. If there is little to no absorption and about equal amounts of scattering at all the visible wavelengths, then the object appears white. If some of the visible light is absorbed by the pigment, then the object appears colored. The scattering in these cases is caused by light falling on pigment particles that have different refractive indices than the surrounding plastic medium (see Fig. 2.6) (Billmeyer and Saltzman, 1981). The amount of light scattered depends on two factors: the difference between the two refractive indices and the particle size of the pigment. The relationship between scattering and these two properties is illustrated in Figures 2.6 and 2.7 (Billmeyer and Saltzman, 1981). It is important to note that these two properties will act independently of each other. For example, the same pigment (same refractive index) with a different particle size will scatter light differently. This means that if scattering properties (particle size) vary from lot to lot of pigment, then the inclusion of that pigment into a plastic medium will also appear different. This is only one of the issues associated with pigment that have an impact on the appearance of an object. Another aspect of pigments that needs to be noted is that they all do not "produce" color by the same physical mechanism. Nassau, in his book *The Physics and Chemistry of Color*, discusses the 15 causes of color. Not all of them pertain to pigments, but a number do. For example, some pigments (e.g., zinc oxide, cadmium yellow, and cadmium orange) produce color because they are semiconductors. When they are illuminated with an energy source (light), electrons move from the valence band to the conducting band through the band gap. Depend-

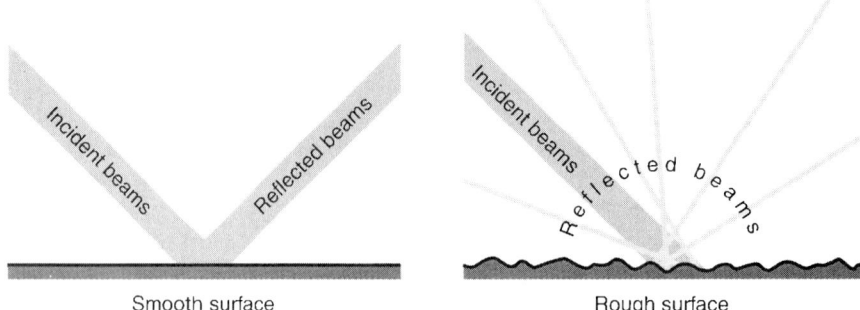

Figure 2.7. Illustration of relationship between two materials and their refractive indexes (Billmeyer and Saltzman, 1981).

ing on the width of this gap, some of the visible light will be absorbed while the rest is reflected back. Cadmium yellow has a band gap of 2.6 eV (electron volts). This gap absorbs energy in the violet and blue part of the spectrum, leaving all others, which leads to its yellow color. Reduce the band gap slightly to 2.3 eV and it absorbs violet, blue, and green, leaving the characteristic color of cadmium orange. The color produced in other pigments will be caused by other methods, such as transition metal compounds or transitions between molecular orbitals of organic compounds (Nassau, 1983). The point is that because these materials interact with light in different ways, the object, depending on its make up, will have a very complex relationship with the light source.

All of the above relationships create a situation where the assessment and control of colored plastic materials are not trivial issues. Just like trying to figure out your phone bill, sorting out problems associated with just the object itself can be quite a task. A check list of questions is provided at the end of this chapter that we hope will be helpful in working through some of these complex issues. The main principle to take from this section on the object is that objects interact with light from the source in multiple ways and the refractive index of the polymer contributes to this interaction.

2.5. THE OBSERVER OR DETECTOR

We will discuss two types of observers: the human observer and the instrumental observer. In visual assessment the observer is the human eye, and in colorimetry it is the instrument receiver. Because the human observer is much more complex, we will start there.

2.5.1. Human Observer: Physiological Response

Given the importance of color, it is amazing how little we understand its nuances or appreciate its power. What is color, anyway? Against all intuition, it turns out to be not a quality of objects but rather an attribute of our brains. Color is a paradox. It exists in light, which to the human eye seems colorless. It does not exist in a rainbow, apple, grass, or paint, which appear colored. Few objects possess inherent color or pigment. Color is seen when light waves reflected off an object meet the eye. An object appears colored because it absorbs some wavelengths and reflects back others. Think of a green leaf. When light hits it, only green rays are reflected from its surface. All other rays are absorbed by the leaf. We classify it as "green" when in reality it is every color but green.

Human vision has laws that may vary from person to person. The laws of physics play a part in our perception of color, but they only provide the starting point for a process that is influenced by the physiology of the eye and the psychology of humans. In broad terms, the physiology of the visual system has been understood for a long time. Only recently however, has progress been made in identifying the substance in the eye that is specifically responsible for sensitivity to see one of the three primary colors: red, green, or blue. Jeremy Nathans of Stanford University identified three genes that enable the eye to see color. The genes direct cells in the retina of the eye to produce three pigments, each sensitive to one of the three primary colors. The existence of pigments had been previously deduced but never before demonstrated.

In humans, light from the sample passes through the pupil of the eye and through the lens to be projected with variable focal lengths on the retina, where, in the case of the normal observer, it is brought into focus. This is similar to a picture that is thrown upon a film by a camera. In the retina at the back of the eye there are two types of photosensitive receptors: rods and cones. The rods, named because of their shape, are responsible for colorless vision in conditions of dim illumination; the cones, operating at higher light levels, are responsible for color perception. Visual acuity is greatest in the fovea, where the density of the receptors is greatest.

The first stage in the transformation of light into neural impulse is a photochemical process that involves the breakdown of various visual pigments that are later resynthesized. One such pigment is rhodopsin, the photochemical substance contained in the rods. This process initiates the electrical message of the rods to be conveyed to the brain. The cones, the color receptors, are further specialized into three types: blue, green, and red. Each type has a different pigment making it sensitive to a different color of light. Their combined power is such that a person with normal vision can distinguish millions of hues, shades, and variations of color. Only when the light and color signals from the eye reach the brain do we "see" color (Fig. 2.8) (Billmeyer and Saltzman, 1981).

2.5.2. Human Observer: Psychological Response

Color is essential to full enjoyment of the world. How we learn to classify and label colors is based largely on our culture. Because colors are not really "out there" but are constructed in our retinas and brains, it stands to reason that culture and psychology play a great role in our perceptions of color. People are drawn to color from an emotional standpoint. Color preferences are rooted in associations arising from our culture, how we were raised, our feelings about ourselves, and what we were taught.

The different wavelengths of light affect our brains differently as well, triggering specific responses in our bodies. For example, the longer wavelengths of red light "excite" the brain and stimulate the heart and nervous system. Colors we wear affect

14   Color as a Science

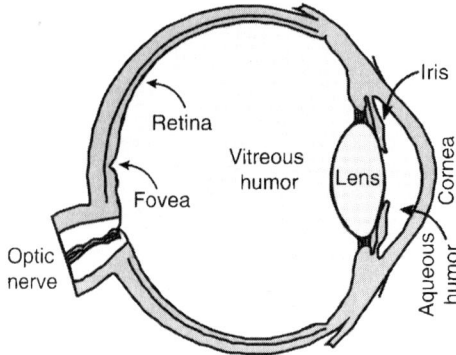

The cross section of the human eye.

Figure 2.8. Cross section of the human eye (Burnham, 1963; Billmeyer and Saltzman, 1981).

our bodies, our minds, our moods. Bright, clear colors help us feel more joyous and light hearted and can actually counteract depression. Blacks, dark grays, and browns and dull, muddy colors tend to depress us, while the green tones are healing.

Studies have been done to determine the effects of colors on humans. For example, when preschool children in New Zealand were placed in a pink room, their physical strength and positive mood increased; the reverse was true of children placed in blue rooms. Blue light, with its shorter wavelengths, tends to "calm" the brain. Green rooms where actors wait before a performance are painted that color because green has a restful effect without the sometimes depressing effect of blue.

Color selections almost always send symbolic messages. Bright, clear colors are positive in nature and action, pure and clear, untainted by ulterior motives. Pastels show immaturity or weakness in the area signified by the color. Dark or muddy colors reveal the negative qualities of doubt, fear, hate, anger, greed, and so on. Wearing these colors will intensify these feelings. Extremely vivid implies an impending lucid state. Colors are indicative of our emotions, each hue having a different meaning taken from the actual observation of clairvoyants who long ago noted the correspondence between certain feelings and the resultant colors visible in the aura.

It is interesting to study some of the meanings related to colors. For example, black denotes the unknown, the mysterious, darkness, death, mourning, hate, or malice, especially when associated with fear or uncertainty. However, in dreams, if the feeling associated with black is joy or happiness, it is thought to imply unmanifested spiritual gifts or qualities. Blue is thought to represent awakened spiritual forces, one who has found his or her life work, also considered the color of honest intentions and mild passions. Brown is earthy or worldly, physically oriented, practical, and materialistic. Tan is associated with purity and noble ideals tinged with doubt, depression, and earthly reason ("I'm only human" or "I must be realistic"). Green is often associated with vigorous growth, renewal and immortality, or youth's inexperience. It is associated with both healing and sickness. Yellow symbolizes hope and a bright future, associated with enlightenment and wisdom. Red is a particularly strong color that commands attention. It is the symbol of activity and power, energy, warmth, talent, and courage. On the negative side it conjures up violence, passion, and anxiety. Purple is a mystical and religious color. Purple partakes of both the integrity of blue and the power of red. Today associations to purple are mainly between faith and spirituality and dreams or superstition. White is purity, the non-color of light. It is associated with innocence. On the other hand, it is also seen as the white flag of surrender or whitewashing or covering up. These have even filtered into our language in expressions such as "green with envy," "seeing red," "having the blues," and so on.

Color is used widely in business and advertising. Advertising provides the best examples of the use of color in its various functions: to attract attention, as a vehicle for encoding information, and as a rich source of symbolism and evocation. Color, working on a subliminal level, is frequently the first thing perceived in an advertisement and can, in a fraction of a second, establish the content of the image and suggest a range of values to the customer.

Color can even prove beneficial for your health. Doctors have long made use of the healing power of light by manipulating wavelengths outside the visible spectrum. X rays, lasers, and radiation treatments have been a standard part of medical practice for years. Some believe that just as lasers can use the power of invisible light to cut through diseased tissue, visible light can be used to alleviate a variety of ailments through its effects on the central nervous system. In ancient times multi-chamber temples of color were used for physical and spiritual rejuvenation. Today color therapists have adapted ancient Hindu belief that the body's internal balance is maintained through seven chakras, or energy centers, which respond to different frequencies of light that we experience as colors.

2.5.3. Human Observer: Color Deficiencies and Color Blindness

About 8% of males and less than 1% of females are born with defective color vision. As was discussed earlier, a person with normal color vision—the trichromat—has three kinds of color-sensitive cones: red-, green-, and blue-sensitive cones. Dichromatism is when only two colors are perceptible and one color is difficult to recognize.

Color blind is a loose term because it implies a complete lack of ability to see color, and total color blindness is rare. Because color deficiencies do occur within the population, it is important to know this for individuals who are matching or evaluating color. The Farnsworth–Munsell 100-hue test is an excellent tool for both color discrimination and analysis of color vision defects. Because the observer is a significant variable in the color equation, it is important to understand their color vision capabilities.

2.5.4. Instrumental Observer

After looking at the complexity of the human observer, the instrumental observer may seem simple in comparison. Colorimetry is an attempt to express numerically that which is seen by the human eye. Today there is virtually no industry in which color does not play a major role and in which a colorimetric system is not used. One of the difficult problems we face today in the plastics industry can be attributed to this fact. This attempt to mimic the human interpretation of color via numeric values as seen through the eyes of a mechanical device is asking for a lot. Most, if not all, the spectrophotometers today do a commendable job of analyzing how objects interact with light. They describe this interaction through the result of a spectral reflectance or transmission curver. However, to our brains this is only a line on a chart and has little to do with color as we see it. A spectral curve cannot describe the beauty of a pink rose or the mystery of deep brown eyes. These are qualities left only to our brain. As we described above, real color exists in the mind of the human observer. Even with all of this in mind, humans appear to have a need to apply a quantity to everything, and so we use the spectrophotometer. Spectrophotometry will be dealt with in detail in another chapter of this book. Here, we want to point out that the difference between the instrumental observer and the human observer is significant.

Nonetheless, plastics manufacturers frequently employ colorimetric techniques for calculating and correcting color formulations and for quality control. Plastics processors and industrial end users are mainly interested in the application of this science in quality assurance. The advantages of colorimetry are obvious. Time is saved in formulating new colors, and there is less need for recoloring work during production. Also, it provides the basis for statistical quality control and for objective discussion between supplier and customer. Unless one considers the previous paragraph, the disadvantages of colorimetry are not so evident at first glance. They include measurements that do not correspond to or even appear to contradict the visually perceived color and the variation of measurements obtained by supplier and customer.

Even today, despite enormous progress made in recent years, the numerical expression of a sensory perception is still subject to certain preconditions, not least because the standard colorimetric systems in use still have faults and limitations. Unfortunately, many manufacturers of colorimeters contribute to the uncertainty with impressive software and by claiming a degree of accuracy to so many decimal places that simply is not possible. Users often trust blindly in the manufacturer's claims, although any number of good books are available that describe the problems of colorimetry.

The lack of correlation of the current mathematical models of converting spectrophometric data to visual sensory perception is a very big contributor to the problems of colorimetry. As we will see later, a number of factors contribute to this correlation problem, the major one being that humans do not see uniformly over the entire visible spectrum. This necessarily means that our model must accommodate this nonuniform behavior, which turns out to be a difficult mathematical problem.

2.6. DESCRIPTION OF COLOR AS A LANGUAGE

Like any foreign language, color has its own terms and definitions. When colors are classified, they can be expressed in terms of their hue (color), brightness (lightness), and intensity (saturation or chroma). Hue is used in the world of color for the classification of pure color, like red, green, blue, or yellow. Brightness refers to how light or dark a color is. Intensity is seen in terms of saturation: vividness or dullness. Each can be measured independently of the others. Figure 2.9 and illustrates how these

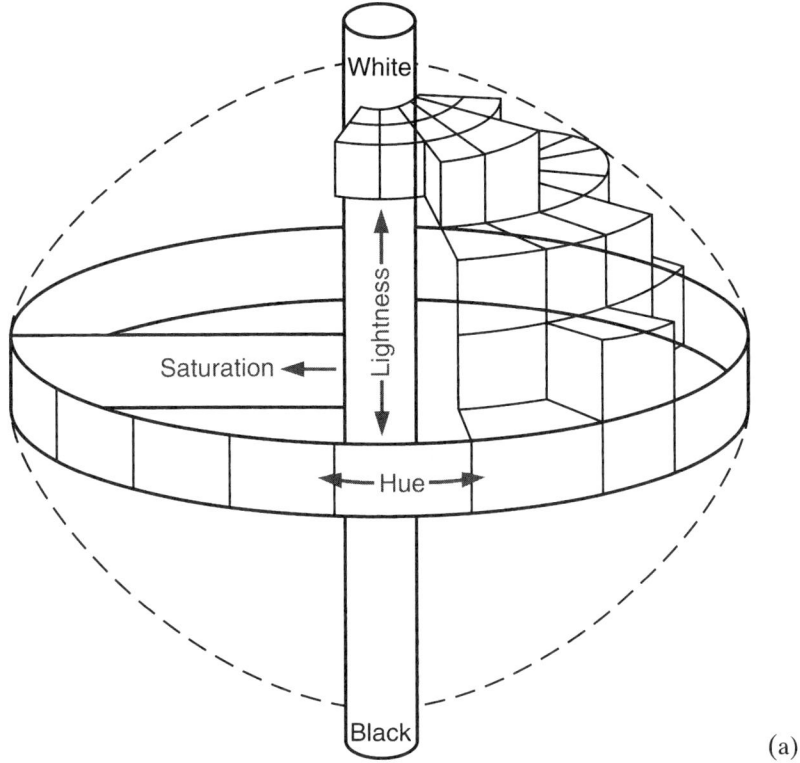

Figure 2.9. Three-dimensional color space (*a*) and color solid (*b*) of hue, lightness, and saturation (Minolta, 1993).

descriptions interact in a three-dimensional system often called a three-dimensional color solid or space (Minolta, 1993).

The everyday names given to shades of color—such as sky blue, aquamarine blue, or turquoise blue—are hopelessly inadequate to meet the needs of industry and science. People can rarely agree on the exact shade the names given to colors describe. One of the most determined attempts to formulate a precise nomenclature for color was made by American artist Albert Munsell. He devised a three-dimensional color system, in the form of a tree, that classifies shades according to perceivable characteristics: hue, brightness, and chroma. The system is referred to as the Munsell system, where 20 colors (hues) are represented by transparent plastic branches. Values (brightness) are indicated by the vertical position on the hue branch. Chroma, or intensity, is indicated by how far the square is from the tree's center. Intensity scales are of three types: (1) tints, the range from a pure hue to white; (2) shades, the range from a pure range to black; and (3) tones, the range from a pure hue to any gray.

By creating scales for hue, lightness, and chroma, we can now measure color numerically. Other methods for expressing color numerically were developed by the Commission Internationale de l'Eclairage (CIE), an international organization concerned with light and color. The two most widely known of these methods are the Y_{xy} color space, devised based on tristimulus values XYZ defined by CIE, and

18   Color as a Science

(b)

If we look for the color of the apple on the color solid, we can see that its hue, lightness, and saturation intersect in the red area.

Figure 2.9. (*continued*)

the $L^*a^*b^*$ color space created to provide more uniform color differences in relation to visual differences. Color spaces such as these are now used throughout the world for color communication.

The concept of the XYZ tristimulus values is based on the three primary colors (red, green, and blue) and that all colors are seen as mixtures of these three primary colors. The tristimulus values are useful for defining a color, but the results are not easily visualized. Because of this, in 1931, the CIE defined a color space for graphing color in two dimensions independent of lightness. This is known as the CIE chromaticity diagram. In this model, achromatic colors are toward the center of the diagram, and the chromaticity increases toward the edges. When we measure an object, the values obtained correspond to a point of the graph. Figure 2.10 illustrates this two-dimensional color space (Minolta, 1993).

In 1964 the CIE changed its recommendation to a different observer. This new observer, called the 10° standard observer, provides a more accurate correlation to visual perception and should be the observer of choice in most, if not all, colorimetry work.

The $L^*a^*b^*$ color space (also know as CIELAB) is presently the most popular color space for measuring object color and is used in many fields. In this color

Description of Color as a Language   19

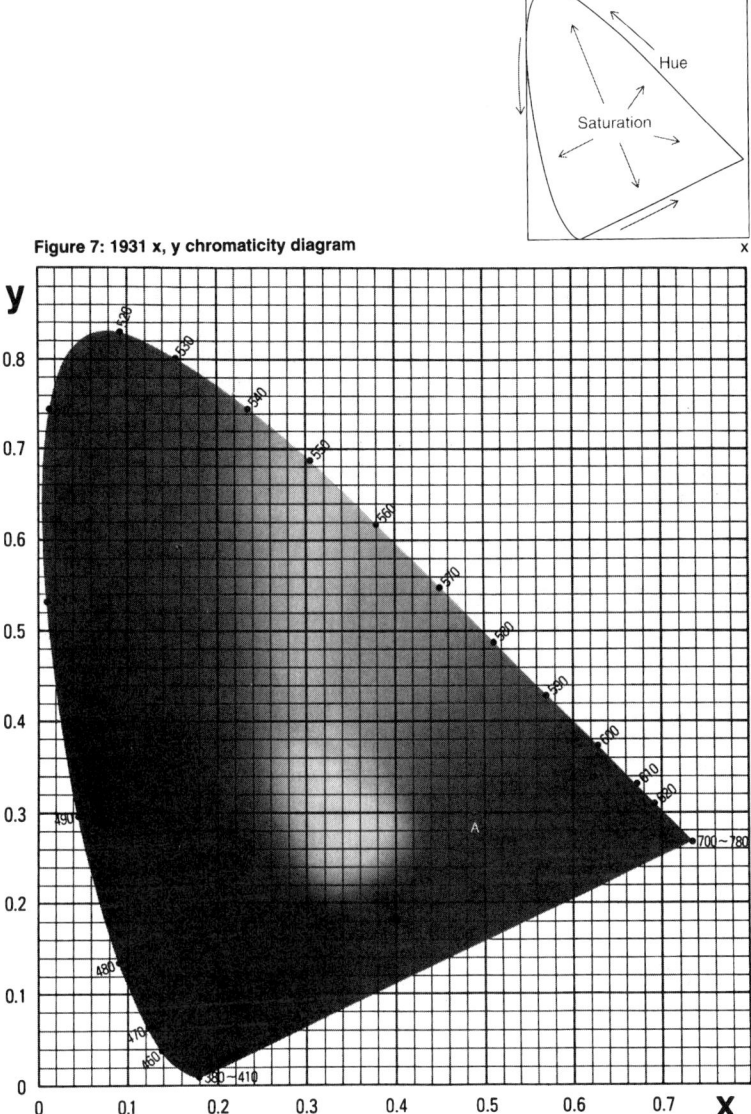

Figure 2.10. The 1931 x, y chromaticity diagram (Minolta, 1993).

system, L^* indicates lightness and a^* and b^* are the chromaticity coordinates. The red direction is represented by $+a$, $-a$ represents the green direction, $+b$ indicated the yellow direction, and $-b$ represents the blue direction. The center is achromatic. Figure 2.11 is a representation the $L^*a^*b^*$ color solid (Minolta, 1993).

The mathematical calculations of the values for XYZ and for the all the above-mentioned color spaces are well documented and do not need to be repeated here. It is important to note that these are all mathematical attempts to simulate

20 Color as a Science

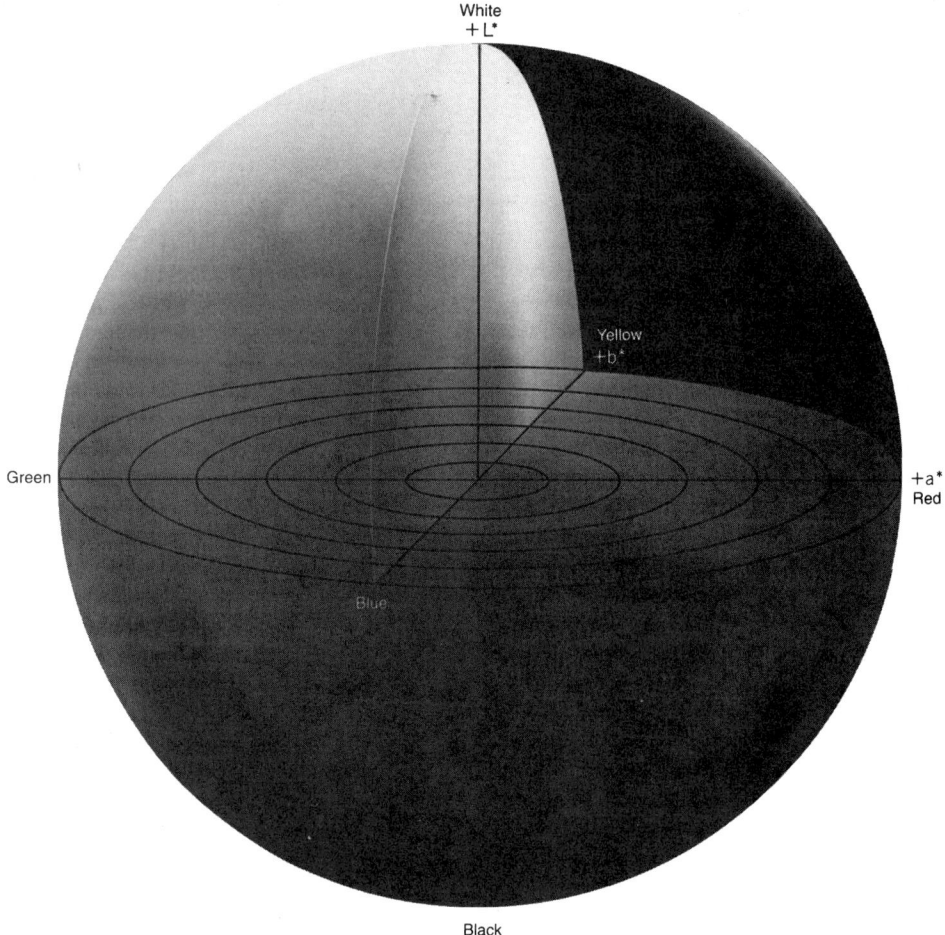

Figure 2.11. Illustration of the color solid for the CIE 1976 ($L^*a^*b^*$) color space (Minolta, 1993).

human color perception. At the writing of this book these attempts are less then perfect, which means "just because the numbers say so, does not mean it is true"!

2.7. APPEARANCE OF COLOR

2.7.1. Surround

Now that we have a basic structure for describing how color appears to an observer, let us consider some additional aspects that affect how we see an object as well as how changing one or more of our triad (source, object, and observer) impacts color. The appearance of an object can be changed by several other phenomena. For example, a color can be made to appear more or less chromatic and/or darker or lighter depending on the surroundings. A white will appear brighter when placed on a dark or black background. This is particularly important in industry because it must be considered when evaluating a sample or object. A sample viewed against two different surroundings will appear different.

2.7.2. Gloss

The gloss or surface structure of an object affects its appearance. An object can be made to look lighter of darker as well as to have slight shifts in hue depending on the surface. This is due to light being scattered and absorbed differently at the surface of the object due to the surface itself. Rough surfaces have a tendency to scatter light more than smooth surfaces and so on. This is an important consideration for industrial applications. If two objects made with the same material and the same process have different surfaces, they will appear to be different colors. This can present considerable problems when attempting to obtain color approvals or evaluating color in a control process. In addition, inclusion of metallic materials into polymers, fluorescent behavior, size of the object, and temperature of the object all have an impact on how humans see color, which only serves to further complicate the problem of color communications.

2.7.3. Color Rendition and Metamerism

What about the situation where the source, object, or observer changes? Because the triad works together to stimulate the brain multiplying their interactions wavelength by wavelength, changing the value of one wavelength changes the response. This means if we view a object under a different light source, our object will appear different. In some cases the object can even appear as a different hue when viewed under two distinctly different light sources. The object is said to have poor color rendition. Similar effects occur if you have the same object and light source and change observers. This is attributed to the fact the different observers can have different spectral response curves, that is, see the visible spectrum differently. Finally, there is the situation where two objects interact with light differently; that is, they have different spectral reflectance or transmission results. When this occurs, the best situation one could hope for is a mismatch. The worst outcome is called *metamerism*! Metamerism is not a "medical term to describe people who are fearful of the metric system of measurement," as one student answered on their exam, but the situation where two objects are seen to be a visual match under one light source but no longer match when viewed under a different light source. As one could imagine this is not a good situation. If both objects do not match under all sets of lighting conditions (their spectral responses are the same at every wavelength of the visible spectrum), then Murphy's law applies: The light source that causes the largest difference is the one your customer will use! In the plastics business metamerism is one of our most difficult problems. Because of the numbers of different polymers and the fact that each polymer set often requires a different set of colorants, metamerism is a high probability. This is almost certainly true when the initial color target is not polymeric but dyed cloth or printed paper.

2.7.4. The Observer

There is a second type of metamerism that is important to discuss and that is *observer metamerism*. This is where two objects appear to one observer to be a match but to a second observer they do not match even when viewed under exactly the same light source. This occurs when the two observers have slightly different responses to the visible spectrum, neither observer being color blind.

2.8. BRINGING IT TOGETHER

In this brief chapter we hope we have been able to establish in the reader's mind that the coloring of plastics materials is not a simple process. However, we would like the reader to know that it is also not an impossible problem. If one takes a sound scientific approach to variables analysis as it relates to color, for the most part the difficulties can be eliminated. As you have seen, there are many variables that must be contended with and these variables do not always act independent of each other. This means we need to define, understand, and control as many variables as possible. We suggest you start with the "simplistic first" theorem, which states: "The most likely reason that your new computer is not working is you don't have it plugged in" (actual data from computer support companies). Start with the simple and work to the complex; it save lots of time and is good, sound scientific thinking. Below are some simple questions to help you remember the basic variables that most often cause color problems. It is by no means all inclusive, for there are times when the solutions are complicated, but this is usually the exception and not the rule.

- Are the light sources the same? Name them, i.e., D-65, D-75 A, B, C, cool white etc.?
- Is the level of Illumination the same in both viewing areas?
- Are the objects the same?
- Are we looking at the same target (standard). How do we know this and can we verify it?
- Are the observers constant or are there different people looking at samples at different times?
- Does the observer have normal color vision (tricky if he or she is a customer)?
- Are we using the same standard observer 2° or 10° in instrumental measurement?
- If Instruments are involved, do they correlate with each other? Can both instruments read the exact same sample(s) and give the same results?
- Are we viewing the samples against the same background?
- Do the samples have the same surface?
- Are the samples being viewed at the same angle?
- Are the samples made from the same polymer? If not, the refractive indexes of the polymers must be considered as a variable. This means the two samples may match at one angle and not at another!
- Is there metamerism involved? Do the samples have the same spectral curves or do they cross each other?
- Are the materials temperature sensitive? Are they thermochromatic?

CHAPTER 3

Colorimetry, Color Specifications, and Production Tolerances by Visual and Instrumental Means

Danny C. Rich

Sun Chemical Corporation
Color Research Lab
631 Central Avenne
Carlstadt, NJ 07072

3.1. INTRODUCTION

In the previous chapter, the principles of the science of color were presented. In this chapter we extend those concepts and transform color science into color engineering. As with all engineering disciplines, color engineering requires a solid basis in metrology. The metrology or measurement of color, known generally as colorimetry, is the first step in the incorporation of color science into the color technology of colored plastics. Just as plastics engineers must deal with the chemical and mechanical measurements of the plastic components, material, and processing equipment, so too, they must deal with the various aspects of the appearance of the final plastic product. This is the realm of the Color and Appearance Division of the Society of Plastics Engineers. Applying the principles of color science to the engineering of colored plastics will take us through basic colorimetry and on into the frontiers of advanced colorimetry. This is the most active of the areas of color science research. The pathway to objective numerical tolerances and statistical process control is not without a few obstacles and pitfalls. Learning how to get around or over such hindrances is one of the goals of this chapter. Like the con-

Coloring of Plastics, Fundamentals, 2nd edition. Edited by Robert A. Charvat
ISBN 0-471-13906-8   Copyright © 2004 by John Wiley and Sons, Inc.

struction engineers of antiquity, who discovered that hand-hewn rock moves far easier on wooden rollers than being dragged along the road, it is hoped that you, the reader, will discover a few new tools or tricks to make your projects run a lot smoother and faster.

3.2. COLORIMETRY

Colorimetry is the quantification of color. It is based on the principles of optical metrology—the science of the measurement of light. The science of color has taught us the number, the kinds of objects, and the optical radiation that must be quantified, and optical metrology teaches us how to make those measurements. Since color is the subject and the radiation is visible radiation, it should be trivially obvious that the first and most fundamental method of colorimetry is with the human eye. In fact, colorimetry with the human eye goes back in time nearly to the beginning of recorded history, when semiprecious stones, wine, and foods were all graded on the basis of their color. So, then, assessing the color and appearance of plastic materials with the human visual system is, quite properly, known as visual colorimetry.

3.3. VISUAL COLORIMETRY

Visual colorimetry is the most basic and the most accurate method of measuring the color of an object. Artists and craftsmen have been practicing visual colorimetry for untold centuries. Visual colorimetry can be as simple as holding a skein of yarn up to the light from the northern sky and judging if the yarn contains the correct mix of dyes. Or it can be as complex as the methods used by the color science researcher during the first half of the twentieth century when they mixed light from three single wavelengths of light (monochromatic) to form a representation of any other color light that they could imagine or generate. Figure 3.1 shows a schematic of a commercial visual colorimeter produced by Donaldson [1] in England that was popular in the 1960s. The most recent commercial visual colorimeter was the VCS-10, produced by Applied Color System [2] in the 1980s. Today, there are no commercial visual colorimeters available in the United States. Recent advances in flat-panel displays may change that in the near future. The brightness of modern flat-panel displays is close to that of a white plastic on the bottom of a commercial color-matching lighting booth [3]. Since flat-panel displays are back lighted with fluorescent lamps, it is quite possible to make a display with multiple sources and more than three primaries, like the two visual colorimeters described above.

    The most significant advantage of a visual colorimeter is that it uses the human visual system to measure the color directly. The most significant disadvantage of a visual colorimeter is that it uses the human visual system to measure the color. Human observers are terribly susceptible to fatigue, to mood changes, to the weather, to just about everything. As a result, while a visual colorimeter is the most accurate instrument for measuring color, it is not the most consistent or reliable. In addition, there can be significant differences between the color perceptions of two

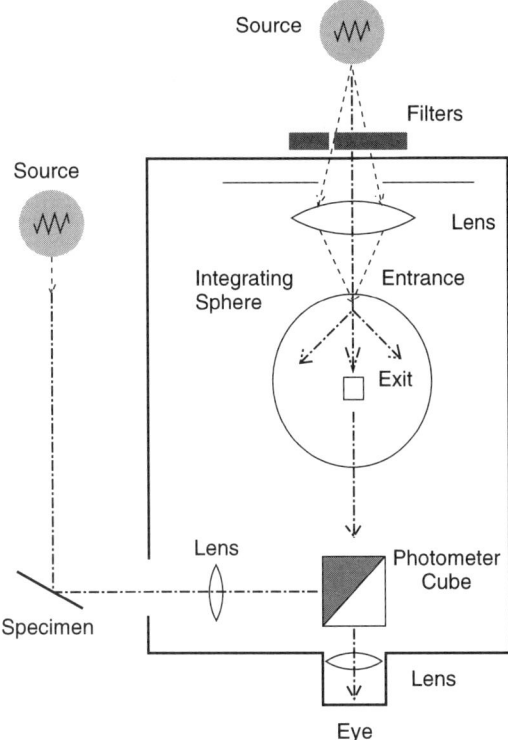

Figure 3.1. Schematic of a visual colorimeter (from Wyszecki and Stiles).

observers, and thus the match points of the two observers will seldom be in close agreement.

3.4. COLORIMETRY BY ANALOG SIMULATION OF VISUAL COLORIMETER

In quality control of colored objects, production requires more consistency than absolute accuracy. After all, there will generally be a product standard having the correct color to which we may reference the color of parts from the current production run. Since the production part and the product standard will be free from the small errors in accuracy, because any small inaccuracies will be the same in measurements of the batch and the standard and thus will subtract out in the different components. What is needed is an instrument with higher day-to-day objectivity than our visual system, and for such an increase in consistency, one must be willing to give up some of the absolute accuracy. This requires an analog simulation of visual colorimetry.

In Chapter 2 it was demonstrated that the visual system requires three things to function correctly: a light source, an object, and a detector. So one designs an instrument with a standard light source, a port for placing a solid sample, and a detector that has the properties of a standard observer. In the visual colorimeter the pri-

26 Colorimetry, Color Specifications, and Production Tolerances

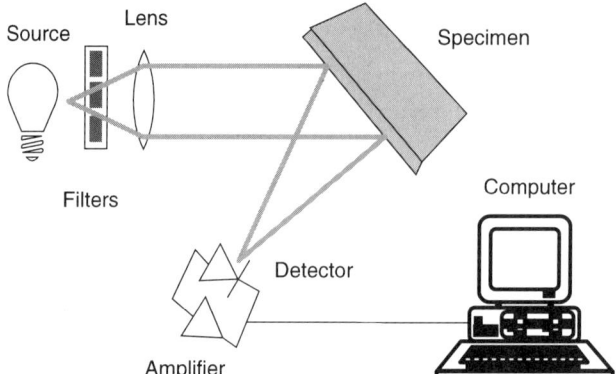

Figure 3.2. Schematic of an analog simulattion of a visual colorimeter.

maries are "red," "green," and "blue" lights. As indicated Chapter 2, in 1931 the International Commission on Illumination (CIE) recommended a standard set of color-matching primaries that have since been renamed to standard observer functions. In the analog simulation of the visual colorimeter, it is desired to illuminate the standard and production specimens with "white" light, so the red, green, and blue (RGB) colored filters are usually placed over the detectors. This is not an absolute necessity and some modern analog systems use red, green, and blue light-emitting diodes (LEDs) to illuminate the specimens. Once the RGB values have been recorded, the data analysis is very similar to that of the visual colorimeter. Using the linear algebra derived by the CIE, an RGB set can be transformed into the now standard CIE XYZ. If the RGB primaries of the colorimeter are not exact linear transformations of the original CIE RGB primaries, the transformation will result in some rather large residual errors. This is the normal case for video cameras and low-cost flat-bed scanners. Errors in color of 15–20 visually noticeable differences can occur. So, the filters in most analog photoelectric "filter" colorimeters are selected to produce as close a match as possible to the CIE spectral color-matching functions. Modern filter colorimeters have absolute errors of 3 visual units or less. Figure 3.2 shows a schematic of a typical filter colorimeter. Figure 3.3 shows the quality of the fit of the product of the light source, filters, and detectors to the CIE color-matching functions, wavelength by wavelength across the visible spectrum.

An analog simulation of the visual colorimeter, such as a good filter colorimeter or a well-designed LED colorimeter, can repeat its readings to about 1 part in 10,000 day after day for years. This provides the plastics manufacturer with the ability to precisely monitor and control the color of the raw materials or finished product. But, today, plastics parts are used widely in combination with other pieces. Take, for example, a warm winter coat. There may be cotton or wool panels, nylon panels, plastic zipper or buttons, painted metal accessories, and so on. Each of these items will have been "colored" using a different set of real primaries (dyes or pigments) due to the chemistry of the coloration process. What might happen to the appearance of that coat as the wearer passes indoors from an overcast daylight into the warm glow of the incandescent lamp in the foyer and then passes into the kitchen

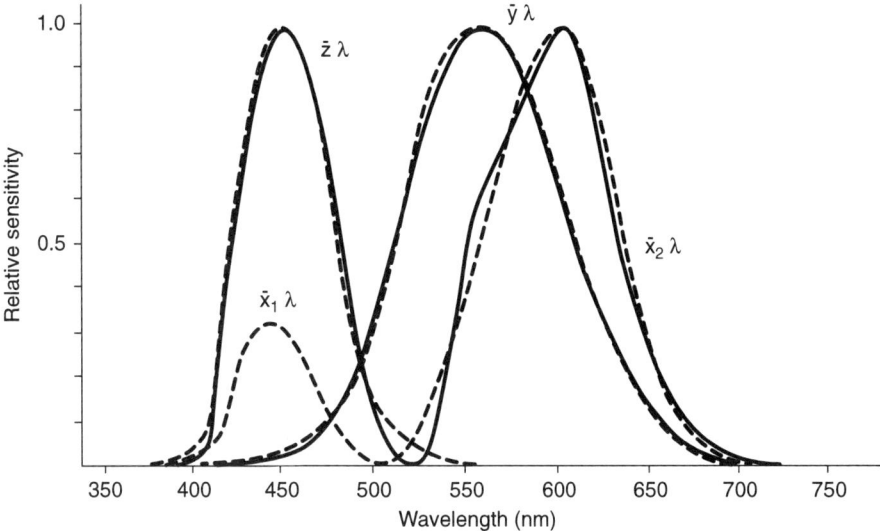

Figure 3.3. Comparison of analog colorimeter fit (solid lines) to CIE 1931 standard observer functions (from Billmeyer and Salzman, 2nd ed.).

lighted with modern compact fluorescent lamps? Color science teaches that under these conditions the potential for metamerism is very great. Visually, we could test for metamerism by changing the quality of the source illuminating the specimen and then redetermining the amount of the primaries needed to make a match. If the numbers were too different, we would reject the match and change the colorant formulation until the difference in the visual colorimetry was acceptably small. The filter colorimeter was designed to optimize the fit of the product of the source, filters, and detectors to one set of CIE illuminant/observer functions, and if we change the light source, we ruin that fit. So, the analog simulation of the visual colorimeter has an additional limitation. It cannot be used to accurately assess metameric matches. The product standard must not just be similar to the production, it has to be identical. To handle the requirements of metameric matches, one must go back to visual colorimetry or develop a new, more sophisticated simulation of the visual colorimeter. What is needed is a digital simulation of the visual colorimeter.

3.5. COLORIMETRY BY DIGITAL SIMULATION OF A VISUAL COLORIMETER

A digital simulation of visual colorimetry is best obtained by characterizing the colored specimen on a wavelength-by-wavelength basis and then computing the required triple product of light source, specimen, and detector using a table of weighting coefficients that represent the CIE standard observer functions, a CIE standard recommended illuminant, and the colored specimen of interest. Just as the filter colorimeter was used to create an analog simulation of the visual colorimeter, it is possible to use a special instrument, the spectrocolorimeter, to create a digital simulation of the visual colorimeter. The instrument still requires a light source, a

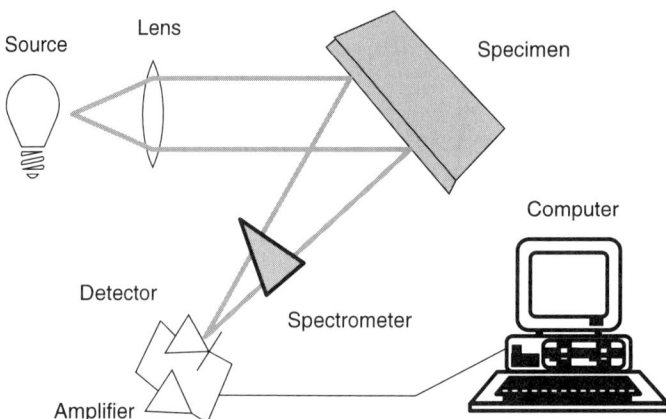

Figure 3.4. Schematic of a digital simulation of a visual colorimeter—a spectrocolorimeter.

specimen port, and a detector. But instead of filters in front of the detector, a spectral analyzer is installed that can select a band of wavelengths centered on a small number of evenly spaced centroids, selecting them either one at a time or in parallel, across the visible spectrum. Figure 3.4 contains a schematic of the digital simulation of a visual colorimeter. The spectrocolorimeter now has all the features that are needed for both accurate and precise color control. The digital simulation is as accurate as the tables of CIE weighting functions and the design characteristics of the instrument itself allow. The only drawbacks to the spectrocolorimeter are its cost and complexity. Advances in optical and optoelectronic technology are quickly removing those drawbacks. In the next section, the characteristics of spectrocolorimeters will be examined in depth.

3.6. SPECTROCOLORIMETRY

Spectrocolorimeters are a unique form of the classic analytical spectrophotometer. Several design changes and design optimization have resulted in a family of instruments that have roots in classic spectrophotometry but also possess new and unique properties. The term "*spectrocolorimeter*" has been the subject of some criticism and disagreement. It was first used in the late 1970s to describe a new series of instruments [4] that looked and behaved like traditional filter colorimeters, returning only tristimulus information. But the design was built not on three or four tristimulus filters but on medium-band interference filters that subsampled the visible spectrum at 16–18 wavelengths. The triple products were computed by a dedicated microprocessor and the tristimulus results presented on a small display, just as in current filter colorimeters. Readings were initiated by pressing a single button on the instrument panel. There were no provisions for spectral plots, readouts, or transfers of the measured spectral data. Prior to this time, digital color measurements were made on traditional recording analytical spectrophotometers equipped with reflectance accessories or on abridged scanning spectrophotometers equipped with analog readouts. Three technical developments led to the change in nature of digital colorimetry [5]. First, the development of holographic gratings that produced the required line structure without the imperfections of a ruling engine and thus a single grating could produce spectra without ghosts and other forms of heterochromatic stray light. Second, it became possible to mount a series of silicon photodiodes onto a single carrier or substrate. This mounting was done by photolithography and resulted in diodes with very precise size and spacing. Third, microprocessor engineering had reached a point where low-cost, digital computers could be built on a single electronic circuit board. Spectrocolorimeters then had solid-state detectors with several decades of linearity that had not been available in low-cost phototubes or photomultiplier tubes. Spectrocolorimeters were equipped with wide-bandwidth spectral analyzers that allowed a lot of light through system at each wavelength, keeping the signal-to-noise ratio high and thus the repeatability low (below 0.5%) while at the same time maintaining very precise position control of the centroid wavelength. The latter was a feature not possible in scanning systems with motors, lead screws, and sine bar mechanisms. The presence of a tightly coupled microprocessor allowed the design engineer to perform many of the fine adjustments numerically rather than optically or mechanically, thus simplifying the design and lowering the cost of the components and the assembly time. Modern spectrocolorimeters have pushed the envelope on this technology even further and exhibit performance characteristics nearly an order of magnitude better than the first instruments of just 20 years ago [6].

3.6.1. Design and Specifications of Spectrocolorimeters

As indicated above, a spectrocolorimeter has three main components: a light source, a spectral analyzer, and a detection system. Each of these main components can be subdivided into additional categories of concern.

Classic analytical spectrophotometers mount the light source on the input to the spectral analyzer, then irradiate the specimen with light of a specific band of wavelengths. That is not the way that the visual system works but it is adequate for measuring spectral diffuse reflectance or reflectance factor. However, if there is any fluorescence present in the specimen, then neither the fluorescence nor the reflectance will be properly characterized. Light in the short wavelengths will be absorbed and then emitted as fluorescence so that the detectors record no loss of light in the absorption region. Thus, all spectrocolorimeters will have polychromatic (white) light incident upon the specimen and will image the reflected or transmitted light into the spectral analyzer. For reflectance factor readings, the spectral quality of the instrument source is not a significant contributor to the quality of the readings. Thus, many small or lower cost instruments will use incandescent lamps or LEDs as the instrument source. For measurements of specimens containing fluorescent dyes or pigments, a source with a spectral quality that conforms to a standard phase of daylight is preferred. Usually that requires the use of high-intensity xenon arc lamps or flash lamps. The flash lamp has the advantage that as it is "on" for no more than a tenth of a millisecond, it contributes little or no heat to the sample surface. Instrument-caused specimen heating has been shown to be a major problem in assessing an instrument's performance properties [7].

In daily experiences, colored objects are illuminated and viewed through a wide range of conditions. The human visual system has the ability to "adapt" to these changes in illumination and view, and the color and appearance of the objects is kept fairly constant. Spectrocolorimeter detection systems are not yet sophisticated enough to simulate this behavior. Therefore, the design engineer must restrict the conditions of illumination and view to a small subset of all possible conditions. The CIE has recommended two geometries for instrumental colorimetry: hemispherical diffuse illumination with near-normal viewing (d:8°) or biconical illumination (45°:0°) [8]. Application of the law of the conservation of energy to optical systems indicates that a reversal of the orientation should have no effect on the results of the measurements. Therefore, 8°:d should be equivalent to d:8° and (0°:45° should be equivalent to 45°:0°. Unfortunately, such a simple geometric description is not adequate to achieve good interinstrument agreement. How diffuse is the "diffuse illumination" must be defined as well as the angular extent of the cones of light in the viewing and/or illumination beams. While, at least in theory, the instrument should be able to characterize a specimen independent of the physical nature of the specimen, this is rarely the case in reality. Thus specimens with significant surface texture or defects are better characterized on a hemispherical diffuse instrument with the surface reflectance included with the body reflectance in the readings. Reports in the literature have shown that even with annular or circumferential illumination schemes, biconical measurements are more sensitive to the quality of the specimen than are diffuse measurements, and small differences in the optical geometry of the illumination and viewing beams result in significant differences in the instrument-to-instrument readings [9,10]. In addition, color measurements intended for use in colorant formulation calculations can provide more control over differences in surface properties, such as gloss, by using a combination of specular (surface reflectance) inclusion and exclusion than by specular exclusion alone using biconical geometry. No instrument geometry can completely exclude the specular reflectance, but biconical instruments have been reported to maintain exclusion of the specular reflectance from the body reflectance over a wider range of gloss levels than hemispherical geometry instruments. The general assumption that biconical readings maintain a higher degree of agreement with the visual evaluation of surface color has not been demonstrated in the literature.

The design and specification of the spectral analyzer are generally more direct than that of the illumination system. The CIE defines the visible wavelength region as covering the range from 360 nm in the near UV to 830 nm in the near infrared (NIR). Few instruments are capable of making readings of reflectance over that range. In practical situations, the CIE indicates that the range may be truncated to 380–780 nm with no significant loss of information. Few commercial spectrocolorimeters cover that range either. Most spectrocolorimeters cover only the basic visual spectral range of 400–700 nm. Some of the high-performance instruments will extend the basic range by 40–50 nm on one end (usually into the near UV or NIR ends, giving a range of 360–750 nm. Reports in the literature by Stearns [11] and in ASTM standards [12] indicate that the spectral sampling interval should be approximately equal to the spectral bandwidth. Thus 16 point instruments that collect data at 20 nm should have a bandwidth of about 20 nm. McCAmy [13] has recommended a value of 80% of the sampling interval for the optimum bandwidth. It has been clear since the time of the Hardy recording spectrophotometer that operating with a spectral bandwidth slightly smaller than the digital sampling interval provides slightly more reliable readings [14]. Modern numerical algorithms provide the ability to subsample or oversample the spectral data and report data at standardized intervals and bandwidths through the use of interpolation and deconvolution procedures. Most of the these procedures are undocumented, being kept as trade secrets by the instrument manufacturers. Deconvolution and interpolation procedures tend to work better in characterizing reflectance properties than transmittance properties. The presence of multiple scattering tends to broaden the absorption bands, making the Nyquist bandwidth generally much larger than 10 nm. Several studies have attempted to identify the need for reducing the size of the sampling interval to 5 nm or less for characterization of the spectral curve [15–17]. The most exhaustive study was done by Fairman [18]. His data showed that going below 5 nm offered no practical advantages, and even the differences between 10 and 5 nm were small, in terms of the effect upon the measurement. It is clear, however, from the work of Stearns [11] and Fairman [18] that numerical integration of the triple product should be performed at intervals no larger than 1 nm. Interpolation of the measurement data to 1-nm intervals from 5–10-nm intervals with spectral bandwidths of 6–8 nm seems to provide spectral data of sufficient similarity to produce compatible tristimulus values at levels that approach the repeatability of the instruments.

The design of the detection system is one of the most standardized aspects of a spectrocolorimeter. With the introduction of tightly coupled microprocessors and the associated analog-to-digital converters, it was discovered that placing all of the analog electronics as close to the detectors as possible resulted in major improvements in stability. Classic analytical spectrophotometers incorporating photomultiplier tubes would place the analog electronics in the same area as the high-voltage supply for the photomultiplier tube. While proper shielding and guarding were used, it is difficult to shield a long wire from all of the effects of motors, switches, and solenoids during the course of a scan. With the new solid-state diodes, high-gain, high-impedance amplifiers could be mounted on a guarded printed circuit directly behind the diodes. This converted the small photocurrent to a voltage and amplified the voltage to levels where environmental noise was below the 1 part in 10,000 required for the photometric scale. Spectrocolorimeters are thus limited, not by the noise in the optical system, but by the analog-to-digital conversion system. Reports on the short- and medium-term repeatability of reflectance factor readings have shown that the statistical distributions of the readings do not follow the central limit theorem, but are significantly more peaked, slightly skewed, and bimodal [19]. The last attribute is an indication that the readings are oscillating between two digital states in the analog-to-digital converter. Most recently, the electronics are being designed such that the preamplifier and current-to-voltage conversion takes place within a single integrated circuit device, a new type of application-specific integrated circuit (ASIC) coined "active pixel or active column video technology [20]. Figure 3.5 illustrates such an integrated detector package [21].

3.6.2. Calibration and Standardization of Spectrocolorimeters

Calibration and standardization are often confused and used interchangeably. Calibration refers to the process of adjusting an instrument or instrument's scale to a known physical standard [22]. Wavelength scales are *calibrated* by comparing the instrument's wavelength scale to a spectral line whose value is known from first

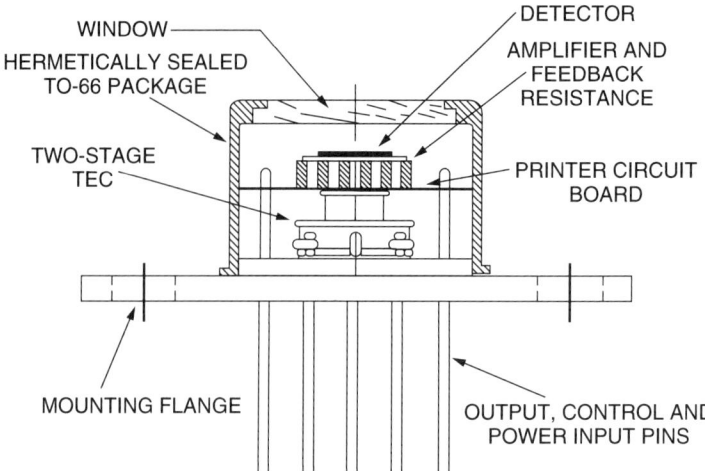

Figure 3.5. Engineering drawing of a solid-state photodetector with integrated amplifier, current-to-voltage converter, and temperature controller. (courtesy PerkinElmer Optoelectronics)

principles or in terms of the fundamental standard of length. An instrument's photometric scale is normally *standardized* against an agreed to standard of diffuse transmittance or diffuse reflectance. The values assigned to that standard may or may not be the fundamentally correct values. Today, fundamental physical standards are maintained for the meter (length), the second (time), the volt (electrical potential), the kilogram (mass), and the candela (the unit of light). There is not a fundamental standard of reflectance. Thus a spectrocolorimeter is never calibrated in the field.

Before any calibration can be performed, one must determine the property to be calibrated. In commercial spectrocolorimeters there are two properties that might be calibrated, the photometric scale and the wavelength scale. Both scales can be adjusted only at the factory. As described above, a calibration requires the measurement of a known physical standard of the property and then adjustment of the instrument scale to agree with the known values. Standards of wavelength include continuous-wave gas discharge lasers, electric arc lamps containing atomic or molecular vapors at low pressures, or hollow-cathode discharge lamps containing metals that can be vaporized by the discharge. In each of these devices, the wavelength of the radiation is known to a very high degree of precision and the natural spectral width of the line is quite narrow. Procedures for utilizing a set of lines from different lamps are well documented in the literature.

The basic definition of reflectance, as used in colorimetry, is the ratio of the light flux *reflected* from a material to the light flux *incident* on the material. Reflectance involves accounting for all of the visible radiation in the system. This is in contrast to a reflectance factor, which is defined as the ratio of the light flux reflected from a material to light flux reflected from a standard material [23]. Material standards of reflectance and transmittance are difficult to obtain. Reflectance, in particular, is difficult since the primary standard of reflectance is the perfect reflecting diffuser, a standard that does not exist in nature. Thus, scales of reflectance must be transferred to physical approximations to the perfect reflecting diffuser using the candela and mathematical corrections to compensate for both the systematic errors in the instrument and in the material standards. In the field, one standardizes the instrument to its working standard, which hopefully has a traceable scale of reflectance factor transferred to it with a kown uncertainty. Random errors in this chain of transfer from the theoretical standard to the instrument working standard are currently the largest sources of uncertainty in the assignment of reflectance factor values to white or neutral materials. The next largest sources of uncertainty are the systematic differences in optical geometry, all of which are currently below the specification levels recommended by the CIE and the International Organization for Standardization (ISO) [24]. This level of uncertainty has been shown to result in differences in the color measurement of the same material by two instruments from different manufacturer [25] or from two different national standards laboratories [26] that are larger than the production tolerances of many plastic components.

3.6.3. Performance Verification of Spectrocolorimeters

There are many sources of errors in basic colorimetry. As indicated above, the level of uncertainty in the transfer stadards can be a major source of error. For high-chroma colors, particularly in the red, orange, and yellow hues, small-wavelength scale errors transfer directly into colorimetric errors [27]. A wavelength scale with a 0.1-nm error in the wavelength scale can result in a 0.1 CIELAB unit color difference error [28]. Similarly, a 0.1% error in the photometric scale transfer can translate into a 0.1 CIELAB unit color difference error for the same high-chroma samples and sometimes even larger color difference errors for neutrals specimens.

When the specimens to be measured contain colorants that are fluorescent, the instrument source must be a reasonably good match to the standard illuminant under which tristimulus values are to be computed. This is one of the reasons that traditional analytical spectrophotometers are not normally used for routine colorimetry. Thermoplastics, which may be pigmented with inorganic or ceramic pigments designed to withstand the high temperatures of the extrusion or molding process, may present unique problems. Some of the inorganic band-gap colorants will exhibit fluorescence in the near infrared (NIR). Modern analytical instruments with NIR-sensitive light detectors will record that NIR radiance as visible range reflectance. This visible–NIR luminescence error may be more of a problem than the UV-activated luminescent pigments that emit in the visible, and this is especially true for instruments with unfiltered incandescent sources with excess radiance in the NIR. For daylight sources, conformance of instrument sources to CIE standard illuminates can be assessed following the procedures given in CIE Publication 51 [29]. Reports in the literature have shown that good-quality daylight simulators can be made by filtered continuous xenon arc lamps [30], by filtered tungsten–halogen cycle lamps supplemented with a UV source [31], such as a phosphor-coated "pen-ray" lamp, or by a properly filtered xenon flash lamp. Unfiltered xenon flash lamps do not produce close matches to any standard phase of daylight.

The geometry of the illuminating and viewing optics has become a critical factor in the differences between instruments. Recent work from the National Physical

Laboratory Teddington, England (NPL), referenced in the previous section, has shown that even national standardizing laboratories within the western European continent can show significant disagreement with each other, even when they have traceability back to one laboratory [26]. In its report, the average differences of 0.5 CIELAB units with a maximum of 2.0 units were observed. This is similar to the size of the differences it reported for a similar intercomparison between industrial laboratories using the state-of-the-art bench-top spectrocolorimeters [25]. However, this is more than an order of magnitude higher than the values reported for the same number of participants spread across three continents (average difference of 0.09 CIELAB units and maximum differences of 0.17 CIELAB units [32]. Thus insufficient control of the irradiance of the specimen or of the viewing environment may result in colorimetric errors two to four times large, than a production tolerance. Currently, high levels of reproducibility and site-to-site agreement on instrumental measurements can only be obtained by purchasing and installing color-measuring instruments from a single manufacturer. The CIE has a technical committee (TC 2-39) working on improving the recommendations for the geometry of color-measuring instruments.

The influence of specimen preparation and presentation cannot be overemphasized. Reports in the literature and in presentations at Annual Technical Conference Regional Technical Conference and other annual technical conferences continue to show that much, if not most, of the random errors and uncertainties in color measurements arise from the specimen itself [33,34]. The human visual system has very poor imaging capabilities, and we have trained ourselves to overlook or ignore many optical artifacts. Spectrocolorimeter are not yet so fortunate. Scratches, nicks, whirls, and inadequate mixing all result in readings that are not representative of the bulk of the materials. The unfortunate thing about this is that the nonuniformity may result in measurements that "look good" when, in fact, the bulk material is off shade, just as the measurements may "look bad" when the bulk material is on shade. If the authors of this book had received a nickel for every time a production specimen's color was measured twice resulting in one reading that was within specification and one that was outside of specification and only the "good" values reported, we would all be living in luxury in some tropical retirement village.

One of the biggest contributors to specimen-dependent measurement error is thermochromism. Thermochromism is the phenomenon where the color of an object is reversibly changed as a function of the temperature of the specimen surface. This temperature effect may be due to changes in the ambient (hot in the summer or cold in the winter) or due to an instrument that conveys large amounts of heat onto the specimen surface from an incandescent lamp or from a positive air flow across hot electronic components. Even standard materials exhibit this effect. The most famous is the British Ceramic Research Association's (now known as Ceram) Ceramic Colour Standards II. Errors of the order of magnitude of 0.1 CIELAB color difference unit per degree Centigrade have been well documented [35]. Commercial specimens are no less susceptible to this effect. This seems to be more prevalent when the colorant has a very high tinctorial strength and a very high temperature stability, an occurrence not uncommon in the manufacture of thermoplastic materials. Considerable care should be exercised in making sure that the surface of the specimen is always at the same temperature when making color measurements.

3.6.4. Application of Colorimetry to Establishment of Product Color Tolerances

The oldest and most unreliable method for establishing and testing product color tolerances is direct visual observation. That being said, we must acknowledge the position of the proponents of visual inspection and their claims that (a) a large volume of colored goods have been successfully produced, visually inspected, and sold over the centuries and (b) customers buy what they "see," not what an instrument measures. Both statements are true and both statements "beg the question." There is absolutely no guarantee that what the inspector sees is anything at all like what the customer sees. In fact, reports in the literature [36] would indicate that the higher probability is that they do not see the same thing. What is true, then, is that well-trained, experienced inspectors will have learned both their own biases and those of each of their customers. That is why the wise manager will do what ever he or she can to supply adequate compensation and perquisites to experienced color inspectors. ASTM D 3134 [37] still has sections in it on how to establish visual tolerances, and D 1729 [38] describes in great detail the specifications and requirements of a visual examination cabinet for the inspection of colored objects. Unfortunately, in the business climate of the twenty-first century it is often not possible to hold back the older employees or to wait for the younger employees to "apprentice" for a decade to develop these skills. The only viable option is to make use of some tool set that will aid the young inspector so as to make him as productive as an experienced inspector. That requires the use of color-measuring instrumentation and some form of numerical tolerances. Numerical tolerances push the borders of colorimetry from basic colorimetry into the realm of advanced colorimetry, where the science is less well defined as it tries to predict what an average observer actually sees.

3.6.5. Methods of Deriving Tolerance Regions from Combined Visual and Instrumental Data

In an almost seminal paper, Berns [39] has described how to use the latest in mathematical and statistical nonlinear optimization to derive custom color tolerance regions using visual (pass–fail) judgments and instrumental measurements of the color of the product standard and the current batch sample. Luo and Rigg [40] have reported that such custom optimization always results in the highest level of agreement (reported as the lowest number "wrong" decisions) between the numerical predictions and the judgments of the experienced inspector. The level of wrong decisions never drops to zero, indicating that visual inspectors either disagree with themselves, a situation documented by Rich et al. [41] in two-outcome, perceptibility judgments, or applied different acceptability criteria to matches for different customers or assumed products. Alston [42] has demonstrated an artificial intelligence based optimization procedure that results in no wrong decisions. To accomplish this, the tolerance regions are made nonsymmetrical, just as is recommended in the visual section of ASTM D 3134. The Alston method has been patented and is available only in certain specific commercial software packages. The biggest disadvantage to both the Berns [39] and the Alston [42] methods is that they require a large number of known batches, both good and bad, from which to develop the statistical inferences that lead to the development of the tolerance regions. The biggest question that remains is how to apply numerical inspections to products produced in multiple repeat batches or produced for the first time?

3.7. EXTENSION OF BASIC COLORIMETRY TO PREDICTION OF COLOR DIFFERENCES

The "Holy Grail" of colorimetry has always been an equation that would accurately and consistently allow single-number shade passing, color control, or product sorting—an equation that could be applied to any color at any time and would predict the visual significance of the difference between two specimens. The limiting value would be anchored either to the visual threshold (just noticeable difference) or to a typical industrial tolerance (whatever that might mean) so that differences greater than some predetermined value would be considered as visually "different" and differences below that value would be considered as visually "identical." The earliest known studies are those of MacAdam [43], Hunter [44], and Nickerson and Stultz [45], dating from the late 1930s and early 1940s. These color difference metrics, so called because they attempt to project a metric or normed vector space onto the affine domain of the CIE tristimulus values, had very little success at reaching these goals until the late 1970s. In 1976 the CIE recommended two equations for study in the hope of bringing some uniformity of practice to the field. The color metrics, known as CIELAB and CIELUV (all capitals are the correct abbreviations), introduced several new concepts to the computation of color differences. Initially, these new formulas were hailed as great advances, providing reasonable mappings of historically accepted color order systems like that of Munsell and at the same time providing reasonable though not particularly good fits to the data of MacAdam and others who had published studies of small color differences. But the ink was hardly dry on the CIE report before industrial researchers began finding problem areas with the equations. As early as 1971 McLaren [46] had been insisting that the Adams–Nickerson formula, from which CIELAB is derived, needed to have significant adjustments to its coordinates as a function of changes in the color of the standard. By 1976 McLaren [47] was showing an equation that, when tested against the known industrial color difference data sets, would provide a percentage of "wrong assessments" about as often as an average shader. His approach required local adjustment of an approximately uniform color space to form even more uniform color tolerance regions. So, even as the CIE committee was finalizing the details of the new CIELAB color metric, Taylor [48] had begun a new study with the retail store Marks & Spensers, McDonald [49] had begun collecting tolerance data in the J&P Coates thread factory, and Rohner and Rich [50] were working with the textile firm AMANN collecting pass–fail data on woven textiles. By the end of that decade the numbers of color tolerance equations were beginning to become unmanageable. J&P Coates [51] had published a formula, Bradford University had published a formula [52], and Marks & Spensers had reported that it had a formula that it would require suppliers to use by purchasing either QC programs or a "blackbox" from Instrumental Colour Systems (now Datacolor International) but would not publish the details. Rohner and Rich [50] also kept their results proprietary inside of the Datacolor,

AG (now Datacolor International) software. Finally, the Colour Measurement Committee (CMC) of the Society of Dyers and Colourists in the United Kingdom made some compromises and published a formula based on the J&P Coates work but incorporating the CIELAB coordinates [53]. This was made a British standard (BS 6923) and soon was adopted by textile associations around the world. The usefulness of such an equation has been demonstrated many times in the textile industry and in some papers from manufacturers of textured, molded polypropylene as well.

The CMC equation has "grown" to be very complex in its form. As can be seen in the equations below, there are many terms that adjust the size of the tolerance region. Initially, these terms were felt to be due to industrial biases between the just perceptible level of color difference and the just acceptable level of color difference. More recent work has shown this not to be the case. In fact, there now appears to be quite good perceptual and physiological reasons for the inclusion of the various terms.

To compute CMC tolerance, one begins with the CIELAB coordinates (L^*, a^*, b^*) of both the standard specimen and the trial (batch) specimen:

$$F_X = \left(\frac{X}{X_n}\right)^{1/3} \quad \text{if } F_X > 0.008856, \quad \text{else } F_X = 7.787\frac{X}{X_n} + \frac{16}{116}$$

$$F_Y = \left(\frac{Y}{Y_n}\right)^{1/3} \quad \text{if } F_Y > 0.008856, \quad \text{else } F_Y = 7.787\frac{Y}{Y_n} + \frac{16}{116}$$

$$F_Z = \left(\frac{Z}{Z_n}\right)^{1/3} \quad \text{if } F_Z > 0.008856, \quad \text{else } F_Z = 7.787\frac{Z}{Z_n} + \frac{16}{116} \quad (3.1)$$

$$L^* = 116 F_Y - 16 \qquad a^* = 500(F_X - F_Y) \qquad b^* = 200(F_Y - F_Z)$$

$$C^* = \sqrt{(a^*)^2 + (b^*)^2} \qquad h^*_{ab} = \arctan\left(\frac{b^*}{a^*}\right)$$

Next the CIELAB coordinates are converted to differences in metric lightness, metric chroma, and metric hue (not hue angle), ΔL^*, ΔC^*, ΔH^*, respectively. By themselves, these parameters define a spherical color difference volume in the rectangular coordinates (L, C, H) with a radius of ΔE_{ab}:

$$\Delta L^* = L^*_{\text{Test}} - L^*_{\text{Std}}$$

$$\Delta a^* = a^*_{\text{Test}} - a^*_{\text{Std}}$$

$$\Delta b^* = b^*_{\text{Test}} - b^*_{\text{Std}}$$

$$\Delta C^* = C^*_{\text{Test}} - C^*_{\text{Std}}$$

$$\Delta H^* = \sqrt{2(C^*_{\text{Test}} C^*_{\text{Std}} - a^*_{\text{Test}} a^*_{\text{Std}} - b^*_{\text{Test}} b^*_{\text{Std}})} \quad (3.2)$$

$$\Delta E^*_{ab} = \sqrt{(\Delta L^*)^2 + (\Delta a^*)^2 + (\Delta b^*)^2}$$

$$h^* = \arctan\left(\frac{b^*}{a^*}\right)$$

38 Colorimetry, Color Specifications, and Production Tolerances

Each component is then adjusted using an equation that depends on the color of the standard specimen. These adjustments terms, S_L, S_C, and S_H, are the complex part of the CMC tolerance equation and the part that changes the color difference sphere into an "ellipselike" tolerance volume:

$$S_L = \frac{0.040975 L^*}{1 + 0.01765 L^*} \quad \text{for} \quad L^* \geq 16$$

$$S_L = 0.511 \quad \text{for} \quad L^* < 16 \tag{3.3}$$

$$S_C = \frac{0.0638 C^*}{1 + 0.0131 C^*} + 0.638$$

$$S_H = S_C (Tf + 1 - f)$$

where

$$f = \left(\frac{(C^*)^4}{(C^*)^4 + 1900} \right)^{1/2}$$

$$T = \begin{cases} 0.56 + |0.2\cos(h+168)| & \text{if} \quad 164° < h < 345° \\ 0.36 + |0.4\cos(h+35)| & \text{otherwise} \end{cases}$$

Finally, some parametric adjustments ($l:c$) are added as well as an overall volume adjustment (cf):

$$\Delta E_{\text{CMC}}(l:c) = \text{cf} \times \sqrt{\left(\frac{\Delta L^*}{lS_L}\right)^2 + \left(\frac{\Delta C^*}{lS_C}\right)^2 + \frac{\Delta H^*}{S_H}} \tag{3.4}$$

In the original CMC papers, these adjustments were considered to be the difference between a just perceptible color difference and an acceptable color difference. More recent study has shown that they are more likely due to the use of textiles or thread samples that have significant amounts of surface texture. Large-scale statistical studies on calendared polypropylene have shown good agreement with the use of the $l:c$ ratio set at 2:1, as is recommended for comparison of textiles and yarns. The figures below show the shape and orientation of acceptability ellipsoids for the standard $l:c$ ratio of 2:1, Figure 3.6 shows the CIELAB a^*, b^* plane ($\Delta L^* = 0$) and Figure 3.7 shows the L^*, C^* plane ($\Delta H^* = 0$). Neither CMC nor any of the other current color tolerance equations show a lightness/chromaticness dependence on hue, so we need show only one constant hue plane to illustrate the shape of the projected tolerance volume. The same is true in the constant lightness plane. There are no lightness-dependent terms in the parameters for chroma and hue.

This now raises the issue of what is the correct value of the $l:c$ ratio for any given material. The CIE Technical Committee TC 1-28 published recommendations on how to handle various parameters that can affect the judgments of color differences as CIE Publication 101 [54]. In that study, the parameter that stood out was the effect of surface texture on the judgments of lightness differences. Small differences in lightness can be assessed for specimens with smooth glossy surfaces but not in samples with textured, patterned, or rough surfaces. Thus studies in the literature

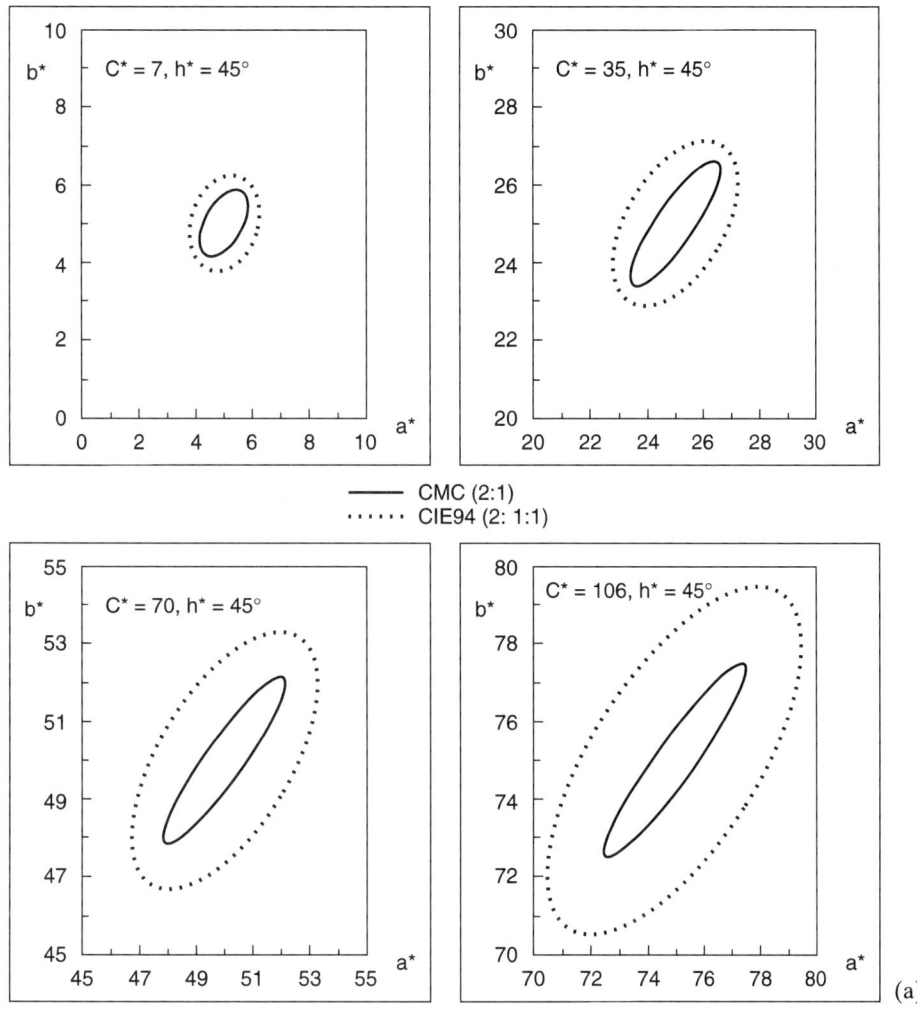

Figure 3.6. Ellipses in the CIELAB a^*–b^* plane for metric hue angles of (a) 45° and (b) 135°.

have reported values for $l:c$ that range from 2.5:1 to 0.5:2 depending on the kind and quality of the materials that were being evaluated. Figure 3.7 illustrates the effect of changing the $l:c$ ratio. It generally only affects the size and shape of the tolerance volume in the (L^*, C^*) plane. However, if the c value is set to any other value than 1.0, then it will affect the size of the chroma tolerance as well. In general, there are very few instances in which the values need to be changed. Since they occur in the denominator of the color tolerance calculation, the numbers serve two purposes. First, they adjust the relative importance of one property to the other. In the CMC tolerance there is an implied hue term so that the ratio is actually 2:1:1 for textiles. During the final preparations of the CMC equation, the optimized equation was tested against both industrial pass–fail data sets, termed acceptability data, and research laboratory data sets, termed perceptibility data. It was seen that the

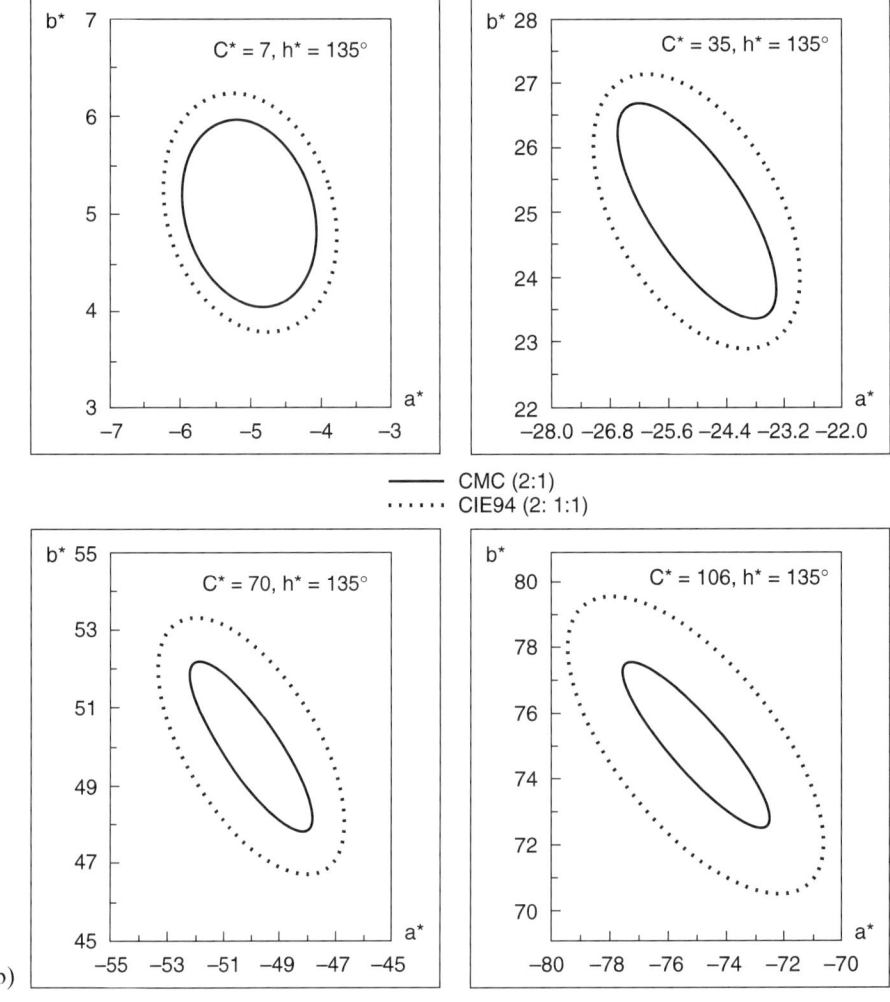

Figure 3.6. (*continued*)

$l:c$ ratios were optimized to different values for the two types of experiments. Unfortunately, what was not investigated was that the research laboratory experiments were done either in paint on paper or inks on metal, since these materials are easier to handle in small quantities. The acceptability data all came from textile production, where surface texture was a significant contributor to the viewing environment. At least one study on textured molded polypropylene [55] found that the 2:1 ratio was an excellent fit to their historical multivariate normal ellipsoids. The second use of the $l:c$ ratio is then to emphasize or deemphasize the importance of the measured property. By setting the $l:c$ ratio to 2:1, one doubles the nominal tolerance on lightness. Datacolor made good use of this feature in its QC software and found that, in general, a production facility tended to maintain the same relative significance for the color difference components across product colors and in many cases across customers. Thus, determining the appropriate $l:c$ ratio would need to

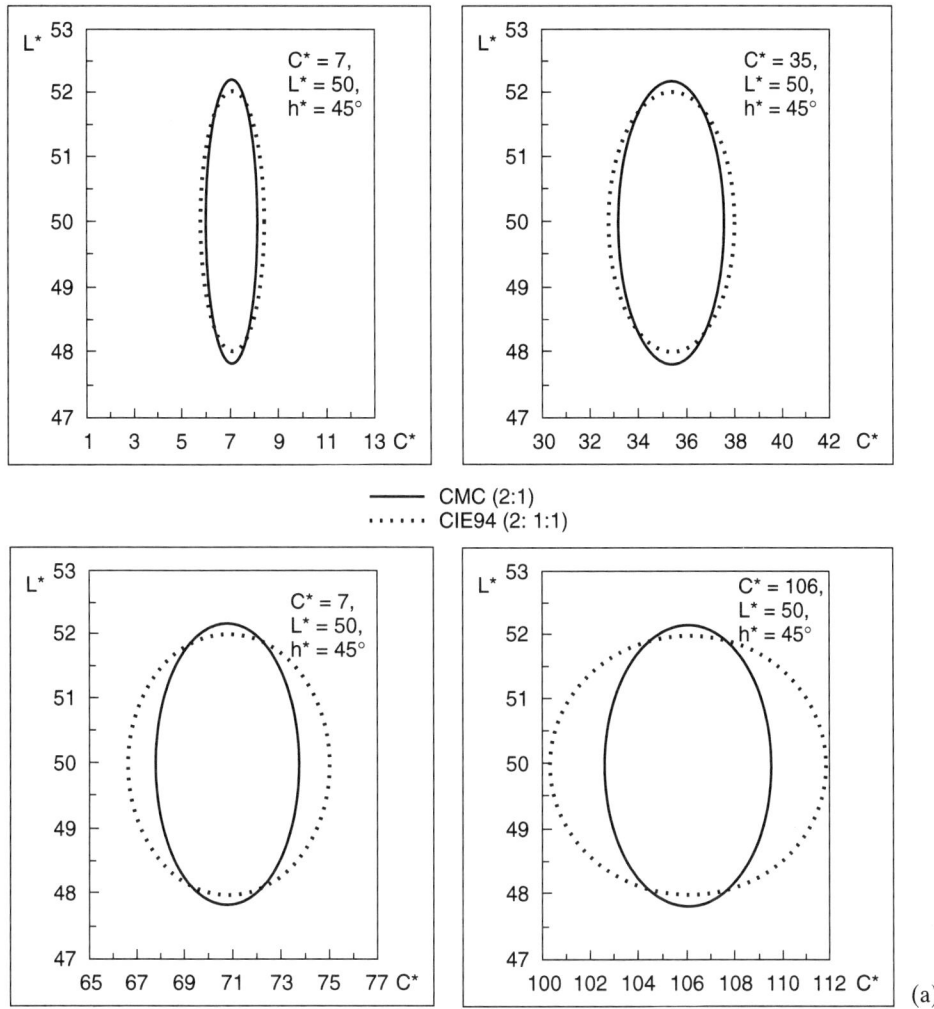

Figure 3.7. Ellipses for CIELAB C^*–L^* plane for metric hue angles of (a) 45° and (b) 135°.

be done only once or at most a few times and then applied to all other events, including new or infrequent production runs. Today, a good "rule of thumb" should be to use an $l:c$ ratio of 1:1 unless some known parametric factor, such as texture, surround, or pattern, would prompt one to use a different factor.

An additional parameter introduced into the CMC equation, the "commercial factor," is a number that changes not the shape of the tolerance solid but the volume of the tolerance solid. The need for this arose because of the definition of "acceptable tolerance" limit. It was desired to have an equation in which the tolerance limit would always occur at the value 1.0. Thus, values below 1.0 would be identified as "acceptable" and values above 1.0 would be identified as "unacceptable," or "pass" and "fail," respectively. But the requirements of automotive interior plastics or computer cabinet panels are much more critical than those of toys or single-use pro-

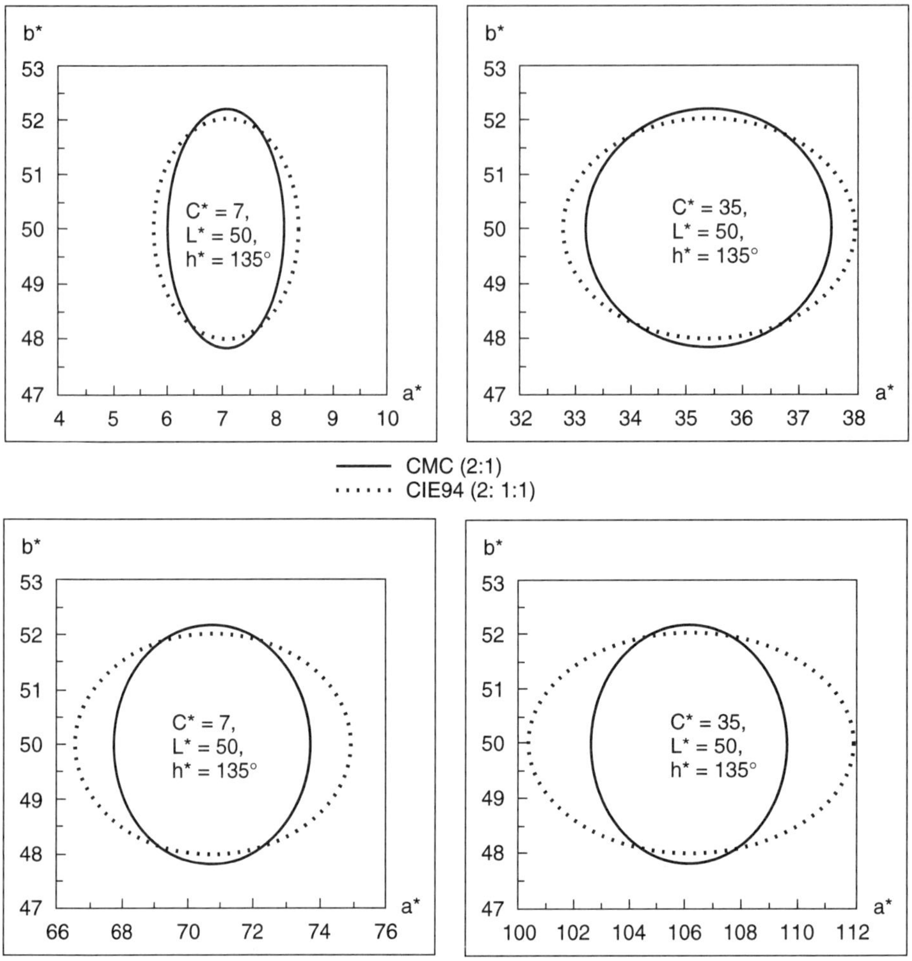

Figure 3.7. (*continued*)

motional items. One could solve this problem in the same way as when using a color difference formula, by setting the tolerance at less than 1.0 for critical items and greater than 1.0 for noncritical items and at 1.0 for typical items. But to keep the single-number shade passing philosophy intact, it was decided to add one more term to the equation and call it the "commercial factor" cf, which would increase or decrease the volume of the acceptance region but the pass–fail decision would still be based on the value 1.0 Great care needs to be exercised when changing this number as it is applied to the volume of the tolerance region as the volume and radii track almost linearly between 0.0 and 1.0.

After observing the rapid acceptance of the CMC equation, the CIE set up a technical committee to look at whether it should issue a recommendation for the CMC equation or derive a new equation. The committee developed a large set of color difference specimens, all produced in automotive lacquer on metal. These samples have extremely smooth surfaces. Visual judgments were performed in a university environment but included experienced color matchers as well as students. When tested against the CMC tolerance equation, the visual judgments showed much more disagreements than other studies. So the committee derived a new equation, finally optimizing it against both the new data and a compilation of data from earlier studies, weighting the newer data most heavily. The CIE equation had a constant S_L term instead of the hyperbolic term of the CMC. It also had only a linear function of C^* in the S_C term and only a linear function of C^* in the S_H term, instead of the complex function h^* found in the CMC tolerance equation. The result of the committee's activity was issued as a recommendation [56] for study until a better formula could be derived and has come to be known as the CIE94 tolerance equation. The statistics of the testing done by the committee clearly showed a performance improvement over the CMC equation. The work is significant for two reasons. It was the first action taken by the CIE since 1976—indicating a renewed interest in color differences and color tolerances on the part of the CIE. It was also the most rigorous numerical optimization reported in the literature. The statistical justification for each step and component in the equation was clearly reported and verified. The committee strongly resisted the temptation to "overfit" the data and included only terms that showed significant statistical improvement in the total sums of squares between the observed and predicted results—in essence, the CIE committee tried to minimize the number of disagreements down to the level of the number of disagreements between observers but no further. The CIE recommendations included much more than just the equations given below. They also set the boundaries within which this equations were considered to be statistically valid. This is something that no one had done before and was a recommendation that was long overdue. These standard basis conditions cover a wide range of appearance-modifying parameters and include the illuminance level of the specimens, the position of the standard and test specimens, the surround and the background conditions, the mode of perception, and the size of color differences. The basis conditions are summarized in Table 3.1. The following equations are used to compute the CIE94 color tolerance values:

$$\Delta E^*_{94} = k_V \left[\left(\frac{\Delta L^*}{k_L S_L} \right)^2 + \left(\frac{\Delta C^*}{k_C S_C} \right)^2 + \left(\frac{\Delta H^*}{k_H S_H} \right)^2 \right]^{1/2} \quad (3.5)$$

Table 3.1. Basis Conditions for CIE94 Tolerance Equation

Illumination	D65 source
Illuminance	1000 lx
Observer	Normal color vision
Background	Uniform neutral gray, $L^* = 50$
Viewing mode	Object
Sample size	>4° subtended visual angle
Sample separation	Minimum possible
Size of color differences	0–5 CIELAB units
Sample structure	Visually homogeneous

where k_L, k_C, and k_H are parametric factors, similar to the CMC $l:c$ numbers; k_V is the volume adjustment to make the predicted volume agree with the visual acceptance volume; and S_L, S_C, and S_H are the location-dependent distortions of the CIELAB color difference metric:

$$S_L = 1 \qquad S_C = 1 + 0.045 C^* \qquad S_H = 1 + 0.015 C^* \qquad (3.6)$$

Immediately after the publication of this equation, the color technologists who had already adopted and endorsed the CMC equations began to test CIE94 to see if it was indeed a superior tolerance metric. Until recently, the issues of the formula for S_L, the complexity of S_H, and whether or not all of the tolerance volumes should point toward the white point have been debated. In both CMC and CIE94 all tolerance regions have their major axis along the lines of CIELAB C^*, but in studies from Bradford University [52] and from the University of Granada, Spain [57], ellipsoids in the purple-blue region (metric hue angles of 290°–325°) are titled away from the direction of the white point. The most recent studies [58] have reported that the S_L function should be neither linear nor hyperbolic but quadratic, centered on a lightness value of about half the difference in lightness between the specimen and the background, a result first reported by Semmelroth [59,60]. It has also been observed that the tritianopic (yellow-blue confusion) lines in the visual system predicts a rotation of the ellipsoids away from the neutral point and toward the location of unique yellow. Studies by Rohner and Rich [50] have also indicated that the fitting of the local nonuniformities may be logarithmic rather than linear or hyperbolic since functions of the form $1/(1 + x)$ are the first terms of the Taylor expansion of $\log(x)$. Using logarithmic adjustments to CIELAB coordinates, they were able to derive a new color difference metric that has been adopted, in a slightly modified form, by the German standards organization, DIN [61].

All of this recent activity and some that has yet to be published are expected to result in further recommendations from the CIE, including a new color tolerance equation that may be termed CIEDE2000[62]. Whatever the actual content of the recommendation, it can be expected that the predictions of perceptible color differences and acceptable color differences will be statistically valid with an error measure that will be generally no greater than the errors made by experienced, professional color matchers.

3.8. CONCLUSIONS AND RECOMMENDATIONS

The precision and accuracy of instrumental color measurements and the associated color tolerance can provide a system of control that equals that of an experienced color matcher. But to achieve this performance, the product standard and production test specimens must be of the highest quality. As was pointed out, more than two decades ago, the weak link in instrumental color control is the specimen. The human visual system can scan a specimen marred by many inconsistencies and the human mind will ignore those features that are not relevant to the job at hand. A analog or digital colorimeter cannot ignore those defects. A spectrocolorimeter is a tool for the color matcher but it is not a hammer. It needs to be treated like the precision instrument it is. A mold-maker would never use his or her tools in a dirty or unsharpened state and a color matcher should never use a tool without knowing that is has been standardized and verified to be in correct operating condition. The second most significant task in making instrumental color control work properly is avoiding "mononumerosis" and "data diarrhea." If the pendulum swings too far either way, the color control system will break down. Colored plastic is more than just a set of tristimulus values. Tristimulus values do not describe or predict what a person "sees"; they describe the set of visual primaries that must be mixed to match the isolated color of the specimen. Appearance attributes such as transparency, shape, texture, and viewing conditions will all affect what we see but are not captured in the tristimulus information. While it is an area of active research, there are no truly good models of color appearance available yet, so trying to collect and analyze too much visual data will only result in mass confusion. Each appearance attribute needs to be identified, characterized with the correct tool, and then an appropriate tolerance applied.

In numerical tolerances "its not the size, its how you use them" that is important. Just making tolerances smaller will not result in an improved product. What is required is learning to live with realistic color tolerances and holding those tolerances batch after batch. If the tolerances are too small, then statistically valid random sampling will be replaced with systematic searches for acceptable readings. Finally, changes to adjustable parameters should be made only after careful review of experimental data that clearly confirm the decision to adjust each parameter.

REFERENCES

1. R. Donaldson, proc. *Phys. Soc. (London)*, **59**, 544 (1947).
2. VCS-10 is a trade name of Applied Color Systems, Inc., Lawrenceville, New Jersey.
3. S. L. Wright, R. W. Nywening, S. E. Millman, J. Larimer, J. Gille, and J. Luszcz, *Sixth Color Imaging Conference: Color Science, Systems, and Applications*, November 17–20, 1998, pp. 100–105.
4. F. W. Billmeyer, Jr., and D. C. Rich, *Plastics Eng.*, **34**(12), 35–39 (1978).
5. D. C. Rich, *Die Farbe*, **37**, 247–261 (1990).
6. D. C. Rich and D. Battle, "Obtaining Excellence in the Color of Decorative Coatings through the Use of Innovative Spectrocolorimeters", paper presented at the Annual Meeting of the Federation of Societies for Coatings Technology. October 13 1994.
7. D. C. Rich and J. M. Jalijali, "Numerical Standards: Are the Materials Ready?, paper presented at the AIC Silver Jubilee Meeting, Princeton, NJ, June 23–24, 1992.
8. *Colorimetry*, Publication CIE 15.2, 2nd ed., CIE Central Bureau, Vienna, Austria, 1986.
9. D. M. Mulholland, *Color and Appearance Div. Newslett.*, **24**(2), 1, 4–5 (1993)
10. T. F. Chong, "Reproducibility of Color Difference Measurements on Textile Samples," in *Proceedings of 1993 American Association of Textile Chemists & Colorists International Conference*, Montreal, Canada, Oct 3–6, 1993, pp. 323–332.
11. E. I. Stearns, *Color Res. Appl.* **6**(2), 78–84 (1981).
12. ASTM E 308, *Standard Practice for Computing the Colors of Objects by Using the CIE System*, ASTM, West Conshohocken, PA 1999.
13. C. S. McCamy, personal communication.
14. F. W. Billmeyer, Jr., "Precision and Accuracy of Industrial Color Measurements," in M. Richter, ed., *Proceedings of the International Colour Meeting, Lucerne (Switzerland)*, Musterschmidt, Göttingen, 1965, pp. 445–456.
15. A. R. Robertson, *J. Opt. Soc. Am.*, **57**, 691 (1967).

16. K. D. Chickering, *J. Opt. Soc. Am.*, **60**, 1027 (1970).
17. W. H. Venable, *Color Res. Appl.*, **14**, 260–267 (1989).
18. H. S. Fairman, *Color Res. Appl.*, **20**, 44–49 (1995).
19. F. W. Billmeyer, Jr., and P. J. Alessi, "Assessment of Color-Measuring Instruments for Objective Textile Acceptability Judgement," Natick Technical Report TR-79/044, U.S. Army Natick Research and Development Command, Natick, MAs, 1979.
20. Active Pixel Technology Primer, Photon Vision Systems LLC, Cortland, NY.
21. Detector schematic provided by EG&G Electro Optics. Division of Perkim Elmer Optoelectronics, 2175 Mission College Blud, Santa Clara, CA 95054.
22. "International Vocabulary of Basic and General Terms in Metrology (VIM), ISO/IDE/OIML/BIPM (1984)," International Organization for Standardization, Geneva, Switzerland.
23. "Standard Terminology of Appearance," ASTM E 284, ASTM, West Conshohocken, PA.
24. D. C. Rich, "Geometric Specifications and Tolerances for Color Measurement," CSIR, Pretoria, South Africa, September 1997.
25. J. F. Verrill, P. J. Clarke, J. O'Halloran, and P. C. Knee, "NPL Spectrophotometry and Colorimetry Club, Intercomparison of Colour Measurements," NPL Report QU 113, National Physical Laboratory, Teddington, United Kingdom, 1995.
26. J. F. Verrill, "Intercomparison of Colour Measurements Synthesis Report," Report EUR 14982 EN, European Commission, Brussels, 1993.
27. D. C. Rich, *J. Electron. Imaging*, **2**(3), 231–236 (1993).
28. "The Relationship between Digital and Colorimetric Data for Computer-Controlled CRT Displays," CIE Publication 122, CIE Central Bureau, Vienna, Austria, 1996.
29. "A Method for Assessing the Quality of Daylight Simulators for Colorimetry," CIE Publication 51, CIE Central Bureau, Vienna, Austria, 1981.
30. G. Wyszecki, *Die Farbe*, **19**, 43–76 (1970).
31. D. C. Rich and J. M. Jalijali, *Textile Chemist Colorist*, **22**(3), 23–28 (1990).
32. D. C. Rich and D. Battle, "Obtaining Excellence in the Color of Decorative Coatings through the Use of Innovative Spectrocolorimeters," paper presented at the Annual International Conference and Exposition of the Federation of Societies for Coatings Technology, October 13, 1994. Atlanta, GA.
33. R. Stanziola, "Color Difference Calculations—Are the Samples Ready?" paper presented at the Inter-Society Color Council Judd Memorial Conference on Color Metrics, February 11–14, 1979. Williamsburg, VA.
34. D. C. Rich and J. M. Jalijali, "Numerical Standards: Are the Materials Ready?" paper presented at the AIC Silver Jubilee Meeting, Princeton, NJ, June 23–24, 1992.
35. J. F. Verrill and F. Malkin, "Thermochromic Behavior of the Ceramic Colour Standard," paper presented at the AIC Silver Jubilee Meeting, Princeton, NJ, June 23–24, 1992.
36. D. C. Rich and J. M. Jalijali, *Color Res. Appl.*, **20**, pp. 29–34 (1995).
37. "Standard Practice for Establishing Color and Gloss Tolerances," ASTM D 3134, ASTM, West Conshohocker, PA.
38. "Standard Practice for Visual Evauation of Color Differences of Opaque Materials," ASTM D 1729, ASTM, West Conshohocker, PA.
39. R. S. Berns, *Color Res. Appl.*, **21**, 459–472 (1996).
40. M. R. Luo and B. Rigg, *J. Soc. Dyers Colourists*, **103**, 126–132 (1987).
41. R. M. Rich, F. W. Billmeyer, Jr., and W. G. Howe, *J. Opt. Soc. Am.*, **65**, 956–959, 1389 (1975).
42. D. L. Alston, "New and Improved Tools for Color QC," paper presented at the Annual Meeting of the InterSociety Color Council, April 24–26, 1994. Troy, Michigan.
43. D. L. MacAdam, *J. Opt. Soc. Am.*, **27**, 294 (1937).
44. R. S. Hunter, *J. Opt. Soc. Am.*, **32**, 509–538 (1942).
45. D. Nickerson and K. F. Stultz, *J. Opt. Soc. Am.*, **34**, 550–570 (1944).
46. K. McLaren, *J. Soc. Dyers Colourists*, **86**, 354–366 (1970).

47. K. McLaren, *J. Soc. Dyers Colourists*, **92**, 317 (1976).
48. P. F. Taylor, *Am. Soc. Quality Control*, **5**, 67 (1977).
49. R. McDonald, *J. Soc. Dyers Colourists*, **96**, 372–377 (1980).
50. E. Rohner and D. C. Rich, *Die Farbe*, **42**(4–6), 207–220 (1996).
51. R. McDonald, *J. Soc. Dyers Colourists*, **96**, 486–497 (1980).
52. M. R. Luo and B. Rigg, *J. Soc. Dyers Colourists*, **103**, 86–94 (1987).
53. F. J. J. Clarke, R. McDonald, and B. Rigg, *J. Soc. Dyers Colourists*, **100**, 128–132 (1984).
54. "Parametric Effects in Colour-Difference Evaluation," CIE Publication 101, CIE Central Bureau, Vienna, Austria, 1993.
55. A. L. Liebeknecht and J. Suthers, "It Takes More Than Right Factors to Make CMC Work," paper presented at the InterSociety Color Council Interest Group II: Symposium on on Industrial Applications of Color, April 19, 1993. Cleveland OH.
56. "Industrial Colour-Difference Evaluation," CIE Publication 116, CIE Central Bureau, Vienna, Austria, 1995.
57. M. Melgosa, J. J. Quesada, and E. Hita, *Appl. Opt.*, **33**, 8069–8077 (1994).
58. "Industrial Color Differences," CIE Technical Committee 1-46, Meeting held in Warsaw, Poland, June 24–30, 1999.
59. C. C. Semmelroth, *J. Opt. Soc. Am.*, **60**, 1685 (1960).
60. C. C. Semmelroth, *Appl. Opt.*, **10**, 14 (1971).
61. K. Witt, "DIN 99" A Poster paper presented at the CIE Quadrenniel Meeting, Warsaw, Pland, June 24–30, 1999.
62. "Improvement to Industrial Colour-Difference Evaluation", Publication CIE 142, CIE Central Bureau, Vienna, Austria, 2001.

CHAPTER 4

Advanced Color Formulation Technology

Robert Olmsted

Gretag Macbeth
617 Little Britain Road
New Windsor, New York 12553

4.1. INTRODUCTION

4.1.1 Computer Color Matching

Computer color-matching systems consist of a spectrophotometer, a personal computer, and color formulation software containing a database of pigments and resins. The system is designed to obtain accurate color formulas more efficiently than manual matching. It serves as a management tool for monitoring formula costs, utilizing waste, and archiving job data. Additionally, it serves as a quality control tool for incoming inspection of raw material.

The database is critical to the accuracy of match predictions and batch corrections. Measuring samples prepared with colorant, resin, and known mixtures of each creates the database. Once created, the database can be edited to meet changes in colorant and resin inventories.

To obtain a formula, samples are measured by the spectrophotometer and the computer searches the database to calculate a formula that most closely represents the color of the sample. Traditional systems rely on opaque or completely transparent samples for reflectance or transmission readings and provide a tristimulus match ($L^*a^*b^*$ values calculated from the reflectance measurement).

Coloring of Plastics, Fundamentals, 2nd edition. Edited by Robert A. Charvat
ISBN 0-471-13906-8 Copyright © 2004 by John Wiley and Sons, Inc.

Newer technology allows for opaque, translucent, transparent samples as well as multiple film thicknesses for both reflectance and transmittance measurements. The newer technology provides a spectral match (matches spectral curve of the sample at each wavelength across the visible spectrum: 360–740 nm). The advantage of spectral matching is its ability to match both color and opacity from a single database.

4.2. TRADITIONAL COLOR MATCHING

4.2.1 Visual Color Evaluation

Color matching is still as much an art as it is a science, and the importance of visual evaluation must not be taken for granted. There still is no substitute for the visual evaluation of color, though it is used more in the quality control function and less in the determination of initial formulas.

Traditionally initial matches required significant trial and error, even with a highly skilled color matcher. In most organizations the color matcher is aided by instrumental and computational methods to supplement visual evaluation. Besides more accurate formulations, the most important advantage to the use of a computer formulation system is the amount of time that it takes to obtain an initial match. By significantly decreasing the amount of time it takes to obtain an acceptable match, the profitability of the end product can increase dramatically.

4.2.2. Visual Evaluation Requirements

For successful visual evaluation of color, there are three important criteria that must be met. The most important requirement is normal color vision. However, the cases where thorough color vision testing is employed are far and few between. By using a test method such as the Farnsworth–Munsell 100-hue test, the level of color deficiency can easily be determined. A controlled viewing environment with standard light sources is also an important part of visual evaluation. Finally, there is no substitute for experience. Today it is becoming more difficult to find sufficiently experienced personnel. Instrumental methods help to reduce the amount of experience required and decrease the learning curve when developing visual evaluation skills.

4.2.3. Additivity Principle

The additivity principle means that if two pigments are mixed in a sample, the total absorption can be found by adding up the individual absorption's of the pigments. Both pigments behave as if the other one was not there. An example of this behavior is described in the following equation using K/S:

$$(K/S)a+b = (K/S)a + (K/S)b$$

where K is the coefficient of absorption and S is the coefficient of scattering.

Experimental data have shown that the additivity principle is not generally valid for a mix of two or more pigments [1]. In Figure 4.1a it can be seen that the addi-

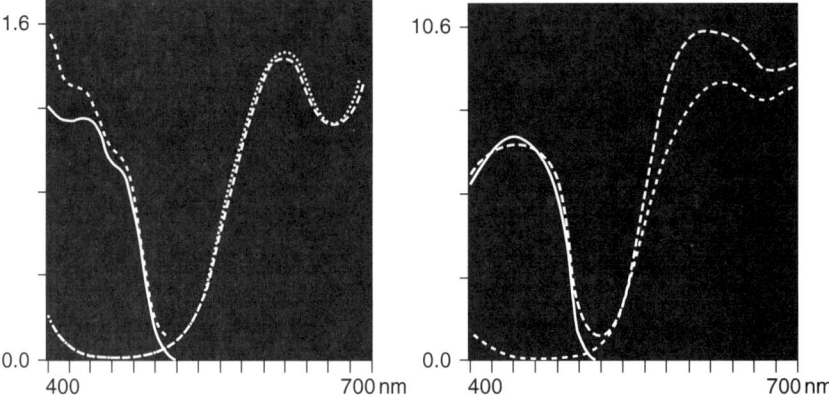

Figure 4.1. K/S versus wavelength.

tivity theory is valid in some cases; but it does not always hold true, as can be seen in Figure 4.1b.

4.2.4. Kubelka–Munk Theory

The Kubelka–Munk theory is used for reflectance measurements and calculations. The original Kubelka–Munk theory described the propagation of light in a stellar or star system. The same equations have been used for the interaction of light with pigment particles in paint, plastic, and ink mediums. Although these applications consider the same visible light, the distance between and the dimensions of pigment particles versus those of the stars are quite different [1].

Originally published in the 1930s by Paul Kubelka and Franz Munk, the Kubelka–Munk equations described the reflectance and transmittance of the sample as a function of absorption and scatter (K and S, respectively). The Kubelka–Munk theory is a two-flux version of the many-flux method of solving radiation transfer problems [2]. Given that the sample must have the same refractive index as air, these equations were not practical for industrial color matching. In the 1940s the Saunderson correction factors were introduced and the Kubelka–Munk equations became more practical for use in opaque systems. Simplifications and assumptions have been made to the original equations, and though these simplified formulas have many limitation, they are the dominant algorithms used in color-matching systems today.

Single- and Two-Constant Systems

Depending on the application, the Kubelka–Munk equations can be divided into two different case—single- and two-constant systems:

1. Single-constant theory assumes that the individual pigments do not significantly contribute to the total scattering of the sample. An example of this theory is the exhausting of transparent dyes into a textile substrate.

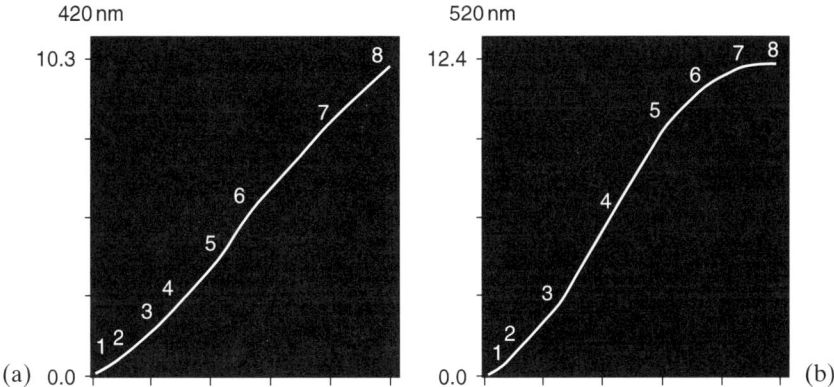

Figure 4.2. *K/S* versus concentration: (*a*) 420 nm; (*b*) 520 nm.

2. If the scatter is assumed to occur from two sources, the colorant(s) and the substrate, it is considered two-constant theory. An example of this theory is the formulation of opaque samples where titanium dioxide is blended with other pigments to achieve color. In this case the titanium dioxide becomes the second source of scatter.

Saunderson Correction

Given the limitations of Kubelka–Munk, a more complex equation was developed by J. L. Saunderson that contrasted the refractive index of the sample to that of air. With the addition of surface or specular (K_1) and internal (K_2) correction factors the equation became more practical for use in opaque systems.

Deficiencies of Kubelka–Munk Theory

Though the Kubelka–Munk theory has proven to be adequate in many applications, it has significant deficiencies that prevent it from being a total solution for color matching. Kubelka–Munk theory continues to be popular because it provides simple analytical equations and reasonable predictions [2].

ASSUMPTIONS OF KUBELKA–MUNK. It is assumed that the colorant layer is sufficient in extent for there to be no light lost from the edges of the layer and that it has uniform composition [2].

Kubelka–Munk reflectance arises from the assumption that the coefficients *K* and *S* are the same for forward and reverse flux. From a many-flux analysis it can be concluded that the angular distribution of the forward and reverse flux is not the same [2].

The Kubelka–Munk theory assumes a linear relation between the colorant characteristic *K/S* and the colorant concentration. In general, it is found that the *K/S* ratio of a component colorant is a nonlinear function of the concentration [3]. This means that it will not be possible to adequately describe the colorant behavior by using a linear relation. Figure 4.2*a* shows the linear relation that results between K/S and concentration. In Figure 4.2b the plot shows a typical nonlinear relationship between K/S and concentration at 520 nm for a series of calibration samples.

For the Kubelka–Munk theory to work, it is assumed that the pigment particles act independently of each other. The net result is obtained simply by adding up the individual actions [1].

4.2.5. Lambert and Beer Theories

The Lambert and Beer theory is reserved for transmittance calculations for very transparent samples. Dating back to the eighteenth and nineteenth centuries, the Lambert and Beer laws state that the absorbence [$\log(1/T)$] for a transparent sample is proportional to the thickness and the concentration of the colorant [4].

The Lambert and Beer laws has been found to be valid at low and moderate concentrations in transparent applications, but it may prove to be inaccurate at higher concentrations. In order for these laws to be valid, the absorption coefficient must be a constant independent of the concentration [5]. Since all colorant layers scatter some light, these equations, even in cases of slightly turbid media, are generally not valid.

4.3. ADVANCED COLOR MATCHING

4.3.1. Turbid-Media Theory

Although Kubelka–Munk is a turbid-media theory, it is limited in its application. It is important to note the advancements that have been made in other areas of turbid-media theory. This will help to determine which will best accommodate the different types of samples we see in today's color industry.

Turbid Media

There are three kinds of optical systems that define turbid media: optically thin, intermediate, and optically thick. Each of these systems can be seen throughout our everyday lives and each is significantly different from the other. Of all the theories that have been developed to handle the turbid media, only one can successfully handle all three optical systems.

OPTICALLY THIN. The scattered light that is observed in optically thin systems is scattered only once; much unscattered light emerges from the sample [6]. An example of such an application would be transparent *dyes* being exhausted into a textile substrate.

INTERMEDIATE. Most of the scattered light in intermediate systems has been scattered many times, but some unscattered light emerges from the sample [6]. A typical application would be a plastics operation that works with pigments and general-purpose polystyrene. Most systems that are typically assumed to fall into the optically thin and optically thick areas are actually intermediate media. Classically, offset printing inks are assumed to be optically thin and screen printing inks to be optically thick. In most cases both of these applications fall into the intermediate-media classification.

OPTICALLY THICK. All the light has been multiples scattered [6]. A paint manufacturer preparing opaque coatings, where titanium dioxide is blended with other scattering pigments to create a color, would be considered an optically thick system.

Application of Turbid-Media Theories [5]

Theory	Optically Thin	Intermediate	Optically Thick
Kubelka–Munk	No	No	Yes
Four flux	Yes	Limited	Yes
Many flux	Yes	Yes	Yes
Doubling	Yes	Yes	Limited
Monte Carlo	Yes	Limited	No
Scattering order	Yes	Limited	No
Diffusion	No	No	Yes

Billmeyer and Richards examined various turbid-media theories for their applicability in the three levels of optical behavior [7]. Of all the theories shown only the many-flux method accommodates all three turbid-media classifications.

Many-Flux Theory

The many-flux theory covers applications with all levels of optical thickness from one mathematical model. By using this model to determine absolute K and S values, the software does not have to define whether white is present in the formulation. All matching is done in one database. There is no need for separate packages that would use Kubelka–Munk single-constant, Kubelka–Munk two-constant, or Lambert–Beer mathematics.

4.4. CONTEMPORARY COLOR-MATCHING SYSTEMS

4.4.1. Many-Flux Theory

Today, the speed of the Pentium processors has allowed for much more complicated algorithms written on the Windows platform, such as computing many-flux calculations and spectral matching routines. New color-matching technology provides sophisticated and intuitive color formulation that is easily learned and used by the expert as well as the novice color matcher.

Using many-flux technology, computer color-matching systems can automatically formulate with or without white, at all levels of opacity, from a *single* database. In addition, satellite systems can offer added value by providing the same high-quality formulation results as a full system, with a low cost and feature limited satellite system.

54 Advanced Color Formulation Technology

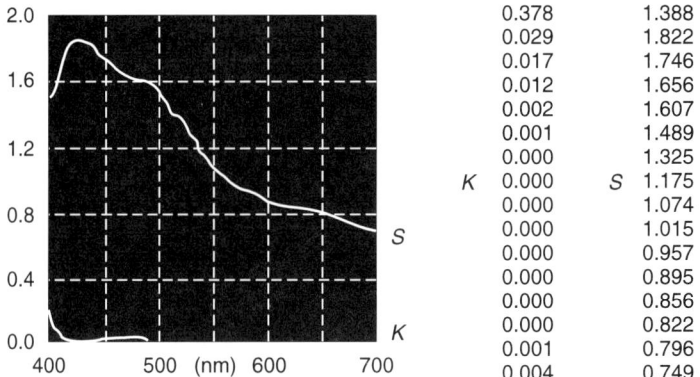

K		S	
0.378		1.388	
0.029		1.822	
0.017		1.746	
0.012		1.656	
0.002		1.607	
0.001		1.489	
0.000		1.325	
0.000		1.175	
0.000		1.074	
0.000		1.015	
0.000		0.957	
0.000		0.895	
0.000		0.856	
0.000		0.822	
0.001		0.796	
0.004		0.749	

Figure 4.3. K and S versus wavelength.

4.4.2. Kubelka–Munk Comparison with Many-Flux Calculations

Two Constant Always Used

Although a number of colorants can have very low absorption (extenders, resins, etc.) or a very low scattering (several pigments, colorants, etc.), there is no such thing as zero absorption or zero scattering. There is no single-constant behavior in the real world; nature is always two constant. Because all samples show two-constant behavior, the calculations are based on two-constant mathematics [1]. Though Kubelka–Munk used two-constant theory in some cases, it is limited due to the simplified equations and the assumptions about sample and colorant characteristics.

All Calculations in Absolute Units

K AND S. Due to the mathematical restrictions of the Kubelka–Munk equations, the pigment K and S data are calculated relative to a reference component (generally the white pigment) [1]. Typically the K and S of the white can be determined with a double measurement (reflectance and transmittance or over white and over black). This is performed on an individual sample or series of samples at constant thickness and varying white pigment volume concentrations. This does allow for the optimization of white pigment loading, though it can result in inaccurate calculations of opacity and/or pigment loading when considered in combination with other colorants.

The many-flux model calculates pigment K and S in absolute units. In addition, the calculation of K and S goes beyond the white pigment to *all colorants* providing accurate opacity and pigment loading predictions. The data in Figure 4.3 show calculated K and S data, where both K and S are variable across the visible spectrum.

K_1 AND K_2. The Kubelka–Munk equation is founded on the premise that once you disperse a pigment in a resin system, there is no further development. The determination of how much light enters a sample and how much exits after diffusion

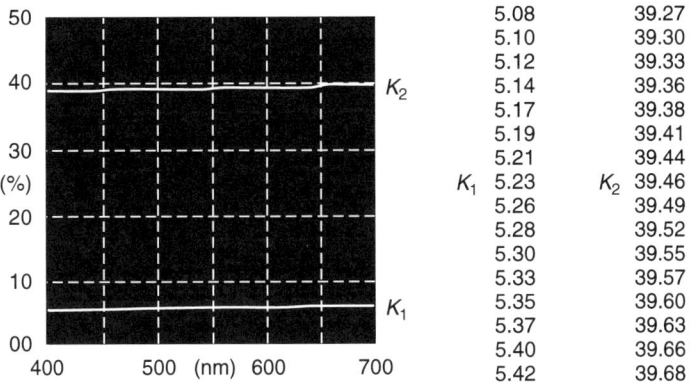

Figure 4.4. K_1 and K_2 versus wavelength.

relates directly to the Saunderson correction factors K_1 and K_2 [8]. However, most available software uses a fixed value for K_1 and/or K_2. In many cases these values are not calculate; the user inputs them (e.g., default values of 4% for K_1 and 60% for K_2). These values are fixed for all wavelengths or they can be a calculated value limited to a single wavelength. These methods are not valid because K_1 and K_2 are dependent on the refractive index of the material, which is wavelength dependent (see Fig. 4.4). Depending on the sample set, using a fixed K_1, K_2 value can lead to inaccurate calculation of absolute K and S. Figure 4.4 shows the variation that can occur in the determination of K_1 and K_2 values at individual wavelengths across the spectrum.

Nonlinear Relationships

Kubelka–Munk theory assumes linear relationships for K/S versus concentration and K/S versus thickness as well as the validity of the additive theory:

1. K (concentration)
2. S (concentration)
3. K (thickness)
4. S (thickness)
5. "additivity"

Many-flux theory treats all functions as completely nonlinear. It does not try to approach nonlinear functions through a piece-by-piece linear approximation.

Additivity Principle Not Used

Software should use a nonlinear function for the relationship between both K and S and pigment concentration. When two or more pigments are mixed into a sample, an interaction is calculated [1].

56 Advanced Color Formulation Technology

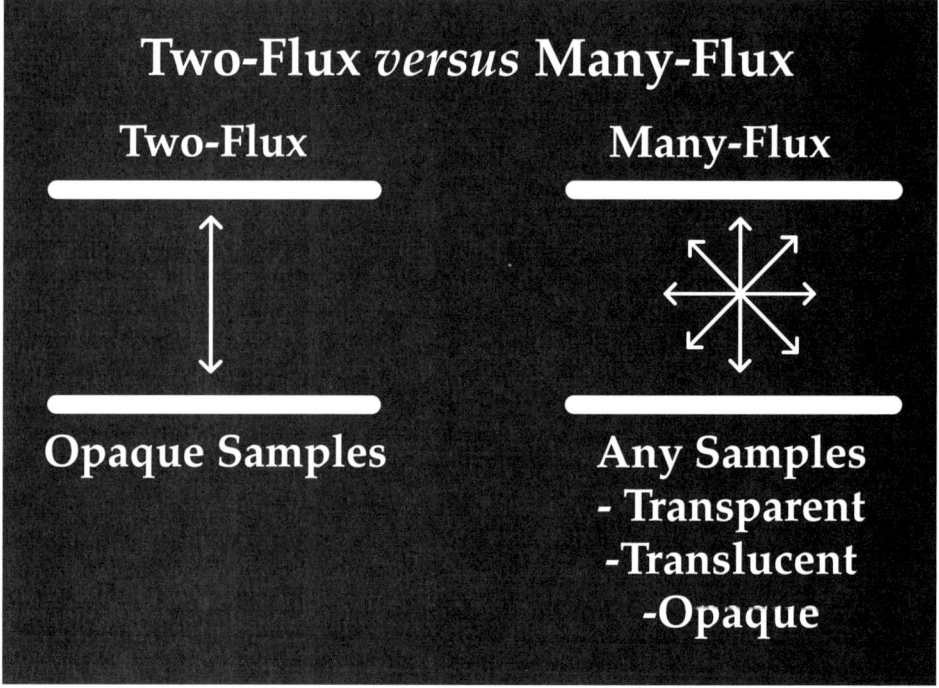

Figure 4.5. Two-flux versus many-flux theory.

4.4.3. Many-Flux Theory Technology

Many-flux theory can be applied to applications that have samples at any level of optical thickness (Fig. 4.5). All the calculations can be done in a single formulation package. The physics that takes place within the colorant/resin matrix demands that directional flux be determined. Many-flux theory considers light flux within the coloran/resin matrix in both an up and down flux (Kubelka–Munk) as well as directional flux.

4.4.4. Spectral Versus Tristimulus Matching

Tristimulus Matching

The typical approach for color formulation uses tristimulus matching routines.

MATCH ROUTINE
Match standard X, Y, Z
  Three equations to be solved
  3 Unknowns = 3 concentrations = 3 pigments

DISADVANTAGES. Because there are three unknowns, there needs to be at least three pigments in the formula. With one or two pigments an exact color match is not possible.

Spectral Matching

Computations are completed through an iteration process. There is no exact tristimulus match, but a best-fit spectral curve match is calculated. This is not a selective spectral match. Formulations are done using a spectral matching routine.

MATCH ROUTINE
Match standard spectral data R_i
    Iteration to achieve best fit

ADVANTAGES
1. There is better pigment selection due to more points of reference.
2. There are no limitations on the number of pigments (minimum or maximum).
3. Metamerism is minimized in calculations (spectral matching ensures quality under all lighting conditions, not just default conditions).
4. In general, there are more accurate formulations.

4.4.5. Dynamic Database

The database is an integral part of a computer color-matching system. It is the characterization of all pigments and resins from which formulas will be calculated. Creating a database involves producing samples of pigments, resins, and known mixtures of each. Once database samples are measured and saved to the database, the database is calibrated.

Contemporary color-matching systems enable sample additions to the calibration database for increased performance. In addition, the sample set is variable for each individual colorant. An application may require only 5 samples for a yellow colorant but 12 samples for a reflex blue to achieve optimum performance. Note that additional samples, are used in the calculation of the absolute K and S data. These samples are *not* a function in a "search-and-correct" calculations.

Complete file calibration uses all selected samples and optimizes the K and S data for all colorants. The results for *all* matches will improve, not just matches that are close in color or formula to the added samples.

Database Sample Set

The following is an example of a database sample set utilizing the many-flux model:

1. Mathematical minimum—2 (2 unknowns and 2 knowns):
    a. Linear relation is always correct (straight line, one sample).
    b. Sample validity cannot be determined.
    c. Concentration is dependent.
2. Nonlinear relation (more than two samples):
    a. Samples that are not correct can be easily identified.
    b. With more samples, there is better characterization.

3. Types of samples for each colorant:
   a. different concentrations with resin,
   b. mixes with white (help to define white using "nongray" samples), and
   c. mixes with black (lower reflectance values help to define the absorption characteristics).
4. Number of samples:
   a. application dependent and
   b. typically 7–10 samples per colorant.
5. Additional samples:
   a. Once the database is complete, it is possible to add more calibration samples to improve performance if necessary.
   b. All that is required is measuring the new sample(s) and recalibrating the database.

SAMPLE CHARACTERISTICS. In many Kubelka–Munk color-matching systems the user is required to present an opaque sample to the spectrophotometer. Depending on the application, this can be done a number of ways. For coatings, the technique of cross-coating several layers of colorant until opacity is achieved is commonly used. The generated sample then has become inconsistent with the typical process thickness. Although this can add error to the formulation, it will still adhere to the limitations of the Kubelka–Munk equation [7].

The samples required for the many-flux system must be at process thickness for the most accurate characterization of each colorant. Unlike Kubelka–Munk samples, the many-flux system can benefit from samples that have not reached complete opacity. The colorant sample's thickness is also used in the calibration process. This is important for applications that continually produce samples at varying thicknesses.

EFFECT OF SAMPLE SET. Table 4.1 shows how the number of samples used to characterize a database can affect the initial formulation. In this example the full set uses nine samples (six mass-tone letdowns, one with white, one with black, and one

Table 4.1. Full versus Limited Sample Set

Sample Set		Resin	White	Black	Yellow	Red	Green	Blue
1.	Actual	90	300	30	60	0	60	60
	Full	90	302	37	61	0	67	46
	Limited	90	301	0	77	24	40	69
2.	Actual	60	120	120	0	60	0	240
	Full	60	118	125	0	53	0	244
	Limited	60	106	155	0	13	0	266
3.	Actual	120	0	0	180	240	0	60
	Full	120	1	0	192	242	0	45
	Limited	120	0	82	199	197	2	0
4.	Actual	300	30	30	0	180	60	0
	Full	300	27	33	0	189	52	0
	Limited	300	25	39	27	168	442	0

with white and black) per colorant and the limited sample uses three samples (one mass tone, one with white, and one with black). The nine-sample database provides initial predictions that are much closer to the actual formula.

4.4.6. Applications

Because basic pigment properties are calculated using absolute units, how they are applied in the matching algorithm add versatility to contemporary color-matching systems.

Reflectance and Transmittance

Both reflectance and transmittance measurements can be applied to the same database. For plastics and translucent liquids and for printing or coating on nonopaque substrates, this is a very important feature. In these applications it is not sufficient to match a standard in reflectance only; a transmission match is equally important. Contemporary color-matching systems can combine both types of matches in one calculation using only one database. To use this capability, it is necessary that the spectrophotometer measure both the reflectance and total transmittance of the sample.

Contrast Measurements

Just as the reflectance and transmittance measurements can be used in one database, contrast measurements can also be applied to a single database (or even combined with reflectance and transmission measurements for the calibration of the database). In this case the two measurements that would be combined are the over white and over black measurements. Typically this technique is applied to coatings and printing inks on paper or screen inks on textiles (white and black cloth).

4.5 SUMMARY: ADVANCED COLOR FORMULATION

In practice, there are many sources of error apart from the inaccuracies of theory. In addition, there is human error, measurement error, batch variation of the colorants, and the nonreproducibility of the coloration process itself [2]. Once a process is under control, the next step is to apply color formulation software. It has been shown that many of the past methods have not proven to be completely viable for the variety of color applications encountered today.

Considering the full scope of many applications, traditional color matching has not provided a total solution for color matching. With the addition of the many-flux theory, and spectral matching, contemporary color-matching systems are the next step in providing a *total* solution for color formulation.

REFERENCES

1. Mark Maes, "Advanced Color Formulation," Colourplas Conference, Manchester, 1996.
2. J. H. Nobbs, *Rev. Prog. Color.*, **15**, 66–75 (1985).

3. G. Wyszecki and W. S. Stiles, *Color Science: Concepts and Methods, Quantitative Data and Formulae*, 2nd ed., Wiley, New York, 1982.
4. Anni Berger-Schunn, *Practical Color Measurement*, John Wiley & Sons, New York, 1994.
5. Deane B. Judd and Gunter Wyszeki, *Color in Business, Science and Industry*, 2nd Edition, John Wiley & Sons, New York, 1975.
6. Fred W. Billmeyer, Jr. and Max Saltzman, *Principles of Color Technology*, 2nd Edition, John Wiley & Sons, New York, 1981.
7. Fred W. Billmeyer, Jr. and L. W. Richards, "Scattering and absorption of radiation by lighting materials," J. Color and Appearance **2**(2), 4–15 (1973).
8. D. Mowery, "A Complete Solution for Computer Color Formulation," Color and Appearance Division RETEL Charlestion, NC Sept. 24–26, 1995.

CHAPTER 5

Visual Color Matching of Plastic Materials

Frances A. Leiby

5.1. INTRODUCTION

The process of defining and describing color is a very complex field and remains the subject of much research and numerous publications. Simply stated, light first strikes an object's surface, where it is modified, before proceeding to the eye, where the object's light or color is received. The eye transmits this received light as neural impulses to the brain for interpretation and subsequent color definition. Thus, color perception is the net result of three interdependent processes [1]. The first process is the modification of light by the colorant(s) at the object's surface, the second process is the perceptual process involving the reception of the light from the object by the human eye, and the third is the psychological process involving the brain's interpretation of the neural impulses from the eye. The physiological process by which light is received and transmitted to the brain as neural impulses is initiated by the responses of the rods and cones located on the retina in the back of the eye. The lens in the eye focuses the light on the retina. The rods have been determined to be responsible for night vision (low-light receptors) and play no part in the color perception process. The cones are thought to be primarily responsible for color vision through a mechanism involving three different types of cones, each responsive to different wavelengths, that is, red, green, and blue. The mechanism of these responses is beyond the scope of this chapter; however, the brain's ability to translate and interpret the eyes' responses to different light wavelengths into a perceived color is a truly amazing process.

Detailed research, aimed at studying the various areas associated with the eye's physiological color response, perception and color measurement, exists today within

---

Coloring of Plastics, Fundamentals, 2nd edition. Edited by Robert A. Charvat
ISBN 0-471-13906-8 Copyright © 2004 by John Wiley and Sons, Inc.

numerous research groups, some of which are spearheaded by Stephen A. Burns [2–16] at the Schepens Eye Research Institute in Boston, Massachusetts; Karl-Heinz Bauml [17–20] at the Institut for Psychologie, Universitat Regensburg in Germany; John A. Watlington at the MIT Media Laboratory TVOT Group; and Michael D'Zmura [21–39] at the University of California at Irvine. Within the contexts of the above research lie two primary objectives: first, to better understand the psychological and psychological responses of the eye and brain to light and color and, second, to develop a system matrix to better define color in order to eliminate, reduce, and even eliminate the "human" influence.

The human factors that influence the visual perception of color relate to the physical health of the eye, the individual's mood, and the individual's ability to discern color. Since color perception results from both the physiological and psychological responses to light, it follows that changes in health due to fatigue, sickness, physical injuries, medication, age, and mood swings all subtly affect the visual perception of color. Then, what characterizes one individual's ability to distinguish subtle color differences relative to another's ability and can this be simply quantified? Generally, these differences in color perception can be due to such physiological differences as heredity, eye injuries, or the aging process. The aging process typically causes the macular fluid in the eye to yellow, which in turns affects the light traveling through this fluid before being received by the fovea at the back of the eye. Thus, physiological differences are the greatest contributing factors that account for the observed differences in color perception between two or more individuals.

The Farnsworth–Munsell 100-Hue test [40] offers a complete set of limit standards combined with standardized sources that are easily administered to an individual to assess and quantify their color acuity. These tests evaluate the ability to discern between graduated subtle shade color differences (i.e., green to red, purple to red, blue to violet, and green to blue), which provide an overall assessment of an individual's ability to discern color nuance within different color spaces. This method is commonly used today to evaluate different individuals' abilities to perceive color differences and, when given on a regular basis, serves as a color reference standard to evaluate color-matching personnel.

Different models have been created to describe color. One of the most common models by Byk-Gardner assigns opponent colors to a three-dimensional plane (see Fig. 5.1). In this model, L, a, b define the color space, where L is a lightness-darkness coordinate, a is a redness-greenness coordinate, and b is a yellowness-blueness coordinate. Figure 5.1 further shows which color in each opponent pair is indicated by positive and negative values of a and b. The positive and negative notations are essential to matching with colorimeters. The opposing coordinates facilitate color discussions and movement decisions. The division of color into an opposing coordinate matrix facilitates the choices for color corrections during the visual color-matching process.

Colorants are typically defined as pigments and dyes that, when added to plastics, define a specific color. The subject of colorants has been discussed in numerous books and papers. However, in the context of this chapter, a dye is defined as a substance that is soluble in the resin system and produces color only by the absorption of light and no scattering of light. Pigments are not soluble in the resin system and therefore must be mixed into the resin by one of many dispersion processes. Factors such as pigment particle size and ease of mixing or dispersion directly influence their

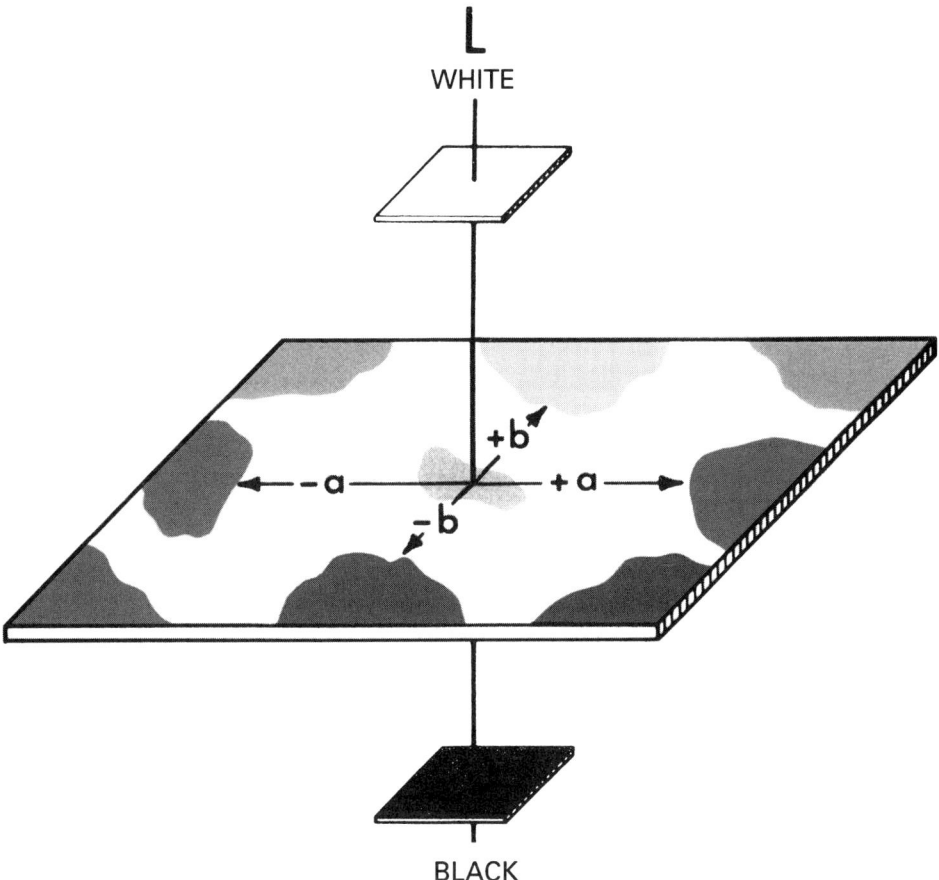

Figure 5.1. Three-dimensional L^*, a^*, b^* color representation.

contribution to the final color of the resin. Furthermore, many pigments will also scatter light with the net effect of increasing the opacity of the color match. The particle size of some organic pigments is so small, though, that they appear to behave as dyes and, thus, are often very difficult to distinguish from dyes in their color effects observed in plastic materials. The masstone of a colorant is the perceived color resulting from the addition of various levels of colorant in a specified polymer. The tint of a colorant is the perceived color resulting from the addition of a low level of colorant in the presence of different levels of titanium dioxide (white pigment). Breaking down a color standard into its various colorant components is just as challenging a task as choosing the colorants for the match to meet the application demands. New pigments and dyes are continually being developed to satisfy today's color demands within the continually expanding colored plastics arena.

Successful visual color matching requires a good understanding of the nature of color, how color is perceived, the influence of the physiological and psychological processes of color perception, the human element variability, and the ability to evaluate or assess an individual's color acuity. Next, it is important to define and discuss the other basic essentials required for successfully visually color matching plastic materials.

5.2. ESSENTIALS FOR VISUAL COLOR MATCHING

The visual color-matching arena is made up of many factors that both directly and indirectly influence the color match. The color standard or master, viewing light source, polymer type, application, thermal stability and light fastness, chemical resistance, opacity, gloss, surface texture, color match tolerance, choice of colorant(s) and additives and their cost, the dispersive mixing process and equipment, and the sample preparation for viewing all require a clear understanding before and during the visual color-matching process. Designing a standard color match request form that lists and defines all of the above factors will clearly define the scope of the target match.

5.2.1. Color Standard

The standard or color master is the single most important factor involved in any match because is serves as the reference to which all match trials are visually compared. It is absolutely necessary to ensure that the "submitted" standard or color master is the "true" standard the customer wants matched. Very often a part piece will be submitted for matching that is not the approved, true standard. Although the submitted piece may be within the color sphere relative to the approved, true standard, the color may lie to one side or the other, making a close match to the approved, true standard impossible. Always attempt to obtain an approved, true standard to use as the reference for future trial matches.

The condition of the standard is equally important. Is the standard dirty? Does it have fingerprints on it? Is it scratched or physically abused? Is the color standard "old"? Try to obtain another color standard if the answer to any of these questions is yes. A current, clean, well-cared-for standard is essential to a successful visual color match. It then becomes equally important that the color standard be stored under optimum conditions. These include storing in the absence of "all" light in a clean, dry environment away from heat along with careful minimal handling, even to the extent of handling only with clean, soft, cotton gloves.

Is the color standard based on the same material as the desired match demands? The color match becomes very difficult if the color standard is based in a plastic material that is different from the desired match plastic. A Pantone standard, an opaque painted paper, is still frequently used as a color standard. However, a Pantone paper standard does not address the issues of opacity, gloss, and texture, which are essential to color matching plastic materials. Similarly, painted metal color standards possess gloss levels that often exceed the range capable for most polymers. This factor also further complicates a visual color match in plastic materials with lower gloss. Thus, if possible, secure the color standard in the same material that the match is desired.

What is the physical form of the standard? Is it textured? What are the thickness and opacity? Is it a flat or stepped plaque or a curved piece from a colored part? Can the gloss of the standard be achieved along with the color match? These questions all refer to the physical form of a color standard to match in an injection, extrusion, or blow molding, or other process applications. Texture affects the visual color perception of the standard. Furthermore, the gloss of the standard is affected by a textured surface of the color standard. Color standards having textured surfaces usually appear "lighter" and are accompanied with a lower gloss than the same color standard possessing a smooth, polished surface that might promote the inherent "high" gloss of a specific base plastic material. Gloss reduction is one of the main reasons for texturing plastic materials, but textured color standards are more difficult to match unless the same base plastic material and mold texture are available for matching trials. Color standards containing curved surfaces interfere with the visual color comparison process by precluding side-by-side comparisons with the color match trials. The perceived color, when viewed across a curved surface, will appear different than the perceived color viewed across a flat surface because of the different angles at which the light hits the surface of each piece. Thus, it quickly becomes evident that the physical form of the standard is just as important as all of the above other factors that affect both the quality of the color standard itself and the ability to achieve a successful visual color match to the color standard.

5.2.2. Light Source

The visual perception of color is directly related to the source of light in which is it viewed. The spectrum of light is dependent upon the source of light. Natural sunlight is different from incandescent light, which is different from fluorescent light. Since the human eye sees only the light reflected and/or transmitted from the colored object's surface, the source of light striking the colored surface plays a major role in how the eye and brain perceive that color. The change in perceived color under different light sources from a colored object is called metamerism. Many pigments and dyes exhibit this property and, therefore, must be used carefully in order to avoid a metameric match, often dictated in the automotive industry. However, if the standard is metameric, then it becomes even more critical to define the light source so as to achieve a good visual color match with that color standard. The customer must agree before matching to a specified light source in which the color standard and trial matches are to be viewed. Otherwise, visual color match correlation between the color standard and the color match will often disagree.

The relative viewing angle of the color standard and the color match trial is also important once the viewing light source has been defined. Remember that the perceived color is that light received by the eye from the surface of the colored object. Thus, if these relative viewing angles differ, then their light reflection angles differ and the resultant gloss angles can differ. Also, the color standard may reflect the direct light while the match trial may reflect indirect light if their relative viewing angles are not the same, and this will also result in a difference in their perceived colors. Suffice it to say that when visually comparing the color standard to the trial match, it is imperative that both objects are viewed from the same angle under the same light source.

The Macbeth Division of the Kollmorgen Corporation [41] has categorized the basic three different light sources, daylight, incandescent, and cool white fluorescent, into a viewing box. Improvements in standardizing these different light sources

5.2.3. Polymer System

Every visual color match in plastic begins with choosing the base polymer system. Each polymer has its own inherent gloss, opacity, thermal and processing stability, additive packages (for thermal and UV resistance), and physical (including rheological) properties that must be assessed prior to starting the color match and maintained after the color match is completed. Is the polymer basically transparent or opaque, whitish or yellowish in color, or glossy in its natural form? Does it require high heat to process? Must the plastic be dried first? Does the base color change after processing the material? These questions require answers before starting a color match to ensure that visual changes noted in the match trials are not the result of the base polymer shifting color due to its thermal processing heat history. Typically, this is why additional thermal and color stabilizers are often added to match trials. A great deal of information on a wide variety of different plastic resins is available today from resin suppliers. This information not only serves as a guide to the thermal processing or mixing of plastic materials, but often includes additional information relating to their thermal stability and rheological properties.

Flame-retardant materials and glass-filled polymers further complicate a visual color match. Opacity factors, a "rough" surface due to glass fiber, and changes in the polymer base color require careful consideration before starting a match. Colorant chemical compatibility with flame retardants must be known. Thermal and rheological properties very often change due to the presence of flame-retardant materials and glass fiber fillers. Glass fiber increases the viscosity of the polymer and thus the shear during mixing that is often further accompanied by changes in the base color of the polymer.

It is very important to understand the plastic material before starting a color match! Know the polymer, its limitations, processing conditions, physical properties, and other attributes. The more information known about the polymer, the more assured and defined is the visual color match.

5.2.4. Application

It is essential to thoroughly understand the requirements of the end-use application for the color-matched part. How does the application determine the color match? What market(s) will be color-matched part service: toys, automotive, cosmetics, packaging, garden equipment, furniture, sports, industrial? Each application limits the available choices of colorants and determines the choices of other additives. Outdoor applications often require UV stabilizers, many of which possess a yellow color. This fact must be factored into the color pallet before starting this type of match. Colorants must have good light fastness to be considered for outdoor applications. Toys and food packaging often necessitate the use of colorants approved by the Food and Drug Administration. High-chroma colorants are targeted by the cosmetic industry to achieve the depth of color often accompanied with high gloss and distinction of image (DOI) required by this industry. The DOI is the ability of the colored part's surface to behave like a mirror; reflected images are perceived on the surface of the colored part, which behaves as a mirror of high resolution when viewed at a specific angle in a specified light source.

The process involved in producing the end-use application is equally important in determining the choices of colorants. Time and temperature determine the process, which in turn often limits the range of available colorants. Rotational molding, blown and cast film, blow molding, injection molding, extrusion, slurry molding, and compression molding all cover a wide processing temperature ranges involving different residence times. The process plays a major role in determining the success of the color-matched application and thus must be thoroughly defined for each application.

5.2.5. Mixing Process

There are many available methods for mixing colorants into plastic materials. Choosing a laboratory method that will most closely match the production process is ideal, but not usually the norm. Colorant dispersions in plastic materials determine the perceived color. Thus, choosing a consistent, reliable mixing method that closely mimics production dispersions is critical to a good visual color match.

Small twin screw extruders, banburies, two-roll mills, and single screw extruders are commonly used today to disperse colorants in color match trials. The injection molding of resin and dry pigment and resin blends is another commonly used method. Each process will often disperse or "develop" the color differently, so it is important to use the same mixing method throughout the complete visual color-matching process. Good correlation between the laboratory color match and the production run is more often obtained when the same piece of mixing equipment is used for both. For example, color matching done on a small laboratory banbury will often scale up very well to a large banbury production situation. Conversely, color matching done on an injection molding machine may not scale up well or correlate to plastic materials processed, for example, on a twin screw extruder. The reasons for this are most likely due to the differences in dispersion or mixing between the injection molding machine and the twin screw extruder. A good mixing process will usually achieve a good colorant dispersion, which will reduce the necessary amounts of pigments that will ultimately reduce the cost of the visual color match. Very often there is not the opportunity to choose optimum laboratory mixing equipment in order to ensure good production correlation. However, a good understanding of the mixing limitations is essential to project and anticipate the "first-time" production run perceived color. Experience plays a strong role in predicting and foreseeing the differences between laboratory-generated color matches and production material perceived color, often before the color match becomes a production situation.

5.3. SCIENCE OF VISUAL COLOR MATCHING

The science of visual color matching is defined as the processes involved in achieving the match requirements, limitations, applications, performance, and processing in a final visual color match. Each step within this process must be clearly defined in order to make the correct choices in subsequent steps. The end result is a "close" visual color match meeting all requirement demands.

The first essential step in color matching is to define the requirements of the color project. A standard form, that is, a color match request (CMR) form, is highly recommended. The CMR form, shown in Figure 5.2, asks the questions needed to define the sphere in which the color match must be accomplished: polymer type, standard, competitive material, performance, volume, cost, and so on. Once the CMR is fully filled out, the scope of the match is defined and the process can be started.

Determining the desired resin, optimal pigments, dyes, and additives and establishing the necessary paperwork to define the color match are the next steps in the process of color matching. An example of a color match work sheet is shown in Figure 5.3. The polymer type, its inherent base starting color, and its processing conditions determine the choices of pigments or dyes to be used. Ultraviolet stabilizers and some thermal stabilizers impart a slight yellow undertone color to some polymers, which must be factored into the base starting color before starting the match. Further complicating the starting base color of a polymer is the fact that some polymers change color during their processing. Polymers that require drying prior to molding or processing require additional preparation before matching in accordance with the manufacturer's recommendations.

The plastic color standard is the reference color control to which all color match trials are visually compared. In some cases, the standard must be returned with the match or after the match is approved. The ideal situation is to keep the color standard, labeled with the customer identification number, the in-house color number, date, and polymer type, if known. Handling, storage, age, texture, gloss, shape, size, and physical form all determine the condition of the color standard both as received and for future reference purposes.

The choices of pigments or dyes in the color pallet that can be used to match the color standard are directly determined by the polymer, its processing conditions, and the application requirements. Pigments and dyes, in addition to their color, can further be classified in several different ways: by its strength; FDA acceptable or permissible; inorganic (heavy metal, non–heavy metal containing); whether it is transparent, opaque, organic, light stable, and/or heat stable; ease of dispersibility; and chemical compatibility. The completed CMR form will automatically determine and limit these choices within the color pallet.

Each colorant is unique in appearance as defined by hue, value, and chroma. Simply phrased, hue is how we perceive an object's color [42]. Hue can further be defined as those graduated color variations that are visually perceived as one moves around the color wheel going from one color to the next color. The value characteristic of color is the degree of "lightness" or "darkness." Finally, the chroma of a color is its vividness or dullness. The three-dimensional color system shown in Figure 5.4 [42] pictorially shows the interrelationships between chroma, hue, and value all of which define a perceived color within a three-dimensional system.

A color matcher can also work within a two-dimensional slice of the three-dimensional color system shown in Figure 5.4. The perceived color trial match is often then described in a two-dimensional arena of "value" versus "chroma," as shown in Figure 5.5. This representation depicts the perceived color in terms of "whiter" versus "deeper," "cleaner" versus "dirtier," and "weaker" versus "stronger." These visual perceptions then direct the color matcher to make

COLOR MATCH REQUEST FORM

CMR NO: _____

CUSTOMER _____	REQUEST DATE _____
_____	DATE LAB RECD _____
_____	REQUESTED SHIP DATE _____
_____	SALES PERSON _____

ATTN: _____

TELE: _____

FAX: _____

STANDARD SUMITTAL FOR MATCHING
- ☐ Pollets
- ☐ Parts(description) _____
- ☐ Other(description) _____

MATCH DATA   Quantity
- ☐ New match for   Chips   Plaque _____
- ☐ Rematch for   Chips   Plaque _____
- ☐ Match for production
- ☐ Concentrate   ☐ Dry color
- ☐ Precolor   ☐ Liquid color

COLOR MATCH IN RESIN:

CUTOMER PROCESS
- ☐ Injection Molding
- ☐ Hot Runner
- ☐ Extrusion
- ☐ Film Extrusion
- ☐ Blow Molding
- ☐ Roto
- ☐ Other

CUSTOMER RESIN PROPERTIES
- Supplier: _____
- Grade resin: _____
- Type: _____
- Color: _____
- Customer code: _____

GLOSS
- ☐ High gloss
- ☐ Low gloss

OPACITY
- ☐ Same
- ☐ Less
- ☐ More

HEAT STABILITY
MAXIMUM OPERATION TEMPERATURE
- ☐ 400 - 450 °F
- ☐ 450 - 500 °F
- ☐ 500 - 550 °F
- ☐ 550 - 600 °F
- ☐ 600 °F +

LIGHT STABILITY
- Indoor
- Outdoor
- Indoor / Outdoor
- ☐ U/V Stabilizer (attach special request)

COLOR MATCH REQUIREMENTS
- Automotive / cosmetics (critical)
- Industrial commercial
- ☐ Identification color only

SPECIAL REQUIREMENTS
- ☐ FDA   Glassfilled
- ☐ NSF   ☐ Flame Retardant

EVALUATION ILLUMINANTS
- ☐ Incandescent / Horizon
- ☐ Coolwhite/ Flourescent
- ☐ D65
- ☐ Outdoor

EVALUATION COLOR SCALES
- ☐ CIELAB   ☐ CMC   ☐ FMC II   ☐ OTHER

Current customer volume: _____   Potential volume of this request: _____

Current competitive pricing: _____   End use/Application: _____

Additional comments: _____

COLOR COST:
- ☐ STANDARD COLOR
- ☐ PREMIUM COLOR- PRICE: $ _____ /lB
- ☐ FOB
- ☐ DLVD

LAB

Assigned Product / Color: _____

Date CMR completed: _____

CMR Approved by: _____

Sample Order information

Price _____

Date shipped: _____

Shipped VIA: _____

Sample Approval: _____   Credit Approval: _____

Figure 5.2. Color match request form.

70 Visual Color Matching of Plastic Materials

COLOR MATCH WORK SHEET

Matcher _____ Date _____ Color # _____
Standard _____ Customer _____ Color _____
Resin _____ End Use _____ GP or FDA _____ Ref. # _____

Material	1	2	3	4	5	6	7	8	9	10

COMMENT: _____

MATCH APPROVED: _____ DATE: _____ DISPOSITION: _____

Figure 5.3. Color match worksheet.

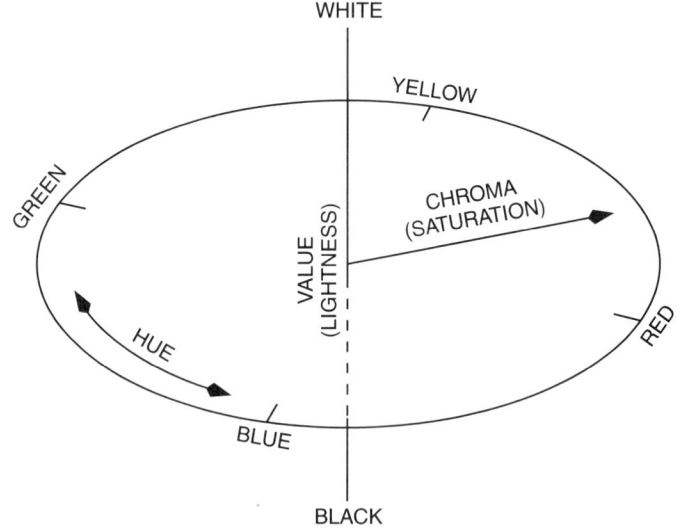

Figure 5.4. Three-dimensional color sphere.

Science of Visual Color Matching   71

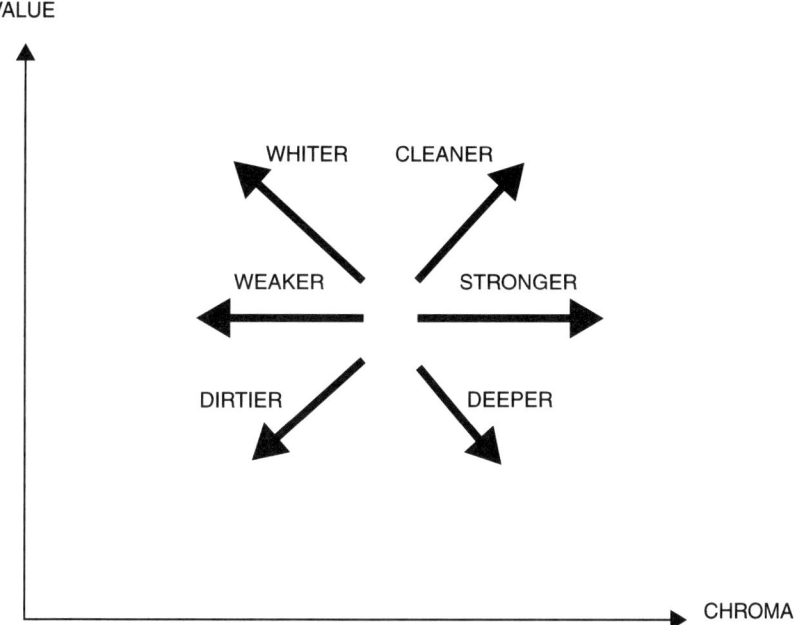

Figure 5.5. Two-dimensional work space.

successive choices and changes in the amounts of colorants as needed to visually match color standard.

Libraries and house files of color masstone, tints, and similar matched colors in different plastic resins can save a great deal of time in defining the color sphere for the target match. Additionally, many colorant suppliers can provide assistance in supplying some of this information. Most colors can be matched with four or five colorants. White and black assist in controlling the light-versus-dark aspects of a match. Choices for the other colorants are usually determined by the information contained on the CMR form as well as the target color. Colorant strength and cost are also important considerations at this point in the matching process. Once the color match color space has been identified, the next steps involve the visual "closing in" on the standard through a series of quantified trials, making the necessary colorant adjustments and reviewing the choices of colorants as needed to fine tune the visual color match within the scope of the CMR form.

Several methods are commonly used today to mix colorants into the plastic resin during the laboratory color-matching process. These options all involve, first, the dry blending of materials and, second, the dispersive mixing of the colorants (and additives) into the melted plastic matrix via one or more of the following techniques: injection molding, milling, twin-screw or, single-screw extruder processing, and banbury processing. Pelletized processed materials are best evaluated in accordance with the application, that is, injection molded, rotationally molded, blow molded, and so on. Direct injection molding trials, however, shorten this step and greatly shorten color-matching times. However, variations in colorant dispersions throughout these different steps often result in differences in the overall color development and, subsequently, the visual color match. It is these differences that often necessitate repeating trial attempts via different methods to ensure visual color reproducibility to avoid excessive costs and production problem situations.

A visual, deductive assessment process that compares each color match trial with the plastic color standard followed by adjustments to the next trial is recommended. One good technique requires the color matcher to ask and answer the following questions after each trial, always keeping in mind all requirements noted in the CMR form:

1. How does the color match trial appear relative to the color standard under different light sources? Is there metamerism?
2. Is the color match trial darker or lighter than the color standard?
3. Is the color match trial bluer or yellower than the color standard?
4. Is the color match trial redder or greener than the color standard?
5. Is it more opaque, brighter, or chromatic than the color standard?
6. Is the standard cleaner or dirtier than the color match trial?
7. Does viewing angle or positioning of the color standard to the color match trial make a difference in the perceived visual color?
8. Are gloss differences between the color standard and the color match trial affecting the way the color is being perceived?
9. Is there a texture difference between the color standard and the color match trial?
10. What colorant(s) component(s) of the color match trial should be increased, decreased, or replaced to fine tune the next color match trial and by how much? Why?

Asking these 10 questions to visually color match will move the color match closer to the standard. Color eye perception, training, and experience minimize the number of color match trials required to achieve the target match. The primary function of a colorant is to impart its color hue to the match, but subtle secondary functions of the same colorant in blends with other colorants often complicate the matching process by lightening, darkening, or shifting the undertone to the red, green, blue, or yellow side of the color match sphere. The ability to predict or estimate these secondary undertone changes parallels experience. Similarly, experience helps determine the amount of change per trial. It is often much easier to overshoot the color in the attempt to bracket the color standard between two or more trials and then adjust accordingly. However, bracketing is not always possible due to unforeseen secondary interactions between colorants, as shown in the following examples.

Example 1: A Gray Color Match

This example shows the influence of primary colorants and their secondary effects in the visual color match of a gray color. A gray match trial appears visually dark, green, and blue relative to the color standard, but the opacity is satisfactory. The single addition of white in the next trial lightens the match but moves the trial even more blue relative to the color standard. Yellow is then tried in the next trial match to compensate for the increase in blue, but the match is now more green. The addition of red is then tried in the next trial match, which succeeds in compensating for the green and blue, but the trial match now appears darker than the color standard. Will successive trials varying the white and red achieve a visual color to the color standard? Is the correct colorant system being used? What is happening to the cost? Succeeding trial matches will answer these questions.

Example 2: A Red Color Match

An opaque, red color match trial is first tried using only red and white pigments. The result is a color that appears visually dark, red, and blue when visually compared to the color standard, but is sufficiently opaque relative to the standard. The addition of more white pigment in the next trial moves the color more to the red and blue side, making the perceived color darker. Decreasing the red in the next trial lightens the match, which is now less red but still too blue. Yellow is then added to try to reduce the blue and further lighten the match, but the perceived color is now too green and dirty. Having tried these approaches, it is now time to consider an alternative pigment approach. A clean, red-shade orange is then introduced in place of the yellow into the colorant system, which results in a lighter, less red, and more yellow match. The color is now bracketed between these match trials; now the red, orange, and white pigments are adjusted to "bring in" the color. Other pigments may be indicated if the color still cannot be satisfactorily matched. Additional reds and clean green or red-shade blues may be indicated to visually fine-tune this color match. At this point in the matching process, trial and error and pigment availability or acceptability (i.e., allowable) determine the subsequent steps necessary to achieve a close visual color match.

Example 3: A Maroon Color Match

A dirty, yellow-shade maroon color standard is to be matched. The colorants chosen for the first trial match are a carbon black pigment and a yellow-shade red. The visual color of this trial match is light red and yellow relative to the standard, but the opacity is equal to the standard. The carbon black is then increased in the second trial match, which results in a darker visual match that is now too green and less yellow. A red-shade violet is added to the third match trial with a reduction in the amount of yellow-shade red, keeping the carbon black level the same. The thinking here is to offset the green without going too red yet still maintain the darkness. However, this match trial is too dark, really green and slightly blue (i.e., no longer yellow). The fourth trial targets a reduction in everything in trial 3. This moves the blue more to the yellow side and the darkness to the lighter side, but the perceived color is still too green when visually compared with the color standard. Trial 5 involves reducing the violet and black and leaving the red, as in trial 4. This trial visually reduces the green and brings the color closer to the standard. Trial 6 repeats trial 5 except the red is slightly increased to achieve a very visually acceptable color match.

Example 4: A Blue Color Match

The color standard is a visually opaque, intense (i.e., high-chroma) green-shade blue. Medium-shade (between red and green) blue and white pigments are chosen for the match. The first trial resulted in a perceived color that was visually light red and blue. All white was removed and a green-shade blue was added along with the medium-shade blue in the second trial. This resulted in a match that was visually too dark, too red, and too blue. Next, half of the level of white from trial 1 was added, the medium-shade blue was totally eliminated, and its amount was replaced with the green-shade blue in trial 3. The lightness and darkness were visually acceptable in this trial, but the match was still too red and too blue. After reviewing the second trial, the fourth trial contained no white, but the medium-shade blue was replaced with a red-shade blue, keeping the green-shade blue but reversing their added amounts. This resulted in a perceived color that is too dark, too red, and too blue but much closer visually. The fifth trial is a repeat of the fourth trial, expect a green-shade yellow is introduced to move the match lighter, greener, and more yellow. Visually, this trial match corrected the color greatly, but the match was still slightly blue. Thus, trial 6 increased the amount of the green-shade yellow, which resulted in a visually acceptable color match to the standard.

Summary

The above examples focus on both the visual and the thought processes necessary to achieve a close visual color match. Constant questioning of colorant choices, trial match reviews, added costs, and colorant amounts along with a "scientific" trial-and-error approach will guide the color matcher to the visual color match. These examples provide some of the insight into this process. An experienced color matcher with a strong knowledge of pigments and dyes and polymers will greatly speed up the process of visually color matching to a color standard. Their visual experience in understanding the secondary effects of multicolorant combinations in different polymers very often guides their choices and minimizes the successive number of trial matches necessary to achieve the visual color match to the standard.

All plastic materials have different degrees of light transmission ranging from transparent to translucent to opaque by nature. Transparent plastic materials do not scatter light as it passes through. A translucent plastic material scatters, reflects, and transmits light, thus creating a translucent or milky visual effect. An opaque material highly scatters light, and although a small portion of light may pass through, the material appears visually opaque to light. The same analogies apply to the addition of colorants to these different plastic mediums. Colorants that dissolve in the plastic material (e.g., dyes) will have little effect on the transparency, translucency, or opacity of the final color match. However, colorants, which do not dissolve are pigments, but must be dispersed in order to achieve their color development which can dramatically affect the transparency, translucency, and opacity of the final color match. Thus, it is very important to understand the visual nature of the base polymer that is to be color matched.

A different approach is taken when visually matching transparent color standards. First, the base polymer must be transparent to light. Second, all colorants to be used in transparent polymers must be themselves transparent to light. This limits the pallet of colorants to mainly organic pigments and dyes. Another critical point in visually color matching a transparent color is the thickness of the color standard or color standard part. This is because as the thickness changes, so does the perceived color due to Beer's and Lambert's Laws that effect the contribution of more or less apparent colorant strength present at the given thickness. Thus, it is extremely important to have the thickness specified before starting a transparent color match. A standards library of transparent, very low concentration, single masstone organic pigments and dyes in the target resin will greatly aid in reducing the times for these matches. Otherwise, the first set of the trials may serve to generate as starting library for this match. Choosing the "correct" organic pigments or dyes will depend upon finding the correct mass tone at very low levels in the specified transparent base resin. Transparent grades of polycarbonates, acrylics, and styrenics have different processing temperatures, different transparencies, and different inherent base colors ranging from yellowish to water clear, all of which determine the choices of possible colorants that can be used to achieve the visual color match. Very often, optical brighteners or other blue-tinting pigments are added to the commercially available base polymers, further complicating the transparent match. Secondary visual color effects become even more important in a transparent color match because there is no opacity to help mask any of the negative secondary color effects. Very often it becomes the norm to change whole color systems before achieving a bracketed match. Thus, the approach to visually color matching transparent color standards requires a good understanding of the available organic pigments and dyes, their secondary color effects, and their primary color strength, hue, and light transmittance.

Certain special colorant types and additives profoundly influence the color-matching process due to their inherent special light effects. Metallic, fluorescent, phosphorescent, pearlescent, and "granite" colorants as well as optical brighteners and ultraviolet radiation and thermal stabilizers further complicate the color matrix. Aluminum flake, commonly referred to as metallic pigment, and pearlescent pigments alter the opacity and color space through the reflective properties due to their particle size and are often adversely affected by white, black, and other colorants. Fluorescent and phosphorescent colorants are also adversely affected by white, black, and many other colorants, increasing the difficulty of the visual color match. Optical brighteners, known to "shift" or "flop" the yellow color space to the blue color space, complicate the match by shifting the base undertone. Granite effects are achieved most commonly through the use of particulate micas of various particle sizes, some of which have specific colorants bonded to their surface to achieve additional visual color effects. Achieving a good visual granite effect color match requires matching both color and granite particle size effects. Good visual color matching is most often the only way to closely match those color standards containing these special effect pigment or additive systems.

The science of visually color matching is complicated. It requires the complete definition of all facets related to the match. The CMR form serves as a constant reference to the color matcher throughout the matching process to ensure that the right colorant choices are made. It is equally important to maintain an accurate written log of match trials and a visual progression of match trials to review throughout the matching process. A periodic review of earlier match trials within the same match project sometimes reveals choices not seen at the time but that might now suggest an alternative direction or approach to investigate in order to achieve the color match. A repetitive, systematic, deductive evaluation process is used to visually compare each color match trial with the color control or standard. Special-effects colorants demand even greater understanding and attention in order to achieve close visual color matches.

5.4. CONCLUSION

Visually color matching plastic materials is the science of formulating with colorants and additives in a specified plastic material to develop a visually perceived color that closely matches the visually perceived color of the color standard as well as all of the requirements outlined on a CMR form. This matching process has many variables that range from the basic physiological and psychological color perception processes of the eye and brain to those choices made along the pathway to achieve the final visually close color match. These variables have been identified and described throughout this chapter, always with the overall focus on clearly defining and understanding each variable and concept before starting and during the visual color match process.

However, a confidence factor does exist for each visual color match. Is the match reproducible in the color laboratory? Is the match the most cost-effective? Is it reproducible in a production situation? Should different colorants have been used that the color matcher was unaware of? All of the tools identified in this chapter support a yes to these questions. One key factor, time, reduces the optimum number of color trial matches to a minimum number. Some color matches are more difficult than others, necessitating more time, and some color matches are just not possible within the context of the existing color standard. The number of pigments and dyes is already astronomically high and will continue to grow, thus prompting the need to always maintain information libraries of existing and new pigments and dyes. A good self-assessment tool for all visual color matchers is to question every visually close color match. This forces an objective review of the completed close visual color match and process. Close visual color matches that comply with all of the CMR requirements most often have a high confidence factor associated with them.

The age of computer technology is here. The software packages currently available for computer color matching have tremendously improved over the past few years and will continue to improve dramatically as current human and computer research continues to search for those color recognition patterns that more closely mimic the human eye and brain processes. So far, the eye still has the lead, but for how much longer?

REFERENCES

1. B. M. Mulholland, *Plastics Eng.* 33–36 (1997).
2. A. E. Elsner, S. A. Burns, and R. H. Webb, *J. Opt. Soc. Am.*, **10**, 52–58 (1993).
3. S. A. Burns and A. E. Elsner, *J. Opt. Soc. Am.*, **10**, 221–230 (1993).
4. S. A. Burns, S. Wu, F. Delori, and A. E. Elsner, *J. Opt. Soc. Am.*, **12**, 2329–2338 (1995).

5. S. A. Burns and A. E. Elsner, "Cone Photoreceptor Directionality and the Optical Density of the Cone Photopigments," in *Vision Science and its Applications*, Vol. 1, Optical Society of America Technical Digest Series, Washington, DC, 1996.
6. S. A. Burns, J. He, and F. C. Delori, "Do the Cones See Light Scatter from the Deep Retinal Layers?" in *Vision Science and its Applications*, Vol. 1, Optical Society of America Technical Digest Series, Washington, DC, 1997.
7. S. A. Burns, J. He, and A. E. Elsner, *J. Opt. Soc. Am.*, **14**, 2033–2040 (1997).
8. Y. Chang, S. A. Burns, and M. R. Kreitz, *J. Opt. Soc. Am.*, **10**, 1413–1422 (1993).
9. S. Wu, S. A. Burns, and A. E. Elsner, *Vision Res.*, **35**, 2943–2953 (1995).
10. S. A. Burns and A. E. Elsner, *J. Opt. Soc. Am.*, **13**, 667–672 (1996).
11. S. Wu and S. A. Burns, *J. Opt. Soc. Am.*, **13**, 649–657 (1996).
12. S. Wu, S. A. Burns, S. Reeves, and A. E. Elsner, *Vision Res.*, **36**, 1573–1582 (1996).
13. S. Wu, S. A. Burns, A. E. Elsner, R. T. Eskew, and J. He, *J. Opt. Soc. Am.*, **14**, 2367–2378 (1997).
14. S. A. Burns and R. H. Webb, "Optical Generation of the Visual Stimulus, in M. Bass, E. W. van Stryland, D. R. Williams, and W. I. Wolfe, Eds., *The Handbook of Optics*, McGraw-Hill, New York, **28**, 1994, pp. 1–27.
15. A. E. Elsner, S. A. Burns, J. J. Weiter, and F. C. Delori, *Vision Res.*, **36**, 191–205 (1996).
16. F. C. Delori and S. A. Burns, *J. Opt. Soc. Am.*, **13**, 215–226 (1996).
17. K.-H. Bauml and B. A. Wandell, *Vision Res.*, **36**, 2849–2864 (1996).
18. K.-H. Bauml, *J. Opt. Soc. Am.*, **12**, 261–271 (1995).
19. K.-H. Bauml, *J. Opt. Soc. Am.*, **11**, 531–543 (1994).
20. K.-H. Bauml, *Perception & Psychophys.* **53**, 338–344 (1993).
21. M. D'Zmura and K. Knoblauch, Spectral Bandwidths for the Detection of Color. *Vision Res.*, **38**, 3117–3128 (1998).
22. M. D'Zmura, K. Knoblauch, M.-A. Henaff, and F. Michel, Dependence of Color on Context in Case of Critical Vision Deficiency. *Vision Res.*, **38**, 3455–3459 (1998).
23. M. D'Zmura and B. Singer, "Contrast Gain Control," in L. T. Sharpe and K. R. Gegenfurtner, Eds., *Color Vision: From Genes to Perception*, Cambridge University Press, Cambridge, pp. 369–385 (1999).
24. M. D'Zmura and G. Iverson, "A Formal Approach to Color Constancy: The Recovery of Surface and Light Source Spectral Properties Using Bilinear Models," in C. Dowling, F. Roberts, and P. Theuns, Eds., *Recent Progress in Mathematical Psychology*, Lawrence Erlbaum, Mahwah, NJ, 99–132 (1998).
25. M. D'Zmura, "Color Contrast Gain Control," in W. Backhaus, R. Kliegl, and J. Werner, Eds., *Color Vision*, Walter de Gruyter, New York: 251–266 (1997).
26. M. D'Zmura, P. Colantoni, K. Knoblauch, and B. Laget, *Perception*, **26**, 471–492 (1997).
27. M. D'Zmura, P. Lennie, and C. Tiana, *Percept. Psychophys.*, **59**, 381–388 (1997).
28. M. D'Zmura and B. Singer, Contrast gain control. In Color Vision: From Genes to Perception, L. T. Sharpe and K. R. Gegenfurtner, Eds., New York: Cambridge University Press, pp. 369–385 (1999).
29. M. D'Zmura, P. Colantoni, K. Knoblauch, and B. Laget, Color transparency. *Perception* **26**, 471–492 (1997).
30. M. D'Zmura and B. Singer, *J. Opt. Soc. Am.*, **13**, 2135–2140 (1996).
31. M. D'Zmura, *Perception*, **25**, 1223–1234 (1996); translation of C. Bergmann" Anatomisches und Physiologisches ueber die Netshaur des Auges," *Seitschrift fure rationelle Medicin II*, **1857**, 83–108.
32. B. Singer and M. D'Zmura, *J. Opt. Soc. Am.*, **12**, 667–685 (1995).
33. M. D'Zmura, G. Iverson, and B. Singer, "Probabilistic Color Constancy," in R. D. Luce, M. D'Zmura, D. D. Hoffman, G. Iverson, and K. Romney Eds., *Geometric Representations of Perceptual Phenomena*, Lawrence Erlbaum, Mahwah, NJ, 1995, pp. 187–202.
34. G. Iverson and M. D'Zmura, "Color Constancy: Spectral Recovery Using Trichromatic Bilinear Models," in R. D. Luce, M. D'Zmura, D. D. Hoffman, G. Iverson, and K. Romney, Eds., *Geometric Representations of Perceptual Phenomena*, Lawrence Erlbaum, Mahwah, NJ, 169–185 (1995).
35. R. D. Luce, M. D'Zmura, D. D. Hoffman, G. Iverson, and K. Romney, Eds., *Geometric Representations of Perceptual Phenomena. Articles in Honor of Tarow Indow's 70th Birthday*, Lawrence Erlbaum, Mahwah, NJ, 1995.

36. B. Singer and M. D'Zmura, *Vision Res.*, **34**, 3111–3126 (1994).
37. M. D'Zmura and A. Mangalick, *J. Opt. Soc. Am.*, **11**, 543–546 (1994).
38. M. D'Zmura and G. Iverson, *J. Opt. Soc. Am.*, **11**, 2389–2400 (1994).
39. G. Iverson and M. D'Zmura, *J. Opt. Soc. Am.*, **11**, 1970–1975 (1994).
40. D. Farnsworth, "The Farnsworth-Munsell 100-Hue Test for the Examination of Color Discrimination," Munsell Color, Macbeth, Division of Kollmorgen Instruments Corporation, New Windsor, NY, 1957.
41. "The Broadest Choice of Standard Light Sources in the World's Most Advanced Color Matching Booth," New Windsor, NY, Macbeth, Division of Kollmorgen Instruments Corp. (6/93), Spectra light II.
42. "A Guide to Understanding Color Communication", August 1980, p. 6, Grandville, MI X-Rite Corporation Brochure.

CHAPTER 6

ISO 9000 and Other ISO Standards, Guides, and Test Procedures

Charles T. Bradshaw

Quality Measurements Corporation
428 Spruce Place
Concord, North Carolina 28026-1023

6.1. INTRODUCTION

Continuous quality improvement provides personal satisfaction that can only be estimated, but the rewards are nevertheless real and definitely obtainable by both large and small organizations. Conformity assessment against standards of the International Organization for Standardization (ISO) has again and again been demonstrated to be a useful tool for improving a supplier's processes regardless of type of products and services provided. In the case of laboratories that color plastics, the data related to physical properties and to color and appearance are normally reported in Certificate of Analysis format, and the validity of those data is of equal importance to the price and the timely availability of the colored products.

The agency issuing the ISO global standards is located in Geneva, Switzerland. Early on, the European Community (EC) adopted ISO standards and guidelines in an effort to remove trade barriers among its member nations. At the same time, it was perceived that the ISO standards erected trade barriers to the United States because registration of a supplier became a necessity for gaining access to European markets. Assurance of quality is obtained through ISO registrars and laboratory accreditation agencies. They perform third-party conformity assessments

---

Coloring of Plastics, Fundamentals, 2nd edition. Edited by Robert A. Charvat
ISBN 0-471-13906-8   Copyright © 2004 by John Wiley and Sons, Inc.

on behalf of customers. Concurrently, with the growing popularity of third-party certification, supplier audits have essentially been eliminated. That goes both for the suppliers' quality systems and for their testing capability, but periodic random sampling and testing for product auditing will continue to be performed in the customers' plant.

6.2. ISO TEST PROCEDURES

In addition to standards and guidelines, the ISO has developed several test procedures for plastics. They are commonly called ASTM equivalents since they comprise a parallel system to ASTM test methods for physical properties, such as tensile and flexural properties, Izod or Charpy impact strength, and deflection temperature under load or Vicat softening temperature. The main difference involved is in the expense of new molding dies that are fabricated to metric dimensions. These dimensions are not just equivalents but are like changing from a 50-lb bag of resin to a 25-kg quantity. Several years ago, "soft conversion" was described as simply relabeling the 50 lb as 22.7 kg, but later "hard conversion" changed to the 25-kg or 55.1-lb package. The hard conversion of dimensions of a tensile type 1A and 1B specimen is $150 \times 10 \times 4$ mm, while for a flexural properties test the preferred specimen measures $80 \times 10 \times 4$ mm and the critical dimension is thickness since it occurs in the formulas as either a squared or a cubed factor. Fortunately, the specification for an Izod bar has the same dimensions as the flexural test bar, but two different types of notch radii may be specified: 0.25 or 1 mm. For the ISO test for deflection temperature under load, when tested in the flatwise position, the same dimensions are used as for the flexural and Izod tests. But when tested in the edgewise position, a larger bar is specified, where the width dimension is actually the critical one although a range of 9.8–15.0 mm is specified.

There are other minor nonequivalencies of ISO test procedures compared to ASTM standards, but for color difference the ISO Test Procedure No. 105 is unique. Those who use Colour Measurement Committee (CMC) procedures—particularly CMC 2:1 Lightness to Color ratio—claim that it facilitates a uniform description for acceptability decisions that is better than any other system in existence. These equations permit the use of a single number tolerance, DE_{cmc}, in a nearly uniform color space. The CMC formula is a modification to the perceptibility CIELAB formula. It is fully described elsewhere in this book, but it deserves some brief notice here because, after all, it is an ISO procedure. The CMC developed the basic British Standard No. 6923, "Calculation of Small Color Differences." Soon afterward, in 1989, the American Association of Textile Chemists and Colorists (AATCC) adopted AATCC Test Method 173, "CMC: Calculation of Small Color Differences for Acceptability." Ford Motor Company indicated a preference for using CMC 2:1 ratio color difference for plastics weathering data for plastics interior trim materials.

6.3. ISO 9000 STANDARDS

Admittedly, there are difficulties in the United States when adopting ISO standards. For one thing, the English language forms used are not the same as in the U.S.

idiomatic usage. For example, the term *supplies* refers both to raw materials and to finished products, except where *customer supplied product* refers to raw materials. Another point of confusion is the common practice in Europe of using a comma to signify the decimal indicator. This occurs not only in tabular data but also in the constants in algebraic formulas. Both may be confusing to Americans, especially when performing a calculation. The American National Standards Institute (ANSI) and the Japanese Standards Institute (JSI) are suggesting the use of a decimal point.

The ISO 9000 series of standards is not just another short-term quality program, here today and gone tomorrow. It begins with a top management commitment as expressed in a corporate quality policy, divisional quality policy, and laboratory quality policy. These brief documents are developed through strategic planning after first developing a vision statement, a mission statement, and appropriate objectives. Quality policies state realistic and attainable goals that are appropriate to the organization and are relevant to the customer. They are mechanisms used to express the goals of the organization, and they are reinforced through management review that is conducted at least annually in order to evaluate the continuing suitability and effectiveness of the quality system. Continuous follow-up must be performed daily by a designated management representative. The proper implementation of documented procedures must also be monitored at appropriate intervals by internal audits performed by a trained auditor who leads an internal audit team. This key function should be independent of the supervisory personnel of the area being audited. Thus, with proper feedback and timely corrective action, the entire operation becomes self-correcting. As a result, the surveillance audits by the third-party auditors and assessors are *no problem*.

6.4. ISO 17025 FOR LABORATORIES

Until recently, the ISO 17025 standard was known as ISO Guide 25 for Laboratory Accreditation. While it covers both calibration and testing laboratories, this discussion will concentrate on how the standard applies to testing laboratories. Laboratory accreditation is an important corollary to ISO 9001/9002 registration. Note that the certification process is called "accreditation" when it applies to laboratories and "registration" when it is for other facilities. Specifically, laboratory assessment examines the appropriateness of test methods, measuring and test equipment (M&TE), and sample preparation. To be effective, a laboratory assessor must be assigned on the basis of his or her knowledge of specific plastics. Next, the M&TE is examined to determine its state of repair and maintenance as well as whether it is properly calibrated and periodically validated through in-service checks. Standard reference materials must be set up for this use. Where possible, they provide an unbroken chain of traceability of measurements to the National Institute of Standards and Testing (NIST) or other recognized national or international agencies.

For color measurement, twice-a-shift "calibration" normally uses just a black light trap and a white standard. This practice cannot provide NIST traceability and no true verification of accuracy is currently available. In early spectrophotometers, a set of colored glass filters calibrated by the National Bureau of Standards (NBS) was commonly used for wavelength and photometric verification. The calibration was limited to transparent samples, of course. It has been claimed that there is no longer a need for such calibration. Modern spectrophotometry incorporates a diffraction grating for a stable monochromatic light source, while the Hardy spectrophotometer was subject to short-term drift due to the mechanical components controlling the slits, mirror, and prisms in the light path. For well over 25 years, quarterly tests have been offered by the Collaborative Testing Service (CTS) of Herndon, Virginia. By subscribing to this program, a laboratory can compare its results with consensus values reported by a large number of participating laboratories. Data are presented with statistical evaluation and the 99% confidence elliptical plots of the chromaticity data, wavelength printout, and specular gloss data.

6.5. QS-9000 QUALITY SYSTEM REQUIREMENTS

In 1994 the automotive industry issued a coordinated set of requirements for suppliers of parts and materials. The first two-thirds of this document incorporates the 20 elements of the ISO 9001 standard. The final one-third of the standard has sections on sector-specific and manufacturer-specific requirements. Chrysler, Ford, and General Motors representatives have standardized the production part approval process (PPAP), quality and productivity improvement requirements, and process planning and tooling design. The final manufacturer-specific section summarizes third-party registration, timing, and other requirements that differ among the four customer groups.

Colored plastics parts and materials are very much a part of QS-9000, and most suppliers have either attained registration or are currently pursuing that goal.

6.6. ISO 14000 SERIES FOR ENVIRONMENTAL MANAGEMENT

In the first year of ISO 14000, it was reported that over 2000 registrations had been issued in Europe, Asia, and the United States. Some are issued on a site-by-site basis, as for Ford Motor Company, whereas others, such as IBM, chose to pursue it on a corporate wide basis. Regulatory incentives combine with a desire to demonstrate improvement in community awareness for these interests. Trade pressure, while present to some degree, does not appear to be the driving force it has been in ISO 9000 and 17025 certifications. The ISO 14000 series is composed of 20 separate documents covering several unique topics including the labeling and life cycle of materials and products. The Environmental Protection Agency (EPA) of the United States may be along way from recognizing the advantages of ISO 14000, but many state initiatives are under way in various parts of the country from California to New Hampshire and from North Carolina to Wisconsin. A group of 10 states have combined efforts in working with the 15 separate federal agencies involved in environmental matters in order to facilitate negotiations, recognizing the adversarial approach commonly experienced by industry. Most large companies are implementing ISO 14000 principles while evaluating the new technology and the economic incentives related to the major investments in equipment. In the United States, third-party registrations have thus lagged behind Europe and Asia. Unlike a manageable number of 20 elements for ISO 9001, ISO 14001 has 62 elements. In addition, there are 5 standards in the ISO 9000 series, but a total of 20 for the ISO 14000 series.

6.7. DOCUMENT CONTROL

Often overlooked are "documents of external origin." In the 1994 revision of ISO 9000, these received special attention. Each new or revised document must be reviewed by authorized personnel before being issued for general use.

For the colorist, communications from the customer regarding color a match are often received verbally or by electronic media. Does the customer want the match to be lighter or darker, redder or greener, bluer or yellower? It is not always the case that communications are expressed in such understandable terms or are correctly transcribed or documented. More often, communications are not clear or are not properly documented.

6.8. QUALITY SYSTEM PROCEDURES

Whatever the type of quality system certification—ISO 9000, ISO 17025, QS 9000, or ISO 14000—one begins with preparing documented procedures and then implementing them. It may be tempting to secure the services of a consultant, but that practice is questionable. A wise consultant acts as a facilitator only. He or she does not actually write any procedure but, instead, provides an outline while the operator or technician closest to the particular operation fills in the step-by-step details. Those individuals who are proficient in word processing can produce a near-finished product, while others need more help. The end result will more likely be a useful document.

6.9. PREVENTIVE ACTION

Before the ISO, a member of first-line supervision often had no effective mechanism for preventing recurrence of a problem. Perhaps a mismatch would be blamed on operator error or misweighing of color ingredients. One asked whether a zero-defects approach could work, but it was usually found to result in "fixing the blame" instead of "fixing the problem." With ISO corrective action follow-up, there is a good chance that such situations would change. When labor-intensive methods repeatedly lead to mistakes, automated equipment is often the answer. When positive results are not forthcoming from training programs, basic math and chemistry training has often been found necessary. When vitally needed materials were found to be missing from inventory, a different approach was needed, such as instituting an effective maximum/minimum ordering system for purchasing practices. The above situations are actual instances where comprehensive review of a color production system results in effective preventative action.

6.10. THE PROCESS AUDIT

Closely related to preventive action is the process audit. Such an audit is performed as part of a laboratory assessment for each test on the proposed scope of accreditation. Performance of a test procedure is observed relative not only to the requirements of the ISO 17025 standard (or ISO Guide 25) but also with respect to performing the ASTM or ISO test method properly. As a minimum, the laboratory must have properly calibrated equipment, the current test method, and trained testing personnel. In addition, a representative number of test demonstrations will reveal whether good laboratory practices are observed in sample preparation, using the correct test parameters and calculating the results. When operators and technicians accept responsibility through such demonstrations, they tend to commit to the quality process. Real improvement occurs with involvement at all levels.

6.11. FUTURE OF ISO 9000 AND OTHER STANDARDS

For colored plastics producers, a combined audit for ISO 9001 or ISO 9002 with laboratory assessment for ISO 17025 may be the preferred method of operation. For supplying the automotive industry, regardless of whether it is as a tier 1, 2, or 3 supplier, QS 9000 is a necessity. The good news is that interest and acceptance of ISO certification has virtually eliminated supplier audits.

CHAPTER 7

Introduction to Colorants

Robert A. Charvat

Charvat and Associates, Inc.
374 Bradley Road
Cleveland, Ohio 44140-1149

The coloring of plastic materials technology has grown by leaps and bounds since the first formal publications over 18 years ago. The first technical conference devoted strictly to the coloring of plastics took place in Rochester, New York, on April 12, 1962. The original use of colorant technology can be said to date back to paintings on cave walls. Until recent times, most color and physical data on colorants have been associated with the coatings or paint industry. In fact, even today, most colorant technical data sheets are heavily biased toward coatings and inks. This is changing as the performance of colorants in thermo and thermoset plastics has become so important to consumer and industrial products. The domain of coloring of plastics is, today, a dynamic and ever-changing technology. New, but mostly improved colorants appear on a regular basis. New and/or improved polymers seem to appear with every monthly issue of plastics technical journals. Keeping abreast of these developments is a challenge to all who have careers in this area of the plastics industry.

Up to this point in the text, the word *colorant* has been used to identify materials that when incorporated into a polymer result in visible color. This use of the word is intentional. The word is generally thought of as a generic identification of any coloring matter. In the world of plastic coloring there are two major categories of colorants, pigments and dyes. It is important to understand the difference. For this text a pigment is defined as a distinct particulate material that remains essentially unchanged during the processing and life cycle of a plastic product. On the other hand, a dye is a colorant that becomes soluble in the polymeric system. This

---

Coloring of Plastics, Fundamentals, 2nd edition. Edited by Robert A. Charvat
ISBN 0-471-13906-8   Copyright © 2004 by John Wiley and Sons, Inc.

does not imply the dye is soluble only in the polymer itself. The dye can just as well be soluble in a component of the polymer system. This suggests that the difference between a pigment and a dye is very clear and complete and no variations exist. Not so! The dividing line between pigments and dyes is not concise. It is possible for a colorant to perform as a pigment and/or a dye in a plastic system depending on a number of issues, such as but not limited to processing. This and other issues will be dealt with in more detail in other chapters of this publication. It can be said that the successful coloring of plastics is a 'no-chemical-reaction event". Basically, if during the processing or life cycle of a plastic product a change in color occurs as a result of the colorants, a chemical reaction has taken place. Therefore, the coloring technology applied has failed. A word of caution here. A visible color change in a product can be the result of failure not directly involving the colorants. Colorants, however, will be the first items indicted for the failure. When this unfortunate event does occur, many times the reaction parameters are unknown. The quick answer is to change colorants hoping for improved performance. This technique, used often, is an answer but not a solution since what actually takes place chemically is never completely understood. The color technologist must always consider the possibility that a visible color change in a product may be the result of a failure *not* directly related to the colorants in the product.

A number of definitions need to be clarified before the reader can completely understand the chapters that follow. These definitions fall into a number of categories and clarify color and colorant properties. The difference between a pigment and a dye has already been made clear.

7.1. COLOR VALUES

Color values refer to what the product or material looks like. For the color technologist, this is a three-dimensional description. First is *hue*, which is described as the color, such as red, blue, or green. The second is *value*, which describes the lightness or darkness of a color. Value ranges from black through shades of gray to white. The third color attribute is *chroma*, which represents the intensity or the saturation of the viewed color. A pile of iron ore would have low chroma. A red fire engine or a red rose would have high chroma. These descriptions are used by the color technologist but rarely heard from the artist.

7.2. COLORANT PROPERTIES

Materials such as colorants can range from highly transparent through varying amounts of translucency to opaque. Dyes will be truly transparent. Plastics themselves can be transparent (water clear) to opaque. These varying opacity levels of colorants and plastics are major factors considered when determining colorant formulations for plastics products.

7.3. STRENGTH AND TINT STRENGTH

Strength and tint strength, which are used interchangeably by industry, are relative, not absolute, properties. Tint strength is a comparative measurement of a colorant in a plastic. Strength refers to how much colorability a given colorant has compared to some other colorant at a given level. This should never be confused with the economics of colorant usage. Cost per unit weight of a colorant is *not* the issue. The total system use cost is the only meaningful evaluation.

7.4. DISPERSABILITY

Dispersability is the separation of pigment particles and clusters called agglomerates and aggregates into their ultimate individual particles *and* completely coating the surface of the individual particles with a vehicle that is usually the polymer. This ideal is rarely if ever obtained. In the real world, the technology exists to separate agglomerates (which are loosely held clusters of particles) into their individual particles. Today's technology does not, and probably never will, have the capability to break apart aggregates (which are very tightly held clusters). The major issue in dispersion is shear energy input to the colorant(s) polymer combination. As an example, if the shear energy is not great enough to break up agglomerates and coat the particles with polymer, no matter how long the shear energy is applied, dispersion will be incomplete. Additionally, if the shear energy is too great, physical damage to the colorant(s) and polymer system may occur. Particle fracture and/or abrasion may take place. This is usually observed as a change in color and/or opacity of the material. It is safe to conclude that the effect of not enough or too much shear energy into a system will give negative results. Surface treatments on organic as well as inorganic colorants is an excellent way to attack the dispersion problem. Most organic and many inorganic colorants are surface treated to improve dispersability and other properties too numerous to itemize here. With an understanding of the above, this discussion should clearly suggest that dyes that are soluble in a system are not subject to any of the parameters of dispersion.

7.5. HEAT RESISTANCE

Heat resistance is the ability of a colorant to remain stable and unchanged during the heat of processing, fabrication, and life cycle of the product. Many color technologists, who should know better, discuss heat resistance of colorants by stating a temperature below which a colorant can be expected to perform satisfactorily. This is only half the issue. The time at the elevated temperature is critically involved. The correct way to state heat resistance is to give a time–temperature combination. To illustrate, the performance of a colorant could be entirely satisfactory at 90 seconds and 175°C (347°F) or 15 seconds @ 300°C (572°C)!

7.6. CHEMICAL RESISTANCE

Chemical resistance could easily be called the heart of colorant stability in plastic systems! When colorants do not perform as expected, there is usually a visible color change. This implies that a chemical reaction has taken place with the colorants. For colorants to perform in a system, they *must* remain unaffected by all the steps of processing and life cycle.

7.7. SOLUBILITY RESISTANCE

Solubility is usually mistakenly categorized as bleed, crocking, transfer, migration, plateout, sublimation, and spue. As an umbrella statement, these descriptions connected to purely solubility include only bleed and transfer. The other items listed are associated with a colorant's ability to be mobile in a plastic system or the effect of a colorant's incompatibility or ability to evaporate or vaporize and then condense on the surface of a plastic or surrounding materials. By definition, dyes are soluble in a plastic system. Dyes must stay in solution and not come out of the plastic during the processing or its life cycle. Inorganic pigments are generally considered to be insoluble. Organic pigments may or may not have solubility in a plastic system. Many events will impact on a colorant's solubility. To illustrate, some, but not all, solubility parameters might be temperature, colorant level in the plastic, and/or the plastic itself.

7.8. LIGHTFASTNESS/WEATHERABILITY

Lightfastness and weatherability are often used interchangeably, but they are *not* the same! Lightfastness is the exposure of materials to high levels or long exposure times of UV light, visible light, total darkness, temperature cycles, and humidity but no direct contact with liquid water. Lightfastness can be generally compared to indoor exposure. Weatherability, on the other hand, is exposure of materials to high levels or long exposure times of UV radiation, visible light, IR radiation total darkness, temperature cycles, direct contact with liquid water, and contact with gases, particulates, aerosols, and other pollutants. Weatherability is easily compared to outdoor exposure where a material is exposed to all the elements of nature and society.

7.9. TOXICITY/ENVIRONMENTAL

Many years ago all colorants were thought of as safe. Not so today! Today, the color technologist, at the beginning of the formulation work, needs to consider all the safety aspects of a plastics formulation, particularly the colorants. The "old timer" at coloring plastics sometimes has a mental road block in separating what is safe and what is safe and legal. There is a real difference. Some colorants have been used for many years with no history of harm to any person or thing. However, because of the chemical makeup of these colorants, they have restricted usage or outright bans in commerce. This is a very dynamic reality to the color technologist. Government regulations can change quickly. Keeping abreast of the prevailing regulation is difficult. As the world becomes more safety conscious and environmentally aware, the colorant palette available to the color technologist become significantly smaller. This makes the colorist's work much more difficult on a day-to-day basis.

The items discussed in Section 8.3 are not all inclusive. The major properties have been addressed in this overview. However, there are additional properties that could become issues for any given situation. It is impossible to cover all the possibilities in this introductory chapter. A close inspection of a specific polymer chapter of interest will give the reader insight on specific properties he or she must address for that polymer.

7.10. COLOR AND PROCESSING STABILITY

The "process" includes not only the cormpounding of the polymer but also the product manufacturing or fabrication steps. Just because a polymer system is able to withstand the rigors of processing and/or fabrication does not necessarily suggest that colorants used in other applications are adequate and workable in a new application. Each new application must be analyzed and designed completely, prototyped, and tested to be sure the proposed design meets all the specifications and goals. These specifications and goals cover all the aspects of the application from processing through fabrication, life cycle, and finally disposal. With a focus on colorants, the usual form, as produced, is a fine dry powder. This fine powder has a particle size ranging from about $0.02\,\mu m$ to over $30.0\,\mu m$ [1]. The lower ranges are where one will usually find organic pigments. Some inorganics *can* have a small percentage of particles in the 30.0-μm range [1]. However, most inorganics will have particle size distributions 100% below 4.0–$10.0\,\mu m$ [1]. The average particle size of a colorant is no more important than the particle size distribution. Generally, the smaller the average distribution, the stronger the colorant. In addition, a narrow particle size distribution will give a cleaner, more intense color compared to a broad particle size distribution, which will give a duller color. This may give a clue why chemically identical colorants from separate manufacturers may not look identical when incorporated into a plastic system. This may be traceable to particle size distribution differences while analysis shows them to be the same chemically. The particle size average and distribution along with the chemical composition of a colorant will have a significant influence on the dispersability of the colorant in polymers, as discussed earlier.

7.11. END-USE STABILITY

End-use stability and performance are the net result of good design and colorant selection. In the case of colorants, end-use stability and performance most often refer to ability to resist attack by chemicals, solvents, heat, weather, and gases and aerosols found in the product environment. The resistance of a plastic product to attack by any of these in a laboratory can produce definitive results. In combination, these elements are difficult to quantify not only in the laboratory but also in actual exposure to them. Thus, testing throughout a project from start to finish is vital to the technical and commercial success of that plastic product. In this chapter the focus is on the role the colorants play in the success.

7.12. ISSUES AFFECTING COLORANT PERFORMANCE

Of the number of items affecting colorant performance, the one most overlooked is the polymer being colored.

The chemical aggressiveness of a polymer, particularly a molten thermoplastic, may quickly degrade some colorants. Many colorants are sensitive to the acidity, alkalinity, or pH of a plastic. Some polymers are chemically hostile. A classic example is molten nylon, which is known to vigorously attack colorants. A different example, but just as critical, is polycarbonates reaction to colorants and other additives. The melt viscosity of polycarbonate can be negatively affected by colorant formulations. Also, some polymers may act as solvents for some organic colorants. The processing temperature can be at a level that fosters the solvency of these selected organic colorants. The level or amount of some organic colorants in a plastic can promote solvency. Like a window of opportunity, at very low and/or high amounts of organic colorants in a polymer, a threshold of solvency or supersaturation may occur. The additives in a plastic system may be the route to solvency. A typical case might be where a colorant formulation may not be soluble in the polymer but may be soluble in a component or additive. The component could be a plasticizer, stabilizer, or lubricant. Some lubricants may change the strength and hue of colorants in a plastic. It is known that some organic and inorganic colorants look quite different when the dispersion in a system is improved or degraded by additives, particularly lubricants. An additive might be a dispersion aid, a UV absorber, antistatic agent, viscosity modifier, antioxidant, flame retardant, blowing agent, biocide, or many other possibilities. These are examples and should not be considered all inclusive.

7.13. COLORANTS

As review, colorants fall into two major categories. The first category comprises pigments, which have two subdivisions, organic and inorganic. Organic pigments are combinations of carbon, hydrogen, oxygen, nitrogen, and sulfur. Chemical analysis of many organic pigments will reveal the presence of a few selected metals. These metals enter the organic pigment chemistry as precipitating agents or as pigment particle surface additives. Inorganic pigments are combinations of metallic or metalloid elements with oxygen, sulfur, or selenium. The second major category of colorants is made up of soluble dyes. Dyes cover a number of chemical identities too numerous to elaborate on in this chapter. They are covered in detail in chapter 12 devoted to solvent dyes. There is a particular observation one can make when visually examining colorant families. In the organic and inorganic chemical classes, a specific chemical family will usually fall into one distinct area of the visible spectrum. As an illustration, quinacridone chemistry produces reds and magentas. Quinacridone chemistry does not make blues and greens. This analogy can be found in almost all the other organic and inorganic chemical families. The dye families, on the other hand, may well produce the entire visible spectrum from one chemical family. The popular anthraquinone chemistry is known for being able to produce almost any color in the visible spectrum. Organic and inorganic pigment families are unable to produce this wide spectrum of colors.

7.14. MAJOR ORGANIC PIGMENT FAMILIES

DIARYLIDE YELLOWS AND ORANGES. Originally called benzidine yellow and orange, this family has six derivations: AAA, AAOA, AAOT, AAMX, and HR [1].

These pigments show varying degrees of bleed and stability with AAA being the least reliable and HR having the best properties. These pigments enjoy great usage in the ink industry, in which the ink vehicle is a polymer. The AAMX type is not nearly as popular as the other types. The AAMX type exhibits color and performance properties quite similar to the AAOA, AAOT, and HR versions.

HANSA YELLOWS. Hansa yellows are quite opaque when compared to the other organic family yellows. Hansa yellows as a family are known as potentially significant bleeders. This limits their use in most thermoplastics. Hansa yellows do have utility in some thermoset plastics. Care should be exercised when contemplating the use of Hansa yellow's in polymers. Hansa yellow's popularity and great usage is in coating applications.

NICKEL AZO YELLOWS. Nickel azo yellow is one of the most mature organic yellow pigments available. It has been used in some cellulosics, PVC, and polyolefins. However, this green-shade yellow has been replaced in most applications by other more stable and efficient organic yellows. This yellow pigment is of more historical interest than of practical use by industry today.

BENZIMIDAZOLONE YELLOWS, ORANGES, AND REDS. These yellows, oranges, and reds are considered one of the high-performance pigments. The yellows tend to be green-shade yellows. Oranges tend to be yellow-shade oranges. The reds are blue-shade reds. Excellent bleed characteristics with good to very good heat stability and fastness to light make these pigments a possible choice for high-performance or engineering plastics applications.

ISOINDOLINONE AND ISOINDOLINE YELLOWS, ORANGES, AND REDS. These premium pigments range from green-shade yellows to yellow-shade reds. They exhibit excellent heat, light, and bleed properties. However, this chemical group is much more popular in coatings than plastics. Therefore, their use in plastics on a regular basis is small since other selections to obtain a color are usually more appropriate and cost effective.

VAT YELLOWS AND ORANGES. This chemical group brings good performance to the coloring of plastics. The vat group includes pigments such as flavanthrone and anthrapyrimidine. Vat pigments have resistance to bleed, heat, light, and weather. These fairly expensive pigments have great utility in coatings, where they produce brighter, cleaner colors than they will produce in plastics systems.

FLAVANTHRONE YELLOWS. This high-performance red-shade yellow is mainly used in coatings. It is thought of as a very lightfast yellow. It provides the same performance in plastics, where durability is required. However, with all the other high-performance yellows available, flavanthrone yellow has enjoyed only partial success in plastics applications.

DISAZO CONDENSATION YELLOWS AND REDS. Disazo condensation pigments provide important performance improvements over traditional azo types. Pigments in this family range from green- to red-shade yellows and from yellow- to blue-shade reds to violets and even a red-shade brown. The condensation chemistry derives the pigments by producing enhanced or very large azo molecules that result in pigments with good resistance to bleed, heat, and light and have significantly better properties than the classical azo group from which they were developed.

QUINACRIDONE REDS, MAGENTAS, AND VIOLETS. This family ranges from a slightly yellow-shade red to a blue-shade red to a violet. In all color ranges the quinacridone pigments provide excellent heat and light stability and in most cases bleed resistance. This family has the potential for bleed problems in styrenics. Quinacridone pigments enjoy great success in coating systems, particularly for automotive finishes. This leads to their being considered for many plastics applications.

THIOINDIGOID REDS AND VIOLETS. This family brings unique shades of deep reds to violets to a limited group of plastics. In general, the thioindigoid group exhibits excellent properties. However, these pigments have some limits in their usage. Even though bleed resistance is excellent, the pigments may be soluble in some polymers such as polyesters. This solubility will most likely have a significant effect on the properties of the thioindigoid pigment.

DIKETO-PYRROLO-PYRROL REDS AND ORANGES. This group of high-performance pigments is the only totally new chromophore to be commercially available for many years. This pigment exhibits excellent bleed, heat, and weather resistance. It enjoys tremendous popularity in the automotive exterior coatings market. Its use in plastics is growing dramatically as applications take advantage of its high-performance properties in plastics.

DIANISIDINE ORANGES: DIANISIDINE. Oranges fall into the classic family of organic pigments. Their chemistry is mature. Dianisidine orange is a yellow-shade orange used primarily in the ink industry. However, for lower end applications it has utility in plastics. The moderate, at best, heat, light, and bleed resistance of dianisidine orange justify its use in less demanding applications.

DINITRANILINE ORANGES. Another limited-use orange from the classical organic pigment group, this yellow- to red-shade orange has low cost along with modest properties. Dinitraniline orange is used mostly in masstone where its modest fastness properties are most applicable. This orange is used mostly in coatings and printing inks and in masstone almost exclusively.

NAPHTHOL REDS. This family of reds is most extensive. The shades range from yellow to blue-reds and even to a red-shade brown. Properties of the naphthol family extend from poor to high performance. Just because a red pigment is labeled as a naphthol does not suggest anything meaningful about its shade and more importantly its performance in plastics. The exact family member and its properties must be clearly identified before using any member of this family. To do otherwise could lead to overperforming in an application or to a complete failure.

RED LAKE C. This pigment tends to be a warm or yellow-shade red. Like many other reds in the classical group, red lake C was, and is, primarily an ink pigment. However, for low-cost products this pigment is very useful in plastic systems. Relatively good heat stability and opacity are an asset. A resistance to bleed is a useful property. This pigment contains a barium salt as part of its chemistry. The presence of barium other than barium sulfate may be a safety issue governing its use.

PERMANENT RED 2B. In the classic group, permanent red 2B is most likely the most used colorant. It ranges from yellow to blue in shade depending upon which metal salt it contains. The strontium salt gives the yellow shade, the barium salt an intermediate shade, and the calcium salt the blue shade. A manganese and magnesium salt variety exists; however, these types have very little utility in plastics since the metals involved stand a good chance of negatively impacting the physical properties of many plastic resins. The colorants in this group are known for having more than modest heat and light stability and for their nonbleed properties. Safety can be an issue with the strontium-and barium-containing varieties but not with the calcium salt species.

PIGMENT SCARLET. This classical organic pigment has been used successfully for years. This blue-shade red is noted for having good heat stability and fair lightfastness. It has been used for years in PVC due to its resistance to bleed in this polymer family. Moderate cost helps its survival from other more stable but expensive red pigments.

ALIZARINE MAROONS. This pigment is an extremely blue-shade red, virtually putting it in violet color space. In fact, many old-time colorists call alizarine maroon the poor man's carbazole dioxazine violet. Moderate heat stability coupled with bleed resistance has supported this pigment for years in modest demand applications.

CARBAZOLE DIOXAZINE VIOLETS. This is a high-performance pigment and is a true violet. This very strong pigment has excellent heat and light stability and is bleed resistant. Unfortunately, this pigment is very difficult to disperse and keep dispersed in most systems. The dispersion issue and the high cost of use sometimes limit its selection even though its color values are remarkable.

INDANTHRONE BLUES. This pigment family exhibits such a very blue-shade red that it can easily be called plum shades of deep blue. This group has very good heat stability, lightfastness, and bleed resistance. This pigment is generally overlooked when deep plum blue is required. Higher cost systems may be chosen, passing up an excellent opportunity to take advantage of this useful pigment type.

PHTHALOCYANINE BLUES. These blues range from very red-shade to green-shade blues. Stability varies over this color range as well due to different crystal structures of the phthalocyanine structure. As a color group, the phthalo blues may be the most popular and most used chromatic colorant. These pigments offer excellent heat, light, and bleed resistance. Nevertheless, the heat stability and lightfastness of the red-shade colorants are not as good as for the greener shades. The red shades are particularly susceptible to solvent attack, which may limit their utility in some liquid systems. Dispersability is an issue with the phthalocyanines. Also, in olefin systems the use of phthalo blues may result in part warping. The solution used today to eliminate this warpage is to replace the phthalo blue with a different chemical-type pigment. This may result in negative economics as well as other physical property and visual difficulties.

PHTHALOCYANINE GREENS. Phthalo greens enjoy the same popularity as the phthalo blues. The shades range from blue-shade greens to the brominated very yellow-shade greens. The blue-shade green is the most stable of the entire phthalocyanine family. The bromtinated or yellow-shade green is quite stable, but not quite as stable as the bluer shades. The phthalocyanine greens have excellent heat, light, and bleed resistance. As with the other phthalocyanines, phthalo greens are difficult to disperse.

7.15. MAJOR INORGANIC PIGMENT FAMILIES

CARBON BLACKS. There is still some difference of opinion as to whether carbon black is organic or inorganic. While some prefer the organic status, carbon black is considered inorganic in this chapter. This is based on the fact that carbon black's properties most closely align with typical or general inorganic pigments. Any debate on this issue will contribute little to using this product to pigment plastics. Most carbon blacks in use today are manufactured by a "furnace" process. In the past other processes were used; however, environmental and economic issues have promoted the changes in carbon black production. Carbon black is one of the most stable pigments in use today. It is literally ubiquitous in plastic products. It is very stable to almost any attack and has uses beyond color values. Carbon black is used to tone or gray chromatic colors. It is the most used pigment for producing grays. A very important noncolorant use is to provide ultraviolet light protection in a number of weathering applications. Carbon black is a major component of many rubber applications, the most obvious being in tires for vehicles of all types.

TITANIUM DIOXIDE. Titanium dioxide, by far, is the pigment of choice for producing whites and pastels/tints and opacifying plastic products. In plastics, rutile crystal titanium Dioxide is the most used material. Anatase crystal titanium dioxide is available but does not enjoy the popularity of the rutile crystal structure. Most titanium dioxide is manufactured by the 'chloride" process. The older, less popular process is the "sulfate' process. Properties, economics, manufacturing, and environmental issues determine the selection for a plastic application. Titanium dioxide comes in a number of modifications that result from numerous organic and inorganic compounds and combinations of both as surface treatments to the titanium dioxide pigment particle. These varieties were developed to produce specific end-use application properties for various plastic products. The heat stability, lightfastness, weatherability, bleed resistance, and chemical resistance of these white pigments are excellent. All the variations available make the choice of titanium dioxide for any given application an important judgment not to be taken lightly.

IRON OXIDE YELLOWS, TANS, REDS, AND BLACKS. This chemical type produces somewhat dull colors at low cost. Where heat stability requirements are at best modest, iron oxides have utility. However, their use in polymers such as PVC, where the presence of iron may be a problem, requires careful evaluation. Heat stability of the iron oxides varies with color. Iron oxide black will have the lowest heat stability. Heat stability increases in order are seen in iron oxide yellows, reds, and finally zinc and magnesium ferrites. The weatherability and chemical resistance of this pigment group vary from modest to excellent. Careful selection for a specific application is required. Iron oxide is ubiquitous in nature. Some natural materials may find their way into plastics application; however, this situation is very rare. Nonetheless, synthetically produced versions of natural iron oxides such as siennas and umbers are available. The utility of these pigments is usually found in paints artist colors and coatings.

CHROMIUM OXIDE GREENS. This pigment is unique in that it is exceptionally stable for heat, chemical and bleed resistance, and weathering even though it is somewhat dull in color. Current commercial products are totally trivalent chromium oxide, which makes the material acceptable in commerce, compared to hexavalent chromium products, which present health, safety, and environmental considerations. Chromium oxide green is the green pigment of choice for construction/outdoor plastic applications. Chromium oxide green should not be confused with hydrated chromium oxide green and chrome green. Hydrated chromium oxide green is cleaner and more blue than chromium oxide green. This pigment contains water of hydration, which makes it less heat stable than chromium oxide green. Chrome green, on the other hand, is a physical mixture of either iron blue or phthalo blue and lead chromate or an organic yellow. Originally, chrome green was only iron blue and lead chromate yellow. Chrome green's original marketplace advantage was as an inexpensive green with modest properties, even though it contained lead and chromium. Years ago, the use of lead-containing pigments was common.

LEAD CHROMATE YELLOWS, CHROME ORANGES, AND LEAD MOLYBDATE ORANGES. These green-shade through orange-shade reds have been a mainstay in plastics for years. However, their use today and into the future is becoming limited due to the toxicological and environmental impact of lead and chromium. These pigments display intense, clean opaque colors. Heat stability, lightfastness, weatherability, and bleed resistance are moderate to very good. A silica-encapsulated variety provides substantial additional heat resistance and small improvements in other properties in plastics. These pigments will most likely continue to dwindle in popularity due to the toxicological and environmental issues revolving around them.

CADMIUM YELLOWS, ORANGES, REDS, AND MAROONS. Cadmium pigments are very important pigments, despite the toxicological and environmental difficulties due to the presence of cadmium and selenium in their composition. These pigments provide exceptional heat resistance, which makes them the choice for engineering plastics and other plastic applications where durability due to heat, bleed, and greases is a major issue. Lightfastness, when properly formulated, can be quite good. Cadmium pigments are noted for having intense, clean, opaque and easy-to-disperse colors. However, their main claim to fame continues to be exceptional heat stability in high-performance plastics. There are two other categories of cadmium pigments that should be identified. The first, an extended variety of these pigments, called lithopones, was originally produced as a coprecipitation of the cadmium color and barium sulfate. This resulted in a lower cost product with very good color values, particularly in full or masstone color. The second variety comprises the cadmium lithopones currently available in the marketplace as physical mixtures or blends of the pure cadmium color and barium sulfate. These physical blends exhibit color and physical properties essentially equal to the true coprecipitated product.

MERCURY CADMIUM ORANGES, REDS, AND MAROONS. These cadmium pigments are similar in shade to the regular cadmium pigments except the selenium in regular Cadmium has been replaced with mercury in the pigment chemistry. The presence of mercury in these pigments has literally removed them from the marketplace due to toxicological and environmental restrictions. The mercury cadmium colors have little commercial value today.

ULTRAMARINE BLUES, VIOLETS, AND PINKS. Ultramarine pigments are noted for their rich colors, moderate heat stability, chemical resistance, bleed resistance, opacity, weatherability, and attractive economics. They are known to be sensitive to an acidic environment, which may have some effect on their utility. Ultramarine pigments are some of the very old and venerated colorants since they were originally naturally occurring materials, mainly in Middle Eastern countries. Sodium is a constituent of ultramarine pigments, which makes it unique as sodium does not usually appear as a component of color pigments.

IRON BLUES. Iron blues are deep green-shade blues. Iron blues are chemically ferri-ferro-cyanide. Over the years, they have been known as Prussian blue, Berlin blue, Paris blue, Turnbull's blue, toning blue, nonbronze blue and Milori blue. The surviving popular name seems to be Milori blue. All have the iron blue chemical structure. In the past iron blue was a favorite of the coatings and inks industry. The utility in the plastics industry was usually found in low- to moderate-cost and low- to intermediate-property-requirement applications. A recent large usage was as the blue component in green plastic trash bags. However, even this application has lost out to phthalo blue in most cases.

COMPLEX INORGANIC PIGMENTS. This group of highly stable pigments had its genesis in the ceramic color industry. Complex inorganic color pigments (CICPs) and complex inorganic pigments (CIPs), as they are known today, for years were called mixed metal oxide pigments (MMO). The MMO nomenclature was dropped by many a number of years ago to avoid the confusion that the pigment group was really just a physical mixture of a variety of metallic oxides and not a unique chemical compound and not tied to the metallic oxides or other compounds used as raw materials [1]. The CIP range from violets to blues to greens to yellows to buffs to browns and blacks. This covers a large part of the visible spectrum; however, there are no oranges or reds in the CIP palette. This is unfortunate, since in today's marketplace oranges and reds with similar properties would be in demand. The CIP group exhibits exceptional heat stability, chemical resistance, bleed resistance, opacity, and weatherability. These are by far the most across-the-board stable colorants available today. The colors are not nearly as intense as other chemical types in the marketplace, which is unfortunate. However, recent advances have made excellent progress in producing pigments that are major improvements over past offerings.

ZINC SULFIDE AND ZINC OXIDES. Both materials are white but do not approach titanium dioxide for use as a tinting pigment or opacifier in plastics. Both materials can have nonpigmentary utility in plastics, such as providing whitening power at much lower abrasion levels than titanium dioxide. Zinc oxide, for instance, not only brings whitening to rubber products, but also performs as an accelerator in the vulcanization process. These products cannot compete directly with titanium dioxide when whiteness and opacity are the only criteria. However, they can play an important role when they contribute to chemical reactions and/or physical properties.

7.16. MAJOR SPECIALTY PIGMENTS

METALLIC PIGMENTS. These pigments are metal flakes of various sizes made up of aluminum, copper, zinc, and/or their alloys. The aluminum flakes develop the silver metallic colors, whereas the copper, zinc, and aluminum alloys produce the gold, copper, brass, and bronze colors. The metal effect is modified, not only by shape and size, but also by perticle size distribution within a given product. Since aluminum, in particular, as a fine dry powder, can form explosive mixtures with air, most of these materials are commercially available in paste or liquid concentrates. In addition, surface treatments of these materials enhance their appearance and performance. Flakes of other metals, such as stainless steel, are used for surface protection purposes such as corrosion resistance and electrical conductivity.

SPARKLE PIGMENTS. These pigments range in colors, sizes, and shapes. These materials are generally made by die cutting multicolored Mylar films. The shapes and colors are only limited by the designer's imagination and the ability to process the pigments into products while maintaining the original physical properties and shapes.

INTERFERENCE/NACREOUS PIGMENTS. This umbrella group encompasses pigments that provide pearlescent and/or multicolored pearl-appearing surfaces. The original pearl pigment was extracted from fish scales. Today, these products have a number of origins too numerous to address in detail in this chapter. The first synthetic pearl pigments were derived from basic lead carbonate and bismuth oxychloride. The popular pigments in the marketplace today are made by depositing very thin layers of materials such as iron oxide and titanium dioxide on the surfaces of mica flakes. The technology is very sophisticated in that the particle size and distribution of the mica flakes is very important to the final appearance of these products. The most crucial technology revolves around the deposition of the iron oxides and titanium dioxides in exactly the correct thickness on the mica flake surface. The exact thickness of the deposited layer is the determining factor for what the observed interference color will be. The thickness of this deposited layer is measured in angstroms, which clearly identifies the difficult technological requirements of the process. The final product has a unique appearance resulting from the varying colors the pigments display depending upon the lighting and viewing conditions.

LUMINESCENT PIGMENTS. This group of pigments covers a number of technologies and markets ranging from color television picture tubes to "glow-in-the-dark" toys. For our purposes in plastics the foremost pigments are the fluorescent and phosphorescent types. The typical fluorescent type is unique, in that it is really a soluble dye that has been incorporated into a polymeric system and then reduced to a discrete pigmentary form. The final commercial product is organic in composition. The functional mode for the fluorescent pigment is to instantaneously convert UV energy into visible-light energy, which reinforces the visible color of the pigment. A wide range of hues are available with this technology. In the case of the phosphorescent pigments, the functional mode is similar except the radiation of the visible light energy takes place over a period of time. This is the glow-in-the-dark feature. The phosphorescent pigments are inorganic materials with limited hues available.

7.17. MAJOR SOLUBLE DYES

Dyes in plastics are becoming more useful every day as the drive to remove metals from plastic product additives such as colorants continue. For an overview, the descriptions are much different. With organic and inorganic pigments, the chemical families tend to fall in specific color spaces and have similar properties. An example would be the organic quinacridone pigment family, watch is basically orange-red, red, and magenta. For the inorganic cadmium pigment family the specific color space is yellow, orange, red, and maroon. This is not the case with soluble dyes for polymers. In most chemical family groups the whole rainbow is available from red, orange, yellow, green, blue, indigo, and violet. There are approximately 40 different polymer-soluble dye entitles to choose from with all the colors just mentioned. This does not turn out to be quite as big problem as it might be. Most polymer soluble dyes used commercially come from just nine major chemical groups. They are anthrapyridone, anthraquinone, azo, nigrosine, perinone, pyrazolone, quinoline, quinophthalone, and xanthene. We can make some general statements to guide evaluations. Most soluble dyes will exhibit poor lightfastness in tints or pastels. On the other hand, fulltone or masstone colors may perform quite well. In fact, the red dye used for automobile taillight lenses exhibits exceptional lightfastness/weatherability in the full color. Another fact is that a specific dye in a polymer system may not be soluble in the polymer itself, which is easy to assume. The dye may, in fact, be soluble in the polymer; however, it should not be overlooked that the dye may actually be soluble in a component of the system. The component may be a lubricant, stabilizer, antioxidant, or any other additive into which the dye will dissolve. Soluble dyes are usually very powerful, delivering great amounts of coloring power at very low addition weights. All this suggests dyes are very useful colorants that have a very large range of performance properties in many colors. Many of these properties bring value to a product; however, some may be negative, leading to the failure of a plastic product, such as bleeding out of the system. It is clear that positive results are available, but caution should be used during product development.

7.18. TESTING AND EVALUATION

For testing and evaluations, one thing should become very clear after summarizing the contents of this book: Testing is imperative to be sure a colored system will meet the demands of compounding the polymer components, processing the polymer into a commercial product, and finally ensuring that the finished product will perform as expected and as long as expected in the hands of the consumer. This means having confidence that all the design and engineering have been done correctly and the testing is successful and valid. Testing has no value if the tests do not accurately reflect the property being evaluated. The fault of most erroneous testing is threefold. First, the test does not measure the intended property. Second, the test is not reproducible. Third, the test is not statistically in control. Therefore, the tests are invalid. Inappropriate tests and evaluations occur all to often and they should not!

7.19. HEALTH AND SAFETY ISSUES

Safety and health along with the protection of our environment are becoming more important to our personal and professional lives. Colorants are subject to intense scrutiny by local, state, and federal government bodies and regulatory agencies. To a certain extent, the world of colorants and color technology is in what amounts to a "catch 22." Most regulations do not focus specifically on colorants; however, the colorants as chemical entities find themselves in categories being regulated in the strictest sense. These issues will restrict the colorist when selecting colorants for an application when it is known the colorants themselves will not be the issue of regulation. These issues are very complex. Dealing with them on an every-day basis is a challenge and well beyond the scope of this volume to adequately address in detail. Whole volumes are available dealing with these issues and they barely touch the complexity of the subject. If we refer back to the testing issue, a few things must be able to be stated with certainty. The most important is that only after significant, correct, complete, and adequate testing can a producer commit a product to commercial-scale production and sale with the confidence the product will perform as planned and promoted.

REFERENCE

1. R. A. Charvat, internal private data, Charvat and Associates Cleveland, Ohio.

CHAPTER 8

Organic Colorants

Peter A. Lewis

8.1. DEFINITION OF A PIGMENT

The largest proportion of organic colorants used in the plastics industry are pigmentary in nature and offer the end user the necessary properties such as lightfastness and thermal stability consistent with the requirements of the end-use application of the colored plastic article. Prior to entering into any discussion relating to organic colorants, it is first necessary to define what is meant by an organic pigment, as opposed to an inorganic pigment and other colorants such as dyestuffs. An organic pigment is now normally recognized as a synthetic product, based upon a backbone of carbon, hydrogen, and nitrogen and as such these organic-derived colorants will produce very little ash if heated above the point of its decomposition.

The term *pigment* may best be defined using the definition proposed by the Colored Pigment Manufacturers Association (CPMA), a definition developed specifically to enable differentiation between a dyestuff and a pigment:

> Pigments are colored, black, white or fluorescent particulate organic and inorganic solids which usually are INSOLUBLE in, and essentially physically and chemically UNAFFECTED by, the vehicle or substrate in which they are incorporated. They alter appearance by selective absorption and/or by scattering of light.
>
> Pigments are usually DISPERSED in vehicles or substrates for application, as for instance in inks; paints, plastics or other polymeric materials. Pigments RETAIN a crystal or particulate structure throughout the coloration process.

As a result of the physical and chemical characteristics of pigments, pigments and dyes differ in their application; when a dye is applied, it penetrates the substrate in a SOLUBLE form after which it may or may not become insoluble. When a

---

Coloring of Plastics, Fundamentals, 2nd edition. Edited by Robert A. Charvat
ISBN 0-471-13906-8   Copyright © 2004 by John Wiley and Sons, Inc.

pigment is used to color or opacify a substrate, the finely divided INSOLUBLE solid remains throughout the coloration process.[1]

8.2. INTERNATIONAL NOMENCLATURE: THE CI SYSTEM

Modern publications that discuss pigments are likely to make use of the system of nomenclature as published jointly by the Society of Dyers and Colourists (SDC) of the United Kingdom and the American Association of Textile Chemists and Colorists (AATCC) of the United States. This system is known as the Colour Index and is a recognized trademark, hence the retention of the *u* in *colour* whenever reference is made to a Colour Index (CI) name or number. This index identifies each pigment by giving the chemical a unique name and number known as the CI name and CI number. In light of the system's uniqueness, the identification of a pigment by mention of its CI name and number will identify the chemical composition of the pigment in a manner acceptable to most government bodies. For example, DNA orange has the CI name of CI pigment orange 5 (PO 5) and the CI constitution number 12075. The CI name for a pigment may be abbreviated as follows:

Colour Index Name	Example	Abbreviation
Pigment blue	Copper phthalocyanine blue, PB 15	PB
Pigment black	Carbon black, PBk 7	PBk
Pigment brown	Brown iron oxide, PBr 6	PBr
Pigment green	Copper phthalocyanine green, PG 7	PG
Pigment orange	Dinitroaniline orange, PO 5	PO
Pigment red	Pyrazolone red, PR 38	PR
Pigment violet	Carbazole violet, PV 23	PV
Pigment yellow	AAOT yellow, PY 14	PY
Pigment white	Titanium dioxide, PW 6	PW
Pigment metal	Aluminum flake, PM 1	PM

The colored organic pigments covered in this chapter all contain a characteristic grouping or arrangement of atoms known as a *chromophore*, which imparts color to the molecule. In addition, the molecule is likely to feature a number of modifying groups called *auxochromes* that alter the primary hue of the pigment in a more subtle way, such as shifting a red to a more yellow shade or a blue to a more green shade while still maintaining the primary hue of red or blue. Possibly the most important chromophore is the azo chromophore (—N=N—), a feature of all naphthol reds, monoarylide and diarylide yellows, benzimidazolones, pyrazolones, and azo condensation pigments.

The term *lightfastness*, used throughout this chapter, refers to the pigment's ability to withstand exposure to light, both natural and artificial, without suffering any visible change in appearance. The most damaging component of light appears

---

[1] Color Pigments Manufacturers Association Alexandria, VA.

to lie in the UV and, as such, a rapid evaluation of a pigment's likely reaction to long-term exposure to light can be assessed using exposure equipment that maximizes exposure to this most damaging wavelength. Many high-performance pigments are exposed under application conditions at specially maintained sites in Florida and Arizona to more fully appreciate their fastness to outdoor exposure and weatherability. Figures given in the literature and commercial brochures for fastness to light are quoted based upon the Blue Wool Scale as described in ISO 105, BSS 1006, and DIN 16525.

8.3. PIGMENT SELECTION

Once the required hue has been determined, the formulator next has to consider what end-use properties will be expected of the pigments chosen to match the chosen hue. It is useless to formulate a colored plastic for an outside application such as garden furniture only to realize that the chosen colors will not withstand exposure to the elements or a color for a blow molding operation where the temperature of the molding exceeds that which the pigment can tolerate. Due attention must be given to the respective manufacturer's technical literature to ensure that pigments have been chosen that will satisfy both the application and end-use criteria for stability to light, solvents, heat, crystallization, acidity, and alkalinity.

Appropriate attention should be given to the fact that the fastness properties of a pigment will be affected by the medium into which the pigment is incorporated. Thus, even though the pigment may intrinsically feature the required properties, it is still of paramount importance that the final pigmented polymer be tested under both manufacturing and application situations in addition to the final end use. Fastness to light and heat are but two of the properties that are affected by the system.

8.4. CLASSIFICATION OF PIGMENTS BY COLOR

8.4.1. Organic Reds

Metallized Azo Reds

Many of the red pigments used within the plastics industry contain the azo chromophore (—N=N—) within the structure of the molecule and as such are termed *azo reds*. A further subdivision is possible into acid, monoazo metallized pigments such as the calcium and barium salts of permanent red 2B (CI pigment reds 48:2 and 48:1, respectively) and nonmetallized azo reds such as pyrazolone red (CI pigment red 38). Typically each of the metallized types contain an acidic grouping such as sulfonic (—SO$_3$H) or carboxylic acid (—COOH) that will ionize and complex with a metal cation such as calcium, barium, or manganese to form an insoluble metallized azo pigment. Conversely, *nonmetallized* azo reds do not contain an anionic group in their structure and therefore cannot form such a reaction product with a metal cation. Manganese-containing pigments are rarely used in the plastics

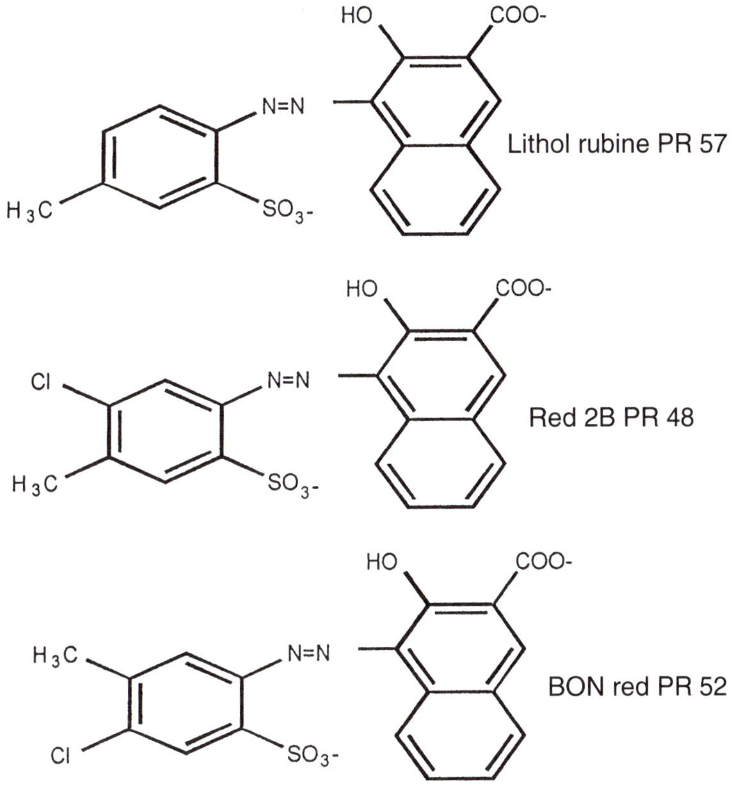

Figure 8.1. Structure of series of metallized azo reds.

industry since the inevitable minute amount of free manganese ions present in the pigment may cause problems when incorporating the colorant into the polymer.

All classical azo reds contain at least one azo group, by definition, and are all produced by a similar reaction sequence involving the chemical reaction termed *diazotization* followed by *coupling*. Diazotization involves reacting a primary aromatic amine with nitrous acid to yield a "diazonium salt," which is then immediately "coupled" to the other half of the molecule to yield the colored pigment as a bright red precipitate.

Figure 8.1 illustrates the structure of a series of metallized azo reds that are of considerable commercial importance and that some use within the plastics industry. Each structure features a molecule based on the coupling of a naphthalene ring structure to a benzenoid structure. Further each molecule contains at least one anionic group, namely sulfonic acid, SO_3H-, capable of complexing with a metal cation. A brief description each of the more common metallized azo reds follows.

LITHOL REDS (BARIUM LITHOL PR 49:1, CI No. 15630:1; CALCIUM LITHOL PR 49:2, CI No. 15630:2). Discovered in 1899, this pigment's major use is in the printing ink industry and finds only limited application within the plastics marketplace. Chemically the pigment is the reaction product from the coupling of 2-naphthol to diazotized 2-naphthylamine-1-sulfonic acid (Tobias acid). The respective metal salts are then produced by reacting the coupled product with either calcium or barium chloride.

The lithol reds are bright reds with high tint strengths and moderate dispersion characteristics; the barium salt is yellower in shade than the calcium salt, which may best be described as a medium red. The use of these pigments in the plastics industry is as a result of their low cost and ready availability. Neither pigment can be recommended for any polymer likely to experience outdoor exposure since the exterior durability of both calcium and barium lithol red is inadequate for such situations. Easy dispersing grades are available that find acceptance in the rubber industry.

PERMANENT RED 2B (BARIUM RED 2B, PR 48:1, CI No. 15865:1; CALCIUM RED 2B, PR 48:2, CI No. 15865:2; MANGANESE RED 2B, PR 48:4, CI No. 15865:4). Discovered by DuPont in the early 1920s the permanent red 2B pigments are azo reds prepared from coupling diazotized l-amino-3-chloro-4-methyl benzene sulfonic acid onto 3-hydroxy-2-naphthoic acid (BON). The barium salt is characterized by a clean, yellow hue as compared to the bluer calcium salt. The barium salt has a poorer lightfastness and weaker tinting strength; however the barium salt will provide better heat stability in certain polymer systems. The calcium salt provides the most widely used blue-shade red.

The use of permanent red 2B pigments in plastics stems from the economy and excellent bleed resistance offered by this series of pigments. The pigments produce clean, bright, strong, and relatively transparent colors. At masstone levels the pigments offer fair light and heat stability, withstanding short exposure to temperatures of 450°F (232°C) as the calcium complex and 475°F (246°C) for the barium complex. These pigments were once often used in blends with inorganic pigments and lead-containing pigments such as molybdate orange PR 104 to produce bright, economical reds, but this outlet has declined in recent years due to the removal of lead from many formulations. The major polymer classes into which permanent red 2B pigments are incorporated are polyvinyl chloride (PVC) and low- and high-density polyethylenes (LDPE, HDPE), and, less frequently, polyslyrenes (PS). The strontium salt CI pigment red 48:3 finds occasional use in polypropylene (PP) fibers, its shade falling somewhere between that of the calcium and barium salts.

LITHOL RUBINE RED (CALCIUM LITHOL RUBINE, PR 57:1, CI No. 15850:1). Made by coupling 3-hydroxy-2-naphthoic acid (BON) onto diazotized 2-amino-5-methyl benzene sulfonic acid (4B acid), this blue-shade red was discovered in 1903 and has found widespread use in the printing ink industry as the process "magenta" in the four-color printing process. Normally used as the calcium salt, lithol rubine is an economical, clean, easily dispersed, blue-shade red with high tint strength. Non-resinated grades are used in the production of clean, bright, blue-shade reds that offer good bleed resistance, high tint strength, and fair lightfastness. Usage of lithol rubine is limited by a rather low heat threshold in that the pigment will only tolerate a maximum temperature of 425°F (218°C) for short exposure periods of the order of 1 min or less before noticeable fade and color degradation will occur. As such, the major applications are in flexible PVC, LDPE processed at low temperatures, and plastisols designed for interior end-use applications calling for an inexpensive red where stability to light is not an issue of concern. Lithol rubine should be used at near-mass-tone levels, that is, with little tinting with white pigments such as titanium dioxide or zinc oxide, in order to maximize the fastness properties of this color.

BON REDS (CALCIUM BON RED, PR 52:1, CI NO. 15860:1; MANGANESE BON RED, PR 52:2, CI NO. 15860:2). Manufactured by coupling diazotized l-amino-4-chloro-3-methyl benzene sulfonic acid onto 3-hydroxy-2-naphthoic acid (BON), these reds first were commercialized in 1910. Characterized by outstanding cleanliness, brightness, and color purity, the manganese salt offers a very blue shade with improved lightfastness as compared to the calcium salt. Both salts form pigments that are intrinsically "hard" and difficult to disperse and as such find little use in the plastics industry. Dispersion equipment such as media mills, containing zirconium beads as the media, or ball mills is generally required to effectively disperse the BON reds.

PIGMENT SCARLET (CI PIGMENT RED 60:1, CI NO. 16105:1, BARIUM SALT). Discovered in 1902, pigment Scarlet is one of the few laked pigments still in production within our modern pigment industry. The pigment is a precipitate of mordant red 9 as the barium salt onto alumina hydrate. Anthranilic acid, 2-aminobenzoic acid, is diazotized in aqueous hydrochloric acid and then coupled onto an alkaline solution of 2-naphthol-3,6-disulfonic acid, R-salt, to form a bright red dyestuff, known as mordant red 9. This dyestuff is then isolated by precipitation with excess sodium chloride and laked on aluminum hydrate by coprecipitation with barium chloride under controlled pH conditions.

Pigment scarlet is a clean, blue-shade red with acceptable heat fastness as compared to the economics of the product. Temperatures up to a maximum of 500°F (260°C) can be tolerated for short periods during the processing of the colorant. The marginal economics associated with the use of this pigment is all that restricts its more widespread acceptance within the plastics industry. The pigment is suitable as a colorant for rubber and finds some use in vinyl plastics and cellulosics. Pigment scarlet shows less tendency toward "plate out" as compared to permanent red 2B when molded in plasticized PVC applications.

RED LAKE (CI PIGMENT RED 53:1, CI No. 15585:1, BARIUM SALT). Discovered in 1902 by K. Shirmacker, red lake C, sometimes also referred to as lake red C, is a yellow-shade red produced by the diazotization of C-amine, 2-amino-5-chloro *para*-toluenesulfonic acid, followed by the subsequent coupling of the resultant diazo onto beta-naphthol. The coupling is carried out under carefully controlled pH and temperature conditions to initially form the sodium salt of the pigment. The insoluble barium salt corresponding to CI pigment red 53:1 is then formed by the replacement of the sodium cation by barium at elevated temperatures. Any rosination that is necessary also occurs at this stage by precipitating the required rosin from solution as the barium rosinate onto the surface of the red lake C pigment.

By far the greatest market for this pigment, especially the rosinated grade, is in the printing ink market, where red lake C is known as a "warm red," due to its distinct yellow shade. Within the plastics industry red lake C offers an economical azo red that finds use in LDPE and polyurethane where the end use allows such a pigment with relatively poor durability properties to be used as the system colorant.

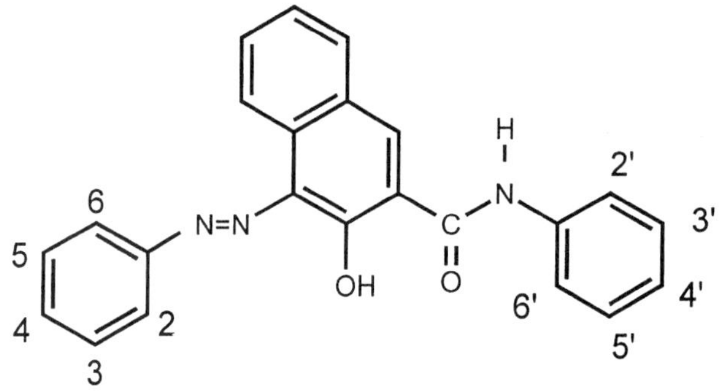

Figure 8.2. Generic naphthal structure.

The moderate lightfastness and poor exterior durability in conjunction with a moderate, short dwell time and heat stability of 500°F (240°C) are the major factors restricting the more widespread use of this pigment.

Nonmetallized Azo Reds

As implied by their description, nonmetallized azo reds do not contain a precipitating metal cation and as such offer increased stability to hydrolysis in highly acidic or alkaline environments as compared to the metallized azo reds previously covered.

PYRAZOLONE RED (CI PIGMENT RED 38, CI No. 21120). The structure of this pigment typifies the nonmetallized azo reds as used in the plastics industry. The pigment is prepared in an aqueous environment by the tetrazotization of 3,3'-dichloro-benzidine followed by coupling the resulting tetrazonium salt onto phenyl-carbethoxy-pyrazolone under controlled pH.

Pyrazolone red is typically soft and easy dispersing. The pigment offers economy with moderate heat, light, and solvent bleed fastness coupled with a high tinting strength. In the plastics industry this colorant finds application in both rubber and flexible vinyl applications under low-temperature processing conditions. In reduced shades the pigment offers poor lightfastness; additionally the colorant does not possess adequate heat stability to allow for its use in most thermoplastics.

NAPHTHOL REDS. Naphthol reds are chemically defined as monoazos of 2-hydroxy naphthoic acid *N*-arylamides without anionic-salt-forming groups. Their individual properties are dependent upon the specific composition of the pigment in addition to the conditioning steps used in their manufacture. As a class, this group of pigments exhibits good tinctorial properties combined with moderate fastness to heat, light, and solvents. Figure 8.2 represents the generic structure that applies to the naphthol reds as available in todays marketplace. Naphthol reds are extremely acid, alkali, and detergent-resistant pigments. Each may be described as a "medium-performance" red since their properties fall somewhere between pyrazolone red and quinacridones, with a cost that corresponds accordingly. Their origin can be traced back to 1880, when attempts were being made to synthesize insoluble dyestuffs in place on cotton fibers.

Only those naphthol reds with additional amide groups in the molecule exhibit the fastness properties required by the plastics industry. Most naphthol reds show some tendency to bleed into the plasticizers in common use within the flexible PVC industry, and only CI pigment red 187 finds minimal use in this polymer. At levels down as low as 0.01%, CI pigment red 170 will still cause bleed problems in highly plasticized systems. Further naphthol reds cannot be recommended for polypropylene when processing temperatures exceed 525°F (274°C) for any appreciable time. For polyethylene, where temperatures during processing are lower, both CI pigment orange 38 and CI pigment red 187 find application. When used at high tint levels, CI pigment red 187 offers a lightfastness rating of 6–7 on the Blue Wool Scale. Although CI pigment red 170 has been used for the coloration of the controlled rheology grades of polypropylene fibers, due attention must be given to blooming that occurs in the deeper shades. Both CI pigment red 170 and CI pigment red 187 are used for the mass coloration of polypropylene fibers, a process where the pigments are introduced to the melt prior to the spinning process:

- CI pigment orange 38, CI No. 12367, is a very yellow shade, bright red with good solvent fastness.
- CI pigment red 170, CI No. 12475, increasingly important as a medium-performance, moderately priced red, is available as both a blue-shade, transparent and a slightly yellower, opacified grade with a lower tint strength as compared to the transparent form. Manufacturing techniques are used to produce the pigment in two crystal phases each exhibiting a unique hue.
- CI pigment red 187, CI No. 12486, is a transparent pigment with excellent heat fastness, moderate durability, and good solvent resistance.

High-Performance Reds

These pigments may be defined as those that will meet the exacting demands of today's environment with respect particularly to the outdoor exposure requirements demanded by climates such as those that occur in Florida and Arizona for as little as two and as long as five years. The high-performance reds considered fall into four basic classes: quinacridone reds and violets, vat dyestuff reds such as perylenes, benzimidazolone reds, and disazo condensation reds.

QUINACRIDONE REDS. These pigments may be described as heterocyclic since their structure comprises a fused ring in which the ring atoms are dissimilar, being a combination of carbon and nitrogen rather than only carbon as we have seen in the previous pigments discussed, as shown in Figure 8.3, where trans linear quinacridone, CI pigment violet 19, is featured. Addition of differing auxochromic groups such as methyl (—CH_3) and chlorine (—Cl) in the 2 and 9 positions around the ring results in the formation of CI pigment red 122 and CI pigment red 202, both described as magentas. The theory behind the superior durability of pigments with the quinacridone structure is that considerable intramolecular hydrogen bonding takes place between molecules through the carbonyl (=C=O) and imino (=N—H) ring atoms.

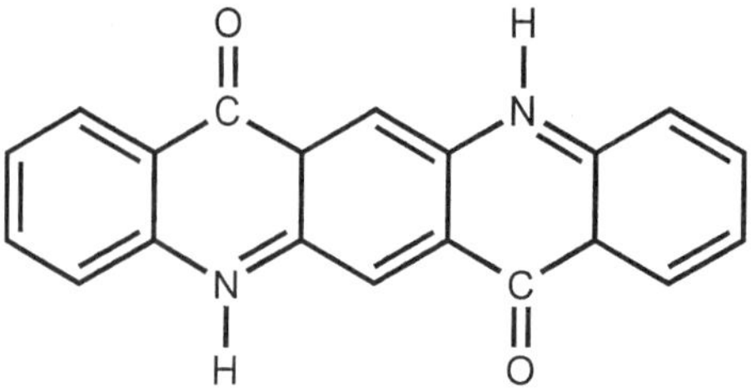

Figure 8.3. Quinacridone PV 19.

Table 8.1. Types of Quinacridone

Colour Index Name	Hue	Comments
Pigment orange 49	Gold	Quinacridone quinone
Pigment red 122	Magenta	2,9-dimethylquinacridone
Pigment red 202	Magenta	2,9-dichloroquinacridone
Pigment red 206	Maroon	Solid solution
Pigment red 207	Scarlet	4,11-dichloroquinacridone
Pigment red 209	Red-yellow	3,10-dichloroquinacridone
Pigment violet 19	Red-yellow, violet-blue	Trans linear quinacridone
Pigment Violet 42	Maroon	Solid solution

These pigments combine excellent tinctorial properties with outstanding durability, solvent fastness, lightfastness, heat fastness, and chemical resistance. Table 8.1 lists those shades currently available commercially.

Quinacridone pigments are manufactured by either the condensation of 2,5-diarylamino-terephthalic acid or the oxidation of dihydroquinacridones followed by subsequent conditioning to produce the product in a pigmentary form of colloidal dimensions. For example, in the first process, the use of 2,5-dianilinoterephthalic acid results in the formation of trans linear quinacridone, PV 19, by a simple ring closure condensation where water is evolved as a by-product. Substitution of the aniline used to produce the diarylaminoterephthalic acid will result in pigments such as PR 122 where ditoluidinoterephthalic acid is used at the condensation stage.

Trans linear quinacridone, CI pigment violet 19, can be conditioned selectively to produce either a red or a violet crystal morph, both of which are stable once they are isolated. While these crystals are hard and relatively difficult to disperse into plastic polymeric materials, they nevertheless are used throughout the plastics industry in both rigid and flexible PVC, LDPE, HDPE, and PS as a result of their exceptional fastness properties and tinting strength. CI pigment violet 19 is regarded as an "automotive grade" pigment and as such is used in Original Equiment Manufacturer (OEM) styles and refinish systems within the automotive paint industry. The other commercially available quinacridones are equally hard to disperse but still find an outlet as plastics colorants. Two such quinacridones are 2,9-dichloroquinacridone, CI pigment red 202, and 2,9-dimethylquinacridone, CI pigment red 122. While all other properties are fully acceptable, it will require the development of "easy dispersing," softer grades specifically for incorporation into

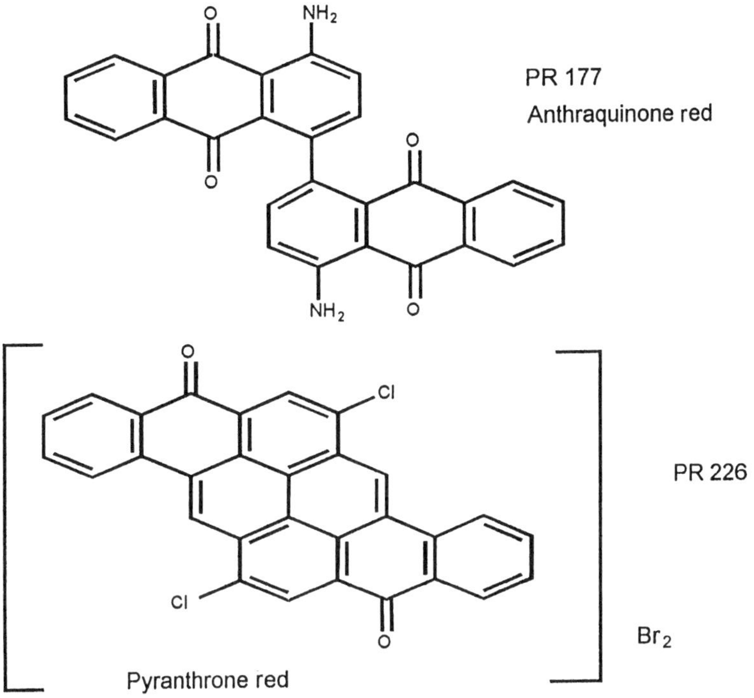

Figure 8.4. Vat reds.

plastic systems before the use of pigments based upon the quinacridone structure becomes more widespread.

VAT REDS. Vat red pigments based upon anthraquinone include such structures as anthraquinone Red (PR 177), perinone red (PR 194), brominated pyranthrone red (PR 216), and pyranthrone red (PR 226), as shown in Figure 8.4. The term "vat pigments" originates from the fact that this class of pigments are related to the vat dyestuffs used originally in the dyeing of cotton. Only their high costs limits the more widespread use of these pigments throughout the plastics industry, with prices in the (1999) marketplace running in excess of $45–$50 per pound. All commercially available vat red pigments have very good fastness to heat, light, acids, and alkalis coupled with excellent weather fastness when exposed in exterior applications. Certain vat pigments have found use in the coloration of thermoplastics, and a few are used for the melt coloration of fibers such as polyester and polypropylene, where costs do not preclude their use. The tendency of several vat pigments to bloom in flexible PVC and polyolefins requires strict end-use testing before such pigments are chosen for these applications:

> Dibromanthrone red, PR 168, CI No. 59300 is a bright, yellow-shade red with excellent fastness properties with no tendency to fade on prolonged outdoor exposure at all depths of shade. However, the pigment tends to bloom in flexible PVC and polyolefins and cannot be recommended for such polymer systems.

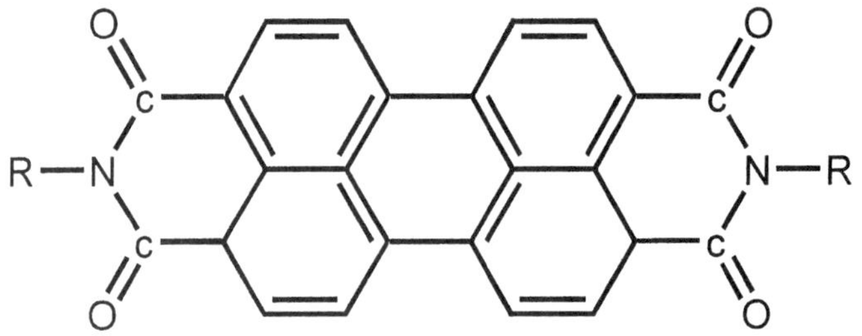

Figure 8.5. Generic perylene structure (R = methyl for PR 179).

Anthraquinone red, PR 177, CI No. 65300 is a medium-shade red with excellent fastness properties. Ideally suited for the coloration of thermoplastics where the end use justifies the cost.

Brominated pyranthrone red, PR 216, CI No. 59710 is a yellow-shade red, classed as a high-performance pigment because of its excellent fastness properties. Neither light nor dark shades will suffer on prolonged outdoor exposure. Again, the pigment finds use in plastics only when the economics can justify such a quality pigment.

PERYLENE REDS. Perylenes may also be described as vat pigments and in fact are the only class of vat pigments that were developed specifically for the pigment marketplace rather than as dyestuffs. Almost all of the perylenes have a structure as shown by the generic formula given a Figure 8.5, that is, they are based upon N,N'-substituted perylene-3,4,9,10-tetracarboxylic diimide. Thus perylenes are derived from the bisimide of perylenetetracarboxylic acid or from perylenetetracarboxylic dianhydride itself, that is, CI pigment red 224. Synthesis begins with the oxidation of acenaphthene to 1,8-naphthalic anhydride followed by its conversion to naphthalimide with ammonia. The naphthalimide is then condensed, under strongly alkaline conditions, to give the perylene-3,4:9,10-tetracarboxylic acid diimide. CI pigment violet 29 is derived directly from the diimide while CI pigment red 179 is prepared by methylating the diimide. Alternatively, the diimide can be hydrolyzed to the dianhydrid, which can then be condensed with aromatic or aliphatic amines to give such pigments as CI pigment red 178.

Perylenes offer shades ranging from yellow-red to blue-red or maroon. They generally possess excellent fastness properties with respect to heat, light, and weather while offering colorants with a high color strength. Perylenes can be used to color most polymer systems, including PVC, LDPE, HDPE, and PP.

BENZIMIDAZOLONE-BASED REDS. This category of reds includes such colorants as CI pigment reds 171, 175, 176, 185, and 208. Benzimidazolon-based reds are azo reds that contain the benzimidazolone structure as part of their make-up. Each of these reds are derived from the NAPHTOL AS molecule as illustrated in Figure 8.6. It is this structure that contributes significantly to the high molecular weight of the pigment and that greatly influences the pigment's fastness properties. All are made via a diazotization and coupling sequence as described earlier.

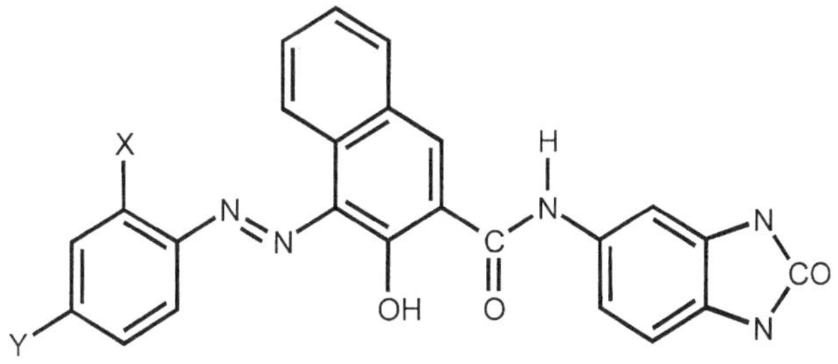

Figure 8.6. Benzimidazolone reds structure (R = methoxy, Y = nitro for PR 171; X = acetyl, Y = H for PR 175).

Benzimidazolone reds are used in the coloring of all types of plastics because of their outstanding heat stability, excellent fastness to light at all depths of shade, and good weatherability. CI pigment red 175 is a highly transparent red with good light-fastness; CI pigment red 171 is also a transparent pigment but with a maroon shade. CI pigment reds 176, 185, and 208 find considerable use in quality printing ink applications but currently have no use in the plastics or fibers industry.

DISAZO CONDENSATION REDS. These types of pigments have been available commercially in Europe since 1957 and in the United States since 1960. Their outstanding fastness properties have resulted in their use in high-performance end-use applications. Figure 8.7 illustrates two typical structures of disazo condensation reds alongside a generic molecule for this class of pigment. CI pigment red 144 results from the diazotization and coupling of 2,5-dichloroaniline onto 3-hydroxy-2-naphthoic acid, which is then converted to the acid chloride and condensed with 2-chloro-*para*-phenylene diamine. Manufacture proceeds via the production of a monoazo carboxylic acid derived from the diazotization and coupling of an aromatic amine with 2,3-hydroxynaphthoic acid. The monoazo carboxylic acid chloride obtained through chlorination is then condensed with an aromatic diamine at both amine sites to give the disazo condensation product. Figures 8.7 and 8.8, showing CI pigment red 242, a bright, yellow-shade disazo condensation pigment with excellent fastness properties, serve to illustrate the size and variation of the structures of pigments within this class.

This class of pigments finds use in the coloration of PVC and polyolefins as a result of their fastness to migration and high heat stability. The small particle size of such pigments as CI pigment red 144 has also resulted in the use of this pigment for coloring synthetic fibers. Recently the disazo condensation reds have found new outlets in the plastics industry as formulators have been replacing cadmium-containing inorganic reds with heavy-metal-free organic colorants.

8.4.2. Organic Blues

Copper Phthalocyanine Blue

The most important and most widely used blue throughout all applications of the pigment-consuming marketplace is copper phthalocyanine blue, CI pigment blue 15,

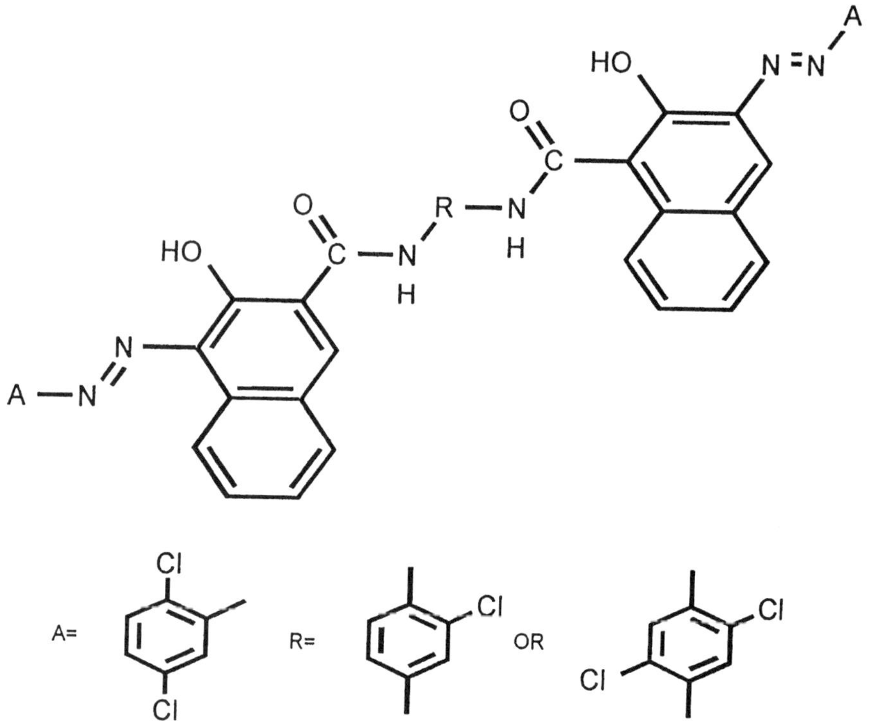

Figure 8.7. Azo condensation reds.

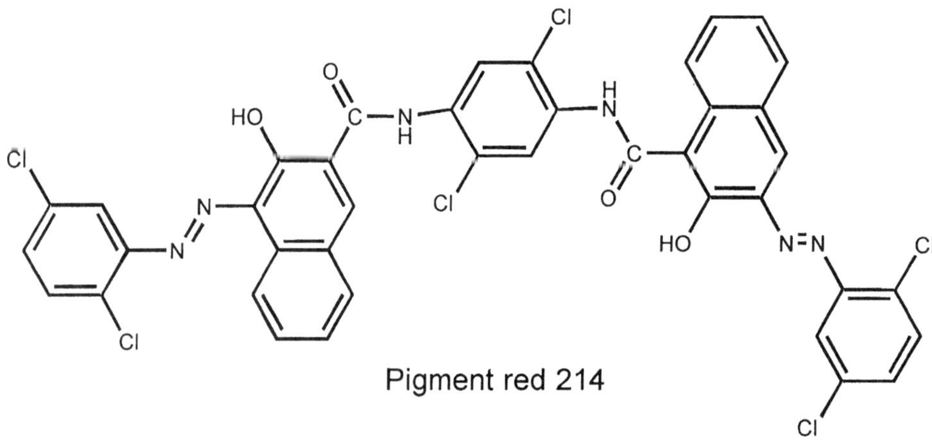

Pigment red 214

Figure 8.8. Disazo condensation red.

Figure 8.9. First described in 1928 by chemists working for the Scottish Dye Works (now part of ICI Americas, Inc.) this pigment has steadily increased in importance to become a product with worldwide significance. The only metal derivative of commercial significance is that of copper, derivatives of other metals having been shown

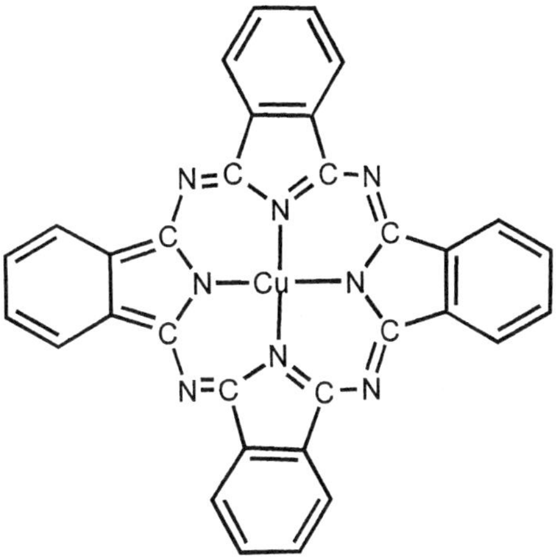

Figure 8.9. Copper phthalocyanine blue PB 15.

to have less desirable shade or fastness characteristics. The metal-free pigment CI pigment blue 16 once found an outlet as a green-shade blue, but its inferior heat stability and its poorer chemical fastness, coupled with a price almost three times that of the copper containing complex, have resulted in a decline in its consumption for all but very special end uses.

Copper phthalocyanine is a pigment that offers brightness, cleanliness, strength, and economy with excellent fastness properties. The only drawback to this pigment is its tendency to change to a coarse, crystalline nonpigmentary form when used in strong solvent systems if the crystal has not been adequately stabilized coupled with the pigment's tendency to flocculate. Another negative is the fact that copper phthalocyanine blues exhibit the phenomenon of surface bronzing when applied at masstone levels and deep tints.

Copper phthalocyanine blues give excellent service in the majority of plastics applications, but formulators are well advised to bear in mind the considerable variation in properties between the alpha and beta crystals, the two major commercially available types. A unique problem area with their use to color plastics is associated with shrinkage and warpage in high-speed injection molding applications and a degree of chemical reactivity as seen with oxidizing and reducing agents.

Copper phthalocyanine is available as two major commercial forms, the alpha and the beta crystal. Other crystal variations such as the "epsilon" crystal PB 15:6 are available commercially but on a very limited basis. The alpha crystal is described as CI pigment blue 15:1 and 15:2 and is a clean, bright-red-shade blue. The beta crystal is described as CI pigment blue 15:3 and 15:4 and is a clean green-shade blue. The beta form is the most stable crystal form and readily resists crystallization. The alpha form, conversely, is the least stable, or meta, form and readily converts to the more stable, green-shade beta crystal. As such, the alpha crystal requires special, proprietary treatments to produce a red-shade product offering stability to both crystallization and flocculation.

Copper phthalocyanine gives excellent performance in most plastic applications. There is, however, considerable variation between crystal types available. The formulator should bear this in mind when choosing a grade for a specific application. Use of any of the unstabilized grades in systems that experience a high heat profile during dispersion or application will result in a shift of shade to the greener side and a loss of strength as crystallization takes place within the unstabilized crystal morph.

CI pigment blue 15, CI No. 74160, is an alpha crystal that offers a red shade and the poorest stability of the copper phthalocyanines and as such is referred to as a crystallizing red-shade (CRS) blue. This grade is most often used in rubber and flexible PVC applications. Care must be exercised during the use of PB 15 since this crystal type has poor heat resistance at temperatures above 425°F (218°C) temperatures at which the pigment will likely revert to the more stable, green-shade beta crystal.

CI pigment blue 15:1 is a modified alpha crystal also having a red shade but with modifications to render the pigment "noncrystallizing" (NC). Most commonly the molecule is lightly chlorinated to the equivalent of one molecule or less of chlorine per molecule of copper phthalocyanine to give "monochlor" blue, identified with the unique CI number 74250. Alternatively, surface treatments can be introduced to the crystal that further decrease the solubility of the pigment and hence confer stability to the structure. This is the most commonly used phthalocyanine blue in the plastics industry due to the high tint strength and exceptional heat stability offered by this clean, red-shade colorant. Temperatures in excess of 500°F (260°C) can normally be tolerated for short cycle times by the stabilized commercial grades available in the marketplace.

CI pigment blue 15:2, described as a "noncrystallizing, nonflocculating" (NCNF), red-shade blue, is an alpha crystal that is stabilized against both flocculation and crystallization using additive technology. Such additives are introduced during the manufacturing or blending operation and are often derivatives of copper phthalocyanine that confer stability to the crystal via a steric hindrance mechanism. Use of this grade is not necessary for most plastic applications, with the possible exception being for solvent cast films.

CI pigment blue 15:3 represents the most stable crystal, green-shade, beta copper phthalocyanine. The use of this crystal modification of copper phthalocyanine blue also ranks high within the plastics industry. The pigment offers excellent lightfastness and a useful heat stability that will vary between 450 (232°C) and 550°F (288°C), depending upon the depth of color required. To maximize the color appeal of PB 15:3, many of the commercially available grades of PB 15:3 are formulated to contain a small percentage of the alpha crystal, a significant factor in determining the heat stability of the chosen pigment. A PB 15:3 that is exclusively the beta crystal will be dull and weak in comparison to the grades containing a small, controlled amount of the alpha crystal. CI pigment blue 15:4 is descriptive of a beta blue that has been modified with phthalocyanine-based derivatives to confer flocculation resistance to the crystal such that it can safely be used in strong solvent systems or systems likely to be exposed to high temperatures. Certain grades of PB 15:4 will exhibit a heat stability of 550 (288°C)–600°F (316°C) in many polymer systems.

Other specialized, more expensive crystal modifications also exist, such as PB 15:5, a red-shade gamma crystal, and PB 15:6, a very red shade epsilon crystal.

The manufacture of copper phthalocyanine is a two-stage process where a crude, nonpigmentary product is first isolated and then the product is conditioned using a variety of techniques to render the crystal pigmentary. Manufacture of the crude proceeds at high temperatures, sometimes under increased pressure and typically using organic solvents as the medium in which the condensation takes place. A phthalic acid derivative is condensed with a source of nitrogen such as urea and a copper salt such as copper chloride in the presence of a metal catalyst such as molybdenum or vanadium. Crude copper phthalocyanine can be characterized as having an average particle size of 15–25 µm, a surface area below $3 m^2/g$, a purity in excess of 92%, and a low color value.

The conditioning stage to produce finished pigment is carried out commercially using one of several processes:

SALT ATTRITION. The crude is ground in a ball mill or a double-arm, sigma blade mixer in the presence of a large excess of sodium chloride and an organic solvent such as diethylene glycol or xylene to give the green-shade, beta copper phthalocyanine blue.

SOLVENT-FREE SALT ATTRITION. The crude is attrited in a ball mill or a double-arm, sigma-blade mixer in the presence of a large excess of salt (sodium chloride) but without using any organic solvents to give the red-shade, alpha copper phthalocyanine blue.

ACID PASTING. The crude pigment is dissolved in a massive excess of concentrated sulfuric acid, from which it is then reprecipitated in a controlled manner by drowning out the acid solution into iced water to obtain the red-shade, crystallizing-grade, alpha copper phthalocyanine blue.

ACID SWELLING. The crude is treated with a smaller volume of less concentrated sulfuric acid than is used in the acid pasting procedure after which the acid-swelled paste is pumped under controlled conditions into iced water to obtain copper phthalocyanine blue as the alpha crystal.

Miscellaneous Blues

INDANTHRONE BLUE (CI PIGMENT BLUE 60, CI No. 69800). Belonging to the class of pigments described as vat pigments, indanthrone blue is a very red shade, non-bronzing pigment with outstanding lightfastness properties. The relatively poor heat stability, with a maximum use temperature of 450°F (232°C) for this pigment, coupled with its high cost, limits the use of PB 60 in plastics.

CARBAZOLE VIOLET (CI PIGMENT VIOLET 23, CI No. 51319). This pigment is a complex heterocyclic molecule that provides formulators with an intense red-shade blue pigment that possesses excellent fastness properties. Only the pigment's relatively high cost and hard nature limit its more widespread use. The pigment is used at very low levels to produce "brighter whites" by imparting a bluer hue to the undertone of the white. Again its use is limited due to its high cost coupled with the

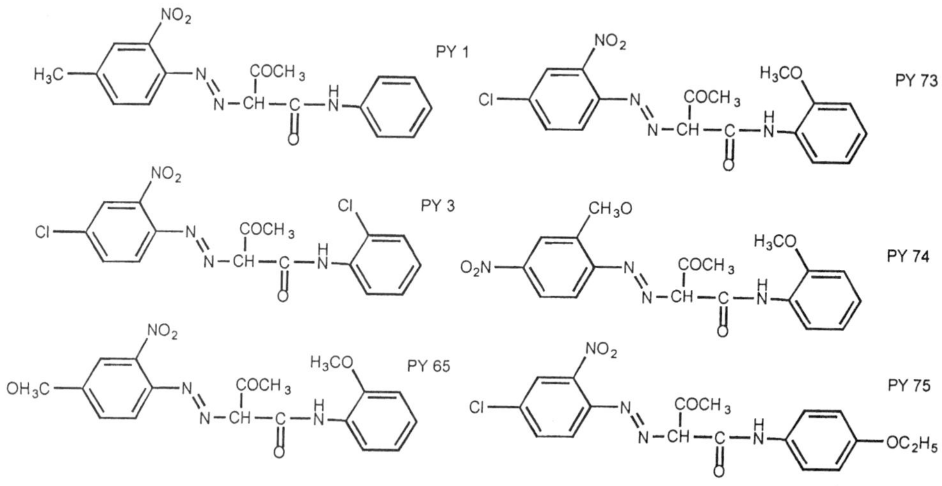

Figure 8.10. Monoarylide yellows.

fact that PV 23 does not meet the highest requirements for fastness to aggressive plasticizers of the type now used in the plastics industry.

METAL-FREE COPPER PHTHALOCYANINE BLUE (CI PIGMENT BLUE 16, CI NO. 74100). This is a green-shade blue, greener than PB 15:3, that has very limited use in the plastics industry. The pigment is reported to be useful in injection molding situations where the molded parts have critical shrinkage and warpage requirements. The crude pigment is manufactured from phthalonitrile, after which it is conditioned using an acid pasting technique. The heat stability of PB 16 is in the range of 475 (246°C)–500°F (260°C).

8.4.3. Organic Yellows

Monoarylide Yellows

These are azo pigments whose manufacture is based upon the diazotization and coupling sequence as mentioned when dealing with azo reds. The structures of the major monoarylide yellows are represented in Figure 8.10. The principal outlet for this class of yellows is in the decorative or architectural paint marketplace. Because of their solubility in many plasticized systems and their tendency to exhibit surface bloom and bleed in most polymer systems, monoarylide yellows find little use in the plastics industry. A possible exception to this generalization is CI pigment yellow 97, CI No. 11767, which has found some use in the coloration of PVC, ABS, polyolefins, and polystyrene. Pigment yellow 97 is a monoarylide yellow pigment produced by coupling of diazo 4-amino-2,5-dimethoxybenzene sulfoanilide to 4-chloro-2,5-dimethoxy acetoacetanilide. This pigment offers improved solvent bleed and lightfastness, especially in reduced shades, as compared to the more widely available monoarylide yellows that find use in the coatings industry.

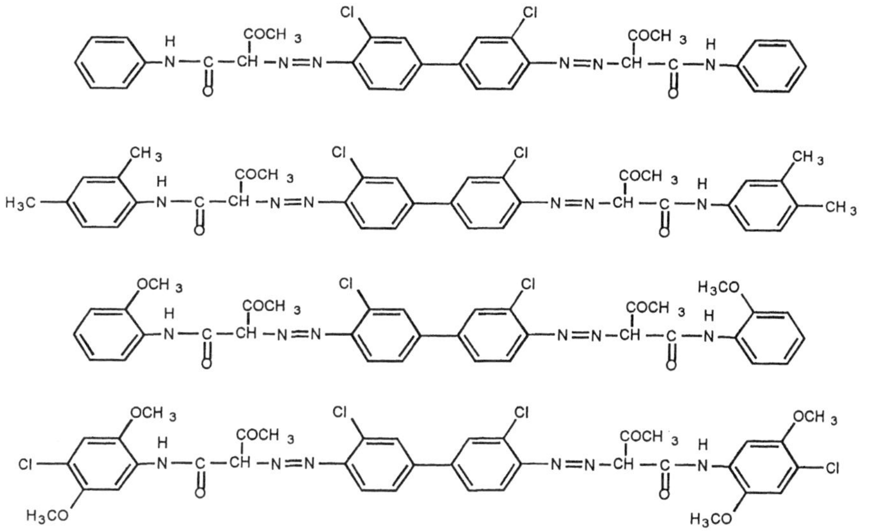

Figure 8.11. Diarylide yellows.

Metallized Azo Yellows: CI Pigment Yellow 62, PY 62, CI No. 13940, Calcium Salt

Pigment yellow 62 is a typical metallized azo yellow that finds use in coloration of rubber, LDPE, PVC, and PP fibers. The pigment is a calcium-based azo yellow produced from the diazotization and coupling of 4-amino-3-nitro benzene sulfonic acid onto acetoacet-o-toluidide (AAOT) followed by conversion of the sodium salt to that of calcium. The use of this pigment within the plastics industry has increased over the past several years as a replacement for diarylide yellows, especially under circumstances that may lead to thermal degradation of the colorant. The pigment is a green-shade yellow with good economy and moderate fastness properties to light, acids, and alkali. The heat stability is rated at a maximum of 475°F (246°C), measured at 0.1% pigment with 1.9% TiO_2 and 5 min dwell time.

Diarylide Yellows

The structures of this commercially important range of organic yellows are shown in Figure 8.11. This figure clearly shows the similarity between each of these pigments, which principally have a backbone structure centered on 3,3'-dichlorobenzidine with modifications to the shade and properties by variation of the coupling component used in the diazotization reaction. Table 8.2 gives a summary of the properties of the major diarylide yellows of commercial significance. Each diarylide yellow offers low cost, reasonable heat stability, and moderate chemical resistance. The major worldwide market for this class of yellows is the printing ink industry. These yellows are approximately twice as strong as the monoarylide yellows dealt with previously; furthermore, they offer improved bleed resistance and heat fastness. Nevertheless, none of the diarylide yellows have weathering properties that would allow for their use in exterior situations and as such should never

Table 8.2. Summary of Diarylide Yellows of Use in Plastics Industry

CI Name	Common Name	Properties
PY 12	AAA yellow	Poor lightfastness and bleed resistance; rarely used in plastics
PY 13	MX yellow	Redder shade than PY 12, offers improved heat stability and solvent fastness; PVC and polyolefins are major outlets
PY 14	OT yellow	Green shade; poor tint lightfastness limits use; PVC and polyolefins are major outlets
PY 16	Yellow NCG	Bright green shade; improved heat and solvent fastness; very limited use in plastics
PY 17	OA yellow	Very green shade; poor lightfastness; improved migration and bleed resistance; PVC and polyolefins are major outlets
PY 83	Yellow HR	Very red shade; improved migration resistance; heat and light fastness over PY 12; PVC and polyolefins are the major outlets

be considered for outdoor applications such as patio furniture. A very limited range of opacified diarylide yellow pigments is available, having undergone an after treatment that has reduced their surface area and consequently given increased opacity that has resulted in these specific types exhibiting improved fastness properties when compared with their nonopacified counterparts. Recent work has shown a tendency for diarylide pigments to initiate decomposition when processed in certain plastic systems at temperatures in excess of 473°F (200°C). Because of this tendency, diarylides should be carefully evaluated at realistic processing temperatures before being approved for the intended application.

Benzimidazolone Yellows

Illustrated in Figure 8.12, these yellows take their name from the fact that each features the 5-acetoacetyl-aminobenzimidazolone molecule within its structure. Additionally, each is an azo pigment with an acetoacetylarylamide nucleus. Their exceptional fastness to heat, light, and solvents is attributed directly to the presence of the benzimidazolone group within the pigment's structure, first described in 1964 and offered to the marketplace in 1969. These pigments are beginning to find widespread use for the coloring of plastics and in the fiber industry where their excellent fastness properties are demanded. Table 8.3 gives a summary of the properties of this class of pigments, all of which are manufactured by the diazotization of the relevant primary aromatic amine and coupling techniques as described earlier in the chapter.

Heterocyclic Yellows

All these yellow pigments contain a heterocyclic molecule within their structure as presented in Figure 8.13. In spite of their apparent complexity, these new high-performance yellows continue to be introduced to satisfy the exacting demands of

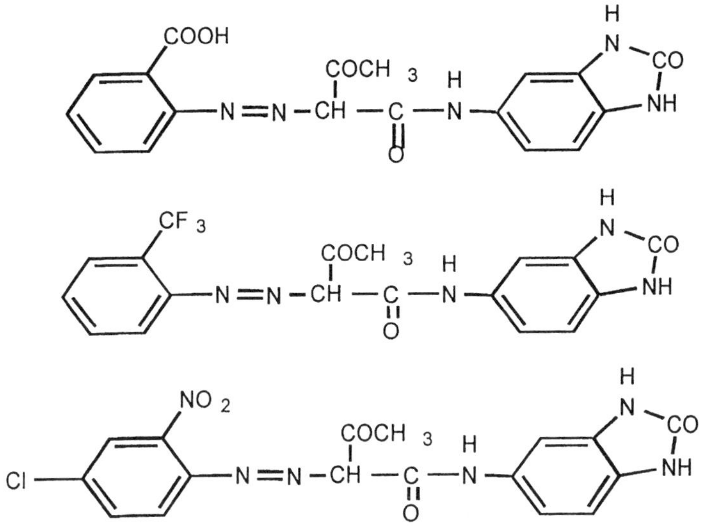

Figure 8.12. Benzimidazolone yellows and orange.

Table 8.3. Summary of Benzimidazolone Yellow Properties

CI Name	Common Name	Properties
PY 120, CI No. 11783	Yellow H2G	Medium shade; excellent bleed resistance and good fastness to light and solvents; offers a heat stability of 520°F; PVC and polyolefins major outlets
PY 151, CI NO. 13980	Yellow H4G	Green shade; good fastness to light and solvents; finds use in PVC and polyolefins; heat stability to 425°F
PY 154, CI No. 11781	Yellow H3G	Redder than PY 151; excellent lightfastness; good solvent resistance; increased use in PVC and polyolefins; not as heat stable as PY 151
PY 175, CI No. 11784	Yellow H6G	Very green shade; excellent lightfastness, good solvent resistance; PVC and polyolefins major outlets; heat stability 375–400°F

the plastic industry and as replacements for the more economical diarylide yellows where it is no longer practical to use these pigment types. Pigments such as isoindoline yellow (PY 139) and quinophthalone yellow (PY 138) are typical examples of such complex, novel chromophores introduced as recently as 1979 and 1974, respectively. All of these pigments find application in high-quality plastics and fiber end-use applications where the end use justifies the price of these high-performance products. Table 8.4 summarizes the properties of the heterocyclic yellows currently available.

Disazo Condensation Yellows

Examples of disazo condensation yellows that are available commercially are CI pigment yellow 93 and CI pigment yellow 128, only the former currently finding use

120   Organic Colorants

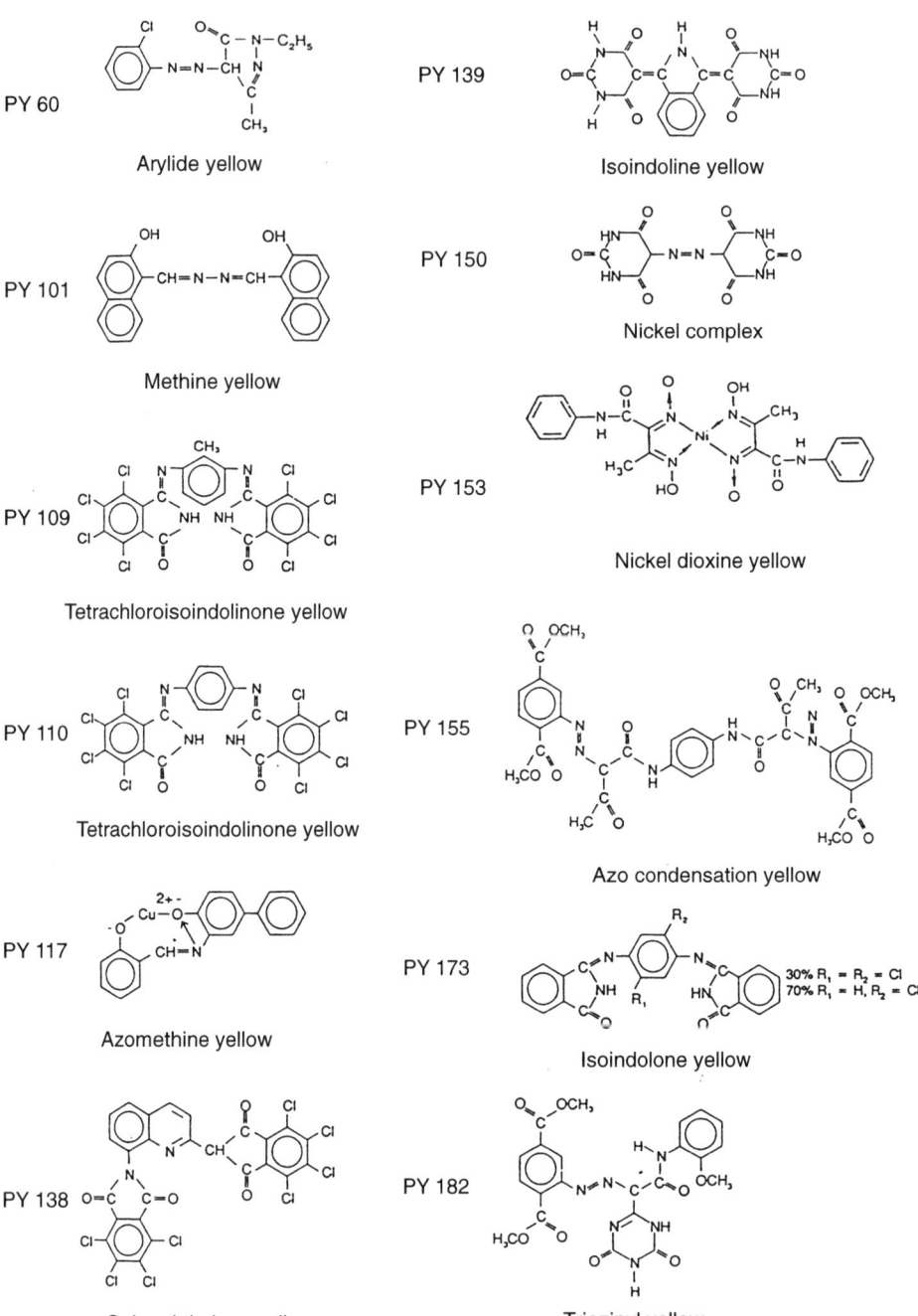

Figure 8.13. Structure of heterocyclic molecule in yellow pigments.

Table 8.4. Properties of Heterocyclic Yellows

CI Name	Common Name	Properties
PY 60	Arylide yellow	Very red shade; moderate light and solvent fastness
PY 101	Methine yellow	Bright yellow; moderate bleed fastness
PY 109	Isoindolinone	Green shade; excellent durability
PY 110	Isoindolinone	Red shade; excellent durability
PY 117	Copper azomethine	Green shade; excellent fastness properties
PY 129	Azomethine	Very green shade; excellent fastness
PY 138	Quinophthalone	Green shade; excellent fastness
PY 139	Isoindoline	Red shade; similar to medium chrome (PY 34); excellent fastness properties
PY 150	Nickel complex	Very green shade; good heat and light fastness
PY 153	Nickel dioxazine	Red shade; excellent fastness properties; poor acid resistance
PY 155	Azo condensation	Green shade; excellent fastness in full shade; both interior and exterior applications
PY 173	Isoindolone	Very green shade; excellent fastness properties in mass tone
PY 182	Triazinyl	Medium shade; excellent fastness properties at full-shade levels; interior and exterior uses

in the plastics industry. Such pigments are manufactured in a unique manner by first producing a monoazo carboxylic acid, in itself a color, by coupling a diazo primary aromatic amine with 2,3-hydroxynaphthoic acid. The acid chloride of this monoazo carboxylic acid is then condensed with an appropriate diamine through both amino groups to give a colored molecule of very high molecular weight. It is not unusual for yellow pigments in this group to have a molecular weight in excess of 1000. Simple variation of the disazo coupling and arylide portions of the molecule may be used to produce a multitude of condensed azo derivatives. Generally the fastness properties and the color intensity of the resultant pigment will increase as the molecular weight of the product increases. As a group these pigments are high-value products, offering the end user good to excellent resistance to such parameters as light, weather, solvent bleed, chemical attack, and heat. The formulator using such colorants has to balance the price of the color with the end-use performance possible. Coloration of PVC and polyolefins is a natural outlet for these yellows because of their fastness to light and heat. Additionally, the fine particle size of many of the commercial grades now available has allowed their use in the spin coloration of synthetic fibers.

8.4.4. Organic Oranges

Table 8.5 lists those orange pigments that have significance in today's pigment industry.

Azo-Based Oranges

These oranges show considerable variation in structure, as can be seen from Figure 8.14. All, however, have the azo chromophore (—N=N—) featured within the

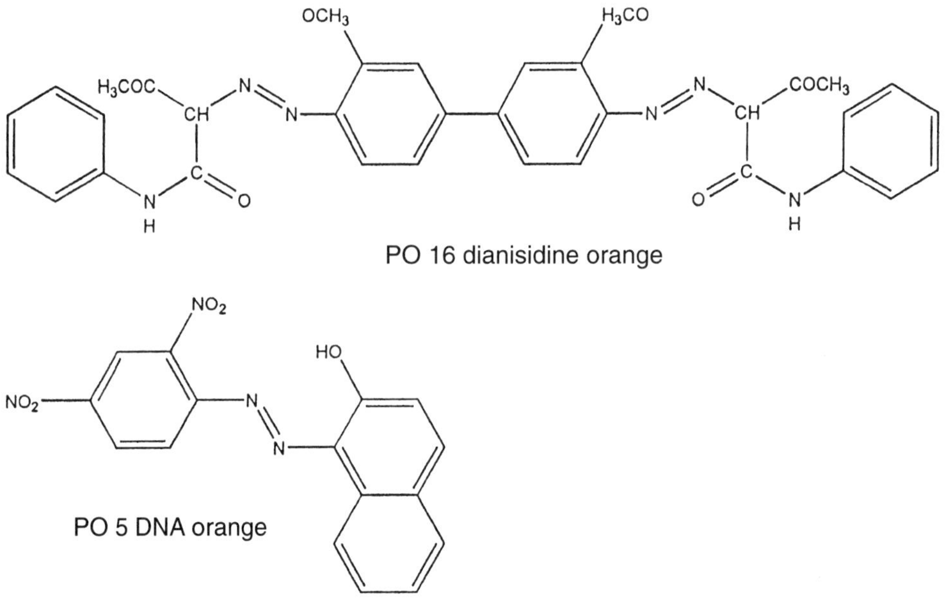

Figure 8.14. Miscellaneous azo oranges.

molecule. In addition, the benzimidazolone oranges are all produced using 5-acetoacetylamino-benzimidazolone as the coupling agent.

ORTHONITROANILINE ORANGE (CI PIGMENT ORANGE 2, CI No. 12060). Prepared by the classical diazotization and coupling technique used for all azo pigments, this pigment is the product of coupling diazo orthonitroaniline onto beta naphthol. Its major outlet is in paper coatings, crayons, and water flexo printing inks. PO 2 is not suitable for use in plastics, where bleed resistance and good solvent and lightfastness properties are of concern.

DINITROANILINE ORANGE (CI PIGMENT ORANGE 5, CI No. 12075). Produced by coupling diazo 2,4-dinitroaniline onto beta naphthol, this pigment offers good lightfastness in full tone and moderate solvent fastness. PO 5 is only used in the rubber and plastic industries in those circumstances where processing temperatures do not exceed 300°F (149°C) and where the levels of plasticizer incorporated into the plastic are minimal.

PYRAZOLONE ORANGE (CI PIGMENT ORANGE 13, CI No. 21110). Synthesized by coupling tetrazotized 3,3'-dichlorobenzidine onto 3-methyl-1-phenyl-pyrazol-5-one, the pigment is a bright, clean yellow-shade product. The pigment finds extensive use in plastics where its economics, high tinctorial strength, and reasonable heat stability (up to 300°F) are acceptable.

TOLYL ORANGE (CI PIGMENT ORANGE 34, CI No. 21115). A diarylide pigment manufactured by coupling tetrazo 3,3'-dichlorobenzidine onto 3-methyl-1-(4'-methylphenyl)-pyrazol-5-one, this orange is a bright, reddish shade offering moderate lightfastness and good alkali resistance but poor solvent fastness. It does find some application in the rubber industry, where strong solvents are not encountered and the processing temperatures do not rise above 300°F (149°C).

NAPHTHOL ORANGE (CI PIGMENT ORANGE 38, CI No. 12367). Naphthol orange is prepared by coupling diazo 3-amino-4-chloro-benzamide onto 4′-acetamido-3-hydroxy-2-naphthanilide to produce a bright red shade orange that exhibits excellent alkali and acid fastness. When used in full shades, this pigment also features acceptable lightfastness; however, at concentrations of only 0.3% the pigment begins to show severe bleed problems in most plastic applications. The pigment does find some use with polystyrenes and the mass coloration of polypropylene fibers.

CLARION RED (CI PIGMENT ORANGE 46, CI No. 15602). A metallized azo pigment manufactured by coupling diazotized 2-amino-5-chloro-4-ethyl benzene sulfonic acid onto beta naphthol followed by reacting this product with barium to yield the barium salt of the pigment, clarion red is not recommended for exterior applications due to its poor lightfastness. CI pigment orange 46 is a very yellow shade azo orange that has a heat stability up to a maximum temperature of 500°F (260°C). This pigment finds minimal use in the plastics industry with its major outlet being the printing ink marketplace.

BENZIMIDAZOLONE ORANGE (CI PIGMENT ORANGE 36, CI No. 11780). The product from the coupling of diazo 4-chloro-2-nitroaniline to 5-acetoacetylamino benzimidazolone, this is a bright red shade orange of high tint strength. In its opacified form this pigment offers excellent fastness to both heat and solvents and gives a hue similar to molybdate orange (PO 104). The major end use of PO 36 is in the paint and coatings industries. The structure of this pigment is shown in Figure 8.12.

Miscellaneous Oranges

Figure 8.15 illustrates the structures of those oranges that fall into a "miscellaneous" category as far as such structures have been declared. Table 8.6 summarizes the properties of this series of pigments, each of which is finding increased use within the plastic and fibers industries.

8.4.5. Organic Greens

Copper Phthalocyanine Green

When a self-shade green is required, rather than a green produced by mixing blue and yellow, copper phthalocyanine is the green of choice. Two pigment chemistries are available. The first is a blue-shade green, CI pigment green 7, based upon chlorinated copper phthalocyanine. The second is a yellow-shade green, CI pigment green 36, based upon halogenated phthalocyanine, where a mixture of chlorine and bromine are used to replace the hydrogens on the copper phthalocyanine blue molecule.

CI pigment green 7 is a blue-shade green made by introducing 13–15 chlorine atoms into the copper phthalocyanine molecule, whereas CI pigment green 36 is a yellow-shade green based upon a structure that involves progressive replacement of chlorine in the phthalocyanine structure with bromine. Figure 8.16 illustrates the proposed structures of the phthalocyanine greens. The most highly brominated product, PG 36, sometimes referred to as "green 6Y" or "green 8G", has an extreme

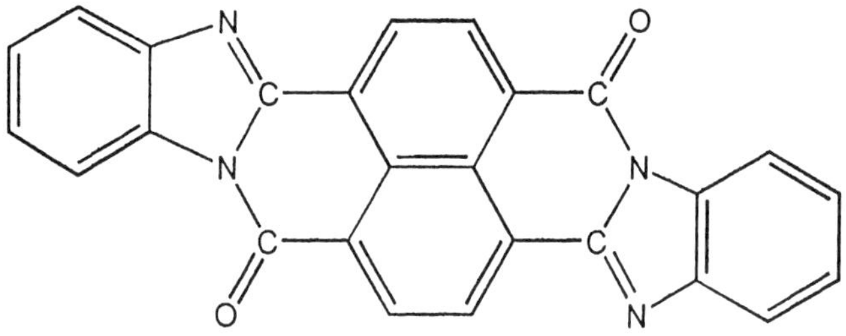

PO 43 perinone

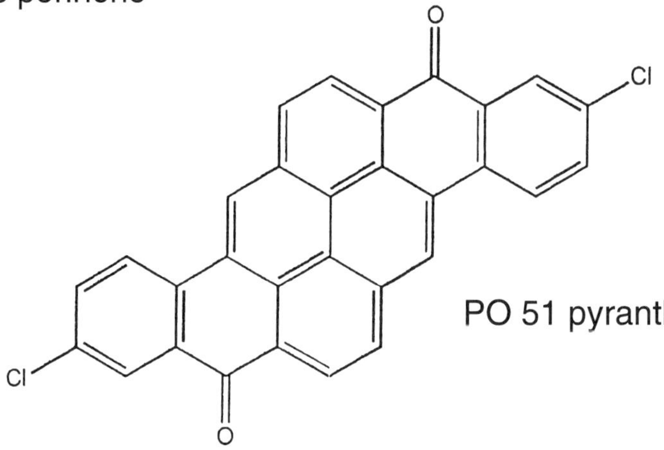

PO 51 pyranthrone

Figure 8.15. Heterocyclic oranges.

yellow shade and contains 12–13 bromine atoms. Both greens exhibit outstanding fastness properties to solvents, heat, light, and outdoor exposure. They can be used equally effectively down to the very palest depth of shades.

Both copper phthalocyanine greens can be processed at temperatures in excess of 500°F (260°C) with little apparent color change, offering even better chemical and color stability than that exhibited by the copper phthalocyanine blues. In general, the brominated PG 36 greens are weaker and more opaque than the chlorinated, PG 7, types. CI pigment green 36 represents less than 5% of the copper phthalocyanine greens used in the plastics industry.

8.5. DISPERSION OF ORGANIC PIGMENTS

Dry pigments comprise a mixture of primary particles, aggregates, and agglomerates that must first be wetted and deaggregated before dispersion forces can take full effect and enable the manufacture of a stable dispersion in the polymer of choice. It is the aim of the plastics compounder to reduce any gross aggregation to the fundamental particle size of the pigment, a value that generally lies in the range of 0.15–0.25 µm.

The process of dispersion requires that air is first displaced from around the particles to fully wet each particle and then to break down any aggregates down to

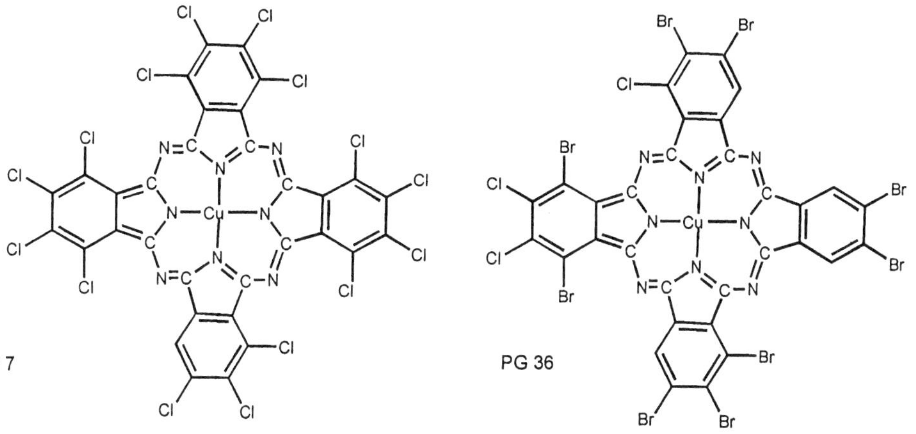

Figure 8.16. Copper phthalocyanine greens PG 7 and PG 36.

their primary particle size. The ease with which this wetting, deaggregation, and dispersion sequence takes place is referred to as the dispersibility of the pigment and varies with both pigment type and polymer system. It is undesirable to grind the pigment, thus breaking down the fundamental pigment particles during the dispersion process.

The wetting or deaeration stage of dispersion is physical in nature, as it requires the polymer system to displace the air associated with the surface of each pigment particle. The dispersion process involves the input of large amounts of energy to the system to fully disperse the pigment and achieve a stable, fully developed, colored dispersion. Many types of dispersion or milling equipment are available that will impart the necessary energy to achieve optimum dispersion. Equipment such as sigma blade mixers, single- and twin-screw extruders, two-roll mills, and Haake and Banbury mixers are found throughout the literature discussing the incorporation of a dry colorant into a polymer system. All are basically highly-energy intensive dispersion and mixing equipment, with variations on the layout and method by which the polymer is contacted with the dry pigment to incorporate the colorant into the system.

8.6. THE HEAT STABILITY PARAMETER

A series of physical and chemical changes can take place with an organic colorant as a result of the pigment being heated above the point at which decomposition becomes an issue or if a pigment is held close to its decomposition temperature for a prolonged period of time. The easiest to follow is a simple chemical breakdown of the pigment molecule as a result of the molecule being subjected to temperatures and times above the maximum advised for the particular class of pigments. This type of breakdown is normally evidenced by a loss of the color intensity of the article being colored coupled with a dull or dirty appearance as pigment decomposition becomes unacceptable. All organic pigments feature a maximum temperature above which decomposition will occur. A second change that can occur is a change in the crystal structure of the pigment, often accompanied by a shift in the hue of the colorant together with a reduction of the apparent strength of the pigment. For example, a red-shade copper phthalocyanine blue, PB 15, will shift in hue from that of a bright, clean red to a duller, green shade if the pigment is exposed to temperatures in excess of 425°F (220°C). This shift is caused by the unstabilized alpha crystal converting to the more stable beta crystal, a change encouraged by elevated temperatures. A change in the pigment's solubility can also accompany an increase in temperature. Such solubility incidents are normally visualized by an apparent increase in strength or color intensity and a shift in shade. Diarylide yellows can exhibit this phenomenon, especially in polymers such as flexible PVC and LDPE, where such surface phenomena as migration and crocking can also result at elevated temperatures. An increased temperature will promote any chemical reaction that may be likely between the pigment and additives in the polymer system. A classic example of this type of situation is with flexible PVC containing both a barium and a cadmium stabilizer system that is being colored with calcium red 2B, PR 48:2. At elevated temperatures, the barium in the formulation will replace the calcium in the pigment with a consequent shift in hue of the colorant to the yellow side of the spectrum.

Process variables while the colored polymer is processed will also contribute to incidents related to color and pigment stability. While it is recognized that most thermoplastic polymers are processed at temperatures in excess of 350°F (177°C), and many are processed in the range from 400 (204°C) to 500°F (240°C), it is critical that the minimum practical temperature is maintained consistent with an economical process when an organic colorant is used in the formulation. The time to which the pigment is exposed to the elevated processing temperature is also critical. Obviously, the rate of any reaction will proceed exponentially with increase in temperature, and considerably more change will occur over a dwell time of 5–15 min than will occur in 1 min or less. The level at which the colorant is incorporated into the polymer may also have an effect on the apparent heat stability. Certain pigments, such as quinacridones, will exhibit better stability profiles when used at deep tones as opposed to light tints. The efficiency of dispersion can also contribute to stability at elevated temperatures. A difference of as much as 75°F (24°C) may be experienced between a pigment that is well dispersed throughout a polymer as compared to the same dispersion where the colorant has not been effectively incorporated into the plastic. The type of polymer will also affect a colorant's stability. This is especially true of such polymers as ABS and flexible PVC, which are themselves sensitive to heat. It is unreasonable to expect a pigment to remain stable while the polymer matrix into which it has been dispersed is changing rapidly.

Thus the choice of organic pigment to be used in imparting color to a polymer system has to be made with full knowledge of the pigment's and polymer's properties. If a pigment or polymer degrades as a result of elevated processing temperatures or environmental exposure, there is little a plastics processor can do to prevent such an occurrence. By careful control of processing variables and diligent choice of pigment, a balanced compromise can be reached between economics and properties that will result in a colored plastic that has a stability acceptable in its end-use application.

CHAPTER 9

Inorganic Colored Pigments

George Rangos

Ferro Corporation
West Wylie Avenue
Washington, Pennsylvania 15301

9.1. INTRODUCTION

Plastic color formulators have relied on inorganic pigments from the inception of the plastics industry. And for good reason. Generally, inorganic pigments provide a great margin of safety in use. They are easy dispersing, heat stable, lightfast, weatherable, insoluble, and opaque and offer good value in use. In this chapter we will attempt to give the reader a broad overview of the myriad inorganic colored pigment types available to the color formulator and of their respective characteristic performance properties.

Colored inorganic pigments may be classified in a number of ways. The following is a variant of a categorization used by the Colored Pigments Manufacturers Association:

Metal Oxides	Complex Inorganic	Metal Salts	Others
Iron oxides	Spinels	Cadmium pigments	Metallics
Chromium oxides	Rutiles	Mercury cadmium pigments	Lusters
	Zircons	Lead chromate yellows	Ultramarines
		Lead molybdate oranges	Gold pinks
		Iron blues	
		Chrome greens	
		Bismuth vanadate	

Coloring of Plastics, Fundamentals, 2nd edition. Edited by Robert A. Charvat
ISBN 0-471-13906-8   Copyright © 2004 by John Wiley and Sons, Inc.

We have shown only those pigment types in each category for which there has been some indicated usage in the plastics industry. For example, within the complex inorganic category, we have identified only 3 of the 14 known structures. The other crystal types are used primarily in glass and ceramic coloration. Within the others category, we will discuss only ultramarines. A separate chapter discusses "effects" pigments.

Over the past 10 years, there has been a dramatic change in the way colored inorganic pigments are used in the plastics industry. Unfortunately for the color formulator, some of the more useful inorganic pigment types from a color space coverage, heat stability, and weatherability perspective—cadmiums, mercury cadmiums, lead chromates/lead molybdates—are fast being regulated out of use. With the demise of these highly chromatic inorganic choices, the color formulator has had to become more knowledgeable about organic colorant performance and more sophisticated in the choice of organic–inorganic colorant combinations to satisfy specific end-use applications. Unlike inorganic pigments, organic colorant performance is often polymer specific, concentration dependent, and TiO_2 reduction dependent. At very low concentrations and/or in high reduction with TiO_2, organic colorant lightfastness, weatherfastness, and heat stability properties can deteriorate rapidly from stated norms. Furthermore, the most widely used inorganic colored pigments—oxides and complex inorganic colored pigments—are lacking in saturation, cleanliness, and/or tinting strength and often cannot achieve certain color space at all or can achieve it only at excessively high concentrations. As a result, organic–inorganic colorant combinations represent a practical way to achieve certain color space and to take advantage of their respective and often complementary performance properties. The organic colorant provides the major component of chroma, saturation, and tinting strength and the inorganic colorant provides the opacity and a baseline of fastness properties that the formulated color may revert to over time.

Now for our survey of colored inorganic pigments.

9.2. IRON OXIDES

9.2.1. Natural Iron Oxides

Iron oxides can be produced from naturally occurring iron compounds or synthetically from iron salts. Natural iron compounds are among the oldest known pigments and today account for about 30% of total iron oxide production. Chemically, they are oxide or oxide–hydroxide compounds of Fe(III). Useful pigmentary forms include α-Fe(OH), yellow goethite; γ-FeO(OH), orange lepidocrocite; α-Fe_2O_3, red hematite; and γ-Fe_2O_3, brown maghemite. The earth contains about 7 wt % of iron oxides. The most important source of natural iron oxides is hydrated aluminum silicate, mined at various locations throughout the world.

Persian Gulf red oxide is derived from Ormuz Island. The hematite deposits contain about 72% Fe_2O_3 and are the highest-chroma naturally occurring red oxide. Spanish red is derived from hematite deposits near Malaga, Spain, and contain about 90% Fe_2O_3. It is less saturated than Persian red and contains carbonates, limiting its use in acidic media.

Raw siennas are derived from limonite ores and are mined in Georgia, Virginia, and Italy. They typically contain 40–60% Fe_2O_3 and 25–45% insoluble siliceous material (Al_2O_3 and SiO_2). Sienna masstones range in hue from a clean light yellow to a darker red-brown or green-brown yellow. Raw sienna tints range from strong and clean to dirty and weak, with no direct correlation to masstone hue. However, the green-brown masstones generally yield a strong, clean tint while the red-brown masstones yield a dirtier tint.

Burnt siennas are reddish brown in masstone with a salmon-colored tint. The difference in hue between raw and burnt sienna results from the conversion of yellow iron oxide to red iron oxide during calcination, during which the water of crystallization is driven off. Burnt siennas typically contain 55–80% Fe_2O_3.

Ochres are lighter in masstone and dirtier in tint than siennas and are mined in Georgia, Virginia, France, India, and South Africa. They are mined from less pure limonite ores. The Fe_2O_3 content in ochres ranges from 20–60%.

Raw umber is derived primarily from Cyprus and contains from 40–60% Fe_2O_3 and 6–25% MnO_2. It has a green-brown masstone and tints ranging from green-yellow gray to blue-gray.

Burnt umber has a dark brown hue that is derived from the conversion of yellow iron oxide to red iron oxide during calcination. The Fe_2O_3 content ranges from 47–63% and MnO_2 content ranges from 12–19%. Siliceous content ranges from 23–28%.

Natural iron oxides are not widely used in plastics because they are coloristically inferior to their synthetic counterparts. They are chemically less pure, less chromatic, less uniform in particle size and size distribution, and 50% weaker. The burnt siennas and umbers are, however, heat stable to about 525°C. The raw siennas and umbers are less suited for plastics, as they begin to lose their water of hydration at temperatures in excess of 200°C.

9.2.2. Synthetic Iron Oxides

About 70% of all iron oxide pigments are produced synthetically. Copperas or ferrous sulfate heptahydrate ($FeSO_4 \cdot 7H_2O$) is the primary source of iron. It is a by-product of the sulfate process for the manufacture of titanium dioxide as well as a by-product of pickling operations in the steel industry. Other sources of iron include ferric sulfate, ferrous chloride, ferric chloride, and the iron oxide slurry from the production of aniline by nitrobenzene reduction.

Red iron oxides are typically produced by one of four methods:

two-stage calcination of $FeSO_4 \cdot 7H_2O$;
oxidation of an Fe(II) salt to Fe(III) and precipitation with alkali from an aqueous solution;
thermal dehydration of yellow goethite, or FeO(OH);
oxidation of the ferrous oxide component of synthetic black oxide, Fe_3O_4.

The final product in all cases is α-Fe_2O_3, but its hue can range from orange to pure red to violet through manipulation of particle size, shape, and surface properties. The four processes yield a range of physical properties. Density can vary from

4.5 to 5.2 g/cm³. Particle size can vary from 0.3 to 4 μm. Refractive index varies from 2.94 to 3.22. Oil absorption varies from 17 to 75 g/100 g.

Yellow iron oxides are produced by one of three methods:

the Penniman–Zoph process of alkali precipitation of ferrous sulfate and oxidation of the resulting hydoxide,

direct hydolysis of ferric solutions with alkali, or

the Laux process of reduction of nitobenzene to aniline with metallic iron in the presence of aluminum chloride or ferrous chloride.

The final product in all cases is α-FeO(OH). The particle size can vary from 0.1 to 0.8 μm, and the resulting hue can vary from a light, strong, clean yellow to a darker, redder, weaker buff. Density is 4.05 g/cm³. The refractive index is 2.3. Oil absorption ranges between 30 and 60 g/100 g.

Red and yellow iron oxides have excellent weathering characteristics; both absorb ultraviolet radiation. They offer good opacity and excellent value in use. Their chroma and saturation, however, are limited.

Red iron oxide has excellent heat stability. It is usable in virtually all engineering plastics.

In contrast, yellow iron oxide begins to lose its water of hydration above 177°C and converts to Fe_2O_3. Consequently, its use in plastics is limited to low-temperature processes. It is well suited for weatherable flexible vinyl applications. At one time, truckloads of yellow iron oxides were used in the coloration of low-density polyethylene disposable green garbage bags.

Neither red nor yellow oxide is recommended for use in weatherable rigid PVC applications. Iron can cause thermal degradation of the PVC during processing and can significantly reduce the weatherability of the PVC compound during exposure.

Use of red iron oxide in plastics has increased as cadmium reds and lead molybdate oranges have been restricted in their use. In combination with the appropriate organic red toners, red iron oxide can be used to cover some of the same color space formerly occupied by cadmium- or molybdate-containing formulations. The red iron oxide component of the organic–inorganic approach also lowers the cost of the formulation, opacifies the formulation, and offers an increment of UV stability to the formulation.

9.3. COMPLEX INORGANIC COLORED PIGMENTS

Complex inorganic colored pigments are solid solutions or compounds consisting of two or more metal oxides. One oxide serves as a host and the other oxide(s) interdiffuse into the host crystal lattice. This interdiffusing is accomplished at temperatures generally in excess of 1000°C. Of the 14 possible crystal structures, the two most important for plastics applications are rutile and spinel. A third structure, zircon, will be briefly mentioned.

The color of complex inorganic pigments results from the incorporation of cations of transition metals into the crystal of the host oxides. The host can be a single oxide such as TiO_2 (rutile) or a mixed oxide such as $TiZn_2O_4$ (spinel) or $MgAl_2O_4$ (spinel). Most of the host oxides can be found as naturally occurring minerals, which are typically colorless when pure. The stability of the host lattice imparts the exceptional thermal and chemical stabilities for which these pigments noted. Typical transition metal cations that are incorporated into the host lattices include iron, chromium, manareganese, nickel, cobalt, copper, and vanadium.

Spinel compounds have a common chemical formula AB_2X_4 and a cubic structure. Of the more than 100 known spinel compounds, only those with oxygen anions are used as inorganic pigments. The following are among those most commonly used in plastics:

Pigment	Colour Index Name	Chemical Formula
Cobalt aluminum blue	Pigment blue 28	$CoAlO_4$
Cobalt chromite blue-green	Pigment blue 36	$Co(Al,Cr)_2O_4$
Cobalt chromite green	Pigment green 26	$CoCr_2O_4$
Cobalt titanate green	Pigment green 50	Co_2TiO_4
Iron chromite brown	Pigment brown 29	$Fe(Fe,Cr)_2O_4$
Iron titanium brown	Pigment black 12	Fe_2TiO_4
Zinc ferrite brown	Pigment yellow 119	$(Zn,Fe)Fe_2O_4$
Zinc iron chromite	Pigment brown 33	$(Zn,Fe)(Fe,Cr)_2O_4$
Copper chromite black	Pigment black 28	$CuCr_2O_4$
Chrome iron manganese brown	—	$(Fe,Mn)(Fe,Cr,Mn)_2O_4$
Chrome iron nickel black	Pigment black 30	$(Ni,Fe)(Cr,Fe)_2O_4$
Chrome manganese zinc brown	Pigment brown 39	$(Zn,Mn)Cr_2O_4$
Chrome Iron black	Pigment green 17	$Fe(Fe,Cr)_2O_4$

Most spinels contain small quantities of other metal oxides as modifiers to alter hue without affecting crystal structure. To illustrate this point, the most common form of pigment green 26 also contains zinc; the most common form of pigment green 50 also contains nickel and zinc; and the most common form of pigment black 30 also contains manganese.

All rutile pigments are derived from the rutile titanium dioxide structure. Rutile forms of complex inorganic pigments are by far the most commonly used in plastics. The following are among the most widely used:

Pigment	Colour Index Name	Chemical Formula
Nickel antimony titanium yellow	Pigment yellow 53	$(Ti,Ni,Sb)O_2$
Nickel niobium titanium yellow	Pigment yellow 161	$(Ti,Ni,Nb)O_2$
Nickel tungsten titanium yellow	Pigment yellow 189	$(Ti,Ni,W)O_2$
Chrome antimony titanium yellow	Pigment brown 24	$(Ti,Cr,Sb)O_2$
Chrome niobium titanium yellow	Pigment yellow 162	$(Ti,Cr,Nb)O_2$
Chrome tungsten titanium yellow	Pigment yellow 163	$(Ti,Cr,W)O_2$
Manganese antimony titanium brown	Pigment yellow 164	$(Ti,Mn,Sb)O_2$
Manganese niobium titanium brown	Pigment brown 37	$(Ti,Mn,Nb)O_2$
Manganese chrome antimony brown	Pigment brown 40	$(Ti,Mn,Cr,Sb)O_2$

In terms of fastness properties, complex inorganic colored pigments are unequaled. Their high calcination temperatures during manufacture impart thermal stabilities up to 1000°C, outstanding outdoor durability, and outstanding chemical resistance.

They are the colored pigments of choice in the coil coatings and vinyl siding industries, where 20- to 30-year warranties are common practice. The most useful chemistries for weatherable rigid PVC applications are pigment yellow 53, pigment brown 24, pigment yellow 164, pigment green 17 (Cr Fe black), pigment brown 30 and pigment brown 29. In addition to their outstanding weatherability and chemical resistance, these chemistries possess beneficial infrared reflectance characteristics that have become increasingly important in the past five years. Among the reasons: darker vinyl siding colors have become more popular and energy costs have skyrocketed. Darker siding colors tend to absorb more heat. Use of these chemistries helps keep surface temperatures lower to prevent heat distortion of the vinyl siding. Lower surface temperatures also tend to keep interior building temperatures lower thereby reducing energy demands in the summer.

In fact, the perceived interest in and need for higher infra-red reflecting characteristics have spurred CICP pigment development, most notably in green and in black color space. In green color space, cobalt-free alternatives to camouflage green (pigment green 26) have been developed that eliminate the cobalt dip in the pigment green 26 reflectance curve. In black color space, modified CrFe blacks have been developed that provide higher reflectance characteristics than the conventional CrFe blacks and pigment black 30s found in many weatherable RPVC formulations. A further elaboration has been the development of high infared-reflecting Cr- and Fe-free blacks. These higher reflecting pigments are finding application outside the weatherable rigid PVC market. In particular, the military has expressed interest in them.

With the demise of cadmiums, complex inorganic pigments are the inorganic pigment of choice in high-temperature engineering plastics applications and obvious formulation companions to very heat stable organic dyes in achieving color space formerly occupied by cadmiums. With the demise of cadmiums and encapsulated lead chromates, their usage has increased dramatically in polyolefins and Acrylonitrile-Butadiene-Styrene (ABS), especially where durability requirements exist, such as in automotive exterior and interior applications. Pigment yellow 53 and pigment brown 24 are especially well positioned here because yellow iron oxide, the most obvious other option to replace cadmium and encapsulated lead chromate yellows, is woefully inadequate in thermal stability.

Again within the context of cadmium and lead chromate alternatives, an increasing use of nickel titanium yellow and chrome titanium yellow is as a TiO_2 replacement in organic colorant–TiO_2 combinations that have been designed to match cadmium or chrome/molybdate color space. In weatherable colorant formulations, use of the rutile titanium yellow pigment in place of Tio_2 might allow a partial replacement or reduced loading of one or more expensive high-performance organic pigments because of the chroma contribution of the titanium yellows and thereby might represent a more cost-effective approach than the original organic–TiO_2 combination. The introduction of the titanium yellow into the formulation might also improve the overall weatherability of the formulation.

From a regulatory prespective, it is the trivalent character of its chromium content that distinguishes chrome titanium yellow from the hexavalent character of the chromium in lead chromate and lead molybdate pigments.

Cobalt blue (pigment blue 28 and pigment blue 36) occupies several niches in plastics applications. It is a viable alternative to phthalocyanine blue in high-density polyethylene molding applications where warping is an issue. It is a viable alternative to phthalocyanine blue in engineering plastics where either solubility issues or thermal stability issues make phthalocyanine usage problematic. It is a viable alternative to phthalocyanine blue at very low concentrations of organic blue where the phthalo might actually solubilize or at a very high reduction with TiO_2 where durability might be negatively impacted (such as in toning a white). To be sure, the high price of cobalt blues and the historic wide fluctuations in cobalt metal prices have done little to endear the pigment to color formulators, especially when a high-performance blue like phthalocyanine blue is available at today's bargain basement prices.

Pigment green 50 is a niche player as well. Its relationship to phthalocyanine green parallels that of cobalt blue to phthalocyanine blue. Its opportunities are further constrained by chromium oxide green, which offers similar but not quite comparable performance properties and a similar but less chromatic and dirtier hue at about one-seventh the price.

Pigment green 26 offers a unique color characteristic—its hue and infrared reflectance make it a key formulation component in camouflage green formulations.

A third structure, zircon, has found limited applicability in plastics. Tetragonal zirconium silicate serves as a host lattice for a variety of transition metals, including vanadium, praseodymium, and iron. Zirconium praseodymium yellow (pigment yellow 159) is a mainstay colorant for ceramic applications and has been used occasionally in high-density polyethylene molding applications as a cadmium and lead chromate replacement where the warping tendencies of organic colorant alternatives disqualified them from use and where the nickel, chromium, or antimony contents of complex inorganic pigments disqualified them as well. The praseodymium yellow has outstanding heat stability but is extremely weak tinctorially, even when compared to the complex inorganic pigments typically used in plastics.

Two of the more limiting qualities of complex inorganic pigments are their relatively low tinting strength and their inherent abrasivity. Their low tinting strength results from their large average particle size, typically 1–4 µm. This limitation impacts in several ways. Their relatively low tinting strength sometimes precludes their being the most cost-effective approach to a target color. Low tinting strength also suggests the possibility of high pigment loadings with resulting deleterious effects on polymer physical properties. In the past five years, significant increases in tinting strength of 20–40% have been achieved for the most widely used calcined inorganic pigments, but they still fall far short of alternative inorganic and organic chemistries. The inherent abrasivity of complex inorganic pigments further limits their use in solution-dyed fiber applications where the fine fiber deniers are easily broken during spinning.

9.4. CADMIUM PIGMENTS

Cadmium pigments represent an exceptionally clean, saturated, tinctorially strong class of yellow, orange, red, and maroon pigments. Their basic composition is cadmium sulfide (CdS) coprecipitated with increasing amounts of zinc sulfide (ZnS) to achieve the greener hues of yellow and coprecipitated with increasing amounts of selenium (Se) to produce the deeper hues of red. The selenium-containing pigments are referred to as cadmium sulfoselenides, Cadmium pigments are available in two forms—lithopone and concentrate (or pure). The lithopone versions contain up to 60% barium sulfate. Coprecipitation of the barium sulfate is a more effective method of extending the pure cadmium pigment than mere physical blending of the pigment with barium sulfate. Lithopones were developed to offer less expensive forms. Cadmium sulfide, zinc sulfide, and cadmium sulfoselenide form hexagonal wurtzite crystal lattices.

Cadmium pigments have been manufactured by both a direct calcination process and a precipitation–calcination process. In the first instance, a mixture of cadmium carbonate and sulfur (and zinc oxide and selenium if the hue to be produced requires their addition) is calcined at 520–600°C for 1–2 h. This direct calcination process is complicated by the volatility of cadmium oxide and selenium, both of which are toxic and require special handling. In the precipitation process, an alkali sulfide solution is added to a solution of cadmium and (in the case of green-shade yellows) zinc salts or to a solution of cadmium and (in the case of deep oranges, reds, and maroon) selenium metal to precipitate the appropriate compound. The precipitate is washed, dried, and calcined at 600–700°C in an inert or reducing atmosphere to convert the precipitated cubic structure to a more stable wurtzite crystal. The calcination conditions control particle size, which ranges from 0.2 to 1.0 μm.

From a performance perspective, cadmium yellows, oranges, and reds represent an almost ideal choice for plastics coloration. They are highly saturated and clean in masstone, strong in tint, opaque, lightfast, chemically stable, and heat resistant to above 400°C. As such, they are technically usable in virtually any of the high-performance engineering polymers on the market today. Cadmiums tend to darken upon simultaneous exposure to UV and moisture. Outdoor durability is therefore dependent upon the polymer's ability to form an effective barrier to moisture vapor transmission. The yellows are not recommended for outdoor exposure in tint, but the oranges and reds have sufficient weatherfastness that they are used as the benchmarks against which organic pigment alternatives are measured. Durability increases as hue moves redder and bluer.

Mercury–cadmium pigments were developed in the early 1950s as a more cost-effective alternative to cadmium sulfoselenides. Mercury replaces part of the cadmium in the cadmium sulfide compound and eliminates the need for selenium. The resulting pigments range from deep orange to a maroon and offer a cleaner, brighter chroma than their cadmium counterparts. Manufacture is the same as for cadmium sulfides, except that mercury salts are added to the cadmium solution that is reacted with the alkali sulfide solution to cause precipitation.

Mercury–cadmium pigment performance is somewhat lower than cadmium pigment performance. They are inferior to sulfoselenides in lightfastness and are not recommended for exterior applications. Nor are they as heat stable as sulfoselenides. Maximum processing temperature is about 320°C. They are quite suitable for ABS but are not sufficiently heat stable for ploycarbonate or polyamide 6.6.

As mentioned earlier in this chapter, cadmium and mercury–cadmium pigments have been largely regulated out of use in the plastics industry, and as a consequence, the color formulator has lost a powerful formulating tool. Organic colorant alternatives to cadmium/mercury–cadmium pigments generally do not possess the latter's versatility; their performance limitations need to be clearly understood. With the exception of bismuth vanadate, inorganic pigment alternatives lack the color space coverage of cadmium/mercury–cadmium pigments.

9.5. LEAD CHROMATE/LEAD MOLYBDATE PIGMENTS

Lead chromate occurs in nature as crocoite, an orange-red mineral. Synthetically prepared lead chromate and its solid solutions with lead sulfate and lead molybdate represent a hue range from primrose yellow to red. The various hues of chrome yellow, chrome orange, and molybdate orange depend not only on composition but also on crystal structure and particle size.

Five chemical compositions constitute these pigment classes. Medium chrome yellow is essentially $PbCrO_4$ and is the reddest hue of yellow. Solid solutions of $PbCrO_4$ with $PbSO_4$ in varying proportions yield primrose and lemon chrome yellows. Their hue difference is also attributable to the their polymorphic nature— primrose yellow is orthorhombic in crystal structure and lemon yellow is monoclinic. The less stable orthorhombic crystal is treated during primrose yellow manufacture to minimize hue shifts and conversion.

Molybdate oranges are solid solutions of lead chromate, lead molybdate, and lead sulfate and range in hue from yellow-shade orange to medium red. Chrome orange is $PbCrO_4 \cdot PbO$ and varies in hue from yellow-shade to red-shade orange with particle size manipulation.

The various compositions of the five pigment types are summarized as follows:

	Primrose Yellow (%)	Lemon Yellow (%)	Medium Yellow (%)	Molybdate Orange (%)	Chrome Orange (%)
$PbCrO_4$	65–71	61–75	90–94	69–80	58
$PbMoO_4$	0	0	0	9–15	0
$PbSO_4$	23–30	20–38	0–6	3–7	0
PbO	0	0	0	0	39
Other	3–8	1–6	4–6	3–13	2

Lead chromates and lead molybdates are produced by precipitation of soluble salts in aqueous media. Various lead sources include litharge, lead nitrate, basic lead actetate, and lead carbonate. The lead carbonate and basic lead acetate are used primarily to control particle size. Other ingredients include acids, alkalis, sodium bichromate, and sodium chromate. Additionally, molybdate orange manufacture involves the use of sodium molybdate and sodium sulfate as raw materials.

Precipitation conditions determine crystal structure. Lemon chrome yellows are precipitated hot with excess lead to form the monclinic crystal. Primrose yellow is precipitated at a lower temperature to form the orthorhombic crystal. Chrome oranges are precipitated under alkaline conditions. The higher the alkalinity, the larger the particle. The precipitation of molybdate orange is complicated by the polymorphic nature of two of the components of the solid solution precipitate: lead chromate and lead sulfate. As with primrose yellow, the crystal is stabilized after precipitation to prevent conversion.

Two particular types of end treatments have enhanced the usability of lead chromates and lead molybdates in plastics applications. So-called predarkened lead chromates have been surface treated with antimony to improve outdoor weatherability. A more significant end treatment has been encapsulation of both chrome yellows and molybdate oranges with silica to markedly improve thermal stability and sulfide staining and chemical resistance as well.

Like cadmium pigments, chrome yellows and molybdate oranges were once workhorse pigments in the plastics industry. Also like cadmiums, chromes and molybdates have come under close regulatory scrutiny in the past 10 years. Of particular concerns are the toxicity issues associated with lead and the carcinogenicity issues associated with hexavalent chromium (the valence state of chromium in these two pigment classes). The argument that the low solubility of these pigment classes suggests minimal health risk to the user has been largely ignored by the regulatory bodies. And again like cadmiums, the inability to formulate with these pigments denies the formulator colorants that are easy dispersing with bright, highly saturated, clean mass tones and strong tints. They offer excellent value in use and very good opacity. Conventional chrome and molybdate heat stabilities range from about 235°C for primrose yellows to about 275°C for molybdate oranges. Among the limitations in use, they are alkali sensitive and susceptible to darkening upon sulfide exposure. Also, the larger particle sized pigments, the chrome oranges and the redder hues of molybdate orange, are susceptible to overgrinding which fractures pigment particles. The chrome orange limitation is not widely relevant because chrome oranges have largely disappeared from manufacture in deference to their cleaner, more lightfast, much stronger molybdate orange counterparts.

The encapsulated versions of chrome yellow and molybdate orange demonstrate improved weatherfastness, chemical resistance, and heat stability. Whereas the conventional chromes and molybdates might be suitable for moderate HDPE and PP processing conditions, the encapsulated types are usable at PP and ABS processing temperatures of up to 300–310°C. The major caveat for use of encapsulated versions is the possibility of cracking the silica encapsulation with excessive grinding and thereby destroying their performance advantages.

9.6. BISMUTH VANADATE

Although patented in 1978, bismuth vanadate yellow ($BiVO_4$) has only recently attracted the interest of the plastics color formulator. A highly saturated, clean, green-shade yellow of high tinting strength with good opacity and lightfastness, it lacked the thermal stability to be of widespread utility for plastics applications. Silica encapsulation of the pigment in the past 10 years has dramatically improved heat stability. During the same period, increasing regulatory restrictions on the use of less expensive encapsulated lead chromates and cadmium yellows have largely neutralized the barrier to evaluation that the much higher price of bismuth vanadate once created.

Bismuth vanadate can be produced by either a wet process or high-temperature calcination. By the first method, an acidic solution of bismuth nitrate, $Bi(NO_3)_3$, is mixed with an alkaline solution of sodium vanadate, Na_3VO_4. The resultant gel is filtered, washed, and converted to crystalline form by low-temperature calcination at 200–500°C. Multiple-phase pigments can be created by coprecipitation of bismuth vanadate with molybdenum, tungsten, or niobium compounds.

In the calcination process, a mixture of Bi_2O_3 and V_2O_3 is milled and calcined at 750–950°C. Wet milling in an alkaline medium of the resultant $BiVO_4$ is recommended to remove unreacted vanadium salts that may adversely affect the pigmentary properties.

Relative to other inorganic pigment alternatives to chrome yellows and cadmium yellows, bismuth vanadate is much more expensive. So, where a complex inorganic yellow or iron oxide yellow will do the job by itself, bismuth vanadate will not likely be a cost-effective alternative. However, it offers much broader color space coverage. It is much more chromatic and much stronger than complex inorganic yellows and iron oxide yellows and therefore more likely to be a stand-alone alternative to chrome yellow and cadmium yellow than the other two pigment classes. It does not have the outstanding exterior durability of nickel titanium yellow, chrome titanium yellow, or iron oxide yellow, but it certainly is sufficient for exterior automotive applications. It is heat stable to about 300°C, sufficient for most polyolefinic and ABS applications. It is not quite heat stable enough for polycarbonate and polyamide 6.6 usage.

A limitation in using bismuth vanadate is the care needed in dispersing the pigment. As with other silica-encapsulated pigments, even partial shearing of the encapsulant will noticeably diminish thermal stability and durability.

9.7. ULTRAMARINE PIGMENTS

Ultramarines are derived from lazurite, a semiprecious stone. Ultramarines can be produced in many shades, the most commercially important being blue, violet, and pink.

Ultramarines are complex sodium aluminates with a zeolite structure. The crystal lattice is approximated by the chemical structure $Na_6Al_6Si_6O_{24}$, with entrapped sodium ions and ionic sulfur groups (S2 and S3). The ionic sulfur imparts the pigment's color.

The composition of ultramarine blue ranges from Na_2O, 19–23%; Al_2O_3, 23–29%; SiO_2, 37–50%; and S, 8–14%. It is prepared commercially from mixtures of china clay, sodium carbonate, silica, sodium sulfate, and a carbonaceous reducing agent such as charcoal or tree rosin. The china clay is preactivated by heating to 700°C before mixing with the other ingredients. The mixture is heated in muffle furnaces to 760–780°C. Within 20h, a green ultramarine is formed. After 40–150h of heating, the pigment is cooled in the presence of oxygen, sulfur dioxide is released, and the pigment hue changes from green to blue. Varying particle size and chemical composition affect the hue. Larger particles, 3–5μm, are characterized by darker masstones and red tints. Lighter masstones with stronger, greener tints are produced with particles of 0.5–1.0μm.

Violet and red ultramarines are produced by treating the ultramarine blue with hydrogen chloride or chlorine gas at 255–260°C or reacting the blue with ammonium chloride. The intensity of the red or violet depends upon concentrations of the reactants and time–temperature durations.

Ultramarine blues, reds and violets are acid sensitive. As such, their thermal and weathering properties are adversely affected in acidic media. Silica-encapsulated versions offer marked improvement in thermal stability and some improvement in weathering properties. The treated versions allow broad polymer usage, even in engineering plastics such as polycarbonate, and provide sufficient durability for automotive interior applications. Weatherability is highly dependent upon the degree of exposure to acid or alkaline conditions.

Ultramarine blue is a particularly effective alternative to phthalocyanine blue in HDPE, where phthalo's warping tendencies limit its use, and in ABS and engineering plastics, where phthalocyanine solubility tendencies at low concentration limit its use. Ultramarine also does not exhibit the shear sensitivity of some phthalocyanine blues. Where weathering and acid sensitivity are not issues, ultramarine blue is a much more cost-effective alternative to phthalocyanine blue than cobalt blue.

9.8. CHROMIUM III PIGMENTS

Trivalent chromium pigments include chromium oxide (Cr_2O_3) and hydrated chromium oxide ($Cr_2O_3 \cdot xH_2O$). Chromium oxide is prepared by calcination, either by reduction of sodium bichromate ($Na_2Cr_2O_7$) with sulfur or carbon at 750–900°C or by reduction of sodium bichromate with ammonium chloride or ammonium sulfate at 700°C. Hydrated chromium oxide is manufactured by hydrolyzing a complex chromium borate, which in turn has been produced by heating sodium bichromate with boric acid in a furnace.

Chromium oxide is used primarily for its fastness properties. It exhibits outstanding outdoor durability, excellent alkali and acid resistance, and exceptional thermal stability. It is very opaque and offers good value in use. These properties make it is a workhorse pigment for the weatherable vinyl color formulator. Its infrared reflectance characteristics resemble chlorophyll, and therefore it is widely used in camouflage green applications. Its major limitations are a limited hue range, low tinting strength and an abrasive quality. Its trivalent chromium content confers compliant status by the Food and Drug Administration (FDA).

Large quantities of chromium oxide find their way into plastics circuitously. Chromium oxide is often used as a raw material in the manufacture of key calcined inorganic colored pigments for the plastics industry, such as pigment brown 24, pigment green 26, pigment brown 29, pigment green 17, and pigment black 30.

Hydrated chromium oxide has a much cleaner, bluer undertone than chromium oxide and also exhibits excellent durability and chemical resistance. Unlike chromium oxide, it is semitransparent and limited in thermal stability to about 280°C. Above 280°C, the pigment loses its water of hydration and reverts to regular chromium oxide, with the latter's duller, dirtier hue. Phthalocyanine green has largely displaced hydrated chromium oxide as a more cost-effective, much higher tinting strength alternative.

9.9. IRON BLUES

Iron blues, or cyanide iron blues, are complex ferriferrocyanide, generally with ammonium, potassium, or sodium cations. They are most commonly produced by a two-step process. First, ammonium, potassium, or sodium ferrocyanide, $M_4[Fe(CN)_6]$, is reacted with ferrous sulfate, $FeSO_4$, to yield $M_2Fe[Fe(CN)]_6$. The latter is digested with hot sulfuric acid and oxidized with sodium chlorate or sodium bichromate to yield the ferric ferrocyanide $M\{Fe[Fe(CN)_6]\}$.

Iron blues have good durability; clean, jet mass tones; and high tinting strength. They have good acid resistance but are alkali sensitive. They are among the more difficult-to-disperse inorganic pigments and have limited heat stability due to the 4–7% water contained in the particle. They tend to decompose above 200°C. Historically, iron blue usage in plastics was limited to low-density polyethylene. The largest end-use application was disposable garbage bags, in combination with yellow iron oxide. They are seldom used today by the plastics color formulator as phthalocyanine blues offer much broader polymer compatibility, performance, and color value.

9.10. CHROME GREENS

Chrome greens are coprecipitates of lead chromate yellow and iron blue. As such, they share the limitations in performance of conventional chrome yellows and iron blues, namely alkali sensitivity and limited heat stability. Consequently, they are at best of limited utility in plastics applications. They range in hue from a very light green to a very dark green.

9.11. RARE EARTH PIGMENTS

The most recent inorganic pigment developments center around rare earth compounds, most notably cerium sulfide compositions. Their development has been driven primarily by cadmium pigment replacement, as the latter have been largely regulated out of use and comparable alternatives still do not exist. (CICP pigments offer satisfactory performance as alternatives to cadmiums but do not get beyond yellow and yellow-orange color space and, as discussed earlier, are deficient in saturation and tinting strength versus cadmium.)

Cerium sulfide pigments were originally designed to cover yellow, orange and red color space and were to be of near-comparable saturation to cadmium pigments. So far, a yellow version has not been introduced and the orange and red versions fall short of achieving the saturation and tinting strength of cadmium. They are also much less cost-effective to use.

Cerium sulfide pigments are produced from hydrated cerium oxide or oxalate and calcined in an oxygen-free, sulfide environment. They are silica-encapsulated to minimize water-reactivity and to improve heat stability and chemical resistance properties. Because of their low relative value-in-use, they are used primarily in engineering plastics and in particular the polyamides where high-performance organic colorant alternatives and other inorganic pigment alternatives are few.

9.12. FDA STATUS

An increasing number of colored inorganic pigments are FDA-compliant. Historically, yellow iron oxide, red iron oxide, black iron oxide, zinc ferrite, burnt umber, raw and burnt sienna, channel carbon black, chromium oxide green, ultramarine blue, cobalt blue and copper chrome black have enjoyed FDA—compliant status, under 21 CFR 178.3297, "Colorants for Polymers". More recently, the FDA has been successfully petitioned with regard to nickel titanium yellow, chrome titanium yellow, and cobalt green under 21 CFR 170.39, "Threshold of Regulation for Substances Used in Food-Contact Articles".

9.13. SUMMARY OF FEDERAL REGULATION OF HEAVY METAL USAGE

A reality of inorganic pigment usage is the restrictive regulatory environment constraining that usage. Below is a table of metals that are regulated by one or more of the indicated U.S. regulatory bodies. Superfund Amendments and Reauthorization Act Title III SARA Title III regulates the toxic inventory and emissions reporting. The Resource Conservation and Recovery Act (RCRA) regulates disposal of hazardous waste. The Occupational Safety and Health Administration (OSHA) regulates exposure to chemicals in the workplace. The Clean Water Act (CWA) limits metal concentrations in water discharge. The Clean Air Act of 1990 (CAA) regulates the abatement of all materials in the air.

Metal	SARA	RCRA	OSHA	CWA	CAA
Aluminum				x	
Antimony	x	x		x	
Arsenic	x	x		x	x
Barium	x	x		x	
Beryllium	x			x	
Cadmium	x	x	x	x	x
Cobalt	x			x	x
Chromium	x	x		x	x
Copper	x			x	
Lead	x	x	x	x	x
Mercury	x	x		x	x
Manganese	x				x
Molybdenum				x	
Nickel	x			x	
Selenium	x	x		x	
Silver	x	x			
Titanium				x	
Zinc				x	

Two initiatives at the state level have also impacted inorganic pigment usage: CONEG and California Proposition 65. The Coalition of Northeast Governors (CONEG) enacted legislation in its member states regulating the amount of cadmium, hexavalent chromium, lead, and mercury in packaging materials disposed of in these states. At least 20 states now have CONEG-type legislation in place and Congress has considered enactment of similar type legislation.

California's Safe Drinking Water and Toxic Enforcement Act of 1986 (Proposition 65) restricts the use and disposal of antimony, arsenic, beryllium, cadmium, hexavalent chromium, lead, and nickel compounds, without specifying limits.

As stated previously throughout this chapter, the pigment types most adversely impacted by this series of federal and state regulations are cadmium, mercury–cadmium, lead chromate, and lead molybdate colorants. Generally, these regulations make a distinction between hexavalent chromium, which is preceived as carcinogenic, and trivalent chromium. It is this distinction that allows the continuing widespread use of chrome antimony titanium yellow while severely limiting the use of lead chromate/lead molybdate pigments.

9.14. SUMMARY OF PIGMENT PERFORMANCE

Included here are two tables summarizing inorganic colored pigment performance. Table 9.1 broadly overviews polymer suitablility; Table 9.2 more specifically delineates performance characteristics.

Some general comments are in order. With regard to color change upon ultraviolet exposure, most pigments show color shift over time. Inorganic pigments typically darken upon exposure, particularly those pigments that contain ions capable of more than one oxidation state, for example, Pb, Hg, Cr, Cu, Fe. In contrast, organic colorants typically fade or shift in hue upon exposure.

Thermal degradation of inorganic pigments generally manifests itself as darkening. Organic pigments also can darken upon thermal degradation. However, two other heat-related color shifts can occur with organic pigments: a change in hue or darkening as a result of high shear stress during typically acceptable processing temperatures (particularly relevant to narrow-gated or hot-runner molding conditions) and a change in hue or saturation as a result of increased solubility of organic pigment in ABS or engineering polymers at elevated processing conditions. Neither of these latter phenomena are relevant to inorganic pigments, which are not shear sensitive (except possibly for silica-encapsulated versions) and not soluble in any of the primary polymer types. Because inorganic pigments are not soluble, they show no migration tendencies as well.

Inorganic pigments are generally easier to disperse than organic colorants. This tendency relates to their large average particle size–about 5–20 times larger. They are not as prone to compaction upon high-intensity mixing and and show less color shift with increasing amounts of dispersing energy.

Table 9.1. Summary of Polymer Suitability

Pigment Polymer	FPVC	RPVC	PS	LDPE	HDPE	PP	ABS	PET	PC	PA 6	PA 6.6
Yellow iron oxide	2	0	1	2	1	1	0	0	0	0	0
Red iron oxide	2	0	2	2	2	2	2	1	1	1	0
Black iron oxide	2	0	2	2	2	2	2	2	1	1	1
Zinc ferrite yellow	2	0	1	2	1	1	0	0	0	0	0
Lead chromate yellow	2	2	1	2	1	1	0	0	0	0	0
Lead molybdate orange	2	2	1	2	1	1	0	0	0	0	0
Encapsulated chromates	2	2	2	2	2	2	2	2	0	1	0
Encapsulated molybdates	2	2	2	2	2	2	2	2	0	1	0

Inorganic Colored Pigments

Table 9.1. (*continued*)

Pigment Polymer	FPVC	RPVC	PS	LDPE	HDPE	PP	ABS	PET	PC	PA 6	PA 6.6
Cadmium yellow	2	2	2	2	2	2	2	2	2	2	2
Cadmium orange	2	2	2	2	2	2	2	2	2	2	2
Cadmium red	2	2	2	2	2	2	2	2	2	2	2
Mercury cadmium orange	2	2	2	2	2	2	2	2	0	1	0
Mercury cadmium red	2	2	2	2	2	2	2	2	0	1	0
Chrome green	1	0	1	1	0	0	0	0	0	0	0
Chromium oxide	2	2	2	2	2	2	2	2	2	2	2
Hydrated Cr_2O_3	2	2	2	2	1	1	0	0	0	0	0
Iron blue	1	0	1	2	0	0	0	0	0	0	0
Ultramarine blue	2	2	2	2	2	2	2	2	2	2	2
Cobalt blue	2	1	2	2	2	2	2	2	2	2	2
Nickel titanium yellow	2	2	2	2	2	2	2	2	2	2	2
Chromium titanium yellow	2	2	2	2	2	2	2	2	2	2	2
Manganese titanium brown	2	2	2	2	2	2	2	2	2	2	2
Cobalt titanium green	2	2	2	2	2	2	2	2	2	2	2
Iron titanium brown	2	1	2	2	2	2	2	2	2	2	2
Chrome iron zinc brown	2	0	2	2	2	2	?	2	2	2	2
Chrome manganese zinc brown	2	1	2	2	2	2	2	2	2	2	2
Copper chrome black	2	2	2	2	2	2	2	2	2	2	2
Chrome iron black	2	2	2	2	2	2	2	2	1	2	2
Chrome iron nickel black	2	2	2	2	2	2	2	2	1	2	2
Bismuth vanadate	2	2	2	2	2	2	2	1	1	1	1

Note: 2 = sufficient thermal stability and limited reactivity with polymer allows broad use, 1 = marginal thermal stability or potential reactivity with polymer restricts use, 0 = generally unsuitable for use.
FPVC, Flexible Polyvinyl Chloride; RPVC, Rigid Polyvinyl Chloride; PS, Polystyrene; LDPE, Low Density Polyethylene; HDPE, High Density Polyethylene; PP, Polypropylene; ABS, Acrylonitrile-butadiene-styrene copolymer; PET, Polyethylene terephthalate; PA, Polyamide; PC, Polycarbonate

Table 9.2. Performance Characteristics of Inorganic Colored Pigments

Pigment Type	Colour Index Name	Heat Stability (°C)	Tinting Strength	Opacity	Lightfastness Mass	Lightfastness Tint	Weatherfastness Mass	Weatherfastness Tint	Acid Resistance	Alkali Resistance	Resistance to Oxidizer	Resistance to Reducer
Yellow Iron Oxide	PIG YLO 42	200	Low	OP	E	E	E	E	P	E	E	F
Red Iron Oxide	PIG RED 101 & 102	350	Low	OP	E	E	E	E	P	E	E	F
Zinc Ferrite	PIG YLO 119	260	Low	OP	E	E	E	E	P	E	E	F
Lead Chromate Yellow	PIG YLO 34	250	Moderate	OP	G	G	F	F	F	P	F	P
Lead Molybdate Orange	PIG RED 104	275	Moderate	OP	G	G	F	F	F	P	F	P
Encapsulated Chrome Yellow	PIG YLO 34	320	Moderate	OP	E	E	E	E	G	G	F	F
Encapsulated Molybdate Orange	PIG RED 104	320	Moderate	OP	E	E	E	E	G	G	F	F
Cadmium Yellow	PIG YLO 35 & 37	425	Moderate	TL	E	E	G*	NR	F	E	F	E
Cadmium Orange	PIG OR 20	425	Moderate	OP	E	E	VG*	VG*	F	E	E	E
Cadmium Red	PIG RED 108	425	Moderate	OP	E	E	E*	E*	F	E	E	E

143

Table 9.2. (*continued*)

Pigment Type	Colour Index Name	Heat Stability (°C)	Tinting Strength	Opacity	Lightfastness		Weatherfastness		Acid Resistance	Alkali Resistance	Resistance to Oxidizer	Resistance to Reducer
					Mass	Tint	Mass	Tint				
Mercury Cadmium Orange	PIG OR 23	350	Moderate	OP	E	E	NR	NR	F	E	E	E
Mercury Cadmium Red	PIG RED 113	350	Moderate	OP	E	E	NR	NR	F	E	E	E
Chrome Green	PIG GR 15	200	Moderate	OP	G	F	G	F	P	P	F	P
Chromium Oxide	PIG GR 17	400	Low	OP	E	E	E	E	E	E	E	F
Hydrated Chromium Oxide	PIG GR 18	280	Low	TL	E	E	E	E	G	E	E	E
Iron Blue	PIG BLUE 27	200	Moderate	OP	F	F	NR	F	F	P	E	E
Ultramarine Blue	PIG BLUE 29	350	Low	TL	E	E	F**	F**	P	F	E	E
Cobalt Blue	PIG BLUE 28 & 36	500	Very Low	OP	E	E	E	E	E	E	E	E
Nickel Titanium Yellow	PIG YLO 53	500	Very Low	OP	E	E	E	E	E	E	E	E
Chrome Titanium Yellow	PIG BRN 24	500	Very Low	OP	E	E	E	E	E	E	E	E
Manganese Titanium Brown	PIG YLO 164	500	Very Low	OP	E	E	E	E	E	E	E	F

Pigment	Colour Index	Max Temp	Tinting Strength	Opacity					
Cobalt Titanium Greens	PIG GR 26 & 50	500	Very Low	OP	E	E	E	E	E
Iron Titanium Brown	PIG BLK 12	500	Very Low	OP	E	E	E	F	F
Iron Chrome Brown	PIG BR 29	500	Very Low	OP	E	E	E	F	F
Chrome Iron Zinc Brown	PIG BR 33	500	Very Low	OP	E	E	E	F	F
Chrome Manganese Zinc Brown	PIG BR 39	500	Very Low	OP	E	E	E	F	F
Copper Chrome Black	PIG BLK 28	500	Very Low	OP	E	E	E	E	E
Iron Chrome Black	PIG GR 17	500	Very Low	OP	E	E	E	E	E
Chrome Iron Nickel Black	PIG BLK 30	500	Very Low	OP	E	E	E	E	E
Bismuth Vanadate	PIG YLO 184	300	Moderate	OP	E	E	E	E	E

Notes:

Heat stability: Indicates maximum processing temperature at typical loading for typical processing times. These values assume no reactivity with any compound ingredients. The nominal heat stability also depends on the ability of the pigment to mask any shifts in polymer color that occur at the upper end of its processing temperature range.

Tinting strength: Assessment is relative to organic colorants, which are typically 5–20 times stronger than the inorganic pigments listed here. Most organic pigments would be rated high to very high by this classification.

Opacity: At typical colorant loadings. At low loadings, many opaque pigments will appear translucent.

Weatherability: E = excellent, VG = very good, G = good, F = fair, P = poor, NR = not recommended. The ratings of the asterisked cadmiums are dependent on the imperviousness of the polymer to moisture vapor transmission. Cadmium yellows in tint and mercury–cadmium pigments are not recommended for exterior exposure. Ultramarine blue (double asterisk) performance is highly dependent on the degree of acid exposure.

Chemical resistance: E = excellent VG = very good, G = good, F = fair, P = poor.

CHAPTER 10

Titanium Dioxide Pigments

Dwight A. Holtzen and Austin H. Reid, Jr.

E. I. du Pont de Nemours & Company
Chestnut Run Plaza, Wilmington,
Delaware 19880-0709

10.1. INTRODUCTION

Pigments based on a titanium dioxide core particle (hereinafter referred to as "titania pigments" are generally recognized and identified as pigment white 6, Colour Index number 77891) and are the preferred white pigments for polymer use [1]. They are unparalleled for the provision of brightness, whiteness, and opacifying power. The advantages of their use are myriad. They possess a high refractive index, thereby providing considerable opacifying power at relatively low loadings. They are nontoxic and are available with a wide variety of surface treatments to better match their advantages with the dispersion and flocculation resistance requirements of a broad spectrum of polymers and liquid systems. They are relatively inexpensive, particularly when compared on a strength/loading basis with other white pigments (vide infra). When properly chosen, titania pigments can provide extensive protection of polymers from the effects of ultraviolet light and the associated generation of destructive radical species in the polymer matrix. Free-flowing varieties, providing ease of addition and of metering, are widely available. Titania pigments do not migrate in a polymer matrix and generally do not require enormous amounts of shear energy for their full dispersion. They are available on a global basis from a number of suppliers, many of whom provide extensive technical service support to their customers.

Other white pigments used less extensively than those based upon TiO_2 for brightness, whiteness, and opacity include lithopone, zinc oxide, zinc sulfide, and

---

Coloring of Plastics, Fundamentals, 2nd edition. Edited by Robert A. Charvat
ISBN 0-471-13906-8   Copyright © 2004 by John Wiley and Sons, Inc.

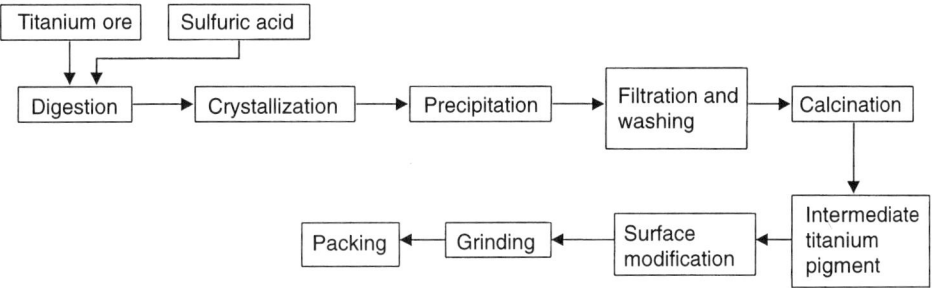

Figure 10.1. Schematic diagram of the sulfate process to manufacture TiO$_2$ pigments.

antimony oxide, among others. These will not be discussed in greater detail. Our initial focus will be on titanium-based materials, as befitting their ubiquitous use in the industry.

10.2. OCCURRENCE AND SOURCES OF TITANIA PIGMENTS

Titanium is the ninth most abundant terrestrial element. Large deposits of titanium in the form of ilmenite (an iron titanate) are found in alluvial deposits formed by the subsidence of ancient seas [2]. This concentration occurs due to the high density of ilmenite compared to typical aluminosilicate-based minerals. The ready availability of ilmenite-based feedstocks has led to their use in the production of titania pigments. Slags, formed during the processing of iron ores containing titanium, often contain appreciable quantities (as much as 95% by weight) of titanium as the oxide. Natural rutile titania (essentially 100% TiO$_2$) is also used as a feedstock in pigment production, although it is generally less available, and, therefore, considerably more costly than many other available ores.

Titania most often appears in the form of one of two predominant crystallographic polymorphs, rutile and anatase. At least seven other distinct phases have been characterized by x-ray diffraction [3], but none are sufficiently commonplace to play a role in pigmentation using titanias. Rutile possesses a higher refractive index than anatase. It is also inherently considerably less photoactive than anatase. For these reasons, rutile has become the predominant polymorph of titania for pigmentary uses in polymers.

The production of titania pigments is carried out using one of two production routes, the "sulfate" or "chloride" processes. In the former, titaniferous feedstocks are roasted and ground and digested in sulfuric acid, impurities are removed, and the resulting titanyl sulfate is hydrolyzed to form a feedstock that is subsequently calcined and subjected to downstream finishing (Fig. 10.1). In the chloride process, titaniferous feedstocks are reacted with chlorine at high temperature in the presence of a carbon source to form titanium tetrachloride and other metal chlorides. The titanium tetrachloride is purified and subsequently burned in high-temperature oxygen to give an intermediate ceramic titania that is then subjected to downstream finishing (Fig. 10.2). In both processes the intermediates are surface treated with a wide range (depending on intended end use) of hydrous inorganic oxides and/or organic materials to generate various performance characteristics.

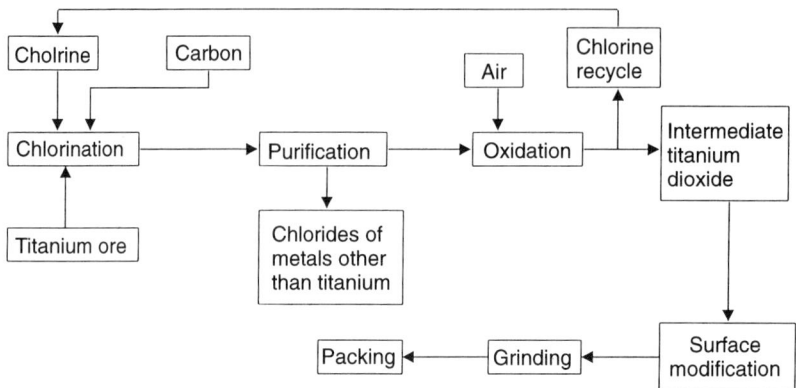

Figure 10.2. Schematic diagram of the chloride process to manufacture TiO$_2$ pigments.

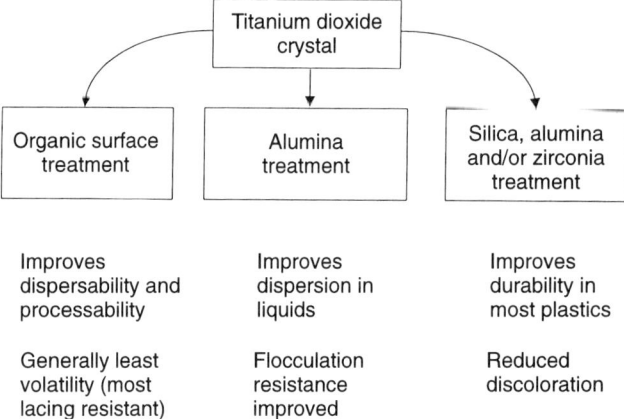

Figure 10.3. Surface treatment of TiO$_2$ pigments. Types and generalized purpose of coating.

10.3. SURFACE MODIFICATIONS OF TITANIA PIGMENTS

As mentioned previously, titania pigments are surface modified in many instances to induce a wide range of end-use behaviors. Hydrous oxide and organic coatings have been in use for many years [4] (Fig. 10.3). Most are precipitated from aqueous solution onto the intermediate materials. The technology involved in this aspect of pigment manufacture is held with the greatest of security by manufacturers, and an extensive patent literature exists [5]. In general, surface coatings are used to enhance both dispersability and, where needed, photostability. Both topics are discussed in greater detail later in this chapter.

Scanning electron photomicrographs of typical rutile titania pigments treated with various organic and inorganic coatings are provided in Figures 10.4 and 10.5.

Inorganic surface treatments are generally placed on the pigment particles in a specific intermediate step of the manufacturing process. Organic treatments can be added at any of a number of addition points throughout the pigment finishing process.

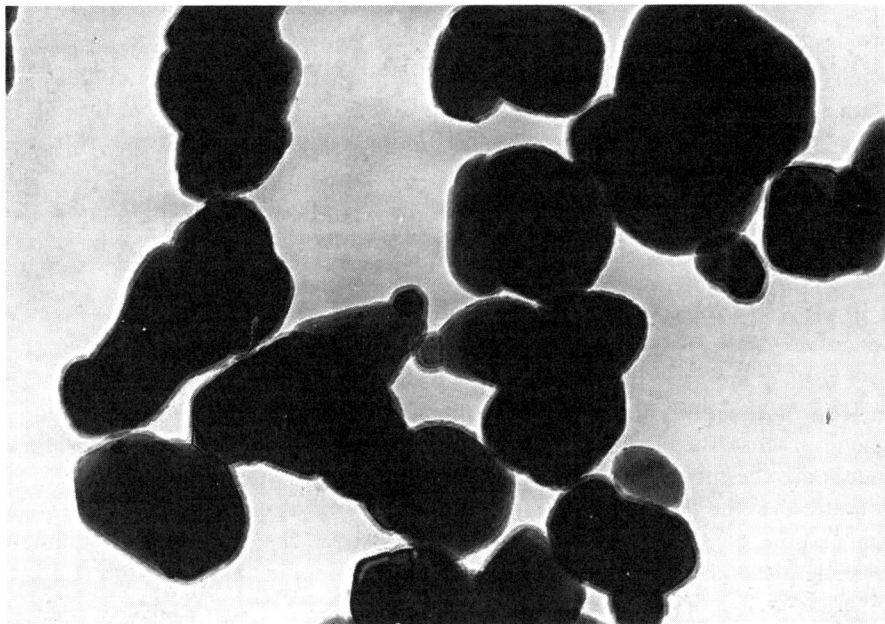

Figure 10.4. Transmission electron micrograph of a silica/alumina-treated grade of titanium dioxide. The surface coating designed to improve durability appears as a shadow around the perimeter of the crystal (300,000× magnification).

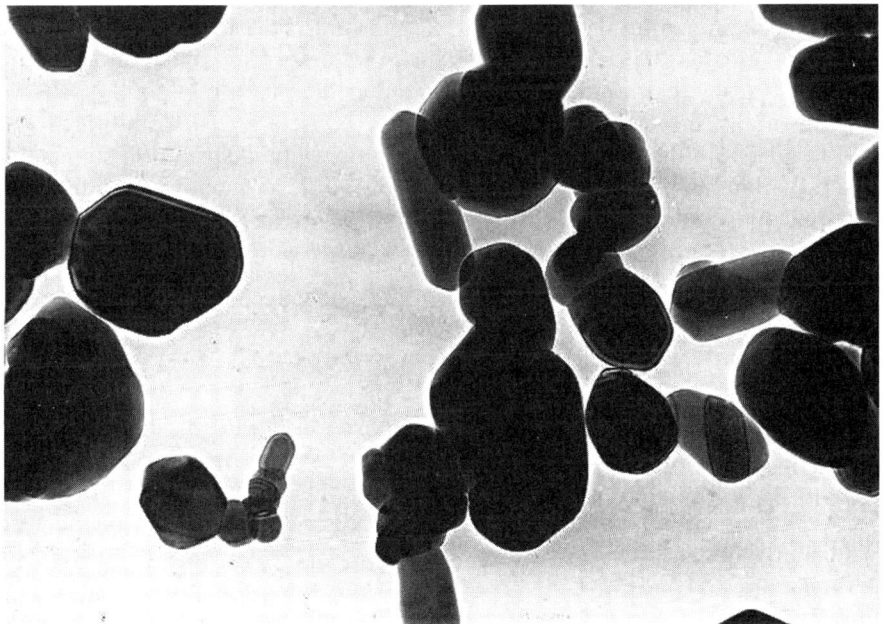

Figure 10.5. Transmission electron micrograph of an organically treated general-purpose grade of TiO_2 (240,000× magnification).

Organic coatings are generally utilized to improve dispersability and flow characteristics of titania pigments. A fascinating array of substances are either used commercially or have been claimed as useful in the patent literature [6]. Organics held in place simply by electrostatic forces [7] as well as those "cemented" to the surface by use of covalent bonds are known [8]. The latter are inert to diffusion through the polymer matrix. Said diffusion can lead to end-use problems that include such phenomena as loss of low-temperature heat seal and poor printability.

10.4. PARTICLE SIZE DISTRIBUTION AND OPTICS

Many excellent articles exist regarding the optical efficiency of titania pigments and the effects on this behavior resulting from changes in particle size distribution [9]. In a general sense, the most important aspects of particle size distribution for titania pigments are the mean diameter of the particles and the percentage of the distribution at either the fine end or large end of the distribution.

Calculations based on Mie theory [10] demonstrate that the optimum diameter of particles intended to scatter visible light should be approximately 0.25 μm. In practice, the theoretical results have been found to hold true [11]. A value of ~0.25 μm is therefore the aim targeted by most manufacturers.

The percentage of particles on the "small" end of the distribution do not scatter light efficiently. In fact, a rapid decline in scattering is noted for particles in the 0.15-μm range, with scattering dropping quickly to zero as the diameter decreases. As such, pigments with an excessive population of particles having diameters in this range will not scatter as efficiently as those with diameters closer to a nominal value of 0.25 μm.

A similar phenomenon is seen at the "large" end of the distribution. However, an additional effect—the degradation of surface gloss—also comes into play, particularly particles with a nominal diameter greater than 0.6 μm or so.

Fundamentally, the most important aspect of particle size distribution is in its control to give both the proper distribution and the same distribution shipment to shipment. Variability in particle size distribution can lead to enormous difficulties in color matching and to unpredictable changes in pigment-loading requirements to obtain equivalent opacity.

10.5. INCORPORATION OF TITANIA PIGMENTS INTO POLYMERS

The methods whereby titania pigments are incorporated into plastic matrices are diverse. Multiple options exist for the plastics processor, including solid concentrates, direct end-use pigmentation, simple dry blends, or the use of "liquid" colorants. A simplified breakdown of how major resin categories are pigmented is shown in Table 10.1.

While the use of concentrate is prevalent in most resin categories, nearly all the TiO_2 used in finished products composed of polyethylene and polypropylene enters by way of solid concentrates. These concentrates typically vary in pigment loading level between 30 and 80% pigment by weight. The manufacture of these solid concentrates is usually done in internal mixers, continuous mixers, or multiple-screw

Table 10.1. Common Routes to Pigmentation with Titanium Dioxide

	Most Common Route	Manufacturing Processes
Polyolefins: LDPE, LLDPE, HDPE, PP	Solid concentrates	Internal mixers, continuous mixers, multiple-screw extruders

Polyvinyl chloride
  Rigid PVC
  Plasticized PVC
	"Dry" blends
	Liquid or paste dispersions
		Intensive high-speed mixers
		Roll mills or disk Impeller mixers

Polystyrene family: GPS, HIPS, SAN, ABS	Solid concentrates, some intensive blends	Internal mixers, continuous mixers, multiple-screw extruders, high-speed dry blenders
Other: acetals, PC, PA, PET	Melt compounding	Predominantly multiple-screw extruders

LDPE, Low density polyethylene; LLDPE, Linear low density polyethylene; HDPE, High density polyethylene; PP, Polypropylene; PVC, Polyvinyl chloride; GPS, General purpose polystyrene; HIPS, High impact polystyrene; SAN, Styrene acrylonitrile; ABS, Acrylonitrile butadiene styrene; PC, Polycarbonate; PA, Polyamide; PET, Polyethylene terephthalate.

extruders. The pigment grade and carrier resin, plus any additives, must be suitable for the end-use application.

Titanium dioxide is usually intensively mixed in high-speed blenders at the end-use level with rigid polyvinyl resins along with the customary additives needed for stability, impact resistance, and processing. Pigment loadings rarely exceed 15% by weight even for products requiring durability on exterior exposure. While many color pigments are introduced in concentrate form, there are essentially no TiO_2 concentrates used in unplasticized polyvinyl chloride (PVC) formulations. Incorporation of TiO_2 into plasticized polyvinyl usually begins with the production of a liquid or paste dispersion of the pigment in monomeric plasticizers or low-molecular-weight polymers. These plasticizer dispersions are normally prepared by premixing pigment, plasticizer, and surfactants followed by high-shear-stress dispersion on three to five roll mills or high-speed disk impeller mills. Selection of flocculation-resistant grades and proper surfactant level are very important to ensure the most economical use of the pigment. Surfactant demand of the TiO_2 pigment should be determined for each formulation. The surfactant demand will vary according to the available surface area of the pigment.

Surfactant demand curves for three different grades with Brunauer, Emmett and Teller Test Method (BET) nitrogen surface areas of 9 and 15 m^2/g are shown in Figure 10.6.

Incorporation of TiO_2 into polystyrene(s), styrene–acrylonitrile, acrylonitrile–butadiene–styrene, and other associated copolymers and alloys is normally by way of concentrates prepared on equipment similar to that used for polyethylene. This concentration step is usually necessary to achieve high-quality dis-persion so color properties are fully developed and physical properties are not compromised.

Titanium dioxide pigments are compounded into the remaining miscellaneous category of resins by many different means, including solid concentrates, dry blends, and liquid concentrates.

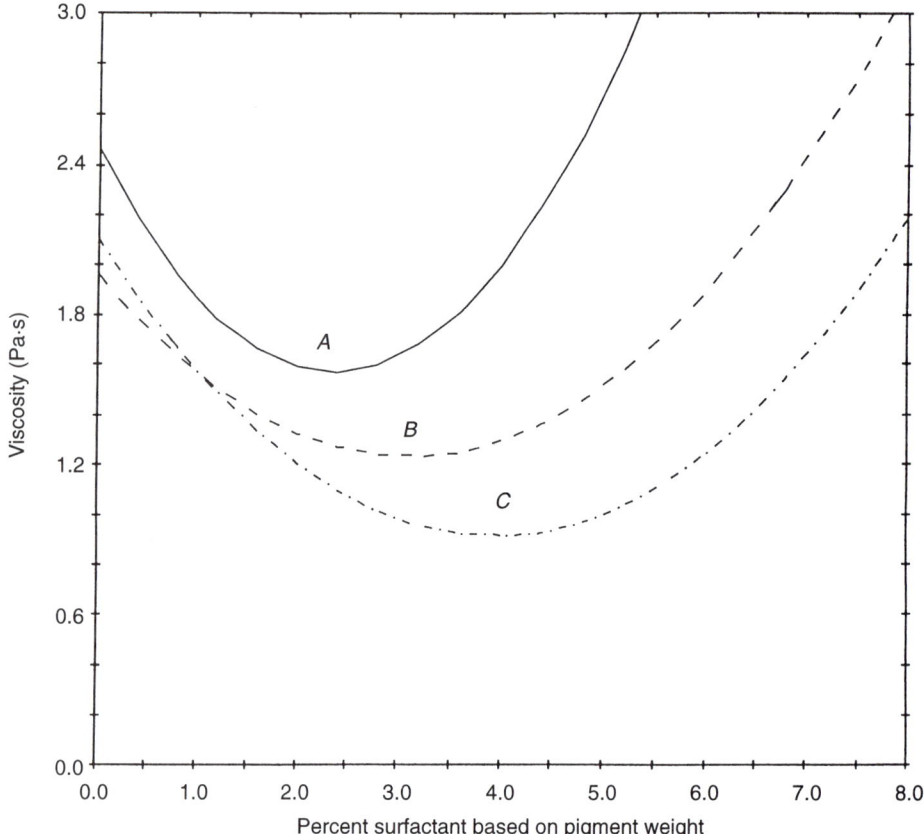

Figure 10.6. Surfactant demand curves of 70% (weight) dispersions of TiO_2 pigments with various levels of alumina treatment. Dispersions were prepared in DIDP plasticizer on a high-speed disk mill. Survactant used was Disperbyk I. Pigment A: titania surface, no alumia surface treatment. Pigment B: 1.5% alumina surface treatment, minimum viscosity achieved at 2.36% surfactant based on pigment weight. Pigment C: 3.0% alumina surface treatment, minimum viscosity at 3.9% surfactant.

10.6. PHOTOCHEMISTRY OF TITANIA PIGMENTS: DURABILITY

As conventional materials are replaced by plastic products, formulating to have the necessary stability to light exposure becomes increasingly important. Selection of the appropriate grade of TiO_2 can have a major impact on balancing the need of improved durability with formulation costs. The grade of pigment, chemistry of the polymer system, activation spectrum of the polymer, and spectral distribution of the actinic radiation must all be considered. What may be an acceptable grade in one situation may have undesirable results under different circumstances. The following gives an overview of the effect that different grades of TiO_2 pigments have on system durability of rigid PVC, polyolefins, and selected aromatic polymer chemistries.

Unpigmented rigid polyvinyl formulations typically have very poor long-term durability. Titanium dioxide pigments are normally compounded into these compo-

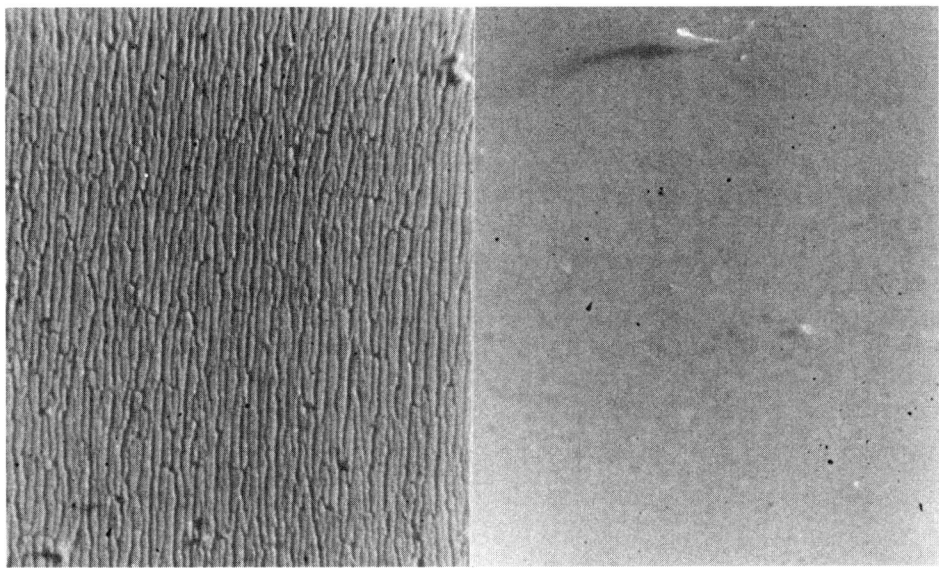

Figure 10.7. Pigment-induced surface cracking, the prelude to free chalking, on exposure to UVA light for 650 h. Exhibits are polypropylene moldings containing 5 wt % TiO_2. Exhibit on the left is a general-purpose grade and the exhibit on the right contains a pigment with a silica/alumina treatment to improve durability.

sitions (such as exterior building products) not only as a colorant but also as a nonextractable, nondecomposing stabilizer. Preferred grades encompass the entire available range of pigment surface treatments depending on the intended use. For example, it may be desirable to encourage controlled surface oxidation to dispose of yellow chromophores that develop during the early stages of PVC dehydrochlorination. This will eventually cause the surface to release degraded polymer and pigment in the form of a chalklike powder. If the "chalking" is not desired, more durable grades such as those encapsulated in SiO_2 should be used.

Polyolefins pigmented with TiO_2, unlike rigid PVC, will typically not exhibit chalking failure on exterior exposure. The normal failure is loss of tensile strength and elongation and the appearance of cracking. In the absence of auxiliary light stabilizers, the degree of photostabilization of the TiO_2 pigment and loading will determine to a large extent the durability. Olefin polymers are essentially nonabsorbing in the ultraviolet portion of the actinic spectrum until significant carbonyl functionality occurs. Studies of the effect of grade on exterior applications such as agricultural mulch films have shown that silica- and some alumina-encapsulated grades can extend the useful life beyond that of unpigmented resin. Surface cracking resulting from ultraviolet light exposure of polypropylene moldings pigmented with organically treated and silica/alumina-treated grades is shown in Figure 10.7.

Polystyrene, its copolymers, and its alloys are used in a wide variety of applications where stability to both direct sunlight and filtered or artificial lighting is important. At wavelengths below 300 nm photolysis [12] of polystyrene can result in unacceptable color development (yellowing) along with reduction of physical properties. Titanium dioxide pigment grades can have a very profound effect on the rate of discoloration depending on the pigment surface treatment and the spectral distribution of the illuminating radiation. In direct sunlight (by which the phenyl ring in the polymer is directly photoexcited), almost all grades of TiO_2 will exhibit beneficial effects, since the pigment is itself a strong light absorber in the ultraviolet. Only with longer term exposure will the grades be differentiated by retention of physical properties. However, in some environments (e.g., office interiors) wavelengths are present that can excite the phenyl group but are subject to attenuation by filtering sunlight through window glass or illumination with fluorescent lighting. On the other hand, the relative intensity of wavelengths that can photoreduce the titanium atom in the pigment are increased. Thus, selection of the appropriate grade of TiO_2 to minimize yellowing on exposure to artificial illumination or filtered sunlight requires a complete understanding of all aspects of the system [13]. The rate of discoloration as a function of grade to interior exposure is illustrated in Figure 10.8 in a polystyrene homopolymer.

The photochemistry of TiO_2 has been studied in tremendous detail over a period of many years [14]. Interestingly, the studies are generally focused on both "ends" of the range of photochemical reactivity. Since titanium readily participates in catalytic reactions upon exposure to ultraviolet light, much has been made of late of the possible uses of this behavior in applications ranging from No_x antipollution systems [15] to the destruction of chlorinated organic wastes [16]. As described earlier, it is just this reactivity that the plastics processor does not generally desire to have in a pigmented system. (So-called photodegradable films, deliberately pigmented with highly photoactive titanias and related photochemically active materials, are obviously excepted from this.) As such, a large body of research exists regarding the study of the reduction of photoactivity of titania pigments [17]. Excellent review articles on the photochemical behavior of titania pigments have been published recently by Diebold [18] and others [19]. A condensed version of the information from these and related articles is now provided.

In oxidation state IV titanium interacts with ultraviolet light and organic materials in the presence of water and oxygen to generate a catalytic cycle that generates organic free radicals and inorganic high-energy species such as the hydroperoxide radical. These free radicals possess sufficient energy to destroy the chemical bonds of a polymeric matrix. The organic moieties are mineralized to carbon dioxide and water during this process. This is the mechanistic basis of the phenomenon known as "chalking" discussed earlier. The "chalk" is the titania remaining after the organic matrix has been converted to carbon dioxide and water. Other effects resulting from the photoprocesses operative in the catalytic cycle(s) are color shift, loss of tensile and impact strength, and dimensional instability.

The mechanism described above and in Figure 10.9 is applicable to the surfaces of pigmented plastics, thin films, and any other system in which water and oxygen are available for reaction at the titania surface. For systems in which water and oxygen are unavailable (e.g., laminate papers), the predominant mechanism of photoactivity is the photoassisted reduction of Ti(IV) to Ti(III). Since Ti(III) is a colored species, a system pigmented with an unstabilized titania will discolor as a function of incident light flux and duration. This discoloration generally appears to be "gray" to the eye. Although these systems can revert upon cessation of light exposure, the initial color performance is not generally regained. As such, photostabilized titanias

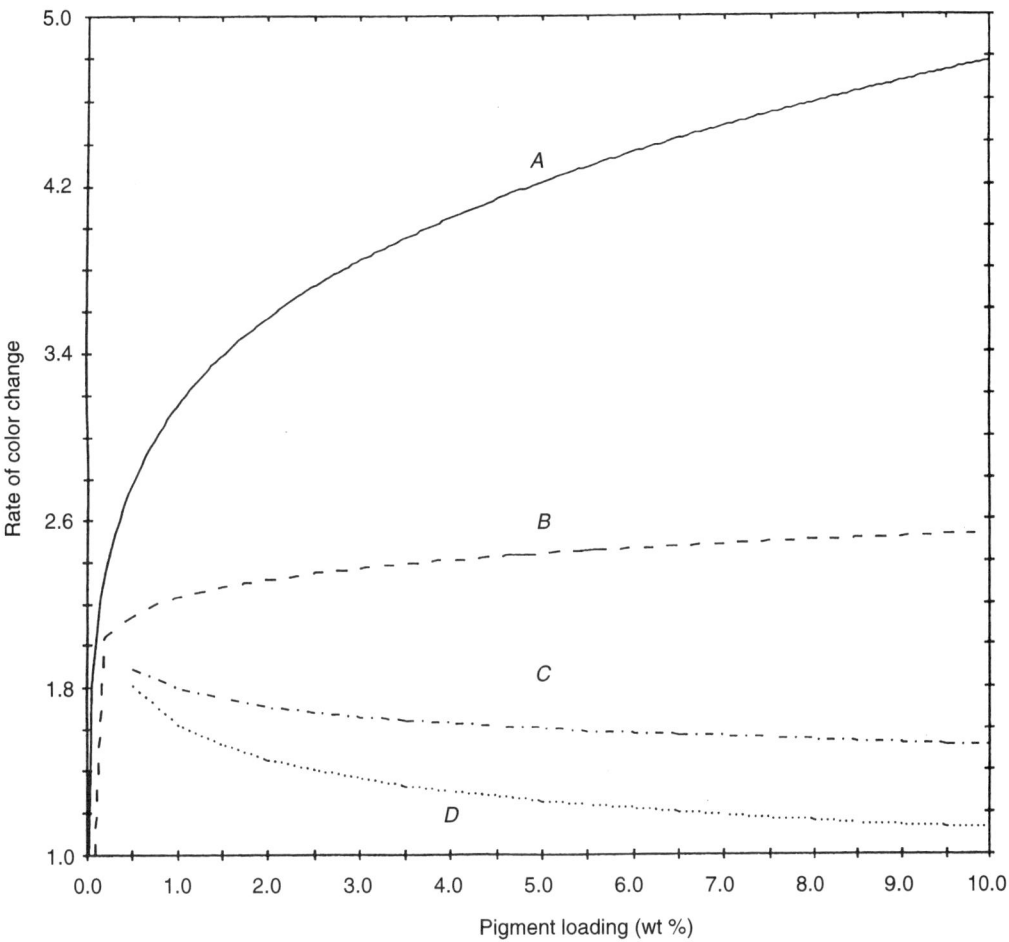

Figure 10.8. South window exposure (Wilmington, Delaware) of crystal polystryrene pigmented with various grades of TiO$_2$. Rate of color change is reported as the change in $b*$ per hour of exposure (multiplied by 1000) from the 1976 $L*a*b*$ color space. Maximum exposure time of 7000h. Pigment A: organic surface treatment. Pigment B: alumina treatment. Pigment C: silica and alumina treatment. Pigment D: silica and alumina treatment.

are required to prevent the decay of color performance even when water and oxygen are excluded from a system.

Of the two major types of titania described earlier, rutile is considerably less photoactive than anatase. It is for this reason that rutile pigments are generally formulated into plastic applications where stability is required. Anatase pigments have been used for "photodegradable" pigmented systems in which sunlight is intended to couple with the very photoactive anatase to rapidly generate radicals and thus convert the polymer matrix into CO_2, water, and other breakdown products (depending on the nature of the film) at an accelerated pace. Photostabilized systems are far and away the larger applications in commerce, requiring stabilized rutile pigments.

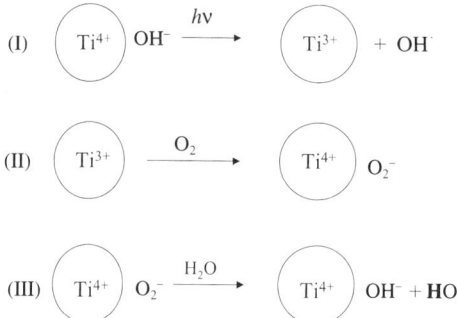

Figure 10.9. Photochemical reactions taking place at the surface of a TiO_2 pigment: (I) photo reduction of the titanium atom resulting in the production of hydroxyl radical; (II) reduced titanium atom reacting with oxygen to form an unstable complex; (III) reaction of the complex with water to produce peroxyl radicals. Both the hydroxyl and peroxyl radicals can react with the polymer matrix and initiate degradation.

Many routes exist for the photostabilization of rutile pigments [20]. Calcination of finished products can be carried out, although there is generally some loss in dispersability and in color performance as a result. The use of redox couples has been touted as a route to photostable systems [21]. The most widely used route to stability is the surface treatment of rutile pigments with hydrous metal oxides, either alone or in combination with organic treatments [22]. In general, coatings of the hydrous oxides of silica, alumina, zirconia, and other elements, often in combination with other inorganic species, are used to reduce photoactivity by what has generally been considered to be an "encapsulation" strategy [23]. A downside to this strategy has been that the most stable materials (i.e., those with the highest levels of oxide coatings) possess sufficient water in the coating matrix to cause "lacing" or "curtaining" when thin films pigmented with these materials are formed. Lacing characteristics of an organically treated grade and a silica/alumina-treated grade are illustrated in Figure 10.10. Recent technological developments minimize this effect [24]. Many options are commercially available for higher gauge and/or lower temperature applications.

Discoloration effects in polymers have been broadly investigated [25]. Titania pigments of certain types can interact very strongly with certain organic compounds, including Butylated hydroxytoluene (BHT) [26], catechols [27], lead-based stabilizers [28], and light stabilizers [29]. These interactions are largely the cause of such phenomena as "yellowing," "pinking," and "graying". Some of these are actually "dark" phenomena, in that they occur in the absence of light and, therefore, arise from a different mechanism than that described earlier. In practice, all of these interactions can be minimized or precluded by the appropriate choice of photostable rutile titania.

The measurement of durability in a given system under conditions of illumination (spectral distribution and intensities), mimicking closely the intended end-use environment, and as determined by changes in appearance and/or physical properties is the ultimate measure of successful formulation in systems requiring photostable behavior. Although many laboratory and field methods exist for the direct measure of photoactivity, these often do not sufficiently reflect the performance that will be experienced in the actual end use.

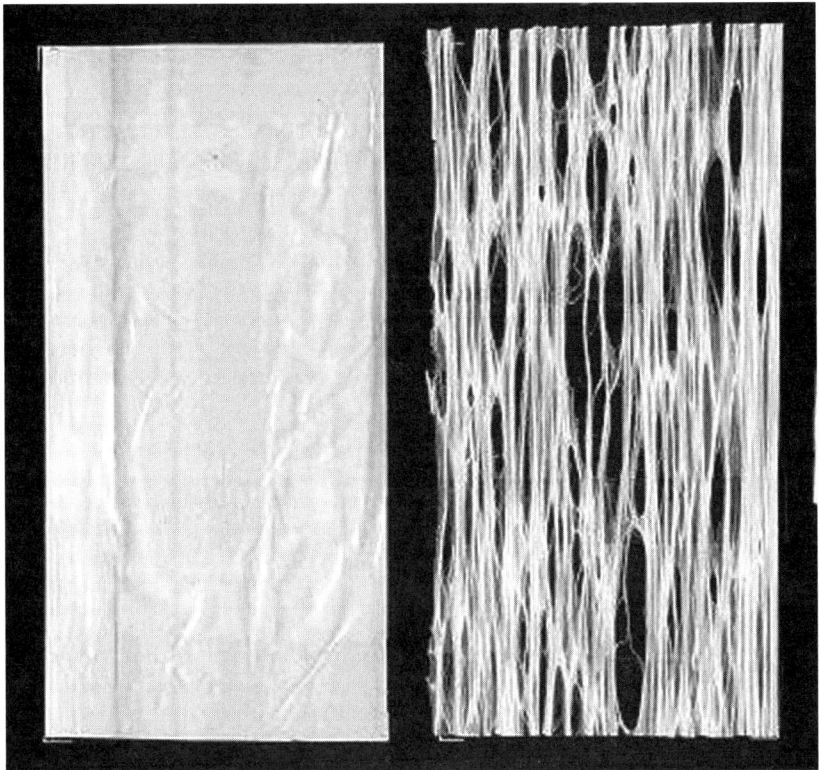

Figure 10.10. Lacing resistance of an organically treated grade of TiO_2 (left) and a silica/alumina-treated grade (right). Films were extruded low-density polyethylene at 1.5 mil thickness containing 15% (wt.) titanium dioxide and cast at 316°C (600°F).

10.7. FUTURE NEEDS

As for all pigments, the need to reduce the amount of energy required for full dispersion is an ongoing concern of titania pigment manufacturers. Further improvements in the uniformity of color performance are also desirable. The flow characteristics of titania pigments, historically considered (and rightly so) to be unacceptable, are rapidly being improved. Novel surface treatment technologies will expand the performance manifold for a broad variety of polymer systems and end-use applications.

REFERENCES

1. L. Leonant, Ed., *Plastics Compounding 1995/96 Redbook*, Advanstar Communications, Cleveland, OH, 1995.
2. J. Barksdale, *Titanium: Its Occurence, Chemistry and Technology*, Roaald, New York, 1966.
3. D. T. Cromer and K. Herrington, *J. Am. Chem. Soc.*, **77**, 4708 (1955); H. Sato, S. Endo, M. Sugiyama, T. Kikegawa, O. Shimomura, and K. Kusaba, *Science*, **251**, 786 (1991).
4. H. B. Clark, in R. R. Myers and J. S. Lang, Eds., *Treatise on Coatings, Vol. 3: Pigments, Part 1*, Marcel Dekker, New York, 1975, Chapter 10; A. S. Ritter, in T. C. Puttnam, Ed., *Pigment Handbook*, Vol. II: *Properties and Economics*, 2nd ed., Wiley-Interscience, 1973.

5. H. W. Jacobsen, U.S. Patent 4,460,655; W. A. West, U.S. Patent 4, Nov. 4 1978, 125,412; W. J. McGuiness, U.S. Patent 3,410,708 Nov. 12, 1968; K. A. Green and T. I. Brownbridge, U.S. Patent 5,203,916 April 20, 1993.
6. J. R. Brand and P. M. Story, U.S. Patent 4,752,340 June 21, 1988.
7. W. Carr, in T. C. Patton, Ed., *Pigment Handbook*, Vol. III, Wiley-Interscience, New York, 1973, p. 29.
8. P. M. Niedenzu, A. H. Reid, Jr., and P. A. Tooley, U.S. Patent 5,562,990 Oct. 8, 1996.
9. J. G. Balfour, Journal of the Oil and Colour Chemists Association (*JOCCA*), June 1990.
10. H. C. van de Hulst, *Light Scattering by Small Particles*, Dover, New York, 1981.
11. P. A. Tooley and Q. K. Le, Conference Proceedings, Society of Plastics Engineers, ANTEC 1995. Boston, MA May 7–11, 1995 Volume III.
12. J. F. McKellar and N. S. Allen, *Photochemistry of Man-Made Polymers*, Applied Science Publishers, London, 1979.
13. D. A. Holtzen, Conference Proceedings, Society of Plastics Engineers, ANTEC 1988, pp. 28–32. Atlante, Georgia April 18–21.
14. S. P. Pappas and R. M. Fischer, *J. Paint Technol.*, **46** (599), 65 (1974).
15. T. Ibusuki and K. Takeuchi, *J. Mol. Catal.* **88**, 93 (1994).
16. M. A. Fox, *Chemtech* Vol 22, 680 (1992); A. Heller, *Acc. Chem. Res.*, **28**, 503 (1995).
17. D. A. Holtzen and A. H. Reid, Jr., Conference Proceedings, Society of Plastics Engineers, ANTEC 1991, pp. 24–29. Montreal, Canada May 5–9, 1991.
18. M. P. Diebold, *Surface Coatings Int.*, **7**, 294 (1995).
19. R. Siddle, *Surface Coatings Australia*, April 1991, p. 6.
20. U. Kaluza and H. P. Boehm, *J. Catal.*, **22**, 347 (1971); F. K. McTaggart and J. Bear, *J. Appl. Chem.*, **5**, 643 (1955).
21. H. W. Jacobson, U.S. Patent 4,461,810 July 24, 1984; B. Barnard and W. T. Laverick, U.S. Patent 4,239,548.
22. L. Chromy and J. Kudela, Journal of the Oil and Colour Chemists Association (*JOCCA*), **52**, 687 (1969).
23. A. J. Werner, U.S. Patent 3,437,502.
24. P. A. Tooley, D. A. Holtzen, and J. A. Musiano, U.S. Patent 5,607,994.
25. D. A. Holtzen, Conference Proceedings, Society of Plastics Engineers, ANTEC 1991, pp. 19–23. Montreal, Canada May 5–9, 1991.
26. D. A. Holtzen, *Plastics Eng.* **43**, p. 42 (1977).
27. K. C. Smeltz, *Textile Chemist and Colorist*, **15**, 1983 (1983).
28. J. H. Braun, *Prog. Org. Coatings*, **15**, 249 (1987).
29. N. S. Allen, J. F. McKellard, and D. G. M. Wood, *J. Polym. Sci.* **13**, 2319 (1975).
30. D. Corless, *Paintindia*, January 1991, p. 83; J. R. Brand, Society of Plastics Engineers Color and Appearance Division RETEC Preprint, 1996; R. Poission, J. Petit, and J. Fischer, *Peintures, Pigments and Vernis*, **40**, 277 (1964); A. Mackor and T. P. M. Koster, *Farbe Lack*, **95**(8), 566 (1989).
31. Cleveland Society for Coatings Technology Technical Committee, Journal of Coatings Technology, **66**(837), 49 (1994); J. H. Colling and T. W. Wilkinson, *JOCCA*, **58**, 377 (1975).

CHAPTER 11

Carbon Black Pigments for Plastics

Scott A. Brewer

Cabot Corporation
157 Concord Road
Billerica, Massachusetts 01821

11.1. INTRODUCTION

The term *carbon black* describes a group of industrial carbons created through the partial combustion or the thermal decomposition of hydrocarbons. Carbon black is unique in that it possesses the smallest particle size and highest oil absorption among the commercially available pigments for plastics. These characteristics help explain carbon black's excellent color strength, cost-effectiveness, and ultraviolet (UV) performance and place it as the most widely used black pigment for thermoplastic applications.

The aim of this chapter is to provide an understanding of how carbon black particle size and shape translate into dispersion quality and other performance attributes in plastics applications. While this chapter focuses on carbon black's effectiveness as a pigment, some attention will also be paid to some of the other plastics performance properties it influences, such as stability against UV radiation, effects on mechanical properties, and electrical conductivity. At its conclusion, this chapter will help the reader select an appropriate carbon black grade for specific plastics applications.

Coloring of Plastics, Fundamentals, 2nd edition. Edited by Robert A. Charvat
ISBN 0-471-13906-8   Copyright © 2004 by John Wiley and Sons, Inc.

11.1.1. Background

The manufacture of carbon black can be traced back for 3000 years, when in ancient China, small lamps covered with ceramic lids were used to burn oil. Smoke from the oil impinging on the lamp lids formed a carbon black used in black inks and lacquers.

The modern version of this ancient method is known as the lampblack process. In this production method, oils are ignited in open shallow pans with a substoichiometric amount of oxygen. The carbon produced passes through a series of settling chambers and filters from which the flocculated black is collected. The procedure of limiting oxygen from burning oil to produce carbonaceous materials, begun with the lampblack process, is the origin of all modern methods of carbon black manufacture.

The oil furnace process is the most common method of production today and is the source of over 95% of the total output of carbon black globally. In this process, a heavy aromatic fraction of petroleum distillate is atomized and sprayed into a furnace preheated to 1200–1900°C. The feedstock vaporizes and decomposes to form carbon black and combustion gases that are immediately cooled with a series of water sprays and heat exchangers to terminate the carbon black reaction and cool the carbon black product stream. The carbon black is separated from the combustion gases in bag filters and is conveyed for further densification either in pelletization processes or in agitator tanks (from which powdered, fluffy black is collected).

The furnace black process is capable of producing a chemically pure, fine-particle carbon black with low volatile content, 1–2%, and pH ranging from 6 to 10, which is suitable for most plastics end uses. This process allows precise control of a carbon black's particle size and shape (or morphology), which ensures uniform color and physical properties in plastics applications (Fig. 11.1).

As formed, carbon black is a fluffy powder possessing low density. The densification process involves the removal of occluded air by agitation and followed by dry or wet process pelletization. In both the dry and wet pelletization process, nearly spherical pellets or beads will form that are typically composed of >99% carbon black and trace impurities such as sulfur. Thus, carbon black is sold as a low density powder or as a pelleted form in pigmenting and other end uses. The choice of a fluffy or pelleted carbon black for dispersion in a given system depends upon the dispersion and handling equipment and end use. For example, pelleted carbon blacks are used most frequently in production of black masterbatch; carbon black powders are typically used to tint chromatic compounds.

11.2. CARBON BLACK MORPHOLOGY

Carbon black is composed of hexagonal planar structures of elemental carbon fused into spherical shapes. These spheres in turn join together to form complex aggregates (Fig. 11.2). Its chemical purity, as compared to other bulk carbons such as cokes and charcoal, makes it a useful pigment for a variety of materials besides plastics, including elastomers, inks, and surface coatings [1]. Carbon black may be applied to create full-shade black plastics or as a tinting pigment to modify the color of chro-

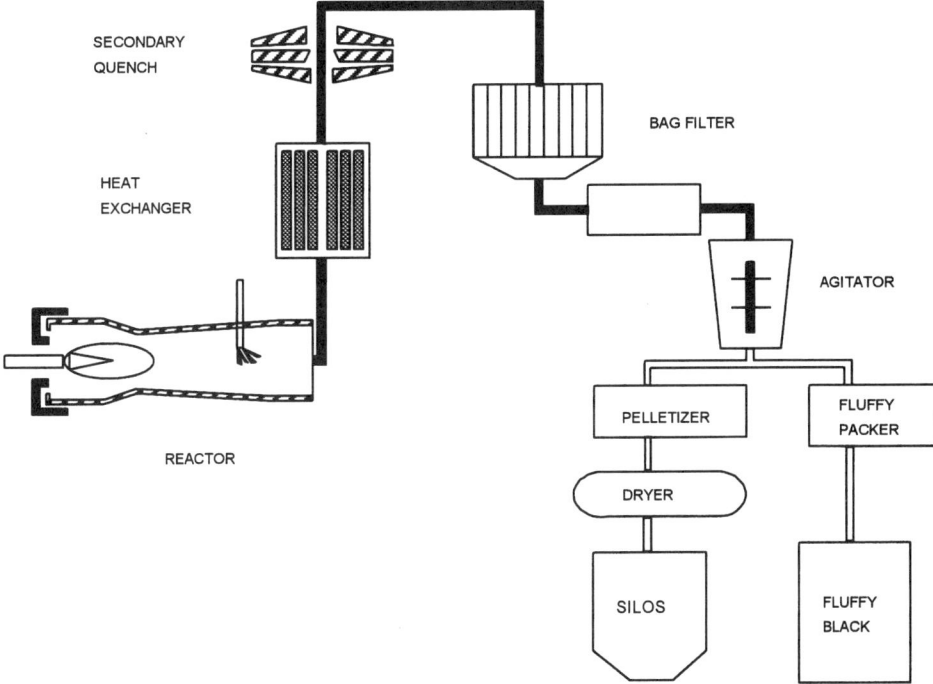

Figure 11.1. Furnace process graphic.

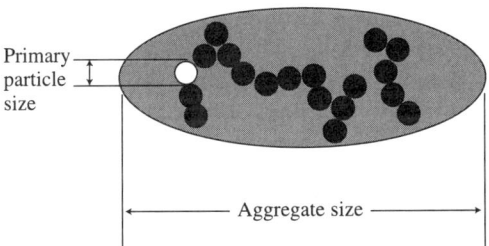

Figure 11.2. Carbon black aggregate.

matic pigments in plastics. As a single pigment, the black color that carbon black imparts to the plastic medium is referred to as "jetness."

In addition to jetness, plastics pigmented with carbon blacks typically exhibit color undertones. Undertone in black plastics appears as a distinct blue or brown-to-orange undertone, depending on the particle size of the carbon black used. In general, in full-shade, black molded applications, fine-particle-size carbon blacks impart a bluer tone. This behavior reverses itself in tints. Large-particle-size carbon blacks impart bluer undertone. Note that the effects of fillers, polymers, and dispersion can alter the typical behavior described above. Tint strength is the relative ability of the carbon black to darken a resin colored with chromatic pigments.

Three variables affect color performance of carbon black in plastics: the particle size and shape or morphology of a carbon black grade, the polymer in which it is

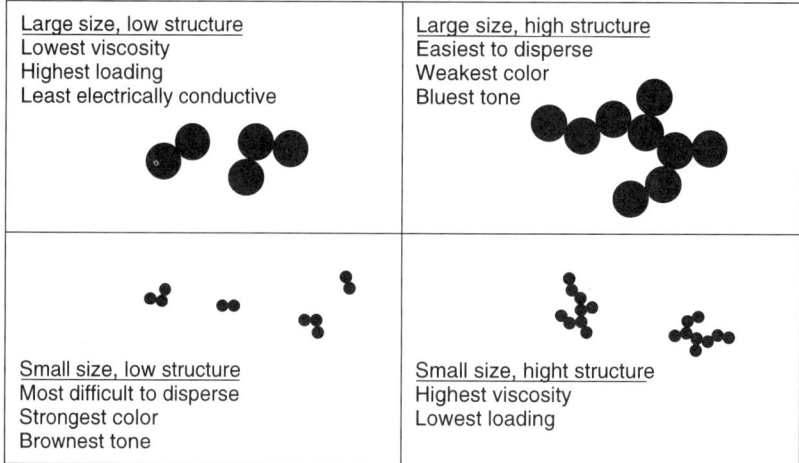

Figure 11.3.

dispersed, and the quality of dispersion in the resin system. Carbon black grades are typically classified on the basis of particle size and structure. Typical commercial carbon black primary particle size measures from 10 to 100 nm in diameter. Above all other properties, particle size can predict the level of jetness that a carbon black will provide after full dispersion in resin. Small prime particles better absorb light because they possess greater surface area per unit weight that is available for the absorption and more efficient forward scattering of visible light. Finer particles, therefore, impart greater jetness and tinting strength [2].

Aggregates are the smallest indivisible units of carbon black. Under normal dispersion conditions, these particles will be broken down further (cracking the primary aggregates can adversely affect application results such as conductivity). The aggregates are constructed from prime particles and are defined in terms of "structure." Structure is determined by the number of particles per aggregate and also by the quantity, branching, and shape that the fused particles form (ultimately described as void volume). High structure refers to an aggregate with numerous fused prime particles and abundant branching and chaining, or higher void volume. Low structure refers to an aggregate possessing comparatively few particles. This attribute is independent of particle size. Aggregates may be composed of fine or coarse particles.

As a rule, low-structure carbon blacks impart greater jetness, but tint structure affects color to a lesser degree than primary particle size. Finally, agglomerates are undispersed clusters of aggregates that are physically bound together. The object of dispersion is to reduce agglomerate size and maximize pigment color development (Fig. 11.3)

11.3. CARBON BLACK SELECTION FOR PIGMENTING

Appropriate selection of carbon black requires careful evaluation of the method of incorporation and dispersion and the application's end-use requirements. Like many

Table 11.1. Particle Size Performance Predictor

Property	Fine Particle, Performance Blacks	Large Particle, Utility Blacks
Mass-tone jetness	Darker	Lighter
Viscosity	Higher	Lower
Dispersion	More difficult	Easier
Wetting	Slower	Faster
Tint strength	Higher	Lower
Conductivity	Higher	Lower
UV absorption	Better	Poorer
Cost	Higher	Lower

pigments, carbon black may often be loaded into a carrier resin at high loadings for delivery as a concentrate or masterbatch. The concentrate is then "let down" into a carrier resin to reach the desired carbon black loading. Depending on end-use requirements, such as UV stability or pigmenting, a compounder will use a final carbon black loading of 0.25–3.0% In some precolor applications, such as styrene polymers, low-dosage, direct addition is used. In these cases, careful carbon black selection and a well-designed dispersion process are prerequisites to good color development.

To achieve best jetness, a formulator would select a grade having very fine particles (high surface area) and a moderate structure (Table 11.1). As mentioned before, this is because small prime particles provide greater overall surface area to both absorb and scatter visible light, and moderate to high structure positively influences dispersion.

Morphology also affects undertone. In full-shade black, fine-particle grades provide a bluer tone, while brown tones are produced by coarse-particle grades. The undertone of a black plastic is a complex function of the scattering and absorption efficiencies as a function of wavelength. Used in combination with other pigments, fine-particle carbon blacks are efficient tinting pigments. However, unlike masstone color where fine particles produce a bluer appearance, it is in tinting functions that larger particles and higher structure increase blue undertone. This tinting behavior often prompts masterbatch and compounders to use two to three types of carbon black: fine particle (<17 nm) for high jetness, 19 nm for UV applications, and coarse tinting blacks (>45 nm).

Achieving a high quality of gloss and low haze in the plastic end product presents more of a challenge. While lower structure grades may give higher gloss, other properties being equal, gloss and haze are very much affected by the quality of dispersion. Also, lower levels of residual carbon black contaminants favor better molded part surface appearance.

11.3.1. Effects of Dispersion on Color

The degree of dispersion in a plastic system is of equal importance to a carbon black's physical properties in terms of color quality and other end-use properties. Dispersion, like color, is strongly influenced by carbon black morphology. Selecting

Table 11.2. Structure Performance Predictor

Property	High Structure	Low Structure
Dispersibility	Easier	More difficult
Wetting	Slower	Faster
Mass-tone jetness	Lower	Higher
Gloss	Lower	Higher
Conductivity	Higher	Lower
Viscosity	Higher	Lower
Loading capacity	Lower	Higher
Tint strength	Lower	Higher

both the most suitable carbon black for the polymer and the appropriate type of dispersion equipment ensure the best processing during dispersion and the best end-use properties. If a carbon black is not completely dispersed in the polymer, the results are reduced jetness and poor physical properties in the end use.

Carbon black prime particle size presents opposing roles in determining color and dispersion quality. While fine-particle blacks can increase jetness through greater absorption of incident light, these grades are more difficult to disperse than large-particle-size blacks. This is due, in part, from the higher dispersion energy needed to displace occluded air with polymer and break interparticle attractions.

Structure also influences dispersion (Table 11.2). The greater volume of high-structure aggregates entangles more molten polymer than low-structure aggregates. Although high structure results in higher viscosities and stiffer compounds, the process of displacing occluded air and breaking interparticle attraction moves forward more rapidly.

11.4. DISPERSION: DEFINITION AND METHODS

As mentioned in the description of the carbon black production process, carbon black agglomerates are formed during pelletization. This increases black density and reduces dust, but for the masterbatch maker or compounder, successful deagglomeration of the carbon black is central to developing excellent end-use properties in plastics. Sufficient energy must be applied to overcome interaggregate attractive forces and approach, as nearly as possible, complete separation of all primary aggregates. Ideal dispersion of carbon black is the condition in which all agglomerates are broken down into primary aggregates, each primary aggregate is separated from every other aggregate, and the surface of each aggregate is completely covered by resin. In practice, this is only possible at low carbon black dosages. A more realistic goal is to estimate a relevant dispersed agglomerate size. For example, $<2-5\,\mu m$ for a fine denier fiber applications.

Carbon black loading can play an important part in obtaining quality dispersion with higher loadings producing superior dispersion, other factors being equal. However, caution should be used to ensure that the concentrate viscosity does not greatly exceed that of the compound polymer—a 3:1 concentrate-to-compound

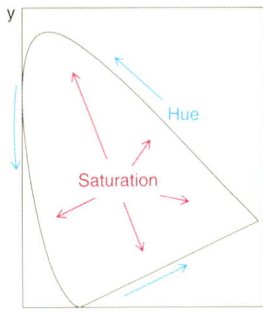

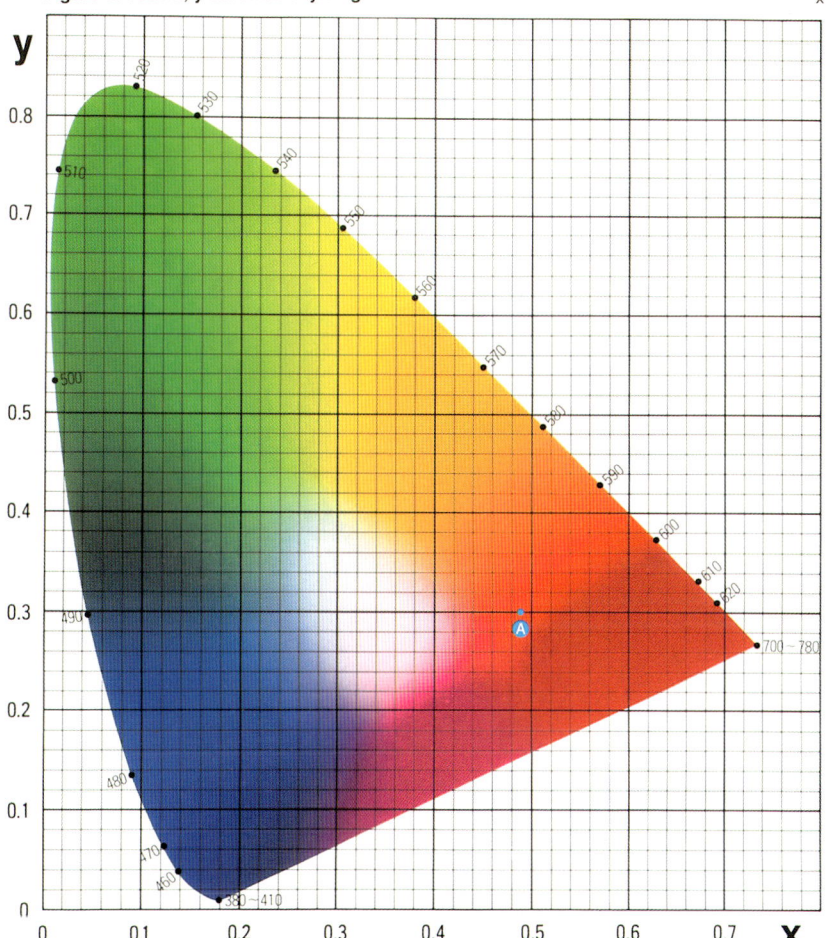

Figure 2.10. The 1931 x, y chromaticity diagram (Minolta, 1993).

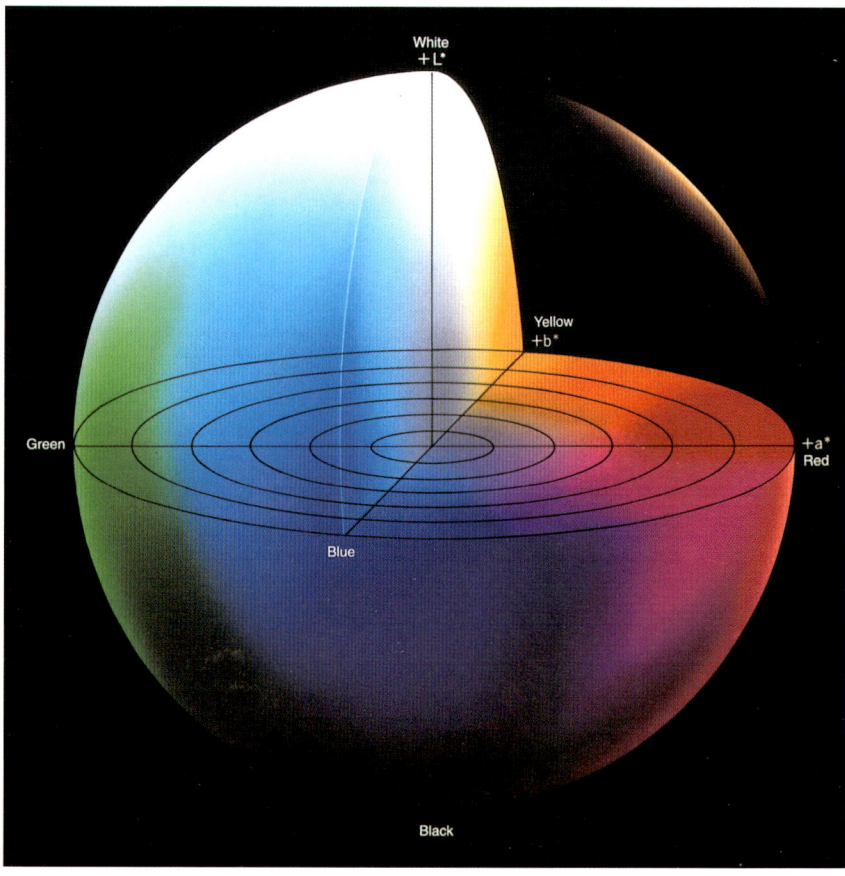

Figure 2.11. Illustration of the color solid for the CIE 1976 ($L^*a^*b^*$) color space (Minolta, 1993).

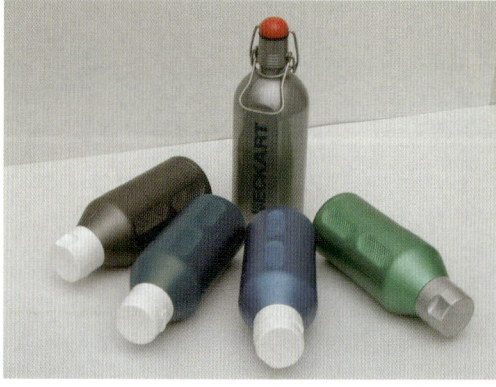

Figure 14.1. Metallic colored moldings.

Atomizer in Operation

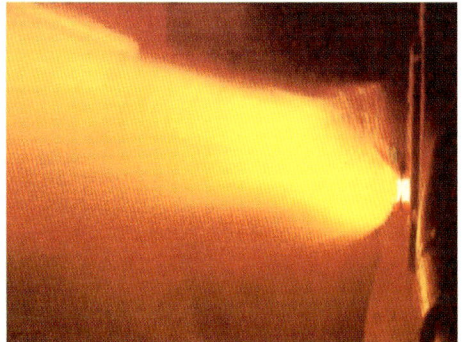

Figure 14.2. Atomizer in operation.

Metallic Pigment Powder

Figure 14.14. Metallic pigment powder.

Figure 15.1. Examples of interference color effects in nature.

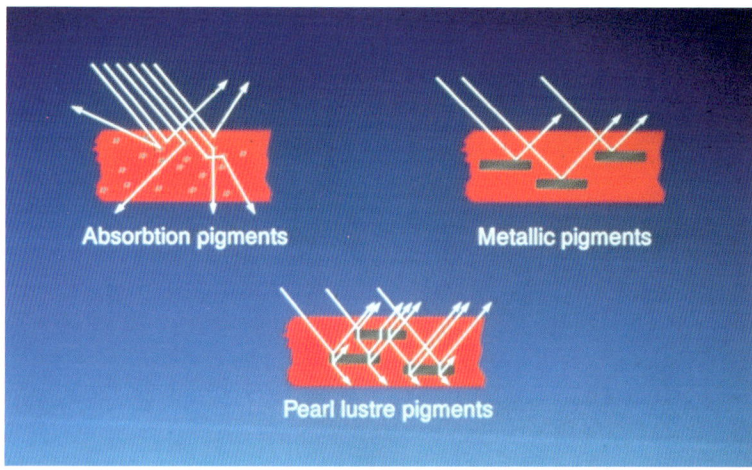

Figure 15.2. Optical properties of conventional pigments, metallic pigments, pearl pigments and pearls.

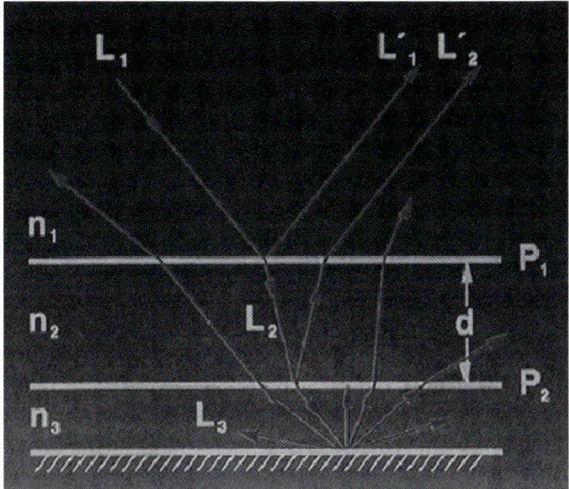

Figure 15.3. Simplified diagram showing nearly normal incidence of a beam of light (L_1) from an optical medium with refractive index n_1 through a thin solid film of thickness d with refractive index n_2 (L_1' and L_2': regular reflections from interfaces P_1 and P_2. L_3: diffuse scattered reflections from the transmitted light).

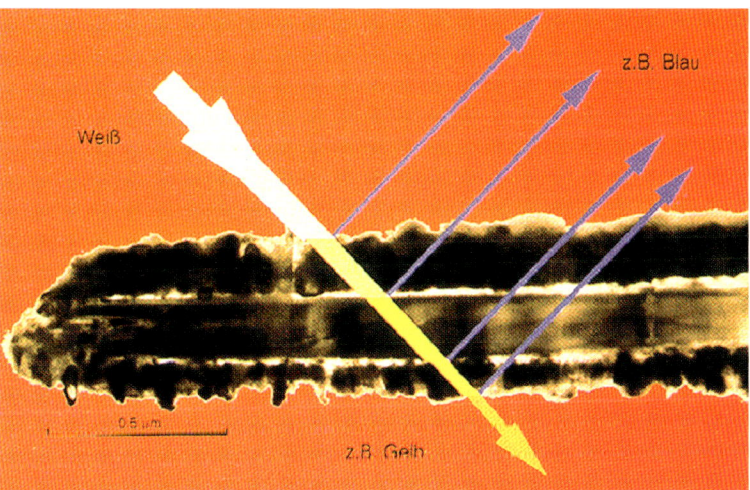

Figure 15.4. Transmission electron micrograph of a TiO_2-mica pigment with the schematic way of light through layers (blue interference effect, yellow transmission).

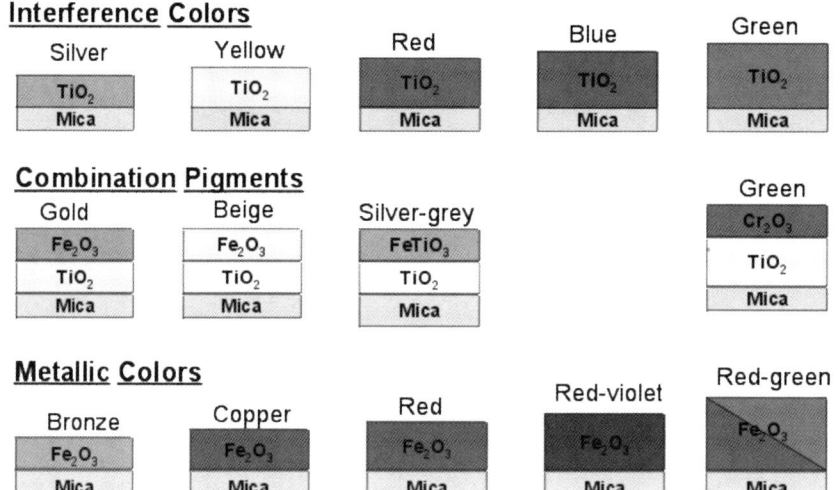

Figure 15.6. Schematic illustration of different metal oxide-mica pigments.

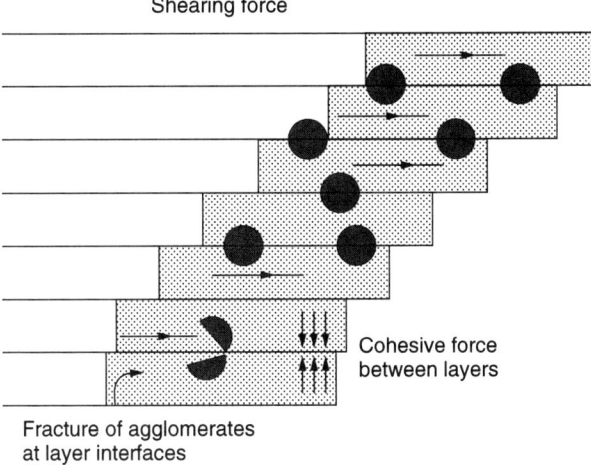

Figure 11.4. Carbon black shearing in plastics.

ratio is appropriate. This helps avoid poor concentrate distribution in the carrier resin.

The use of the concentrate or masterbatch method has proven to be one of the most effective methods of incorporating carbon black into a resin. A masterbatch, as defined in the plastics industry, is a resin containing a high concentrate (20–50%) of pigment and other fillers that is subsequently reduced in a compound resin to accommodate the end-use requirement.

Traditionally, additives have not been found to significantly enhance the dispersion of carbon blacks in plastics. But more recently, for polyolefins, plastics formulators have applied polyacrylate ester waxes as dispersing agents [3]. Note that waxes and lubricants appear in various forms in black masterbatch formulations. The gains that such additives may offer in dispersion quality must be carefully weighed against the additive's effect on a compound's end-use processing such as molding or its end-use performance such as UV stability. A well-designed dispersion process will bring about the best development of color and physical properties.

The concept of shearing force is central to the process of carbon black melt compounding. A fluxed plastic matrix may be considered as a stack of layers that, when subjected to shearing stress, tend to slide along each other. The greater the resin's cohesive force, the greater energy required for shearing. Shearing force acts on carbon black agglomerates at the interfaces of layers, tending to fracture them into primary aggregates. To achieve the best carbon black dispersion, formulators select polymers and equipment to provide the highest possible shear. Note that the need to maximize shear to disperse black must be balanced with the need to avoid polymer degradation (Fig. 11.4).

11.4.1. Effects of Dispersion on Polymer Performance

As mentioned earlier, dispersion quality also plays a strong influence on color. If carbon black is not completely dispersed, in a 100× magnification light micrograph,

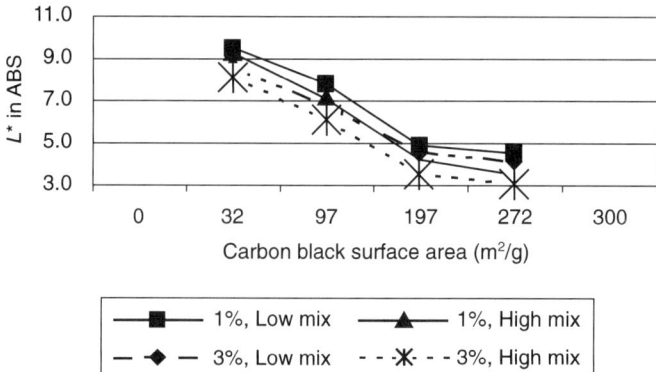

Figure 11.5. L^* for 1 and 3% loaded Acrylo-butadiene styrene copolymer (ABS) resin.

agglomerates containing between 10 and 100 aggregates may be observed (<2% dosage). The effect is to hide some carbon black surface from incoming light, thus increasing the amount of reflected light and making the overall material lighter. Poor dispersion also can result in streaks or other discoloration in the plastics end product, as well as affect how the plastics appearance will age—particularly if it is exposed to UV light.

The effect of mixing intensity on color development can be seen in Figure 11.5. The two curves marked with open and solid circles are L^* values of injection-molded parts made from the low- and high-intensity mixing processes containing 1% carbon black. The high-intensity mixing afforded a jetter (lower L^*) color than the low-intensity mixing procedure. The color "gap" between the low- and high-intensity mixing increased with increasing carbon black surface area. Previous work has shown that the energy required to disperse a carbon black increases with increasing surface area, primarily because of the high van der Waals attraction forces between the primary aggregates.

Molded parts prepared at 3% carbon black loading are also shown in Figure 11.5. The relationship between low- (open triangles) and high- (solid triangles) intensity mixing schemes is similar to that for 1% loaded parts. As one would expect, the 3% black loaded parts are darker.

11.5. CARBON BLACK SELECTION FOR OTHER PROPERTIES

Besides color, other requirements of the end product, such as UV protection, conductivity, reinforcement, or a combination of these, influence selection of carbon black.

11.5.1. Carbon Black for UV Stability

More recent studies have been conducted by Cabot Corporation to understand the performance of various carbon blacks with different morphologies in thin-gauge

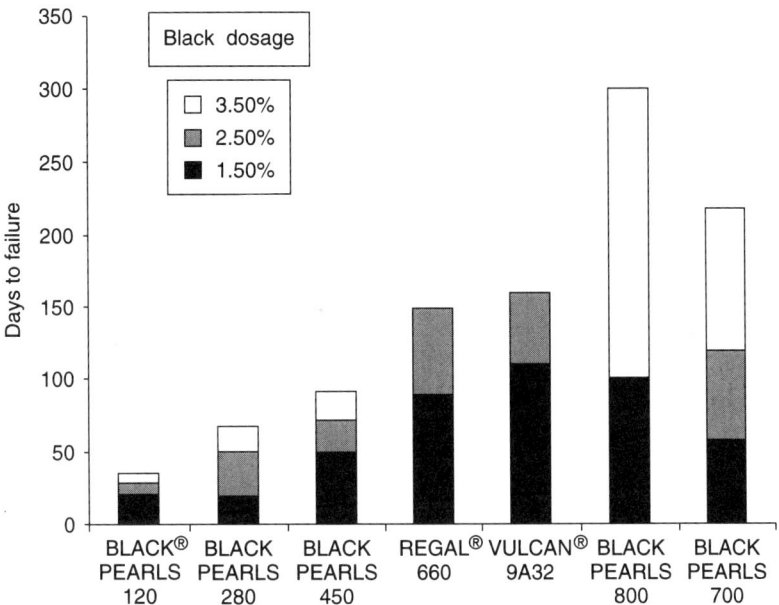

Figure 11.6. QUV-A weatherometer testing (75-μm sample thickness) relationship of days to failure (50% elongation) to carbon black type and loading.

(75-μm) Lineur Low Density Polyethylene (LLDPE) film and to develop a model that predicts the most effective carbon black morphology for maximum weatherability in plastics applications. Cabot's study also examined the effect of various loadings (1.5, 2.5, and 3.5%) on weathering performance under natural accelerated outdoor exposure, xenon arc weatherometer exposure, Q-Panel UV-B radiation (QUV-B) exposure, and carbon are exposure.

Cabot's work established that UV stabilization increases with decreased primary particle and aggregate size. This is a function of the increase in surface area of carbon black that is available to absorb the light. Note that UV stabilization tends to level off for particles smaller than 20 nm due to the lesser degree of back scattering (toward the incident light) demonstrated by the accompanying smaller primary aggregates. Since finer aggregates exhibit less scattering than coarser aggregates, a greater amount of forward-scattered UV light will threaten the stability of the polymer.

QUV-B Weatherometer Testing

Figure 11.6 illustrates the relationship in UV stability between large-particle-size carbon blacks, midrange-particle-size carbon black, and small-particle-size carbon blacks at a given carbon black loading (1.5, 2.5, and 3.5%). The small-particle-size grades (ELFTEX® TP, VULCAN® 9A32, BLACK PEARLS® 700, and BLACK PEARLS 800) provided extended days to reach 50% of tensile elongation at break, signaling value in improved UV stabilization in an artificial weathering test environment.

168 Carbon Black Pigments for Plastics

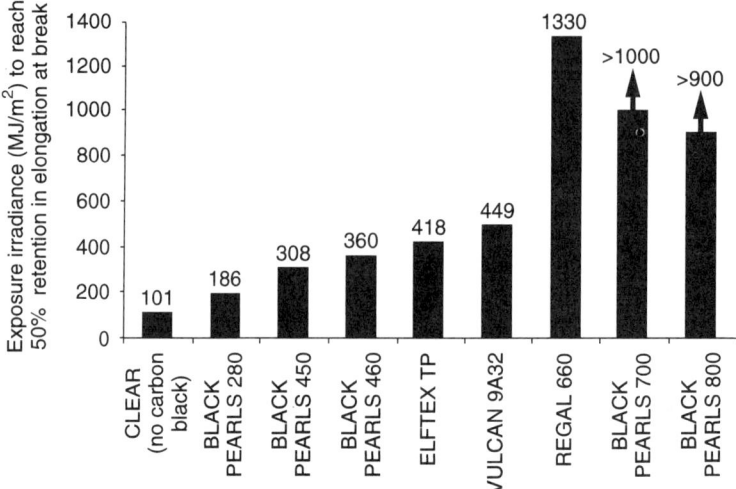

Figure 11.7. Outdoor weather testing, Phoenix, AZ: 75-μm sample thickness, 1% carbon black loading in LLDPE.

Outdoor Weathering: Phoenix, Arizona

The evaluation of samples weathered outdoors revealed distinct grade-family performance groups (Fig. 11.7). Large-particle-size grades (United 120 and BLACK PEARLS 280 carbon blacks) exhibited early deterioration, as evidenced by irradiance for 50% elongation at break. Slightly finer particle size blacks (BLACK PEARLS 450 and 460 carbon blacks) showed slight improvements. Even finer particle size blacks (VULCAN 9A32, and ELFTEX TP carbon blacks) exhibited higher exposure limits to 50% percent elongation at break. The smallest aggregate diameter blacks (BLACK PEARLS 700 and REGAL 660 carbon blacks) exhibited the highest tolerance to exposure irradiance levels—the former never reaching failure within the testing period shown. The sample that did not contain carbon black (clear) failed at an extremely fast rate versus samples containing any carbon black grade.

Effect of Carbon Black Dispersion on UV Stability

Dispersion quality directly impacts UV absorption efficiency by affecting the final carbon black particle size in the polymer. An optimal dispersion is one that evenly distributes carbon black throughout a polymer dispersed down to the smallest carbon black unit (i.e., an aggregate). A poorer dispersion will result in agglomerates of two or more aggregates. Agglomerates, which are larger than aggregates, result in increased scattering and reduced absorption of UV radiation. This explains how the level of dispersion directly influences both the absorption and scattering of UV radiation and, ultimately, the retention of the polymer's inherent physical properties (Fig. 11.8).

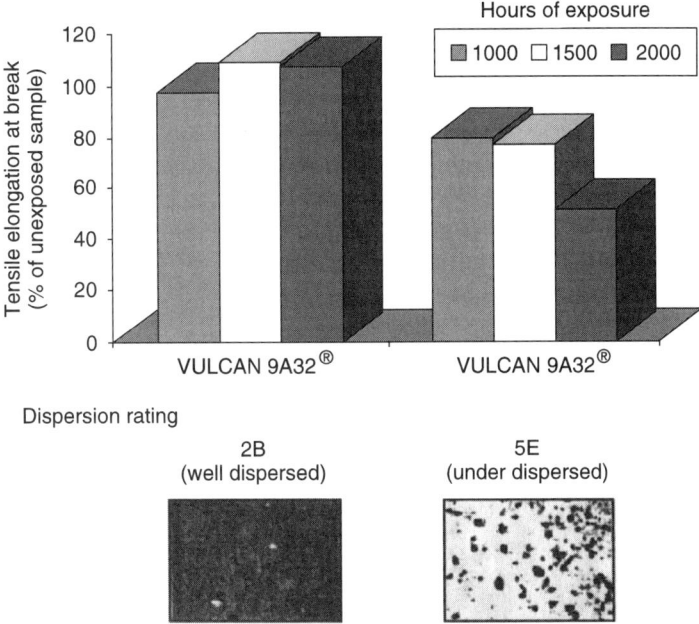

Figure 11.8. Tensile elongation in a carbon arc weatherometer: impact of carbon black dispersion; 2.5% carbon black loading in LLDPE.

11.5.2. Carbon Black for Conductivity

Carbon black is far less resistant to the flow of electricity on its surface than the plastic resins in which the black is dispersed. Therefore, carbon black can be used to lower the resistivity of plastics, imparting antistatic, semiconductive, or conductive properties. End uses for conductive carbon blacks range from material handling bins and device carrier tapes in the electronics industry to fuel system components to semiconductive strand shielding for power cable.

To achieve acceptable conductivity, one usually selects, the finer the aggregate size carbon blacks, other factors being constant. As a result, more aggregates per unit weight of carbon black are distributed throughout the fixed volume of resin. This means smaller distances between aggregates and greater ease of electron transfer from one aggregate to another. Therefore, fine particle size (small aggregates) will lower electrical resistivity when dispersed in plastics, other properties being constant. With regard to electrical resistivity, higher structure creates a more irregularly shaped aggregate, which in turn provides more potential paths for electron transfer on the surface of carbon black through the plastic. This translates into decreasing resistivity with increasing structure when dispersed in plastics. Note that in applications that require moderate conductivity and excellent physical property retention, new specialty grades may be best suited to deliver such a balance of end-use properties. This comparison of dosage, conductivity (usually measured as surface resistivity), and mechanical properties is illustrated in Figures 11.9 and 11.10. The loading level of a carbon black grade is an important factor in the conductivity of the compound. In general, higher loadings increase conductivity; however, because

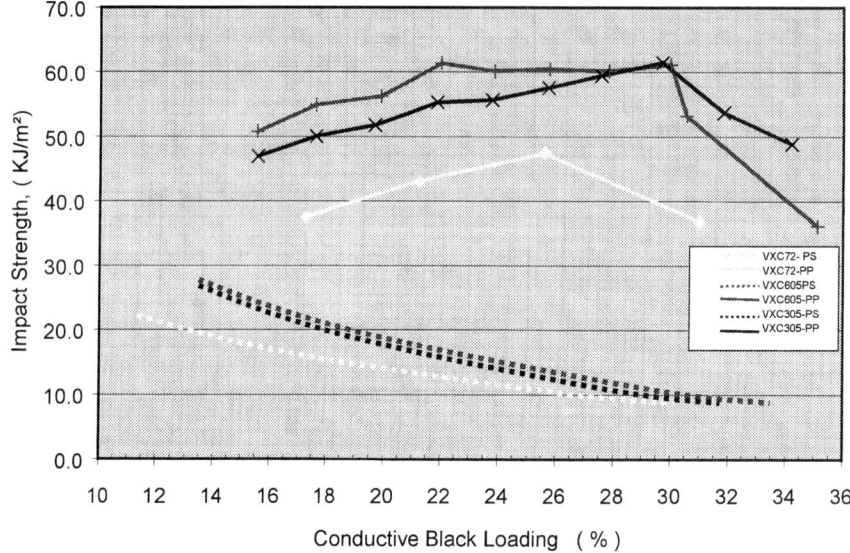

Figure 11.9. Dart impact vs. dosage.

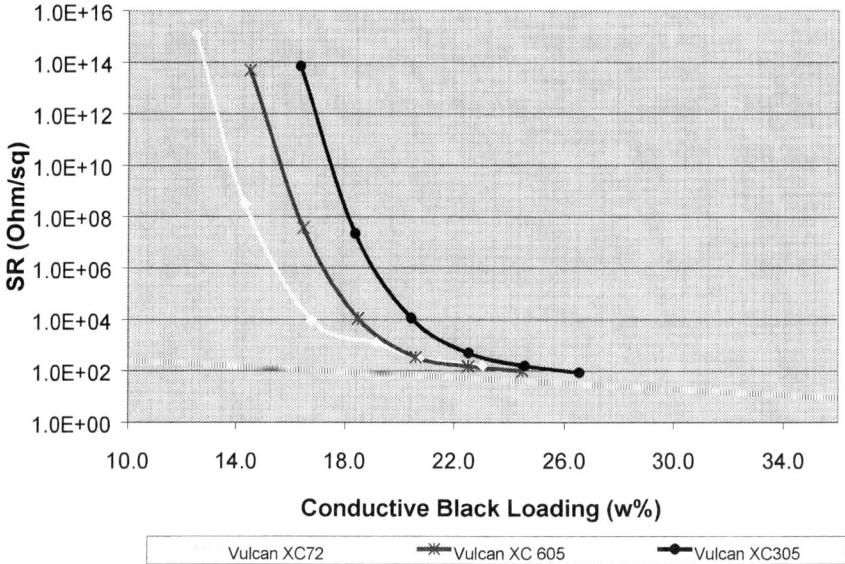

Figure 11.10. Resistivity vs. dosage.

carbon blacks vary significantly in their inherent conductivity, some grades require higher loadings than others to achieve the same level of conductivity.

11.5.3. Carbon Black for Physical Properties

Because of its relatively fine particles and high surface area, carbon black has a much more significant effect on compound physical properties than do other

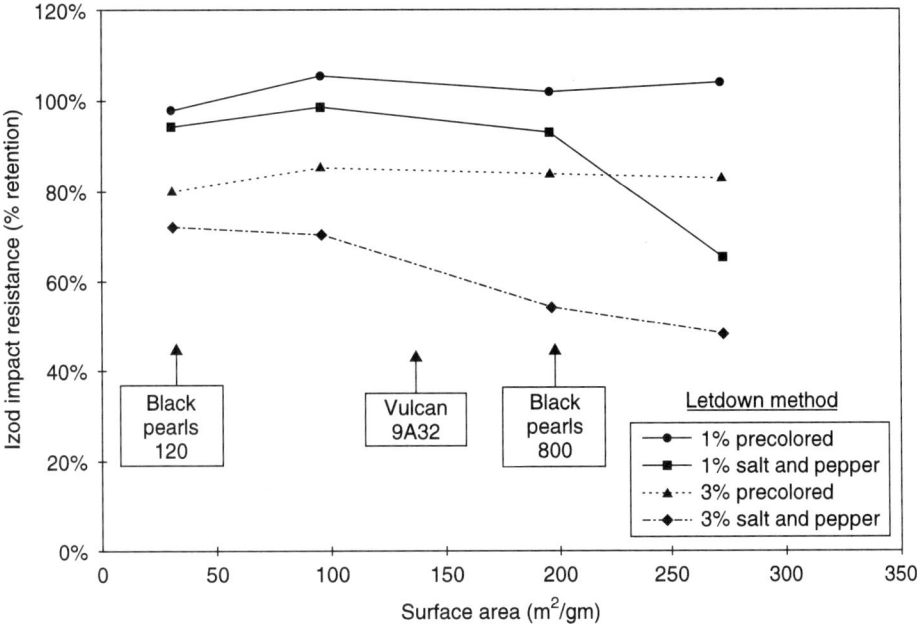

Figure 11.11. Dosage and letdown effects on Izod: injection-molded ABS.

mineral fillers. Carbon black tends to increase hardness, tensile modulus, tensile yield strength, and brittleness temperature. The extent of these effects is magnified by higher carbon black loadings, but other factors play an important role as well.

Tensile Strength

In the case of carbon black as a filler, the ultimate tensile strength of a plastic compound results from interfacial effects between the carbon black and the resin. These interfacial effects are enhanced by the presence of a higher concentration of carbon black aggregates (higher loading) and by decreased aggregate size.

The quality of dispersion also affects stress distribution. In a well-dispersed system, the carbon black–resin interfaces are well distributed, which helps to distribute stress. Poor dispersion can result in larger agglomerates acting as stress concentrators causing a decrease in tensile strength.

Impact Strength

A property of great importance in applications for various carbon blacks in plastics, impact strength will be illustrated with ABS. The physical properties of ABS compounds are extremely sensitive to both quality of dispersion and aging. In this case, as indicated by Figure 11.11, precolored compounds show unproved dispersion over salt and pepper blends mixed in the holding machine. Better dispersion yields better impact retention.

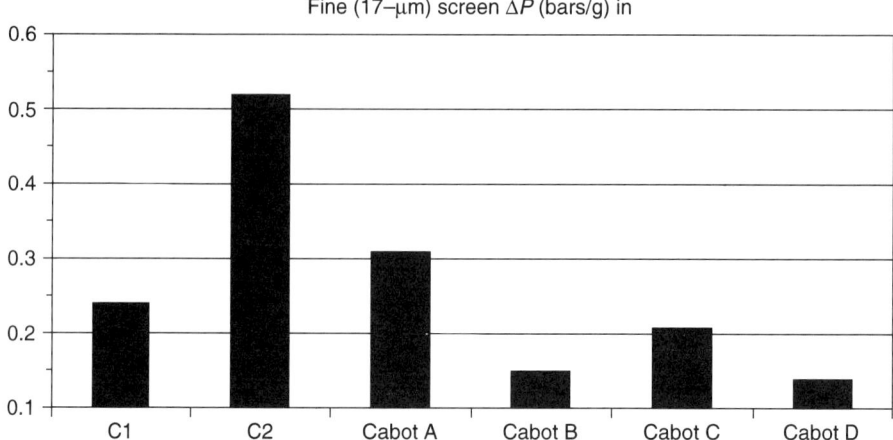

Figure 11.12. ΔP results for a range of carbon blacks.

11.5.4. Carbon Black Effects on Resins: Possible Pitfalls

The comparative chemical cleanliness, residual contaminant level, and control over the properties of carbon blacks produced through the furnace process cannot be overstated. Carbon blacks containing high levels of residue and chemical contamination may adversely affect end-use performance. For plastics used in sensitive applications such as fiber and for some engineering resin formulations, raw material specifications require low levels of chemical and residual contaminants.

Residual Contamination

Especially important is that carbon black possesses low levels of residual contaminants. Residual impurities alone within a carbon black can adversely affect an application's dispersion properties, processing, molded surface characteristics, and color. Impurities can be defined as moisture, residual carbonaceous residue, sulfur, and certain ionic species.

For example, in the fiber-spinning process, filter cartridges are employed to remove the contamination that can block the small spinneret orifices and/or cause fiber breakage during spinning. However, smaller contaminants that pass through the filter cartridge can still cause considerable fiber breakage, particularly for fine-denier fiber. When the pressure buildup is excessive, productivity suffers and the filter cartridge must be replaced and the clogged filter cleaned. The changing and cleaning of filter cartridges represent significant costs to fiber spinners. For carbon black to process efficiently through these filters, it must possess very low residual contaminant levels.

The presence of contaminants in carbon black is typically measured by incorporating a sample into a resin and extruding the concentrate through a screen pack (e.g., 325 mesh or 44 µm). Residual contamination is indicated by the pressure rise behind the screen pack over a period of time (ΔP), although factors such as dispersion quality can affect this test. The fine-screen ΔP data shown in Figure 11.12 illustrate that within grades considered low residue, significant differences can exist in screen pack performance. Low ΔP translates to better processing in fiber and many other carbon black–plastics end uses.

Melt Viscosity (Processability)

Carbon black has a very significant effect on the melt viscosity of a plastic color concentrate. The key factors operational here are surface area, structure, and loading. Increasing any of these will add to melt viscosity of a resin, which in turn affects its dispersion quality and processability. Viscosity is the property of resistance to flow that the body of a material exhibits, with a higher viscosity indicating greater resistance to flowing. Color concentrate manufacturers and masterbatch formulators typically load 30–50% carbon black in various carrier resins for subsequent letdown to realize the 0.5–3.0% typical for most coloring applications. Users prefer masterbatches with high carbon black and low carrier content, but increased loading also increases viscosity, causing a greater disparity between the viscosities of masterbatch and the neat resin. The results of this disparity can be a poor distributive mixing that negatively affects final color, dispersion quality, and performance in the end product.

In plastics, relative viscosity can be defined as the ratio of the viscosity of a concentrate to that of the neat carrier polymer at the same temperature. In this use, relative viscosity predicts relative jetness—defined in the same manner—more accurately than the carbon black loading in a concentrate. A mismatch between masterbatch and letdown viscosities causes incomplete mixing and is the reason that relative jetness can decrease with an increase in relative viscosity [5].

Compound Moisture Adsorption (CMA)

Moisture absorption refers to a material's incorporation of airborne water vapor from the air. A carbon black that possesses low CMA enables compounders to extrude under higher temperatures while minimizing the surface defects caused by entrapped water escaping through the plastic. Since carbon black is a high-surface-area material, it tends to be hygroscopic. The level of surface area largely determines the amount of moisture absorbed by carbon black during storage and transit. In selecting grades of carbon black for formulating plastic compounds, it is well to keep this fact in mind because the presence of moisture can sometimes present problems in processing. In some moisture-sensitive applications, predrying of all compound ingredients may be advantageous.

11.6. CONCLUSION

This chapter was designed to tie the details of carbon black particle size and shape to its performance in thermoplastic applications. For example, selection of fine-particle-size, low-structure carbon blacks will translate into high jetness for automotive exterior molding applications. Also, important in high-value applications such as fiber or thin-wall molding are requirements for carbon black cleanliness and purity. Successful application of carbon black begins by defining the needs for the application. Then, to ensure best performing carbon black is selected, the black supplier and end user must engage in a good dialog that required balance of price and performance are met. Figure 11.13 summarizes the carbon black performance attributes that need to be considered when selecting carbon black for a plastics application.

REFERENCES

1. Dannenburg, Carbon Black, p. 1, Part 2.
2. 53rd ANTEC, M. C. Yu, M. A. Bissell, and R. S. Whitehouse, "The Effect of Carbon Black Dispersion on Polymer Performance," society of Plastics Engineers ANTEC Boston, MA May 7–11, 1995 Volume 3 pg 3246. Cabot Corporation, 1995.
3. Baltimore RETEC, A. Knebelkamp, H. Buskies-Keup, G. E. Hahn, and W. Schafer, "High Performance Dispersing Waxes for Polyolefin Color Concentrates," Technical Service Polymer Additives, 1997.
4. "The Role of Carbon Black Morphology In Reducing UV Oxidation of Linear Low Density Polyethylene Films," Cabot Technical Report S-115, 1997.
5. M. C. Yu, J. Menashi, and D. J. Kaul, "Carbon-Black Morphology: Its Effect on Viscosity and Performance," *Plastics Compounding*, November/December, 1994.

CHAPTER 12

Soluble Dyes

Craig Weadon
Polysolve, Inc.
Crown Point, Indiana 46307

Walter Martin
Rainbow Quest Color Consulting
504 Liathirwood Court Virginia Beach, VA 23462

12.1. INTRODUCTION

Dyestuffs are used in clear plastics to provide clean, bright colors and transparent shades where optical transmission is important, such as acrylics, crystal polystyrene, and polycarbonates. Dyestuffs are also used in other resins to impart strong, clean colors. In the presence of titanium dioxide strong inexpensive pastels are produced. This phenomenon occurs because the dyestuffs absorb certain wavelengths of light while transmitting other wavelengths. No scattering of the light occurs since the dyestuffs are transparent and therefore do not scatter light.

A definition for plastics dyes that has evolved in practical use is somewhat more useful than traditional ones; namely, a dyestuff is a dyestuff when used in a given polymer because it is soluble in the polymer. In contrast, pigments are therefore defined as not being soluble in the polymer being colored.

Dyestuffs are easy to disperse and usually are readily soluble within the polymer matrix. Dyes that have been designed for use in plastics generally are oil or solvent soluble, although some are only resin soluble. While many dyestuffs are water

---

Coloring of Plastics, Fundamentals, 2nd edition. Edited by Robert A. Charvat
ISBN 0-471-13906-8   Copyright © 2004 by John Wiley and Sons, Inc.

soluble, these generally are not used to color plastics unless the molded objects are to be dip dyed. Certain applications such as carpet fibers and sunglass lenses are produced by coating or bonding of the dyestuff to the external surface of the polymer. Some water-soluble dyes are used as Food Drug & Cosmetic (FD&C) certified colorants, but they are usually prepared as precipitated lakes on barium sulfate or other inorganic carriers. Various applications using dyestuffs will be discussed later in this chapter.

At this time a review of the history of dyestuffs and their use is appropriate, along with a brief review of the development of plastics.

12.2. HISTORY

Berries and leaves of certain plants have been used for centuries to provide the colors necessary to make wool, cotton, flax, silk, and other fabric fibers with an attractive shade differing from the natural colors of the materials. These colors were usually boiled with the formed fabric or yarn to impart color. Additional processes such as cooling the solution and/or the addition of lime, carbonate (chalk), or vinegar was found useful to set the color onto the fabric and impart some degree of fastness.

Dyes were also prepared from certain marine animals, such as the Tyrian purple (royal purple) color used by the Roman emperors. Production of these dyes was very limited in quantity and they were very expensive. Only a limited amount of material could be made from the collected shelled animals. This typically was reserved for royalty and persons of rank.

Large agricultural plots were necessary to grow the plants used to make dyestuffs. Extensive labor was necessary to harvest those plants and extract the limited amount of available dyestuffs, making the dyes very expensive to use.

Initial work on the manufacture of coal tar derivatives started in England and Scotland. Charcoal was made from limited oxidation of wood. An oily residue known as tar was found in pools at the bottom of the charcoal pit. When coke was produced for the growing steel mills of England, this same residue was found and named coal tar. Around 1850, experiments with this residue yielded a multitude of products during distillation. We know one of the components of the residue as kerosene, which led to lamps for lighting homes and public buildings. Since a large amount of material was created and burning seemed the only way to get rid of it, chemists searched for ways to use the by-products found in the coal tar residues. William Henry Perkin is credited with being the first chemist to conjure up a brilliant dyestuff from the black, heavy coal tar. Perkin had been looking for a means to synthesize quinine for medical purposes when he accidentally discovered the first coal-tar dyestuff. This heralded the growth of dyestuff chemistry and started a new industry in England. During the London World's Fair of 1862, it was declared the displays of colored fabrics were something of a victory parade for the English dyestuffs industry.

France did not produce dyestuffs synthetically but had a large industry based on naturally occurring dyestuffs. Large plots of land in France—thousands of acres—were set aside to grow madder. The resulting dyestuff products in 1868 had such a wide market that nearly 80 million francs of income was gained annually from the sale of the bright red color, most of it from export sales. Development of the synthetic dyestuff industry in England was so rapid that, by 1888, it was no longer profitable to grow madder since the dyestuff synthesis cost only a fraction of the price of the naturally grown material. The synthetic materials produced in the laboratory were of equal or greater purity and at 1% of the cost of the naturally occurring dyestuff. To the dyers in the Orient, the source of the color that rendered their fezzes and garments was immaterial, but the cost of the garment was paramount, and synthetic dyestuffs significantly reduced its cost.

Alizarin and mauveing were the next dyestuffs synthesized, followed by literally hundreds and hundreds of brilliant colors. Large industries for production of synthetic dyestuff developed in England and Germany. Anthracene, an intermediate product involved in dyestuff chemistry in the early years, was available only from England. Germany developed its branch of the dyestuff industry only by importing large amounts of material from England.

Germany's three big chemical companies, Bayer, BASF, and Hoechst, took up dyestuff manufacturing in the mid-1800s. Hoechst provided dyes used in the leather industry, generally acid of direct dyestuffs. While attempting to improve the dyeing, the chemists made an interesting discovery. A precipitated form of aldehyde green was made and was used to dye a silk gown for Empress Eugenie. Prior to that time, green dyestuffs for silk all appeared blue under the gas lighting used in building. This new green gown remained green! The green dye became known as "empress green" and became the base for a new and extremely profitable market for dyestuffs. A few years later it was replaced with methyl green, which in turn was replaced by malachite green. These dyes were produced by the company, which eventually became Hoechst.

For as long as 5000 years before commercial synthetic dyestuffs were made, indigo had been used as a coloring material. For centuries, indigo had been produced from vegetable matter grown only in India. Adolph Bayer was awarded the Nobel Prize for his laboratory synthesis of indigo in 1879. Production of synthetic indigo started within several years and replaced the natural grown dyestuff.

Artificial silk was first produced from cotton waste in the early 1900s. Three Englishmen are credited with discovering how to produce viscose (rayon) from a cellulose solution using wood and woody materials. During World War I, this process was used to make guncotton (by nitrating the cellulose) and other explosives. The rayon was also used as artificial silk. Special dyes, now known as acid dyes, had to be developed to color this product.

To this point in time, polymerization as a process was applied only to refined naturally occurring materials. It was not until the early 1930s that chemists mastered polymer synthesis from small molecules.

Polyvinyl chloride (PVC) and polyvinyl acetate (PVA) are considered to be the first synthetic polymers created. Safe-handling cellulose acetate soon replaced explosive cellulose nitrate. Polyacrylonitrile and polyamides (Nylon) soon followed. American companies such as DuPont pioneered the development of plastics. England was responsible for the early development of polyester polymerization.

Polyethylene and polystyrene were developed in England in the late 1930s but not commercialized until the end of World War II. the copolymerization led to the development styrene-butadiene rubber (SBR). In turn, when added to transparent polystyrene, SBR rubber improved the brittleness of the polystyrene while imparting impact strength; however, a milky, hazy polymer that had lost its transparency resulted. Copolymerization led to the development of other polymers, such as Acrylonitrile-butadiene-styrene (ABS) and others.

The 1950s gave rise to the polypropylenes and polycarbonates. Other engineering resins such as Polyetheretherketone (PEEK) and polysulfones were developed later.

12.3. PLASTICS COLORING

With the development of man-made fibers, a methodology was needed to color these products. Many of the prepared dyestuffs for natural fibers did not impart color to the synthesized polymeric fibers. Hoechst (Germany), ICI (England), and DuPont (America) were the pioneers to develop dyestuffs to successfully color the man-made fibers. The dyestuffs used were either coupled to the polymer or chemically bonded to the polymer in the case of reactive dyes. These dyes are generally applied to the fiber in a water solution containing the dye and chemicals designed to assist in the process.

Solid molding resins needed dyestuffs with much greater heat stability than solution-dyes for fibers, plus they required solubility in the polymer matrix to successfully color the polymer internally. The development of molded and extruded products established the need to color these products efficiently, economically, and in quantities never previously imagined. Dyestuffs have played an important role in this process.

Different resins are processed at widely varying temperatures. The polymers may also contain acidic or basic groups that might interact with the dyestuff. This process chemistry must be carefully considered when selecting dyestuffs for use with plastics. The dyestuff must remain within the polymer and not migrate to the surface. This phenomenon, known as crocking, is the main reason why dyestuffs are not used to color polyethylene and most polypropylene resins, since few color compounds are soluble in these resins. Crocking also occurs with many fiber color formulations and is a standard test method employed in fiber testing. Crocking can also occur when a dye soluble in a given polymer is used in an amount greater than the resin can dissolve and contain.

12.4. CATEGORIES OF DYES

The dyestuffs used in plastics generally fall in three categories: solvent, disperse, and vat dyes. Solvent dyes are just that—soluble. However, solvent dyes may be soluble in different media, for example, oils, aromatic solvents, aliphatic solvents, waxes, and alcohols. The dyes used to color plastics, as we have discussed, are usually totally organic in constitution and are soluble in aromatic and/or aliphatic solvents and may be oil soluble. The important property here is the ability to dissolve readily in polymers. Alcohol-soluble dyes generally find use only in cellulosic polymers and are relatively insoluble in styrenics, acrylics, or polycarbonates. Another group of solvent dyes are metal–organic molecules consisting of an organic structure attached to a chromium atom. While these dyes are soluble in typical lacquer solvents and have extremely good lightfastness and transparency, they are usually rather weak in strength and are expensive, with their use in plastics limited to fibers where their fastness and transparency outweigh their cost.

Disperse dyes were originally developed for incorporation into certain cellulose acetate fibers. Applications of disperse dyes include the coloring of nylons, polyesters, and acrylics. Since many of these dyes were also designed to be mixed—"dispersed"—with a dispersant, care must be exercised in avoiding dispersed versions of these dyes, since the dispersants used are not meant to be incorporated in plastics and seriously diminish the brightness and transparency of the dye. Disperse dyes in their pure form are often also classified as solvent dyes, carrying both Colour Index names. Many disperse dyes developed for polyesters are finding use in other polymers.

Vat dyes fall mainly into two classes of chemicals: indigoids and anthraquinoids. These dyes are insoluble in water but may convert from a dye to a pigment in plastics in the presence of certain oxidizing agents. These colorants may also revert to a colorless "leuco" state during the processing of polymers at elevated temperatures. Often an observer will confuse this colorless state by thinking that the color has "burned out" at the elevated temperature when it has merely changed form. The use of additives to prevent the oxidation can help to protect the colorant. Acidic salts employed in emulsion polymerization can also drive this leuco conversion or insolubilization. This shifting of color is utilized in solution dying. Care must be exercised in the use of some of these vat dyes, as they are extended and buffered by the addition of salt compounds when manufactured; this can cause corrosion of metal equipment or plug extruder screen packs during the compounding process.

Tables 12.1–12.3 show polymers generally suitable for coloring with dyes (Table 12.1), polymers that may be suitable for dye use (Table 12.2), and polymers unsuitable for dye use (Table 12.3).

Table 12.1. Polymers Suitable for Use of Dyes

Styrenics	Polyesters: transparent and filled
Crystal polystyrene	Rigid PVC
Impact polystyrene	Polysulfone
Styrene acrylonitrile copolymer (SAN)	Polypropylene, foamed only, some dyes
ABS	
Acrylics	Thermosets
Butadiene: styrene	Epoxies
Polycarbonate: clear and filled	Polyesters
Nylons, including filled	Phenolics

Table 12.2. Polymers That May be Suitable for Coloring with Dyes

Thermoplastic polyurethanes
Acetals: copolymers only
Thermoplastic rubbers: some only
Thermoplastic olefins: some only
Cellulosics: only certain dyes

180 Soluble Dyes

Table 12.3. Polymers Not Suited for Dye Coloration

Plasticized flexible PVC
Polyethylene, all types
Polypropylene, except foam
Most rubber
Ethylene vinyl acetate
EMA—Ethylene-methyl acrylate

Table 12.4. Chemical Classes of Dyes for Plastics

Azos—monoazos and diazos	Xanthenes
Anthraquinones	Phthalocyanines
Quinolines	Methines
Perinones	Triarylmethanes
Perylenes	Aminoketones
Indigoids	Azines

12.5. FLUORESCENT DYES AND BRIGHTENERS

Fluorescent dyestuffs are finding increased use in producing eye-catching colors used in promotional materials, safety applications, and toys. These dyes reemit visible light when they absorb light of near-ultraviolet wavelengths. The intensity of the reemitted light is not always readily observed by the human eye, leading to the use of these dyes to brighten duller conventional dyes. When strong ultraviolet sources or "black lights" are used to illuminate the object, strong visible light will be emitted by the object when little or no illumination is apparent.

The presence of light-scattering mineral pigments such as titanium dioxide or iron oxides will quench the flourescent effect. Similarly, excessive concentration of the fluorescent acts to prevent reemission of light and is a waste of money.

Fluorescent brighteners are used to overcome the natural yellowness of some plastics. The brighteners are usually colorless. They function by absorbing wavelengths of light in the near-UV range and emitting the light in the blue-violet region. Care must be exercised in the application of these materials due to their yellowing when exposed to ultraviolet light. They are also used to reinforce the effects of other fluorescent dyestuffs by emitting light at a wavelength that the fluorescent dye absorbs. This greatly increases the amount of emitted light from the fluorescent dye.

12.6. CHEMICAL CONSTITUTION

Commercial dyestuffs are drawn from the number of different chemical classes. Table 12.4 shows some of these classes. Of these classes, the azos and anthraquinones are the most numerous among the commonly used dyestuffs. A few important dyestuffs exist in each of the other chemical classes.

Azo dyestuffs are generally characterized by high strength, fairly high brightness, and low and low to medium stability. They are used mostly to color relatively low-temperature, high-volume polymers. A wide range of colors is available, with the notable exception of blues and greens.

Anthraquinone derivatives, the "coal-tar" family, exist in almost all colors, dominating the blues, green, and violets and least present in the yellows and oranges. Many anthraquinones are very stable to light and heat exposure. Most of the dyes used in high-temperature engineering resins are anthraquinones since this class of dyes dates to the very beginning of synthetic dye chemistry. Many of the commonly used fluorescent dyes are also anthraquinones, with xanthenes and indigoids representing most of the remainder.

12.7. FASTNESS PROPERTIES

With the exception of solubility in organic solvents, the properties of dyestuffs are the same as other coloring materials. Water solubility is important when extrusion pelletizing of plastics is employed, so no migration will occur to contaminate metal surfaces or reduce the effectiveness of the coloring material.

Lightfastness is usually checked using a source of intense UV light often coupled with environment—simulating conditions such as humidity and salt spray. The plastic manufacturer must decide the criteria used for evaluation, depending on the expected use of the product being manufactured. Many individuals agree lightfastness of a particular colorant must equate to a minimum of 100 h of exposure without noticeable fade or shift in hue. Usually tints or tintones are exposed as the changes are more readily seen when titanium dioxide is present. This is probably because the exposure only occurs close to the surface of opaque polymers. Concentrations of dyestuff in the test pieces are usually as follows: 0.01% dyestuff, 1.0% titanium dioxide, and 98.99% resin.

Heat application may result in sublimation or melting of the dyestuff or may bring about decomposition. It is best to express the acceptability to heat in three distinct areas: (1) the temperature up to which the dyestuff is stable, (2) the melting point, and (3) the temperature where sublimation may occur.

It is important to note that not all dyestuffs behave the same when used in different applications or different polymers. Similarly, the hue of the dyestuff may change when the dyestuff is used in different plastics due to its solubility in that resin as well as factors created by the inherent color of the polymer. Residual volatile components in the polymer as well as catalysts and other additives have a distinct effect on the property of the dyestuff. Trace contaminants in the dyestuff can also have a profound effect on the dyestuff performance in different resins.

Length of time of dyestuff exposure to heat at a given temperature is also a critical factor when evaluating heat stability.

Migration of the colorant from the plastic to food-containing products requires intensive testing to assure that the migrated colorant is not harmful to humans or other animals. The chemistry of the dyestuff and its purity are both important in this respect.

Polymers of the same chemical composition generally behave the same when dyestuffs are added as colorants. However, the spectral curves are not always the same due to the additives used in polymerization. Changes in concentration as well as solubility affect color strength in polymers. Solution dissolving of the dyestuff should not be used as the means to check dyestuff performance in plastics. Masstone coloring will also mask the impurities present as well as distort the spectral adsorption curve. A phenomenon known as dichroism exists with dyestuffs. This is the change in appearance of the dye in high concentrations versus its appearance in low-concentration tints. This is mostly noticeable in the red and red-pink dyestuffs but can also occur when comparing the dry, highly concentrated pure material compared to its application use level in other colors.

12.8. PRACTICAL CONSIDERATION

No amount of printed material can substitute for proper definition of requirements, testing of component materials in the application, and careful formulation. Recommendation of a particular dye in a given application may be rendered useless by the use of a different set of functional additives in a color concentrate, for example.

Some of the most common pitfalls are as follows:

> Poor definition of requirements may lead to inadequate stability or overformulation for needs that do not really exist. In a competitive marketplace, of

Table 12.5. Dye Use in Common Thermoplastics

Color	Dye Chemical Type	GPPS	LDPE	HDPE	PP	ABS	Acrylic	Flexible PVC	Rigid PVC
Violet	Anthraquinone	W	N	N	L	W	W	L	L
Blue	Anthraquinone	W	N	N	L	W	W	N	L
Green	Anthraquinone	W	N	N	L	W	W	N	L
Yellow	Anthraquinone	W	N	N	L	W	W	N	L
	Azo	W	N	N	L	N	W	N	N
	Pyrazolone	W	N	N	N	W	W	N	N
	Pyridone	W	N	N	N	L	W	N	N
	Quinoline	W	N	N	N	W	W	N	N
	Quinophthalone	W	N	N	N	W	W	N	N
	Xanthene	W	N	N	N	L	W	N	N
Red	Anthraquinone	W	N	N	L	W	W	N	L
	Azo	W	N	N	L	L	W	N	N
	Benzopyran	W	N	N	N	W	W	N	N
	Perinone	W	N	N	N	W	W	N	N
Orange	Azo	W	N	N	L	L	W	N	N
	Perinone	W	N	N	N	W	W	N	N
	Polymethine	W	N	N	N	W	N	N	N
	Methine	W	N	N	N	W	W	N	N
	Indigoid	W	N	N	N	L	W	N	N
	Thioxanthene	W	N	N	N	L	W	N	N
Black	Azine	N	N	N	N	N	N	N	N

Abbreviations: General purpose polystyrene, (GPPS); LDPE, HDPE, low- and high-density polyethylene; PP, polypropylene.

Table 12.6. Dye Use in Engineering Thermoplastics

Color	Dye Chemical Type	Acetal	Nylon	Cellulosic	Polyester	Polycarbonate	Fluoropolymer
Violet	Anthraquinone	N	W	L	W	W	L
Blue	Anthraquinone	N	W	L	W	W	L
Green	Anthraquinone	N	W	L	W	W	L
Yellow	Anthraquinone	N	W	L	W	W	L
	Azo	N	N	L	W	N	N
	Pyrazolone	N	N	N	W	W	N
	Pyridone	N	N	N	W	W	N
	Quinoline	N	N	L	W	W	N
	Quinophthalone	N	L	L	W	W	L
	Xanthene	N	L	L	W	W	N
Red	Anthraquinone	N	W	L	W	W	L
	Azo	N	N	L	W	N	N
	Benzopyran	N	L	N	W	W	N
	Perinone	N	W	L	W	W	N
Orange	Azo	N	N	L	W	N	N
	Perinone	N	W	L	W	W	N
	Polymethine	N	N	L	W	W	N
	Methine	N	L	N	W	W	L
	Indigoid	N	N	N	W	W	N
	Thioxanthene	N	N	N	W	W	N
Black	Azine	N	N	N	N	N	N

Table 12.7. Dye Use in Thermosets

Color	Dye Chemical Type	Epoxies	Polyesters	Phenolics	Silicones	Polyurethane
Violet	Anthraquinone	L	L	N	N	L
Blue	Anthraquinone	L	L	N	N	L
Green	Anthraquinone	L	L	N	N	L
Yellow	Anthraquinone	L	L	N	N	L
	Azo	N	N	N	N	N
	Pyrazolone	N	N	N	N	N
	Pyridone	N	N	N	N	N
	Quinoline	N	N	N	N	N
	Quinophthalone	L	L	N	N	N
	Xanthene	N	N	N	N	N
Red	Anthraquinone	L	L	N	N	L
	Azo	N	N	N	N	N
	Benzopyran	N	N	N	N	N
	Perinone	L	L	N	N	L
Orange	Azo	N	N	N	N	N
	Perinone	L	L	N	N	L
	Polymethine	N	N	N	N	N
	Methine	N	N	N	N	N
	Indigoid	N	N	N	N	N
	Thioxanthene	N	N	N	N	N
	Azine	N	N	N	N	N

course, overformulation leads to overpricing, which usually leads to loss of business to the competitor.

Properties may be taken for granted. If a dyestuff is used in a food-packing application, proof must exist for nonmigration in the given application for colorants not specifically approved. What works in one situation may not work in another.

Additives may be used carelessly.

Materials may be blended without regard to possible interactions.

The list of caveats is endless. With that caution in mind, Tables 12.5 and 12.6 give a compilation of dyes used in a variety of thermoplastic resin types. They are listed by color and within color range and by common chemical types employed. Table 12.7 provides the same information for thermosets. In the tables, dye types marked with a "W" are widely used, those with an "L" have limited use, and those with an "N" should not be used.

REFERENCES

1. *Colour Index*, rev. 3rd ed. Vol. 1–6, Society of Dyers and Colourists, 1975. Bradford, England.
2. *Technical Manual of the American Association of Textile Chemists and Colorists*, Vol. 39, AATCC, 1963. Research Triangle Park, NC.
3. M. Ahmed, *Coloring of Plastics*, Van Nostrand Reinhold Company, 1971. New York, NY.
4. E. Bäumer, *A Century of Chemistry*, Econ Verlag Gmbtt Düsseldorf, Germany, 1968. Druckerei, 1968.

CHAPTER 13

Photochromic and Thermochromic Colorants

John J. Luthern

University of Akron
2174 Thomas Road
Hubbard, OH 44225-9785

13.1. PHOTOCHROMIC MATERIALS

There are many types of photochromic and thermochromic materials. Due to the introductive nature and consequent brevity of this work, only selected subject areas will be examined. This work is intended to introduce product designers, chemists, color matchers, and laboratory technicians to the special colorant area of photo- and thermochromics.

In 1899, Marckwald discovered that two separate organic compounds, 2, 3, 4, 4-tetrachloro-1-(4H)-naphthalenone and benzo [c]-1, 8-naphthpyridine hydrochloride, developed color in sunlight and lost this color when placed in the dark. The study of photochromic materials had begun.

Photochromic substances are able to transform between two states that possess different absorption spectra. The change, in at least one direction, is induced by the absorption of electromagnetic radiation. The change in the other direction is induced by radiation of a different wavelength and/or a thermal mechanism. This behavior is depicted in Scheme 1.

This chapter will classify and review photochromics according to the mechanism whereby the material manifests its photochromic behavior. This chapter is not intended to be a comprehensive review of every photochromic family; excellent works that explicitly deal in great detail with photochromics are already available.

Coloring of Plastics, Fundamentals, 2nd edition. Edited by Robert A. Charvat
ISBN 0-471-13906-8   Copyright © 2004 by John Wiley and Sons, Inc.

Scheme 1.

$$A \xrightleftharpoons[h\nu'/\text{heat}]{h\nu} B$$

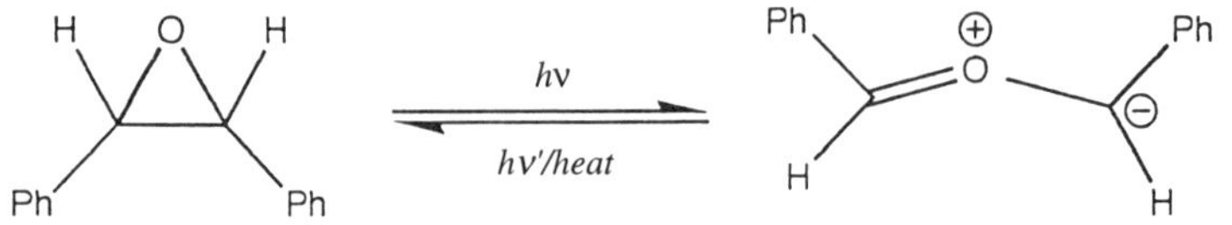

Scheme 2.

The discussion here is limited to the following topics that concern photochromic classes that are potential plastic colorants: chemical types or classification, manufacture, activation processes, attributes, and industrial usage.

We begin by examining photochromism involving heterolytic bond cleavage. Compounds belonging to this group undergo an electromagnetically initiated heterolytic bond cleavage to generate a colored structure. Specifically, this bond cleavage is a result of UV radiation absorption that converts the colorless dye structure to a colored structure, which is sometimes called the merocyanine form because of its structural similarity to merocyanine dyes. In all compounds belonging to this family, charged ions are formed as a result of UV radiation absorption.

1,3-Electrocyclization is a photochemical reaction capable of generating photochromic species. An example of a 1,3-electrocyclization reaction involving a monocyclic aryloxirane is the interconversion of *cis*-stilbene oxide and the corresponding carbonyl ylide, shown in Scheme 2 [1].

Note that the stereospecificity of the reaction in Scheme 2 follows the orbital symmetry predicted by the Woodward–Hoffmann rules [2]; the oxide and ylide interconvert via a disrotatory mode. As is the case with all three-membered heterocycles mentioned here, UV irradiation of the heterocycle generates a highly colored ylide intermediate. Unfortunately, the stability of the ylides derived from monocyclic oxiranes is poor and photochromic behavior is evident only upon irradiation at low temperatures (77 K) [3]. This drawback has been somewhat circumvented in a few cases by annulation of the ylide functionality, which increases its stability. For example, 5-oxabicyclo[2.1.0] pentane develops a purple color when irradiated at 253.7 nm at room temperature [4]. This reaction is shown in Scheme 3.

Indenone oxides and cyclopentadienone oxides are two types of polycyclic oxiranes that generate stable ylides upon UV irradiation at ambient temperature [5]. The photochromic behavior of an indenone oxide is depicted in Scheme 4.

Another class of three-membered heterocyclic compounds that exhibit photochromic activity via 1,3-electrocyclization reaction is the aziridines. For the most part, similar to the oxiranes, these materials require low temperatures to demonstrate photochromic behavior [6]. Also, some of the photochromic transformations are not completely reversible. However, several room temperature stable ylides have been reported for aziridines possessing the general form of Scheme 5 [I], where X and Y may be a mixed combination of nitro, methoxy, and hydrogen groups. The

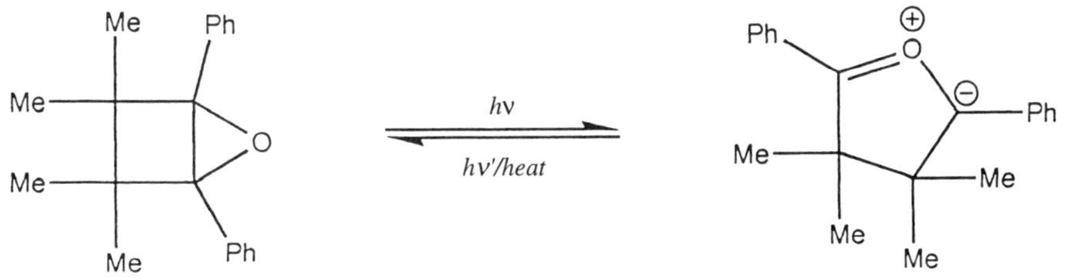

Scheme 3.

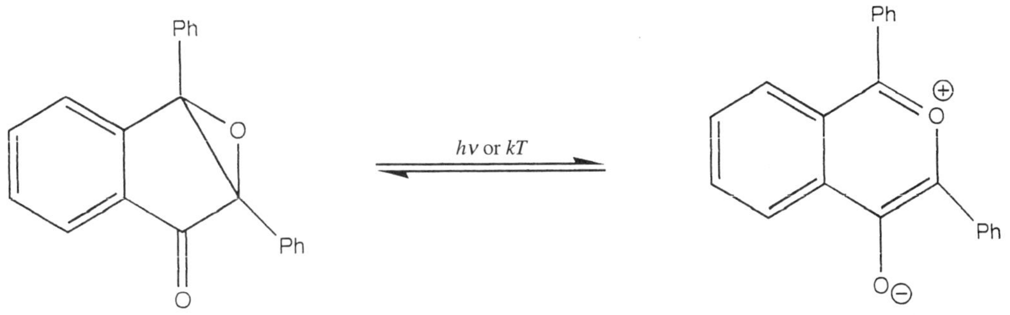

Scheme 4.

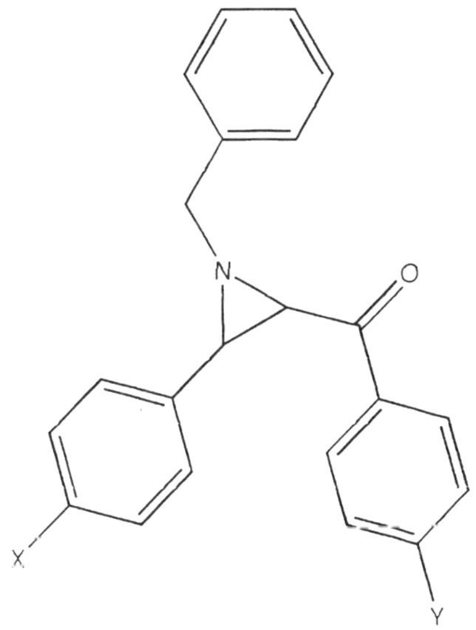

Scheme 5.

oxalic salts of certain bicyclic aziridines generate a red-colored protonated azomethine ylide under UV bombardment. The red-colored ylide is stable for weeks in the absence of visible radiation [7]. These materials have been used in the manufacture of photochromic windshields and glasses.

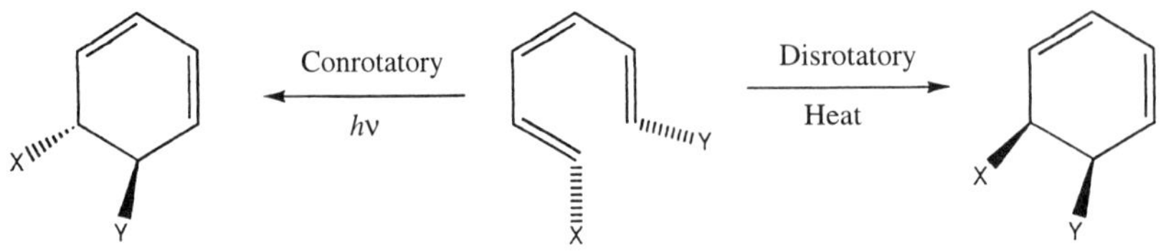

Scheme 6.

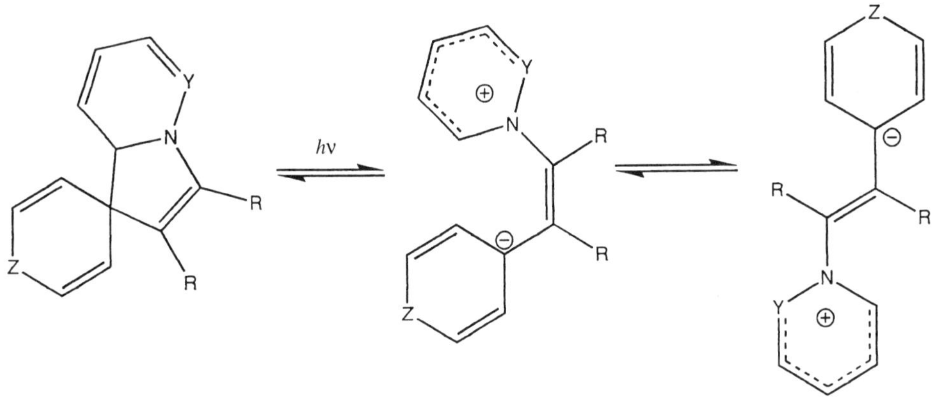

Scheme 7.

The next subclass of photochromic compounds are those based on 1,5-electrocyclization. An example of $4n + 2$ 1,5-electrocyclization is given in Scheme 6.

In hydrocarbon $4n + 2$ 1,5-electrocyclization reactions, only irreversible reactions take place (i.e., ring closure in the case of eight-membered rings and ring opening in the case of seven-membered rings). However, reversible photochromic 1,5-electrocyclization reactions have been achieved by creating molecules whose open-chain and cyclic forms possess similar energies. This new class of photochromic compounds are polycyclics that possess a δ^2-pyrolline unit [8]. These compounds are called the spiro[1,8a]dihydroindolizines (DHIs). The DHIs develop a strong color when irradiated with UV-A or short-wavelength visible radiation [8, 9] (up to approximately 440 nm). This color is due to the E and Z zwitterionic (e.g., betaine) isomers produced by the irradiation of the cyclized DHI. The photochromic reaction is shown in Scheme 7.

In the literature, the DHI's are claimed to be as efficient a photochromic class as the fulgides and anils. However, DHI's are rapidly degraded by singlet oxygen and their use requires an oxygen barrier of some sort. When dissolved in certain polymers, these materials exhibit photochromism for a duration of 600–5000 cycles [10].

Symmetrically substituted DHI's may be synthesized by one of the following two alternatives: When powerful electron-withdrawing groups such as —CN are incorporated into the DHI, it is necessary to irradiate the appropriate spiropyrazoles in a heterocycle/CH_2Cl_2 mixture [11]. Typical heterocycles are the azines, whose members include pyridines, pyridazines, quinolines, and isoquinolines.

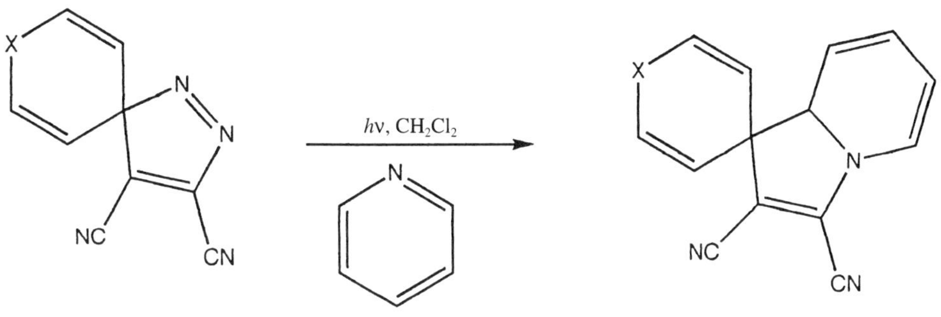

Scheme 8.

Yields are between 40 and 70%. An illustrative example of this synthesis is given in Scheme 8.

Three alternative routes for DHI synthesis involve the reaction of azines with spirocyclopropenes [12], retro-1,5-electrocyclization [13], and via diazo-compounds [14].

The DHI's may be absorbed into the surface of a polymer. For example, CR 39 (diethylene glycol-bis-alkyl-carbonate) with absorbed DHI is used as a light filter [15]. The DHI's may be solution cast with certain polymers. Examples of photochromic plastics prepared this way are poly (methyl methacrylate), poly (n-butyl methacrylate), copoly (vinylidene chloride-acrylonitrile) (e.g., SARAN F), polycarbonate, and polystyrene-butadiene (e.g., Panarez).

In terms of durability, the DHI's possess greater photoresistance than the spiropyrans and less photoresistance than the spirooxazines.

The spiropyrans (e.g., spirochromenes) are important members of the heterolytic cleavage class. Spiropyrans have been one of the most intensively studied photochromic classes [14]. The spiropyrans consist of two orthogonal and electronically isolated heterocyclic halves. In the inactivated, or colorless, state, the two halves share a common spiro linkage to an sp^3 hybridized carbon atom. On exposure to UV radiation, the C_{spiro}–O bond cleaves in a heterolytic fashion, and the two halves of the molecule adopt a virtually planar geometry. Increased planarity allows for pi-cloud overlap between the indoline and pyran parts of the molecule, and this merocyanine-like structure has its first electronic transition in the visible region and consequently exhibits color. The open merocyanine form may exist as one of eight stereoisomeric forms; however, the transoids are energetically favored due to the lack of steric hindrance present in the cisoid isomers. The thermal barrier to isomer interconversion is approximately several kilocalories per mole [16]. The photochromic behavior of a typical spiropyran, 6-nitro-1′,3′3′-trimethylspiro[2H-1-benzopyran-2,2′-indoline] (6-nitro-BIPS) is shown in Scheme 9.

Spiropyrans are prepared via a condensation reaction between an ortho–hydroxy formyl compound and a methylene indoline material (i.e., Fischer base) [17]. See Scheme 10. Also, the corresponding iodo-indolinium salt of the Fischer base may be employed in place of the Fischer base to expedite the purification procedure.

In the case of a typical spiropyran, the coloration–bleach cycle may be repeated anywhere from 100 to 1000 times. Typical values for the molar absorptivity of the colored merocyanine form range from 10,000 to 50,000 L/mol/cm. A few spiropy-

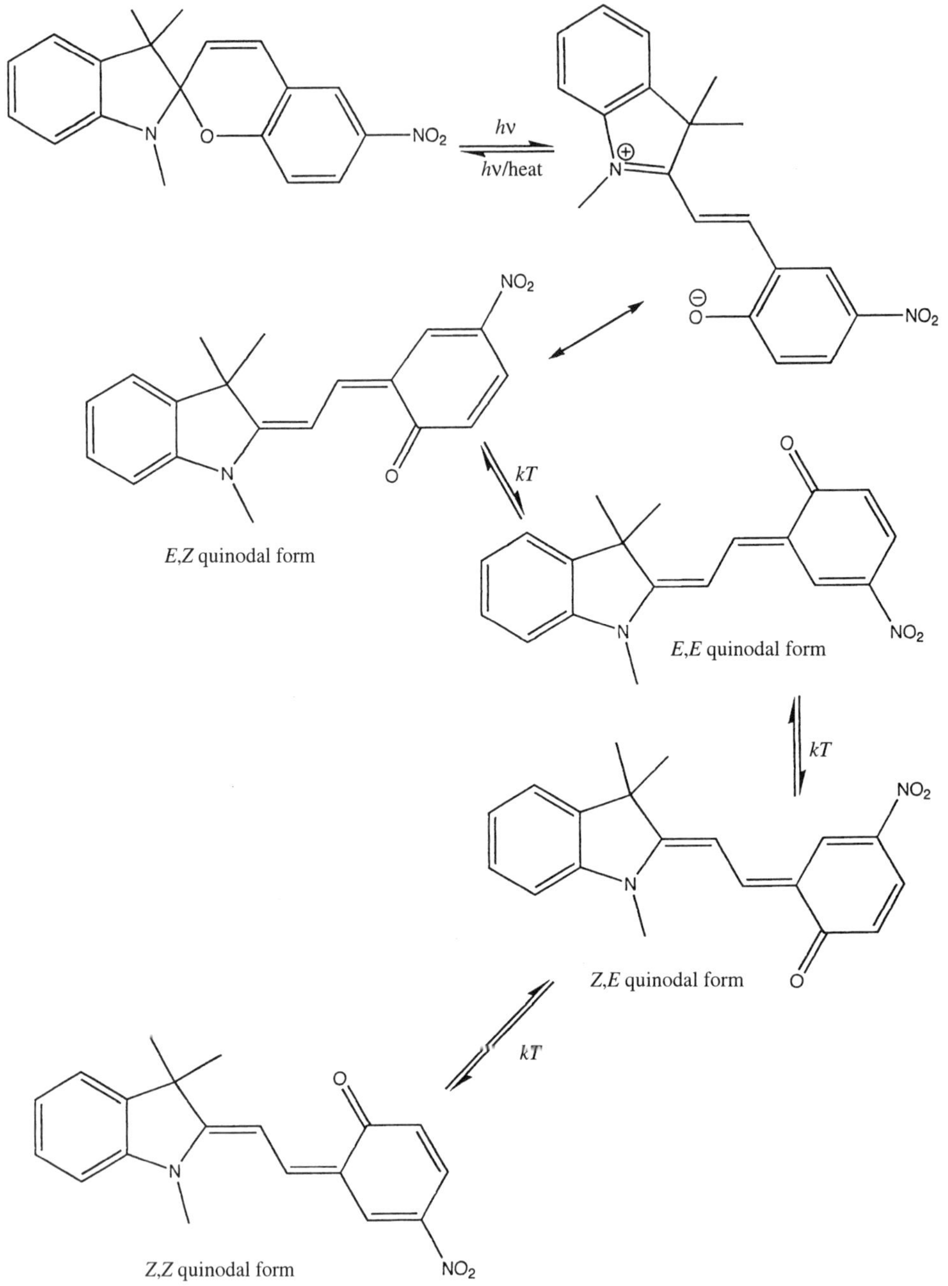

Scheme 9.

rans, among other types of photochromic substances, exhibit a phenomenon called "reverse photochromism," which occurs when a spiropyran solution possesses the colored merocyanine form in the absence of UV light. This colored form is bleached by the action of ultraviolet light [18].

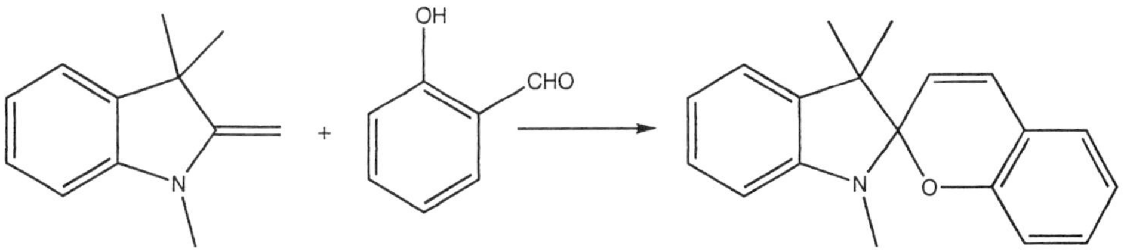

Scheme 10.

The spiropyrans are not photochromic in the solid state but do exhibit photochromic behavior when dissolved in any of the following: solvents, gels, films, plasticized resins, and plastic solids.

Another very important member of the heterolytic cleavage classification is the spirooxazines (SOs). The chemical structure of the SOs is similar to spiropyrans except for the substitution of the pyran ring with an oxazine ring. Also, the photocoloration mechanism is similar to that of the spiropyrans. First, the SO must be dissolved in a solvent or polymer matrix to be able to demonstrate photochromic activity. Prior to UV exposure, the inactivated, and thus uncolored, SO exists as a spiro compound with two electronically isolated halves. When exposed to UV radiation, the spiro C–O bond heterolytically ruptures and a greater degree of conjugation per molecule is made possible. The open merocyanine form can exist in any one of three mesomeric forms: a resonance form, a keto form, and a zwitterion form [19]. A general diagram of SO photochromic behavior is presented in Scheme 11. Spirooxazines can be prepared by a condensation reaction between an ortho-nitroso aromatic alcohol and an alkylidene heterocycle. For example, equimolar amounts of 1-nitroso-2-naphthol and Fischer's base are refluxed in toluene for about 2 h. After purification, the photochromic product is 1,3,3,-trimethyl-spiro [indoline-2,3-[3H] naphth[2,1-b] [1,4]oxazine (NISO) [20].

The thermal decay rates of the colored forms, which correspond to the amount of time required for a UV-activated photochromic to lose its color, typically range from 0.02 to 1.51 mol^{-1} s^{-1}. However, this decay rate is highly influenced by temperature and the nature of the polymeric host. For instance, at 0°C a SO possesses a thermal fade rate of 2.67×10^{-2}, and at 50°C the same molecule demonstrates a thermal decay rate of 3.3221. In fact, the thermal bleach reaction of some SOs ceases at –60°C.

Spirooxazines are utilized in solvent or solid solutions such as gels, films, coatings, and plastics. They are applied to plastics by any one of the following techniques: internally cast with catalyzed monomers (specifically acrylics with diazo catalysts), thermoplastic injection molding (a limitation: SOs are stable at 200°C for only a few minutes), coatings, solvent and aqueous dyeing, and vapor–liquid transfer [22]. They are utilized extensively in plastic photochromic eyewear. In addition to eyewear, photochromic optical plastics are used in decorative, glazing, and lighting applications.

Another major category of photochromics are those whose photocoloration mechanism involves concerted electrocyclization reactions. Members of this group include the fulgides and the chromenes.

The chromenes are important photochromic compounds that are structurally related to their parent compound, 2-H-chromene, whose photochromic behavior is

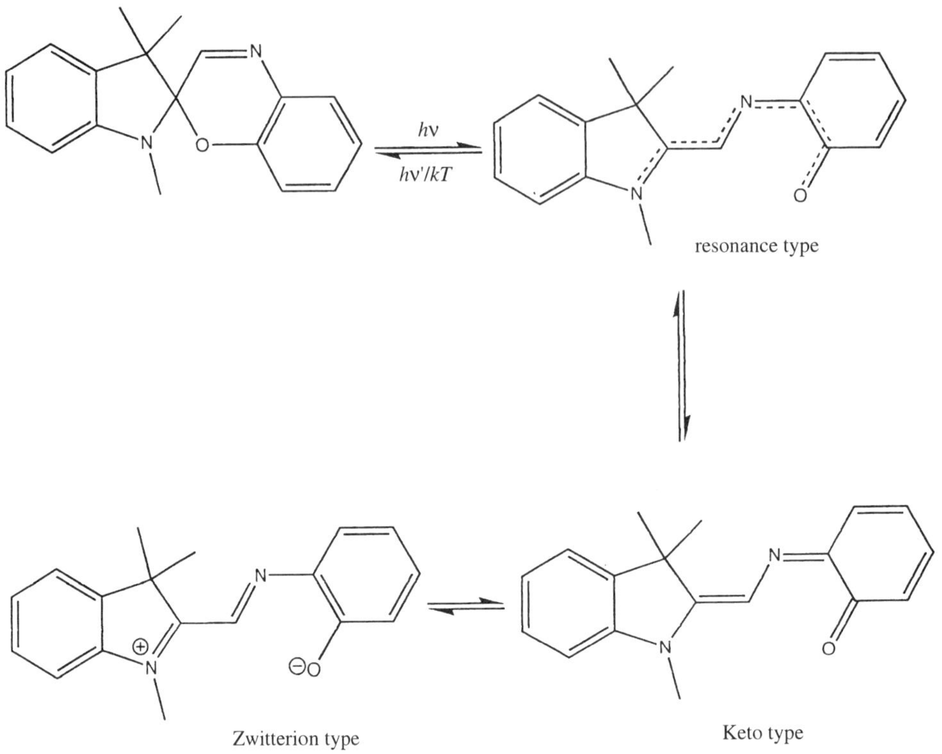

Scheme 11.

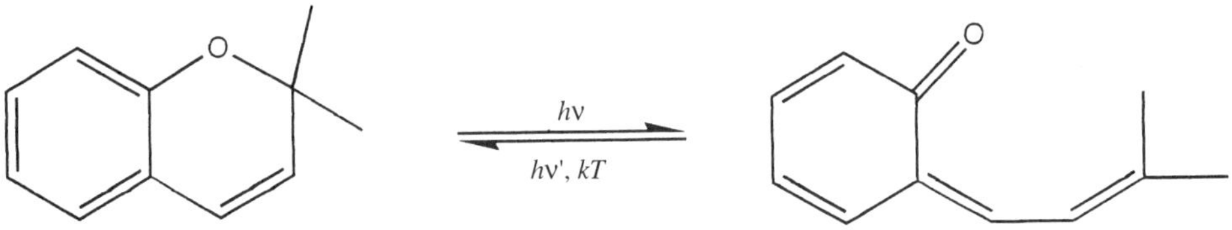

Scheme 12.

depicted in Scheme 12 [23]. 2-*H*-chromene and its simple derivatives have been known to be photochromic since the work of Becker in the mid-1960s [24], for example, the photochromic behavior of the chromene derivatives 2,2-diphenylchromene and 6,6-diphenylnaphtho(2,1:2,3)pyran-4-*d* [25] has been studied. Van Gemert has synthesized numerous, structurally complex chromene-related molecules called diaryl naphthopyrans [26]. These molecules possess a high quantum efficiency for photo-induced coloration in the near-UV region and lack the structural features that caused the rather poor fatigue resistance of the earlier chromene compounds.

Upon activation, the diarylnaphthopyrans typically become red, orange, yellow, or some hue in combination of these three colors, although the cooler colors of photoactivated diarylnaphthopyrans are also commercially available. Within certain

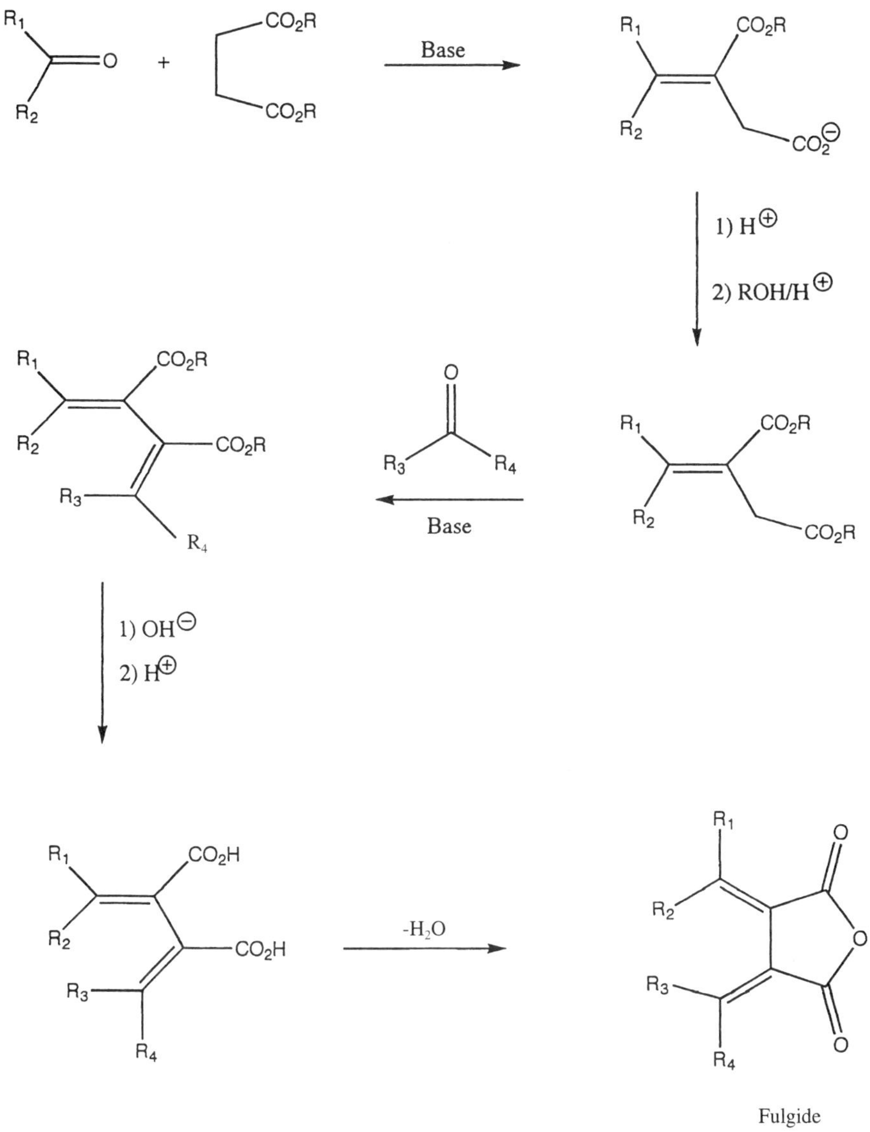

Scheme 13.

limitations, the diarylnaphthopyrans may be processed by virtually any thermo plastic processing method, provided the maximum temperature does not substantially exceed 200°C for long periods of time.

The fulgides were first synthesized by Stobbe early in the twentieth century. Because these compounds usually crystallized as shiny, bright crystals, Stobbe named them fulgides from the Latin word *fulgere*, which means to glisten and shine. However, these materials were not commercially viable until the early 1980s, when fulgides possessing both durability and good photochromic behavior were synthesized. These materials are synthesized via a series of Stobbe condensations, as shown in Scheme 13 [27].

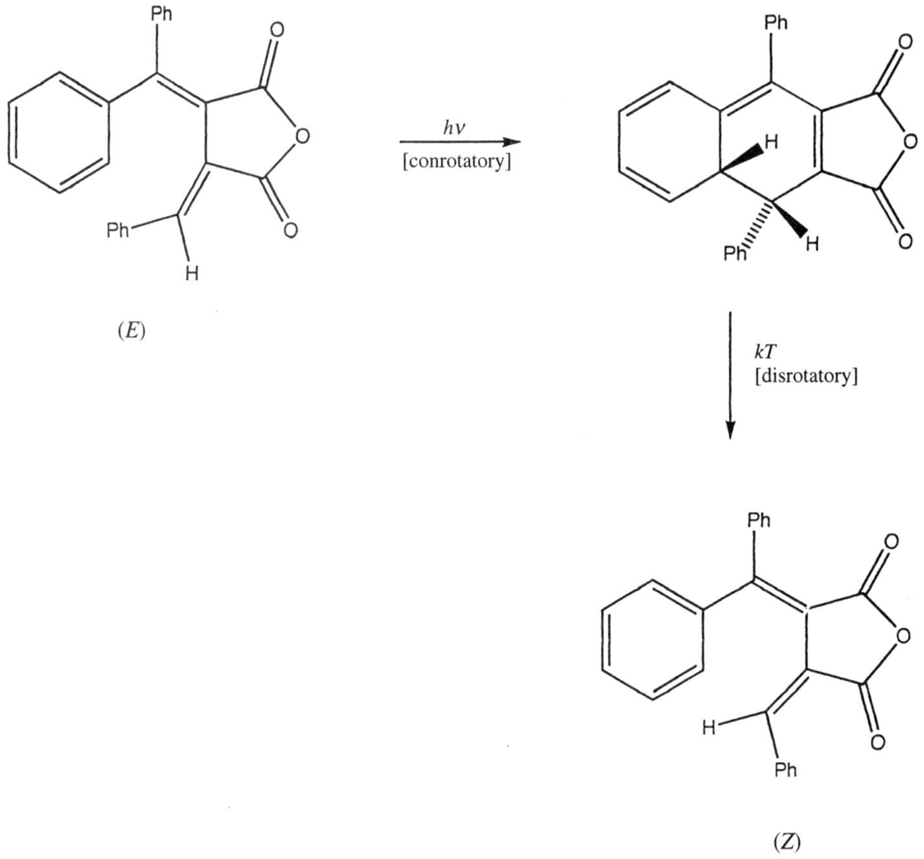

Scheme 14.

The photochromic behavior of the fulgides is due to a photochemical cyclization reaction that generates the colored specie, a 1,8a-dihydronaphthalene. The addition of heat initiates a thermal electrocyclic ring-opening reaction during which the colored specie is transformed into the original fulgide [28]. These reactions are shown in Scheme 14. The photochemical cyclization and thermal electrocyclic ring opening proceed via a conrotatory and disrotatory mode, respectively, and therefore are in agreement with the Woodward–Hoffman selection rules [29].

In the inactivated state, fulgides are yellow or orange crystalline compounds. On exposure to UV radiation, the fulgides develop a red, orange, or blue color. The fade rates of the fulgides vary from less than a minute to many hours [30]. At least one fulgide exhibits polyphotochromism, in which UV irradiation of the appropriate fulgide generates two different 1,8a-dihydronaphthalenes that possess different visible absorption spectra. Specifically, one isomer is red and the other blue. Irradiation of the mixture at longer wavelengths within the visible region will selectively open the ring of the blue form [30].

Applications include eye glasses, spatial light modulation [31], optical waveguide construction [32], nonlinear optical switching [33], inks, and paints.

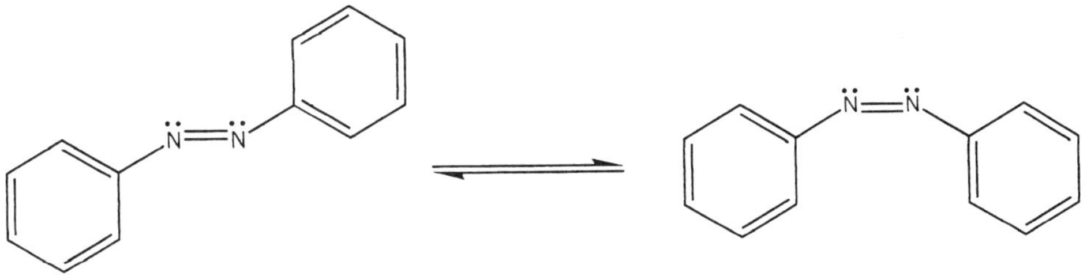

Scheme 15.

The next class of photochromics examined here are the compounds that undergo cis–trans isomerization. In this family, the trans isomer is converted to the cis isomer by electromagnetic radiation. The conversion of the cis to the trans isomer is due to absorption of radiation of a different wavelength, including infrared radiation. In general, the thermal isomerization follows first-order kinetics and the reaction constants may vary from two orders of magnitudes of liters per mole per second to 2×10^{-7} L/mol s [34]. Also, the absorption bands of both isomers are broad and overlap one another at virtually every absorbing wavelength. This phenomenon prevents the 100% photochemically induced conversion to either the cis or trans isomer. Among the classes of compounds that undergo cis–trans (e.g., E–Z) isomerization about an ethylenic bond are the stilbenes, the indigo-type compounds, azo compounds, and the polyenes. The E–Z isomerization of nonpolar stilbenes is catalyzed in either direction by the absorption of visible light and therefore will not be covered here, although the literature on this subject is extensive and enlightening [35–40]. Ethylene linkages containing 1,2-aryl groups with polar substituents exhibit electromagnetic absorption within the visible region, although the hue is strongly dependent on substrate polarity.

Azo compounds undergo a cis–trans isomerization reaction when exposed to UV, visible, or infrared radiation [41]. This cis–trans rearrangement is shown in Scheme 15. In more recent literature the trans isomer is called the E form and the cis isomer is called the Z form. The E form absorbs radiation in the UV region and rearranges to the Z form. This transformation is accompanied by a deepening of the original color, which itself is due to the thermal equilibration of the cis and trans isomers. The Z form rearranges to the E form upon the absorption of infrared or visible radiation.

Simple azobenzenes are yellow, orange, or red. In the absence of oxygen, azobenzenes demonstrate good lightfastness.

Azobenzenes have been utilized to measure the free volume in polymers and the speed of polymeric segmental motion [42, 43]. Azobenzenes that are covalently bonded to a polymer backbone may influence various properties of the macromolecule. Photoisomerization of such substances will cause changes in wettability [44], viscosity [45], solubility [46], membrane properties [47], and swelling properties [48].

Another major class of photochromic materials comprises those that undergo tautomerism via hydrogen transfer. Two subclasses in this category are the triazoles and methyl salicylates. Upon UV absorption, the molecule is promoted to an excited state where the contribution of charge transfer structures is augmented. This augmented charge transfer character increases the acidity of aromatic hydroxy compounds as well as the basicity of a carbonyl oxygen. In molecules containing both

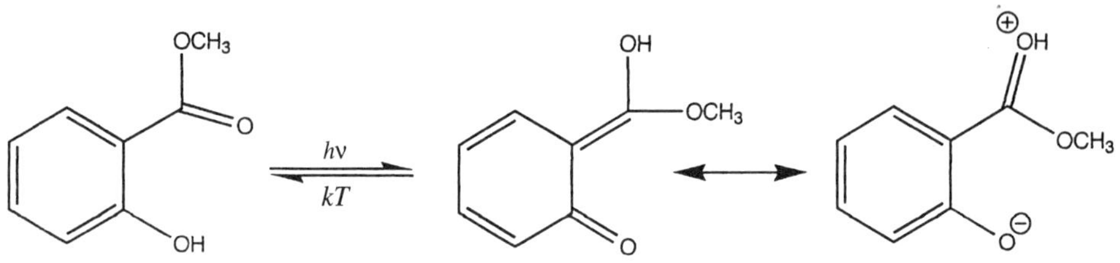

Scheme 16.

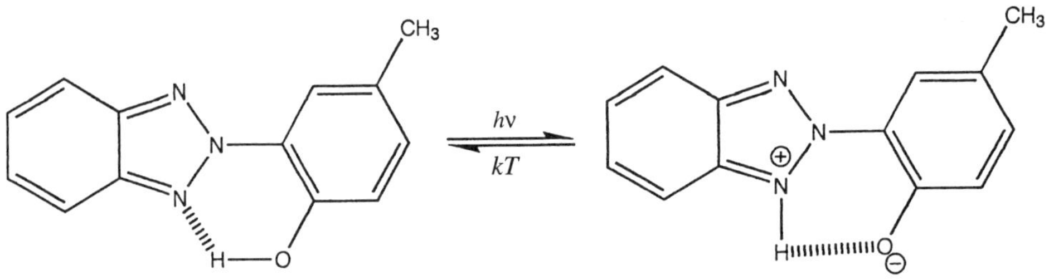

Scheme 17.

types of functional groups, the net result is a facilitation of intramolecular proton transfer. The photoexcitation of methylsalicylate is depicted in Scheme 16.

In addition of methyl salicylate, the hydroxyphenyltriazoles are efficient UV absorbers. These UV absorbers quickly convert the absorbed radiation to ground-state, thermal vibrational energy. When used as compounding ingredients, these compounds are valuable stabilizers that serve to prevent photodegradation of both polymeric substrate and other additives. The photochromic behavior of 2-(2'-hydroxy-5'-methylphenyl)benzotriazole is shown in Scheme 17. Because the deactivation of the excited states of the triazoles and salicylates does not involve visible light absorption, no color change occurs during the tautomerization process.

13.2 THERMOCHROMIC MATERIALS

Thermochromic substances are materials that exhibit reversible color changes within certain transition temperature ranges. The following is a description of one type of thermochromic system. Thermochromic composition type systems usually consist of at least two components: (1) an electron-donating chromogenic material and (2) an electron acceptor. When isolated from one another, the electron donator and electron acceptor are colorless or faintly colored materials. Upon mixing, a strongly colored complex is formed. In the absence of other ingredients the color change is irreversible. The result of the mixing of the two aforementioned components is a permanent image whose lifetime is dependent upon the photo and oxidative resistance of the colored complex formed by the two components. This is the basic principle that operates thermal copying systems.

When a reversible color change is desired, it is necessary to include an additional component. This component is referred to as a solvent in earlier patent literature and as a desensitizer in more recent patents. The melting point of the desensitizer determines the temperature range in which the color change will occur. When the desensitizer melts, the electron donor and electron acceptor of the colored complex dissociate and the color fades. At temperatures below its melting point, the desensitizer does not affect the colored complex.

In practical applications of reversible thermochromic systems, it is difficult to maintain the necessary intimate proximity among electron-donating chromogenic substance, electron donor, and desensitizer. For example, it is possible for one component to evaporate from the system during either processing or storage. Also, application of this thermochromic system to molded plastics will generate thermochromic parts; however, without specific modifications, upon cyclic temperature changes, the repeated liquification of the system will likely cause bloom at some point in time and ruin the part. In addition to the aforementioned disadvantage, the plastic will also possess a rather large color change temperature range. Both undesirable phenomena may be avoided by first microencapsulating the thermochromic system and then incorporating the microencapsulated material into the plastic part.

According to a patent involving the preparation of thermochromic thermoplastic fibers [49], microencapsulated thermally color changeable materials are unsatisfactory because of the following reasons. After incorporation of the microcapsules into the fiber: (1) the resulting fiber is unstable, (2) the microcapsules may be separated from the fabric by rubbing or bending the fabric, (3) the lightfastness of the thermochromic material is poor, (4) the surface of the fiber is uneven, and (5) the uneven fiber surface causes diffuse reflection of incident light, which causes a whitening of the fiber. The problem may be circumvented by constructing the fiber from two separate phases. One phase, phase A, contains a microencapsulated thermochromic system and a non-microencapsulated thermoplastic material with a melting or softening point of 230°C or lower. The other phase, phase B, is a thermoplastic, fiber-forming material such as polyester or polyamide. The two phases are processed in such a manner so that phase A is surrounded by phase B. Phase B functions as a smooth, protective, fiber-forming coating for phase A.

Reversible temperature-indicating devices have been constructed that utilize complex salts such as $Ag_2Hg_2I_4$ and Cu_2HgI_4 [50]. However, the industrial usage of these compounds is limited because of a hysteretically unstable coloration\ decoloration fluctuation. In other words, as a hot, thermochromic complex salt is cooled, the resulting color change does not occur within the same temperature range in which the color change appeared upon heating. As far as the author knows, the number of complex salts that undergo a thermochromic color change below 100°C is limited to only a handful of compounds. For example, in the case of $Ag_2Hg_2I_4$, as the temperature increases from ambient temperature upon heating, a color change occurs at 50°C. Also, Cu_2HgI_4 undergoes a color change from red to brown at 70°C.

Within certain restrictions, unsaturated polyester resins are made thermochromic by the addition of a methanolic Co(II) chloride solution. The unsaturated ester, $CoCl_2$ solution, peroxide catalyst, and accelerator are combined, and the resulting composition is cured at low temperature. The cured resin is colorless at 10°C but deepens in color as the temperature rises, until a blue color is attained at 40°C. This color change is reversible; however, the thermochromism is destroyed by heating to temperatures above 70°C [51].

Recent research efforts [52] involved the synthesis of certain thermochromic antimony and arsenic compounds. It was discovered that the antimony compounds exhibited a gradual color change from green to yellow over the range −200°C to ambient temperature. The arsenic compounds exhibited a color change from yellow to red over much smaller temperature ranges, typically over a 1–6°C range. All color change temperature ranges for the arsenic compounds occur between 116 and 295°C. The color change temperature ranges correspond to the melting point range of the particular compound. A major disadvantage concerning the use of these compounds is that they rapidly degrade when exposed to either moisture or oxygen. A U.S. patent issued to Japan's Agency of Industrial Science and Technology [53] describes an inorganic substance that reversibly changes color from pale green at temperatures above −130°C to pale red at temperatures below −130°C. The thermochromic substance is a composite of strontium and manganese oxides. The composite is synthesized by combining $SrCO_3$ and MnO_2 and heating to 900–1200°C for 15 min in the presence of air. Suggested use of these substances include the visual temperature detection of cryopreservation vessels and temperature and leakage detection in pipelines transporting low-temperature liquids. The materials would be utilized as labels or paints.

Another major classification of thermochromic materials are the liquid crystals. Liquid crystals are organic materials capable of existing as a crystalline solid or as a liquid phase. While in the liquid phase, certain long-range ordering of the molecules occurs. There are two broad classes of liquid crystal phases: (1) the smectic mesophase, in which the molecules form primarily a laminar structure, and (2) the nematic mesophase, in which the long geometric axes of the molecules are parallel. A subclass of the nematic phase, although sometimes categorized as a class unto itself, are the cholesteric liquid crystals, also called the cholesteric mesophase. This mesophase possesses long-range helical order in addition to the linear order of the nematic mesophase. The cholesteric liquid crystals are optically active (chiral) compounds that adopt a planar, uniformly twisted helical structure. The pitch is defined as the distance in which the helix undergoes a 360° turn. Thin films of cholesteric liquid crystal exhibit strong optical activity and selective light reflection. The film may reflect UV, visible, or infrared radiation. When the selectively reflected light is part of the visible portion of the electromagnetic spectrum, the film exhibits color. This selective light reflection and optical activity are due to the helical, long-range order of the cholesteric molecules, and both are directly affected by the pitch of the helix. Often this pitch is strongly temperature dependent, and consequently, the observed color is also highly temperature dependent. Cholesteric liquid crystal films may change color within a temperature range of 10°C. The color change may be positive or negative; that is, as the temperature increases through the color change temperature range, one may observe a positive color change, such as violet to red, or a negative change, such as red to violet. Cholesteric liquid crystals are thermochromic within the approximate temperature range of −10 to 2000°C. The cholesterics are relatively expensive compounds that are easily ruined by chemical contaminants.

Cholesteric liquid crystal films are applied by painting or spraying a solution of the liquid crystal onto a surface. Cholesteric liquid crystals have been used to measure microwave energy, infrared light (such as in surface thermography), visible

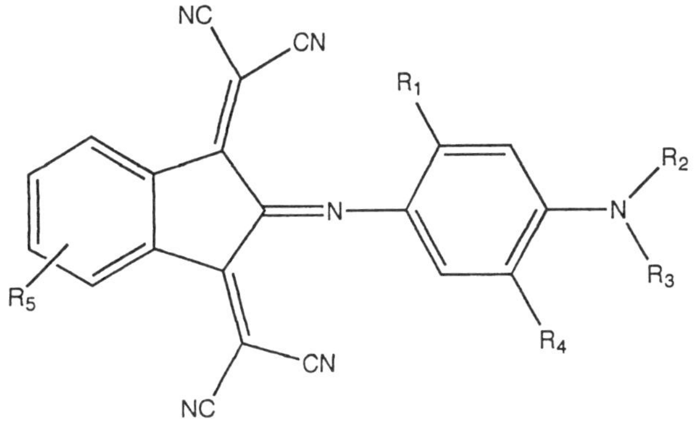

Scheme 18.

light, and laser output and as detectors of small amounts of pollutants in the atmosphere [49].

The following soluble dyes fall into the category of inherently thermochromic dyes. These materials are molecules that are chemically altered by the addition of heat and display a concurrent color change. Researchers at Akzo Nobel have synthesized several infrared-absorbing azamethine dyes that possess the general formula of Scheme 18.

These dyes may be dissolved with virtually any film-forming polymer and a solvent, then cast into single or multilayer films. Because these azamethine dyes undergo an irreversible thermochromic change, it is crucial that the drying temperature is kept below the temperature at which the thermochromic reaction takes place [54]. Ciba-Geigy has recently investigated irreversibly thermochromic azolyl compounds [55].

The previously mentioned photochromic spiropyrans also exhibit thermochromic activity upon melting [56–58]. The melts are usually purple, red, or blue [19] and the color change is reversible; however, the stability of the spiropyran/merocyanine composition is low at the thermochromic temperature transition range. As a consequence, the useful thermochromic lifetime of the spiropyrans is relatively short. Other molecular types that exhibit both photochromic and thermochromic activity are the indenone oxides, anils, spirooxazines, and chromemes.

Several research groups are currently investigating the inherent thermochromism of various polymers. Certain polysiloxanes exhibit reversible thermochromic activity [59, 60]. The thermochromic behavior of these macromolecules is due to order–disorder conformational changes that accompany a particular temperature change. This transition perturbs the electron delocalization of the silicone backbone and results in a shift in the absorption maxima in the UV–visible range.

The poly(3-alkylthiophenes) exhibit a reversible thermochromic change that is due to a transition between low-temperature and high-temperature solid-state structures. The thermochromic mechanism involves the conformation of the alkyl group, which is dependent upon the temperature. At low temperatures the alkyl side chains adopt a fully extended, staggered conformation. As the temperature increases, the population of gauche conformations in the alkyl side chains increases and causes structural disordering in the originally planar, thiophene backbone. This perturbation of the backbone's conjugation length affects the polymer's electronic structure and generates a large shift in the absorption maxima in the material's UV–visible spectra [61, 62].

REFERENCES

1. T. DoMinh, A. Trozzolo, and G. Griffith, *J. Am. Chem. Soc.*, **92**, 1402 (1970).
2. R. B. Woodward and R. Hoffmann, *Angew. Chem. Int. Ed. Engl.*, **8**, 781 (1969).
3. R. S. Becker, R. O. Bost, J. Kolc, N. Bertoniere, R. Smith, and G. W. Griffin, *J. Am. Chem. Soc.*, **92**, 1402 (1970).
4. D. R. Arnold and A. Karnischky, *J. Am. Chem. Soc.*, **92**, 1404 (1970).
5. G. H. Brown, *Photochromism, Techniques of Photochemistry*, Vol. 43, Wiley-Interscience, New York, 1971.
6. A. Padwa and L. Hamilton, *J. Heterocycl. Chem.*, **4**, 118 (1967).
7. A. M. Trozzolo, A. Sarpotdar, and T. M. Leslie, *Mol. Cryst. Liq. Cryst.*, **50**, 201 (1979).
8. H. Durr, *Angew. Chem.*, **101**, 427 (1989); *Int. Ed. Engl.*, **28**, 413 (1989).
9. H. Gross and H. Durr, *Angew. Chem. Int. Ed.*, **18**, 216 (1982).
10. H. P. Jonsonn, Thesis, Universitat des Saarlandes, Saarbrucken, 1985; P. Spang, Thesis, Universitat des Saarlandes, Saarbrucken, 1985.
11. H. Gross and H. Durr, *Angew. Chem. Int. Ed.*, **21**, 216 (1982).
12. G. Hauck and H. Durr, *Angew. Chem. Int. Ed.*, **18**, 945 (1979).
13. H. Durr et al., DOS P 35 21 432.5 (1985).
14. H. Durr et al., DEOS P 35 21 432.5 (1988).
15. H. Starkeather et al., *Ind. Eng. Chem.*, **47**, 302 (1955).
16. M. Aoto et al., *MRS Int. Mtg. Adv. Mats.*, **12**, 219 (1989).
17. E. Inoue et al., *Kogyo Kagaku Zasshi*, **71**, 1228 (1986).
18. R. E. Fox, "Research Reports on Test Items Pertaining to Eye Protection of Air Crew Personnel," Final Report on contract AF41 (657) 215 April (1961) AD440226.
19. R. C. Bertelson, "Photochromic Processes Involving Heterolytic Cleavage," in G. H. Brown, Ed., *Photochromism—Techniques of Chemistry*, Vol. 3, Wiley-Interscience, New York, 1971.
20. H. Ono and C. Osada, U.K. Patent 1, 186, 987, 1970.
21. N. Chu, *Can. J. Chem.*, **61**, 300 (1983).
22. J. Crano et al., "Spiroxazines and Their Use in Photochromic Lenses", in C. B. McArdle, Ed., *Applied Photochromic Polymer Systems*, Blackie & Son, Glasgow and London, 1992.
23. Padwa et al., *J. Org. Chem.*, **40**(8) (1975).
24. R. S. Becker, J. Michl. *Am. Chem. Soc.*, **88**, 5931, 24 (1966).
25. J. Cottam and R. Livingstone, *J. Chem. Soc.*, 6646 (1965).
26. B. VanGemert et al., *Mol. Cryst. Liq. Cryst.*, **246**, 67 (1994).
27. W. S. Johnson and G. H. Daub, *Org. Reactions*, **6**, Ch.1 (1951).
28. R. J. Hart and G. Heller, *J. Chem. Soc. Perkin Trans.*, **1**, 1321 (1972).
29. P. J. Darcy, R. J. Hart, and H. G. Heller, *J. Chem. Soc. Perkin Trans.*, **1**, 571 (1978).
30. J. Whittal, "$4n + 2$ Systems: Fulgides," in H. Durr and H. Bouas-Laurent, Eds., *Photochromic Molecules and Systmes*, Elsevier, Amsterdam, 1990.
31. C. Kirby and I. Bennion, *IEE Proc.*, **133J**, 98 (1986).
32. I. Bennion and A. Hallam, *Radio Electron. Eng.*, **53**, 313, (1983).
33. R. Cush et al., *Electron. Lett.*, **23**, 419 (1987).
34. G. M. Wyman, *Chem. Rev.*, **55**, 625 (1955).

35. R. H. Dyck and D. S. McClure, *J. Chem. Phys.*, **36**, 2326 (1962).
36. D. Gegiou et al., *J. Am. Chem. Soc.*, **90**, 3907 (1968).
37. D. Schulte-Frohlinde et al. *J. Phys. Chem.*, **66**, 2486 (1962).
38. E. Lippert et al., *Angew. Chem.*, **73**, 695 (1961).
39. G. Fischer et al., *J. Chem. Soc. B*, 1156 (1968).
40. C. M. Anderson et al., *J. Am. Chem. Soc.*, **72**, 1263 (1950).
41. H. Zollinger, *Azo and Diazo Chemistry*, Wiley-Interscience, New York and London, 1961.
42. J. G. Victor and J. M. Torkelson, *Macromolecules*, **21**, 3491 (1988).
43. C. D. Eisenbach, *Photographic Sci. Eng.*, **23**, 183 (1979).
44. K. Ishihara et al., *J. Polym. Sci. Polym. Chem. Ed.*, **21**, 1551 (1983).
45. M. Irie et al., *Macromolecules*, **14**, 262 (1981).
46. M. Irie et al., *Macromolecules*, **16**, 210 (1983).
47. T. I. Anzani et al., *J. Chem. Soc. Perkin*, **2**, 903 (1985).
48. K. Ishihara et al., *J. Polym. Sci. Polym. Chem. Ed.*, **22**, 121 (1984).
49. Tanaka et al., U.S. Patent 5, 153, 066, assigned to Kuraray Co. and Pilot Ink Co., Japan October 6, 1992.
50. T. Kito et al., U.S. Patent 4, 723, 810, assigned to Pilot Ink Company, Japan, March 22, 1988.
51. R. W. Heseltine and J. B. Dawson, U.S. Patent 3, 723, 349, assigned to Pilkngton Brothers, Liverpool, England, March 27, 1973.
52. A. J. Arduengo, U.S. Patent 4, 710, 576, assigned to E. I. Du Pont de Nemours, Wilmington, DE, December 1, 1987.
53. M. Kuroda and M. Araki, U.S. Patnet 5, 165, 797, assigned to Japan's Agency of Industrial Science and Technology, Tokyo, November 24, 1992.
54. P. de Wit et al., U.S. Patent 5, 426, 143, assigned to Akzo Nobel, Arnhem, Netherlands, June 20, 1996.
55. W. Fischer et al., U.S. Patent 5, 481, 002, assigned to Ciba-Geigy Corp., Tarrytown, NY, January 2, 1996.
56. J. H. Day, *Chem. Rev.*, **63**, 65 (1963).
57. P. Rumpf and O. Chaude, *Compt. Rend.*, **233**, 1274 (1951).
58. O. Bloch-Chaude et al., *Compt. Rend.*, **240**, 1426 (1955).
59. W. Welsh, Antec' 93, Conference Proceedings, New Orleans, LaMay, 1993, Vol. 1, pp. 298–300.
60. E. Karikari et al., *Macromolecules*, **26**(15), 3937 (1993).
61. K. Tashira et al., *Synthetic Metals*, **55–57**, 321 (1993).
62. M. Ahlskog et al., *Synthetic Metals*, **57**(1), 3830 (1993).

CHAPTER 14

Metallic Pigments

Bernhard Klein
ECKART-Werke
Kaiserstrasse 30
Fürth, D907763 Germany

Translated by
Hans-Henning Bunge
Eckhart Americas, L.P.
32644 Luke Road
Avon Luke, OH 44012-1646

14.1. INTRODUCTION

Metallic pigments belong to the oldest known class of special-effect pigments (Fig. 14.1). In early Egyptian times gold dust, a by-product in the production of leaf gold, was incorporated into varnishes and used to beautify wood and stone. Gold dust was also used in letterpress inks for expensive book editions. With increasing demand, it became of interest to use less expensive pigments such as silver, tin, and copper and its alloys. Before the end of the nineteenth century these less expensive pigments based on copper and brass started to be produced on a large scale mainly for use in inks.

At the beginning of this century copper alloys dominated the metallic pigment market. As soon as aluminum pigments could be mass produced, around 1920, they quickly became very popular. They combine chemical inertness, low density, and ease of handling together for use in a wider range of applications. In addition to

---

Coloring of Plastics, Fundamentals, 2nd edition. Edited by Robert A. Charvat
ISBN 0-471-13906-8 Copyright © 2004 by John Wiley and Sons, Inc.

Photos of Plastic Parts

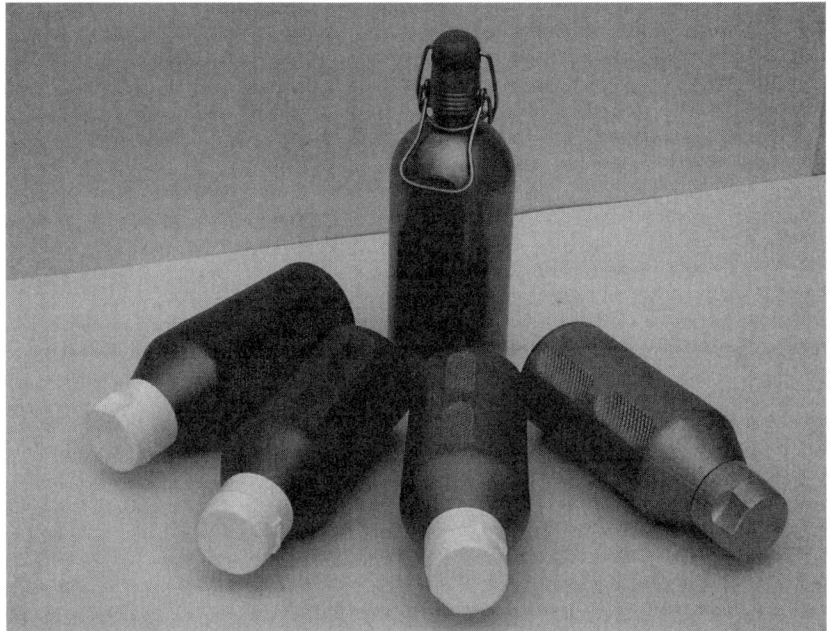

Figure 14.1. Metallic colored moldings.

their optical effect, they displayed various technical advantages [1, 2], that increased their demand. These advantages included electric and thermal conductivity, increased density of plastics modified with metallic, electromagnetic shielding, and corrosion protection of metal structures.

14.1.1. Difference between "Regular" Pigments

Pigments generate light scattering at the interface of the particle and the vehicle. This is also true for most special–effect pigments. However, the difference in metallic as compared to pearlescent pigments is that the light is scattered by a larger pigment surface area in the form of oriented reflection. Metallic pigments act as microscopic mirrors. Therefore, the metallic effect is only possible at the presence of lamella light-reflecting particles and strongly is dependent on the angle between the light source, particle orientation, and observer. Pearlescent pigments demonstrate their special effects through light interference and reinforcement.

14.1.2. Optical Effect

Metallic pigments are mainly used to achieve a metallic appearance of a nonmetallic object. Especially at the end of this century the imitation of metal became important, and starting in the 1950s colored metallic effects in automotive coatings became very popular all over the world.

14.1.3. Nonoptical Effects

With the increasing importance of plastics in this century the nonoptical effects of metallic pigments gained importance. One application is the complete optical barrier function (at a certain minimum thickness in the micrometer range) of metallic pigments for visible infrared (IR) and ultraviolet (UV) radiation. This includes the protection against heat as well as light, particularly in the UV range. For these applications mainly aluminum is being used. For protection against the shorter wavelength of nuclear radiation, pigments based on heavy metals such as lead are applied.

Also gaining in importance are the electrical and thermal conductive properties that can be achieved by adding metallic pigments to otherwise well-insulating plastic. The good conductivity of metallic pigments permits their particles to serve as an antenna by increasing the absorption of microwaves of plastics. To optimize performance, the following factors need to be coordinated: particle size, dielectrical properties of the substrate and wavelength. This is important if these factors approach the electromagnetic resonance of the absorption at its peak.

Other important topics are the antistatic properties and electromagnetic shielding (EMI). Adding special metallic pigments to the plastic can in some cases generate both effects. The main influencing factors are in this case the pigment surface, the specific metallic surface, and the pigment volume concentration.

14.2. PRODUCTION

14.2.1. Raw Material

The raw material for producing metallic pigment is copper and its alloys and part aluminum ($\geq 99.7\%$). These metals are bought as bars weighing a number of pounds each and are therefore mostly reduced in size to ingots. These are generally melted down and then sprayed (i.e., atomized) with the help of gas or converted to thin foils (Fig. 14.2). For the production of pigments atomized material is preferred. The shape of the particles and their size distribution can vary depending on the process, the chemical reactivity of the metal, and the gas used for the spraying process. For certain applications, foils are preferred. The shape of the atomized material is strongly influenced by the processing gas. Air will cause oxidation of the metal surface, particularly for aluminum, and prevent the formation of the thermodynamically more favorable spherical shapes (Fig. 14.3). When using inert gases like helium, the particles will approach an almost perfect spherical form before they harden. These spherical particles can be used for either high-quality pigment flakes or a special types of spherical metal pigments.

14.2.2. Pigment Production

The formation of metallic pigments takes place in different types of milling equipment. The type of mills, the quantity and size of the milling media, as well as the volume in the mill influence the pigment particle characteristics. In addition, pro-

Atomizer in Operation

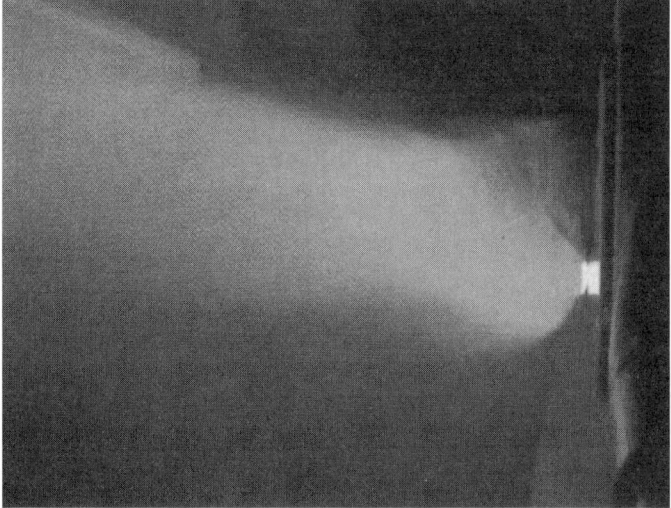

Figure 14.2. Atomizer in operation.

Atomized Aluminum Powder

Figure 14.3. Atomized aluminum.

cessing variables such as milling time, temperature, and rotating speed of the mills influence to a large degree the appearance of milled material and represent the know-how of the metallic pigment manufacturer (Fig. 14.4).

The milling process is done wet as well as dry. The wet milling process has improved the safety of the manufacturing process, particularly for aluminum pigments. A dust cloud of small aluminum particles can be explosive!

Spherical Aluminum Powder

Figure 14.4. Spherical atomized aluminum.

By using modern dry milling processes for aluminum pigments in the presence of inert gases and also various precautionary steps during their final application, their handling has become safe.

The same techniques also permitted the development of pigments with a wider range of optical properties.

Dry Milling

For the dry milling process, the atomized metal is separated in suitable fractions to which fatty acids are added to prevent the fusion of the particles (Fig. 14.5). The milling is done either in a continuing process, and the oversized particles are fed back into the mill, or by the batch method. The particles are separated by size after the grinding process is completed.

Wet Milling

In the classical method atomized aluminum is dispersed in mineral spirits and in the presence of fatty acids. The atomized aluminum is flattened out or actually ground by the milling process (Fig. 14.6). The resulting flakes offer the typical metallic appearance.

The slurry is then separated by particle size and the mineral spirits squeezed out by filter presses. The resulting press cake contains 70–90% metal pigment and undergoes the following additional steps:

- Removal of the remaining mineral spirits results in pigment powder.
- Addition of aromatic solvents or mineral spirits results in the typical pastes applied in coatings (primary pastes).
- Exchange of the mineral spirit for other solvents or plasticizer results in mineral spirit–free material in the form of dust-free crumbs.

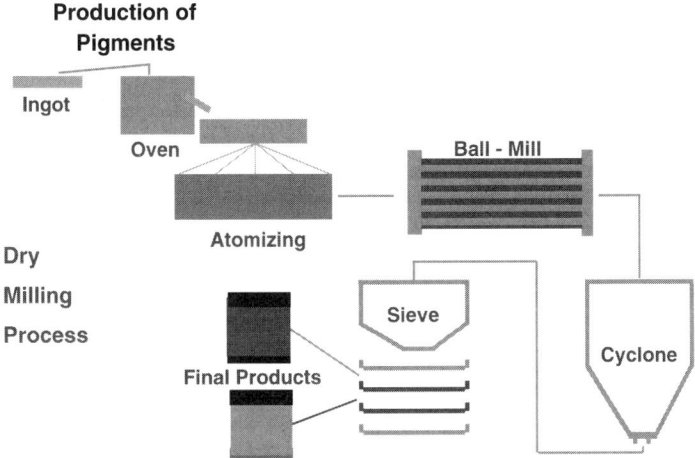

Figure 14.5. Production of metal pigments, dry process.

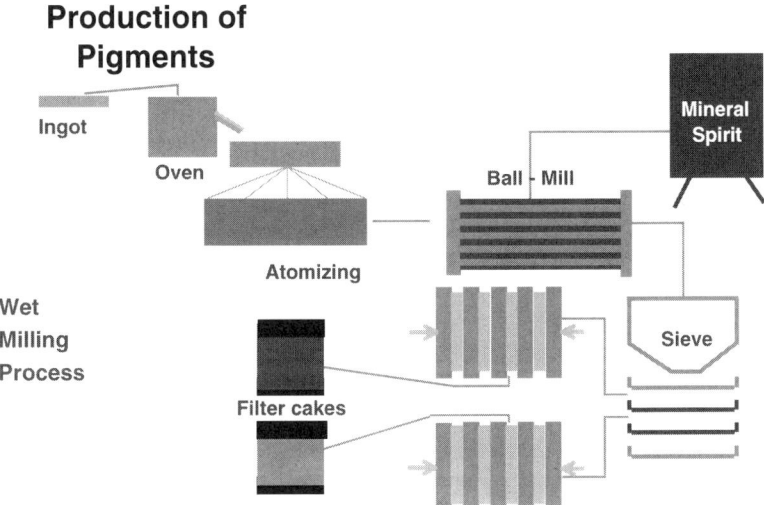

Figure 14.6. Production of metal pigments, wet process.

Mills can be distinguished between the commonly used horizontal classic types and the more effective modern versions based on a stirring action. The latter are also known as pearl mills and are used for more difficult to disperse pigments. The traditional hammer mills are being phased out because of their unsafe operation.

Pellet Production

Since the middle of the 1980s metallic pigments have been available as pellets, which offers a number of advantages. They are safer to ship than solvent containing pastes or potentially explosive powders. Because they contain less carrier, the volume of

208   **Metallic Pigments**

Figure 14.7. Ball mill for wet grinding.

Metallic Pigment Pelletizing

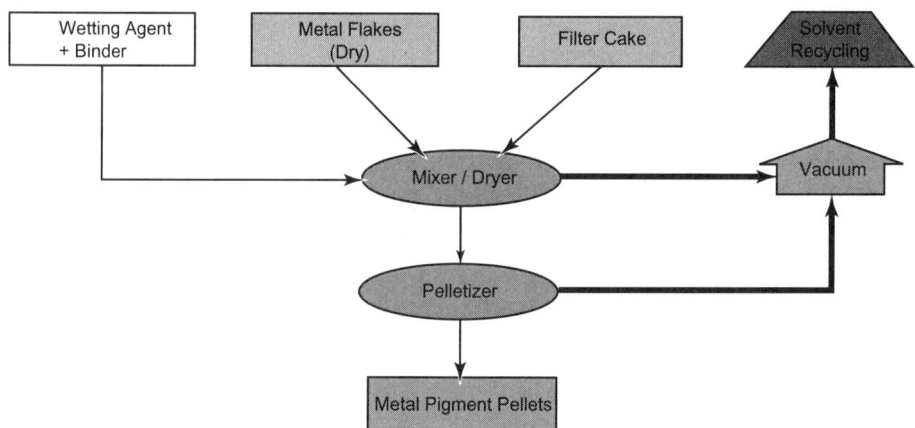

Figure 14.8. Metallic pigment production, pelletizing.

pigment to be handled is reduced as well as the amount of foreign material introduced into the final system.

The first pellets contained only 70% metallic pigment while pastes were already available with 80%. More recently developed pellets contain 80–90% pigment, which can in general be universally applied in different resin systems similar to powders. These concentrated pellets display optimized pigment dispersion. This is an advantage for the concentrate maker because equipment is designed for optimum shear to achieve the best possible dispersion of regular colorants, which however will damage the flake of the metallic pigments.

The pelletizing process is illustrated by the flowchart in Figure 14.8. The pellets can be prepared from press cake or pigment powder. In some processes temporary flow agents are added, which will evaporate.

Cornflake Pigment

Figure 14.9. Aluminum flake, corn flake type.

Cutting Procedure

A totally different method is used to produce very large size flakes or glitter. The raw material is polished aluminum foil; for very brilliant products, aluminum is deposited on plastic film by vacuum metallizing. In a second step the metal surface is coated with protective clear ink; to obtain colored glitter, colored transparent ink containing dyes or pigments is applied.

The film is then cut in two directions to obtain square or rectangular flakes. For flakes with a more integrated pattern (stars, hearts, etc.), the foils are stenciled. This process results in more waste.

14.3. PIGMENT SHAPES

14.3.1. Irregular Shapes (Corn Flake)

Irregular-shaped pigments are produced from irregular-formed atomized material (Fig. 14.9). They have a high aspect ratio (diameter/thickness) and excellent opacity. Their irregular borders contribute substantially to the light scattering, which causes a certain amount of a grayish haze.

Very fine aluminum pigments display a very grayish silver effect if they do not have a chance to orient themselves parallel to the surface, as is often the case in injection-molded parts. They are preferably applied in film and coextrusion processes because of their very high opacity, ability to orient, and reflectivity of light.

For especially coarse aluminum the pigment manufacturer tries to achieve very thick flakes to improve their resistance to shear during processing in plastics. To achieve the desired metallic appearance with these coarse pigments, manufacturers require a higher pigment concentration.

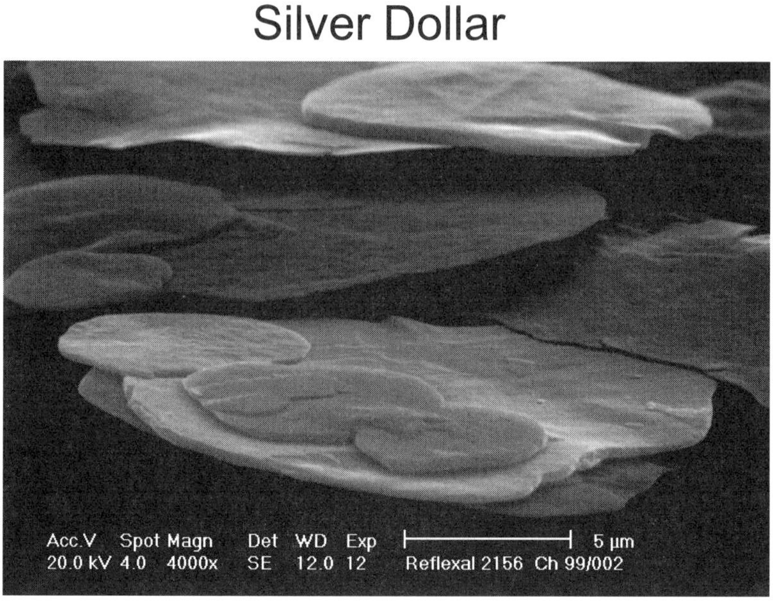

210 Metallic Pigments

Figure 14.10. Aluminum flake, silver dollar type.

14.3.2. Lenticular Shape (Silver Dollar)

Lenticular-shaped aluminum pigment are obtained by selecting spherical atomized aluminum with a narrow-particle-size distribution (Fig. 14.10). Modifying the grinding conditions ensures that only flakes are formed and the aluminum is not actually reduced in size. These pigments have a smooth border and contain only a small amount of fines. As a result, they are very brilliant and have a clean shade because the grayish cast caused by light scattering and fines is removed. Frequently they have a low aspect ratio (diameter/thickness) reducing their ability for orientation in plastics.

14.3.3. Spherical Shapes

A special class spherical pigments are made by polishing selected atomized material (Fig. 14.11) [3]. Their optical properties differ totally from metal flakes, because only a very small area (one point) of their surface reflects light like a mirror. A plastic molding colored with spherical pigments therefore appears gray with a scattered sparkling effect. A useful application is only possible in very transparent plastics of relatively large thickness to achieve a better three-dimensional effect. However, this means that these pigments do not have the same metallic effect as are commonly associated with aluminum pigments.

It was hoped that spherical pigments would avoid the isotropic effect along the flow and weld lines and their the appearance caused by flakelike pigments. This expectation by the manufacturer and user, which led to their development, was not fulfilled.

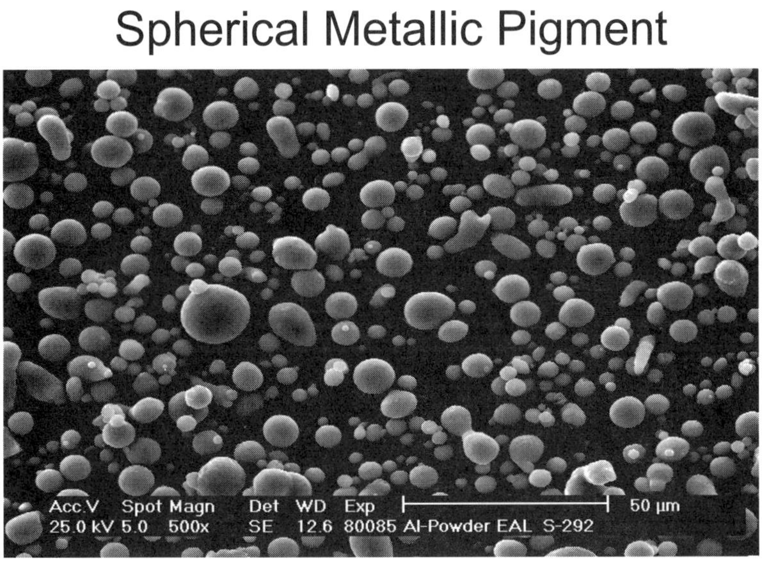

Spherical Metallic Pigment

Figure 14.11. Spherical metallic pigment.

The spherical shape has additional drawbacks. Because a sphere has the smallest possible surface area of all geometric shapes, its ability to develop opacity/hide hide is low. Another difference is in settling behavior. A sphere will settle in a straight line while flakes move back and forth like a leaf. This effect can mainly be observed in reactive resin systems where enough time during the fluid stage is available to result in concentration differences of the aluminum in different sections of the part.

14.3.4. Glitter

Glitter-type pigments can be produced in various shapes with narrow tolerances (Fig. 14.12). In one type of application they are sprinkled onto a gluey surface since they are relatively well protected by their own coating. For application in plastics they have one distinct disadvantage. The protective coating of the more economical versions are based on polyurethane or epoxy resin. For the more expensive products clear, very transparent, highly crosslinked epoxy resins are used, which have low flexibility and low heat resistance. The maximal heat resistance of the economical glitters is 230°C and that of the expensive version 250°C depending on color. The resistance to molten, very polar resin systems such as polyamide leaves much to be desired. They are therefore only suitable for low-melting plastics such as polyvinyl chloride (PVC), low-density polyethylene (LDPE), (Thermoplastic polyurethane), and (unsaturated polyester) resins and can only be incorporated with very little shear. Only if the coating does not contain any colorant can a separation from the aluminum surface be tolerated because the coating is then invisible. For this reason these pigment types will find no further mention in the following section, in particular since the classical aluminum pigments increasingly offer an equivalent and more economical replacement.

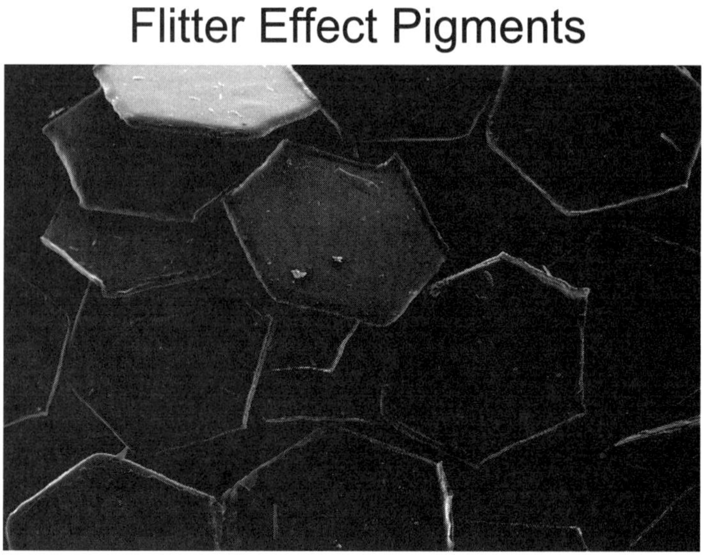

Flitter Effect Pigments

Figure 14.12. Metallic glitter.

14.3.5. Surface Treatment

Because the metallic effect is strongly influenced by the pigment surface, its modification has long been a concern. Also, the need for improvement against chemical and physical influences forces the pigment manufacturer to constantly modify the surface treatment to optimize its performance. This includes mechanical aftertreatments as well as physical or chemical applied layers.

Polishing

An important method to increase the gloss of metallic pigments is polishing. The main aims of polishing the pigments are to eliminate any surface unevenness and reduce the scattering of light resulting in high directional reflectance.

Inorganic Treatment

The most popular inorganic treatments are those with silica [4], phosphates [5], molybdate [6], and chromate [7], all of which serve mainly to reduce the chemical reactivity of the metallic pigments (Fig. 14.13). This includes corrosion inhibition of the pigment and inhibition of its often catalytic effect as well as a change in the physical surface characteristics, for example, where chargeability is important.

A totally different method is being used if the aluminum pigments are coated with colored metal oxides. By this method very heat resistant, colored metallic pigments can be produced of various shades depending on the applied metal oxide. These shades are less intense and less brilliant [8] if compared to aluminum pigments colored with organic colorants [9].

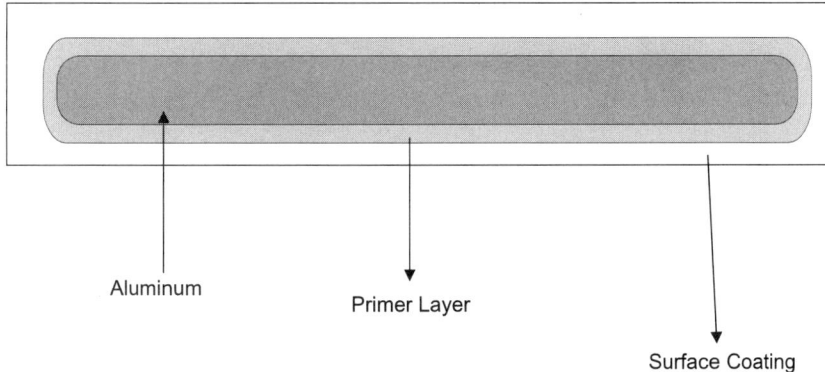

Figure 14.13. Coated flake.

Organic Surface Treatment

The organic coating can best be visualized as a paint coating of metal flakes with systems drying by solvent evaporation or crosslinking for more demanding applications. For some high-performance pigments more than one coat is applied. The first layer is usually adhesion promoting and is then top coated with a highly crosslinked, very resistant lacquer [8].

14.3.6. Supply Form

For the end user the appearance of the pigment is of interest as well as how it handles under specific conditions. For this reason for many years a variety of versions have been available that combine ease and safety of handling. Lately product cost has become more influential than in the past when selecting supply forms. However, special effects also play a very strong role which often makes cost secondary if it means gaining a market advantage.

Powder

Initially metallic pigments were mainly supplied as powders for reasons mentioned earlier (Fig. 14.14). In particular, for aluminum powder, with its high affinity for oxygen, safety became a major safety issue.

Metal powder is not easily incorporated into plastics without damaging the flakes. The poor flow characteristics of flakes make it more difficult to feed them into an extruder. Also they stay suspended in air for long periods, which presents a safety hazard.

Pastes

Pastes are manufactured either by dispersing the powder into a liquid carrier or by increasing the solid content of the ground pigment slurry by various methods.

Metallic Pigment Powder

Figure 14.14. Metallic pigment powder.

PRIMARY PASTES. For the pigmentation of conventional coating systems usually pastes based on mineral spirits are used. To these presscakes produced by the wet grinding process (Section 14.2.2.), aromatic or aromatic-free solvents are added to adjust the aluminum pigment concentration to approximately 65%.

SECONDARY PASTES. A considerable step forward was achieved with the development of secondary pastes. They are manufactured from dry ground or wet ground pigments from which all mineral spirits are removed. They are replaced with carriers such as special solvents, low-viscosity binders or plasticizers. By this method from a few base pigments a wide spectrum of pigment pastes can be prepared tailor made for specific needs. Their pigment concentration ranges for application in coatings or inks from 60 to 75% and for plastics from 80 to 90%.

Pellets

The latest advances in pigment preparations are pellets (Fig. 14.15). They represent the consequential step from the secondary pastes. The main difference is that at least part of the carrier is solid at room temperature or is converted to the solid state by chemical or physical means. The most common carriers are waxes, oligomeric homo- and copolymers, and surfactants.

14.4. PROCESSING

The incorporation of metallic pigments into thermoplastic or duroplastic resins differs little. In this section, only unique differences related to a specific material

Aluminum Pellet

Figure 14.15. Aluminum pigment pellet.

will be mentioned. The best effect of all metallic pigments is achieved in very transparent, clear resins.

It should also be noted that most metal detectors used during production will respond to very low metallic pigment concentrations and therefore need to be turned off, which in turn requires special handling.

14.4.1. Masterbatches

All ingredients of a formulation, such as polymers, additives, extender and reinforcing material, and colorants as well as black and white pigments, should be plasticized and dispersed before the addition of aluminum pigments. The aluminum pigment in the form of predispersed pellets or pastes is incorporated into the batch to obtain a uniform blend with no additional dispersion.

In a multistage process the metallic pigment should only be added during the last phase. When using a twin-screw extruder the metallic pigment should be added, if possible, to the front third section. The screw should be designed not to shear any more at that point. If the blending becomes a problem, a static mixer with little shear is recommended. Also the increasingly popular Buss Kneader [10] can be used for producing metallic batches because it can be adjusted to generate little shear but good mixing [11]. High shear causes the flakes to be destroyed or bend. Both increase the gray haze and loss of brilliance.

14.4.2. Compounding

As in Section 14.4.1, all ingredients are predispersed and the already dispersed pellets or pastes of the metallic pigments are then blended with the compound to

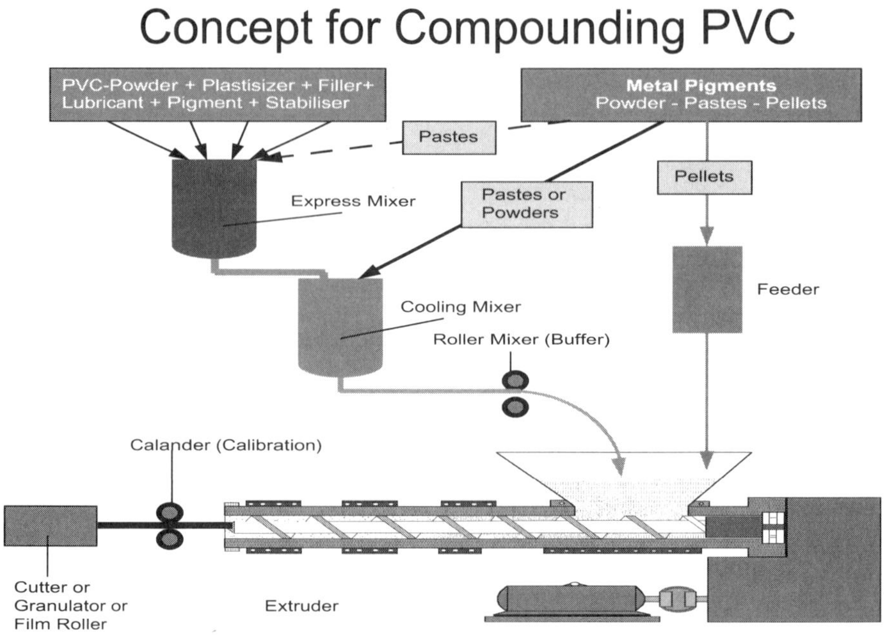

Figure 14.16. Concept for compounding PVC.

form a uniform mix. In a multistage process the aluminum needs to be added during the last stage. When using a twin-screw extruder, it should be added in the last third of the barrel, possibly through the vent opening. If a homogenous mix cannot be achieved, the blend can be mixed further in a low-shear blender. Sometimes increasing the temperature in the middle section of the extruder helps.

The pigmentation of Polymethyl Methacrylate (PMMA) and ultrahigh-molecular-weight polyolefins presents familiar difficulties. Due to the high difference in viscosity between the pigment concentrate and these polymers, it is sometimes necessary to use special masterbatches based on high-molecular-weight carriers.

14.4.3. Polyvinyl Chloride Coloration

Because of its importance and different handling requirements, PVC will be discussed separately. Polyvinyl chloride powder is mostly dispersed in a high-speed blender together with fillers, colorants, possibly a plasticizer as well as a stabilizer, and other additives. During this process the material will reach temperatures of 120°C and above. Metallic pigments should be added at the end and homogeneously distributed by the slower running follow-up blending units (e.g., cooled blender, two-roll mill, extruder) (Fig. 14.16).

The PVC Emulsion types (containing up to 2.5% emulsifier) and those containing filler require a higher pigment loading because of their opacity. Since the emulsifiers are mostly alkaline, it is recommended to use silica-encapsulated metallic pigments to ensure consistent quality of the final product.

When extruding PVC, it is often preferred to meter the masterbatches or monobatches of the metallic pigment directly into the extruder.

The universal pigment concentrates offered by some pigment manufacturers contain 50% to more than 95% metallic pigment. They can be added directly for a final pigment loading of about 1% and will in most cases keep the processing cost down and result in a brilliant product.

14.4.4. Continuing Coloration

To a continuing process the pigment dispersions are added directly to the plasticising equipment by volumetric or gravimetric measuring units. Further blending can then be done on a single- or twin-screw extruder.

The second most popular method is dry blending of batches. The blend of metallic concentrates is fed directly into the processing equipment. When using pigment powder of middle to coarse particle size, mixing can become a problem at a high pigment loading. The addition of 0.5–1% adhesive promoting additives can sometimes help to bind the metallic pigment to the resin. Because pigment concentrates (pellets) as well as solvent-free pastes are already well dispersed, the plasticizing action of the processing equipment is usually sufficient to achieve a uniform blend. In critical cases the pastes are first diluted (1:1) with the polymer before reducing them to the final concentration.

Injection Molding

To optimize the metallic effect when using an injection-molding machine, a higher screw speed, a longer plasticising period, sufficient backpressure, and a suitable screw design are prerequisites. The pigments are added as ready-to-use compounds, masterbatches, or powders.

The use of screens as well as torpedoes can improve the blending process. The use of hot-runners can be critical because the needle gate can cause material of different temperatures to merge, which will cause flow lines. A proper mold design can avoid this effect. Also, the size of the gates should be sufficiently large to avoid their clogging with coarser pigment particles.

(Co)-Extrusion

By this process it is possible achieve metallic effects similar to painted surfaces. Its popularity is therefore likely to increase in the near future. Specifically, extruded semifinished product displays optical qualities that are in some cases superior to a painted surface.

Calendering

For the calendering process pastes or pigment powder can be used under strict observance of safety regulations (e.g., explosion-proof areas). When using pellets, the material should first be plasticized to avoid the danger of imprints by the mill rolls. The plasticized roll, which forms on top of the mill rolls, ensures a good blending of the pigment lengthwise. For a good general distribution the plasticized material should be folded a few times.

When using gold bronze pigments the trimmed material should not be reprocessed because their increased exposure to heat oxidizes the pigment and will affect the shade. In addition, formation of Cu ions causes the development of HCl, which in turn results in cumulative double bonding of the PVC affecting its performance.

Blow Molding

Both injection blow molding and pinch tube processes are especially suitable for metallic pigments, because the expansion of the plastic during the blowing process causes further orientation of the metallic pigment parallel to the surface.

To optimize the metallic effect, the conditions outlined in Section 14.4.4 under Injection Molding should be maintained. The pigments are added as compound, masterbatch, or powder. The use of static mixers or special mixing screws can improve the homogeneity of the blend. For the pinch tube process the high melt viscosity should be considered when preparing the masterbatch.

Vacuum Thermoforming

Vacuum thermoforming is very suitable for metallic effects. For this application extruded, coextruded, and calendered plates with a highly reflective surface are being used. The already oriented metallic pigments are further oriented during the forming process. The difference in expansion of the surface segments will improve the optical appearance further. Depending on the process, products based on coextruded material with a clear surface layer can approach the optical effect of coatings with superior gloss. When using weather-resistant plastic and colorants with excellent outdoor durability, high-quality vacuum-formed parts can be obtained for outdoor use, including automotive applications [11].

Insert Molding

Insert molding, or in-mold decoration, helps overcome the flow lines common to metallic colored intricate injection-molded parts. By this process a film with a uniform orientation is thermoformed and fused with the surface of the injection-molded part.

Films are either coextruded [12] or printed on the back of a very transparent material. The binder of these screen inks or gravure inks has to have a high melting point or crosslinking to form an elastic film to avoid being replaced by the hot injected polymer. It is also possible to coat the backside of the film with an adhesion promoter, which prevents direct contact between the printed area and the polymer. This coating can also improve the adhesion between the film and the injection-molded part [13].

Laminating

Laminating lends itself to the manufacture of resilient products with a large flat surface area and excellent surface properties. These products can be converted to their final form by thermoforming and insert molding. This method avoids the use of paint with its negative side effects, that is, the influence of solvents on the physical properties of the injection-molded part (tension and plasticizing effect). The lamination process also circumvents the optical problem of flow and weld lines.

Coating and Printing

These are special technologies only somewhat related to plastics but with substantial importance for the use of metallic pigments. The most common resins used for these applications are PVC, thermoplastic and thermosetting polyurethane, and acrylics. Silicone resins play only a minor role in the application of metallic pigments. The coatings are usually prepared with a high-speed mixer or dissolver. In both cases the shear should be kept to a minimum. Particularly when using gold bronze pigments, special attention should be paid to the pot-life and aging stability. In some cases aggressive chemicals in the coating system affect the metallic pigments.

The film is continuously coated from the front or backside by rolls or knife coating. For small-volume items frequently gravure or a roller coater is used. All these applications result in an even orientation of the metallic flakes parallel to surface. The frequently used spraying technique has a disadvantage in this case because, due to its high viscosity, the metal pigment shows only little orientation.

In very large volumes a roller application of a melted coating is used. This technology is closely related to the calendering of film and their lamination onto a textile carrier, as described in Section 14.4.4 in the discussions on calendering and laminating.

Pressing

The pressing processes do not provide any dispersing action for the metallic pigments. The premixed components are only brought into the desired form and then sintered (melted). The homogenous appearance therefore depends totally on the preblend of all powdered components.

Reactive Resin Systems

In reactive systems it is important to consider the chemical interaction between the metal and the resin components. In many cases the catalytic properties of the metal can alter the reactivity of the resin systems.

In certain cases the acidity Melamine Formaldehyde resins (MF resins) or alkaline Polyurethane (also known us PU) resin (PUR resin) and alkali hardener (epoxy resin) can corrode or even totally dissolve the metallic pigments. In case of bronze pigments a color shift is possible. For these applications special metallic pigments are recommended, which are protected by an organic or inorganic coatings [14].

14.5. APPLICATION PROBLEMS

The chemical behavior of metallic pigments and their physical appearance require some specific ways as to how they are incorporated and applied. They are mostly distinctly different from conventional methods used for handling nonmetallic pigments.

14.5.1. Incorporation

Metallic pigments should be incorporated with as little shear as possible to preserve their flakelike structure and to ensure an optimum metallic effect.

Mixing Time

The mixing time should be as short as possible not exceed 10–15 min. If the mixing unit is equipped with a heating/cooling system, the metallic pigments should only be incorporated during the cooling stage. If this is technically not possible, they should be added just before the cooling phase. When using pellets, the mixer should be heated to a temperature of at least >100°C and long enough to obtain a homogeneous blend.

Maximum Mixing Temperature

When incorporating bronze pigments, the following temperatures should not be exceeded:

Regular bronze pigments	190°C
Heat-stable bronze pigments (silica coated)	260°C

For aluminum pigments the temperature is less restrictive. In this case their chemical reactivity needs to be considered, particularly with halogen-containing polymers. In extreme cases, if halogen is released, a strong reaction can be anticipated (Wuerz synthesis [15]).

Humidity

In general, metallic pigments are not hygroscopic. The plastic resin should be dried together with the pigment at 60–80°C where necessary. In critical cases (e.g., polyester) the drying can be done under vacuum.

Dwell Time

The dwell time of regular bronze pigments in the processing units should not exceed 5 min. It is recommended the shortest processing time for optimum gloss, brilliance, and homogeneity be determined.

Pigment Loading

The recommended pigment loading (in relation to the polymer) for plastic compounds is as follows:

Pigment Type	Size (µm)	Percent Loading
Fine aluminum	5–30	0.01–1.0
Medium fine	30–60	0.5–2.0
Coarse	>70	2.0–5.0
Fine gold bronze	8–30	0.3–1.0
Coarse gold bronze	30–80	1.0–3.0

It is recommended to optimize the loading for each specific item. Too much pigment not only increases the cost but may also reduces the impact resistance.

Dispersion

Problems regarding dispersion or uniform blending can be solved by

- replacing the standard screw with a mixing screw,
- reducing resin viscosity,
- using static mixers or torpedo inserts,
- extended mixing or plasticizing time, and
- using double-screw extruder or Buss-kneader in place of single-screw units.

Increasing the backpressure, the torque, or even the plasticizing time can also improve the dispersion.

14.5.2. Molding

The anisotropy of the pigment particle causes orientation along the flow pattern. If the flow pattern is parallel to the surface, an increased brilliance is achieved even to a degree that a flop effect can be observed, as in painted objects. On the other hand, flow disturbances and the merging of resin of different temperatures are frequently accentuated by metallic pigments through flow and weld lines.

Reduction of Flow Lines

- Flowlines are reduced by large-size metallic pigments and high pigment loading.
- The higher the viscosity of the polymer the less is the pigment orientation, the fewer the flow lines.
- Dark shades, particularly those containing carbon black, mask the flow lines.
- A large gate opening and a high injection velocity create turbulence in the mold, which reduces the visibility of flow lines. A film gate offers the best results and a bar gate is preferable to a pin gate. For tubelike moldings a cone gate is of advantage. However, they have the disadvantage that the molded part requires cleaning, while submarine gates usually do not cause this problem.
- Optimizing the wall thickness and placing the gates at the most advantageous location when designing the mold will often eliminate optical imperfections or reduce their visibility.
- Large parts with long flow paths require many gates. Here cascading gates (sequential injection molding) will exclude weld lines.
- For complex items or a mold with multiple cavities and if optical perfection is required, counterflow injection molding or the similar SCORIM[1] process will help to eliminate flow lines.
- A higher resin temperature and a heated mold will prolong the flow time of the plastic, including layers close to the surface. This will enhance the disorientation of the aluminum flakes and thereby the optical appearance of the molded part.

---

[1] SCORIM stands for Shear Controlled Orientation in Injection Molding, a technology offered by British Technology Group.

222 Metallic Pigments

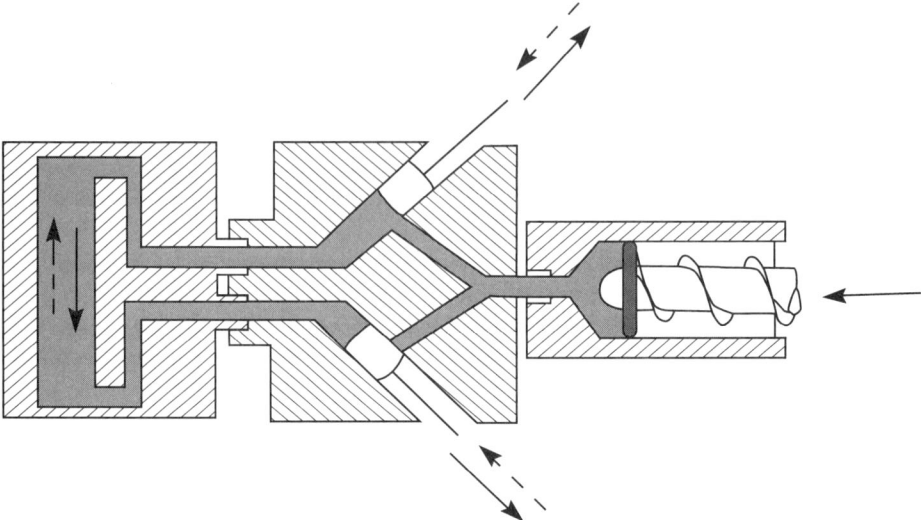

Figure 14.17. SCORIM, phase A.

- Often, by combining two or more of these remedies, it is possible to obtain acceptable parts, even in difficult cases.

THE SCORIM PROCESS. It is possible to attach the SCORIM unit to the injection molder to manipulate the filling process of the mold so that in the area of the weld lines the resin is mixed [16]. This will improve the structural strength substantially and at the same time the unwanted orientation of metallic pigments is prevented. This process typically requires a 10–15°C higher temperature of the resin and the mold. The process is based on a prolonged flow ability of the polymer.

The SCORIM unit is placed between the nozzle of the injection molding machine and the mold (Figs. 14.11–14.13). The gates (i.e., the sprues and runners) are located so that all potential weld lines are placed between them from the flow point of view.

With this technology a substantially improved surface quality can be achieved of injection-molded parts containing lamellar pigments and fibers.

During phase A the main portion of the plasticized polymer is injected into the tool. Only then does the SCORIM unit start its function (Fig. 14.17).

At the end of this injection process (80–90% filled), the two pistons will hydraulically separate the mold from the nozzle. The missing amount of polymer is now added from the small cylinders, countering the phases in the mold and the different molten Section 5 of the plasticized plastic is moved back and forth two to three times over the theoretical weld lines. After completion of this mixing process the cylinders are synchronized (phase B) (Fig. 14.18).

Through a compression and decompression action the density of the mold contents is increased, and the more intimate contact with the mold walls improves the heat reduction. The surface of the molded part is substantially improved and voids as well as sink marks disappear. The improved dissipation of the heat reduces the cycle time up to 15% depending on the polymer.

During the holding pressure phase (phase C) the two cylinders compensate for the shrinkage of the plastic because the screw is hydraulically disconnected from the mold (Fig. 14.19) [17].

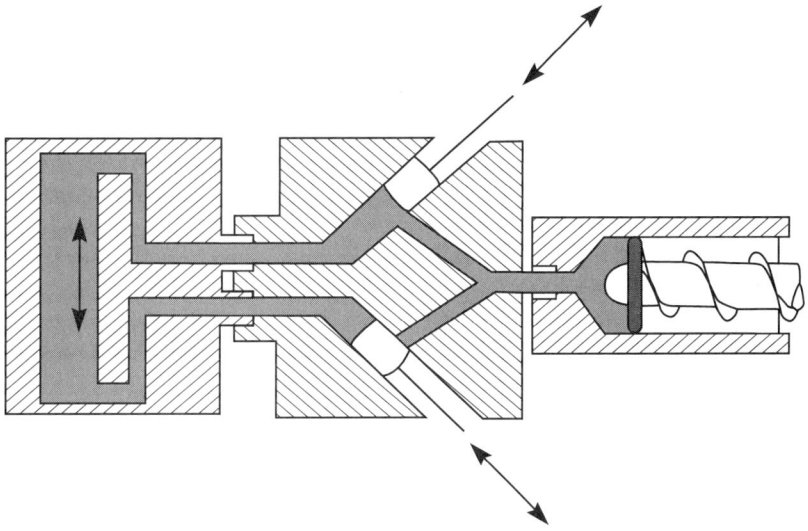

Figure 14.18. SCORIM, phase B.

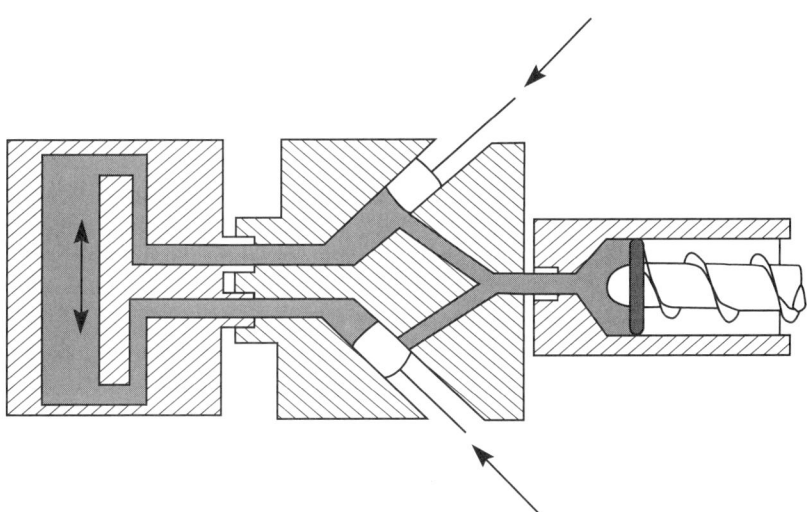

Figure 14.19. SCORIM, phase C.

Starting at the end of phase A, the mold and the injection molding machine are separated, allowing each polymer to be plasticized according to its specific needs.

MOLD DESIGN. The design of the mold tool is very important when using metallic pigments [18]. This includes the mold, the design of the sprue bushing their number and locations. Also critical for the appearance is the relation of wall thickness and flow pattern. If possible, the weld lines should be transferred to the back of the item or to their edges. A rough surface of the mold not only improves the

Table 14.1. Technical Application

Antistatic, electrical conductivity, electromagnetic interference	Electronic units, parts to be electrostatic coated, boxes and encasements for explosion proofed area for the chemical industry
Increase of density	Writing utensils, weights, toys, stands
Light protection (optical barriers, uncolored light filters)	Film, awnings for light protection, technical spectacle frames, coextruded multilayered films
Substitutes for metal and cast iron	Engine parts, tool boxes, valve caps, oil pans, intake pipes, cooling units, ornaments
Microwave absorber	Sectioned polyethylene Terephthalate (PET) parts with varying heat demand, evaporation promoter
Thermal conductivity	Reduced cycle time, substitution of heat conducing parts
Thermal radiation protection	Foams for thermal protection of buildings, heating pipes, thermal barrier, films for roof insulation, shutters for heat protection

Table 14.2. Decorative Application

Utensils for the office, kitchen, and household	Writing utensils, cans, boxes, handles, and bowls
Electrical units	Vacuum cleaner, switch boxes, handles, knobs, electrical kitchen utensils
Chassis parts for cars	Hub caps, covers, bumpers, spoilers, handles, rails, trims
Electrical units for the home	Plastic casings for the audio and video units, remote controls
Metal and cast iron substitutes	Ornaments, automotive parts as valve caps, oil pans, intake tubes, cooling units
Furniture parts	Vacuum-formed or extruded border strips, table tops
Jewelry	Clips, bracelets, necklace
Toys	Balls, model trains, toy cars
Sports and recreational items	Sky boots, boats, bottles, luggage, bicycle frames and accessories

demolding process but also dampens the bright glitter effect of larger aluminum particles, thereby contributing to a more uniform optical appearance.

The location of the nozzle should also be carefully planed because the different flow velocity in various areas within the cavity can appear as flow lines. The extrusion blow molding process with the accumulator unit can also be the reason for flow lines. Usually the incoming melt is separated by the ram, which causes a flow line in the back. This can sometimes be moved to the parting line of the tool, where it becomes almost invisible. In cases where this is not feasible, it may be possible to create a spiral action (turbulence) when filling the accumulator system. The resulting flow patterns are then mixed as they exit, creating the parison.

Shaping Process

When filling the mold, a high flow velocity has a favorable effect on the uniformity of the surface of the molded part. At the same time the metallic particles are exposed to a higher shear force that can cause a reduction in brilliance.

When thermoforming coextruded sheets, the thickness and the viscosity of the upper layer in relation to the pigment particle size and its thickness influence the surface appearance. Of course, the degree of elongation and tight bends are critical.

Properties of Molded Part

As mentioned earlier, pigmentation with metallic pigments, particularly with large particles noticeably reduces the elongation properties and impact resistance. This is caused by their cutting edges.

When filling the mold during the injection molding process, for highest brilliance the flow pattern should be parallel to the surface of the area to be viewed. This permits preferential orientation of the flakes in this area and therefore a better reflection of light.

14.6. TYPICAL USES OF METALLIC PLASTICS

Tables 14.1 and 14.2 give a brief overview of the application for metallic pigments.

REFERENCES

1. H.-H. Bunge and B. Klein, *Injection Molding Int.*, **Nov./Dec.**, 38 (1996).
2. J. D. Kerr and B. Klein, *Kunststoffe Plast. Eur.*, **Aug.**, 34, 1130 (1995).
3. WO 94/02551.
4. R. Besold, *Paint & Ink Int.*, (3) (1994).
5. WO 95/04783.
6. EP 0653 465 A1.
7. US 5.028.639.
8. DE 4.030.727.
9. EP 0668 392 A2.
10. I. F. Johannaber, *Kunststoffmaschinenführer*, Seite 690–691, Carl Hanser Verlag, München, Germany, 1992.
11. H. Kappacher, *Kunststoffe*, **86** (4), 388 (1996).
12. A. Grefenstein, *Kunststoffe*, **87** (10), 1332 (1997).
13. E. Buerkle, *Kunststoffe*, **87** (3), 320 (1997).
14. DE 2.630.731, DE 4.030.727, US 5.028.639, WO 95/04783.
15. Langenbeck/Pritzkow, *Lehrbuch der organischen Chemie*, Verlag T. Steinkopf, Dresden, Germany, (1996).
16. UK 8.531.374.
17. *BTG London*, "News Release SCORIM," 1995–1997.
18. H. H. Bunge and B. Klein, *Kunststoffe*, **86** (9), 1342 (1996).

CHAPTER 15

Pearlescent Pigments/Flakes

Gerhard Pfaff

Merck KGaA, Pigments Division
Frankfurter Strasse 250
Darmstadt, Germany

Joachim Weitzel

EMD Chemicals Inc.
Merck KGaA Darmstadt Germany
Hawthorne, New York

15.1 INTRODUCTION

Pearlescent pigments are synthetic or natural occurring pigments which are used to achieve a lustrous, brilliant or iridescent color effect by light interference. Pearlescent pigments are platelet-like and have a much larger particle diameter than conventional absorption pigments. Conventional color pigments particles are in the sub-micron size, whereas the particle diameter of pearlescent pigments are normally between 1 to 200 microns in size.

Pearlescent pigments were originally used to simulate the appearance of natural pearls, which is why these pigments are called "pearl luster pigments". The visual effects generated by reflection and transmittance of light by thin multilayer films are not restricted to pearls and sea shells. Other wide-spread, eye-catching effects shown by nature include bird feathers, fish scales, gems, minerals, insects (Fig. 15.1). The investigation of the principles of nature's construction mechanisms and architecture shows that these brilliant colors are based on structured biopolymers or laminated structures made by biomineralization [1].

Coloring of Plastics, Fundamentals, 2nd edition. Edited by Robert A. Charvat
ISBN 0-471-13906-8   Copyright © 2004 by John Wiley and Sons, Inc.

Figure 15.1. Examples of interference color effects in nature.

Natural pearls and nacreous shells have been used for decorative applications for several centuries. The history of pearlescent pigments traces back to the year 1656, when the French rosary maker Jaquin isolated a silky lustrous suspension from fish scales (pearl essence) [2]. This suspension was applied to small beads to create artificial pearls. There were several trials over the years to produce improved synthetic pearl colors with either organic or inorganic substances. A strong demand for pearlescent effects evolved from the growing coatings and plastics market. The breakthrough for pearlescent pigments occurred with the invention of metal oxide coatings over a mica platelet [3]. Today, more than 80% of the world market for pearlescent pigments are mica based products. Pearlescent pigments are used to simulate the mother of pearl effect, iridescent (rainbow), or metallic effects. It is possible to obtain brilliance or two-tone color and luster flops (change with viewing angle) in transparent color formulations. The most interesting applications today are plastics, coatings, automotive paints, printing inks, and cosmetic formulations.

The optical properties of pearlescent pigments are characterized by the phenomenon of light reflection or interference. Metallic luster, interference, or interference reflection may be observed depending upon the transparency, absorption, thickness and/or layer structure of the pigment particles. A fundamental definition of nacreous, also known as pearlescent, pigments has been formulated by Greenstein [4]: "Thin platelets of high refractive indexes which partially reflect and partially transmit light". There exists the comprehensive term for pearlescent pigments consisting of materials that do not completely absorb visible light (e.g. TiO_2) or that contain materials with a high refractive index and additional absorption characteristics (e.g. Fe_2O_3) [5, 6]. The terms "nacreous pigments" and "pearlescent pigments" are frequently used in a parallel manner.

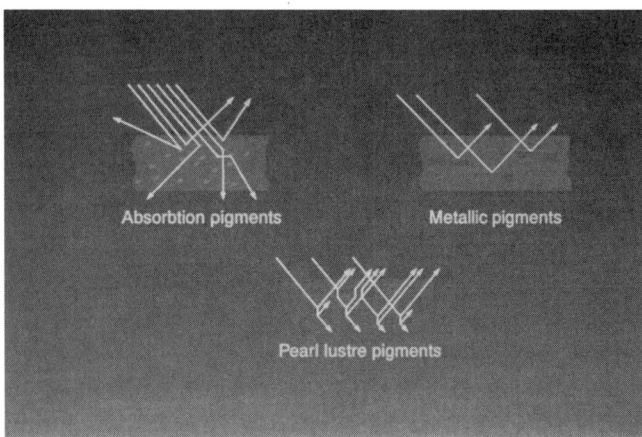

Figure 15.2. Optical properties of conventional pigments, metallic pigments, pearl pigments and pearls.

Figure 15.2 shows a comparison of the optical properties of conventional and effect type pigments. Conventional pigments interact with light by absorption and/or diffuse scattering. Effect or lustrous pigments are comprised of either pearlescent or metallic particles [3, 4]. Metallic effect pigments consist of thin metallic flakes (e.g. Al, Cu/Zn bronze) which act as small mirrors that reflect almost all the incident light. The mirror luster of these pigments can be very brilliant. Pearlescent pigments try to simulate the luster of natural pearls. These consist of alternating, transparent layers of materials with a high and a low refractive index. Small thin platelets of pearlescent pigments with a high refractive index can be oriented in parallel alignment in a matrix with a lower refractive index, e.g. a lacquer system, a plastic, or a cosmetic formulation. If the distances of the different layers or the thickness of the platelets are of appropriate size, interference or pearl luster effects are generated. The pearl luster effect is soft, and appears to come out of the depth and is characterized by multiple reflections. Synthetic pearlescent pigments are transparent or partially light-absorbing platelet-like crystals. They can also consist of a multilayer structure, where the layers have different refractive indices and light absorption behaviors.

Table 15.1 shows an overview of pigments with luster effects. Effect pigments can be classified in metal platelets, oxide-coated metal platelets, oxide-coated mica platelets, platelet-like mono-crystals and cured thermochromatic liquid crystals and crushed PVD-films (Physical Vapor Deposition).

15.1.1. Optical Principles

The optical principles of pearlescent pigments are shown in Fig. 15.3 in a simplified manner where nearly normal incidence without multiple reflection and absorption occurs. At the interface P_1 between two optically different materials with the refractive indices n_1 (e.g. lacquer film) and n_2 (e.g. single layer of a thin crystalline platelet), a part of the light beam L_1 is reflected (L_1) and a part is transmitted (i.e., refracted; L_2). The intensity ratio between L_1 and L_2 depends on n_1 and n_2. In multilayer arrangements like in pearl or pearlescent and iridescent materials, each interface

Introduction

Table 15.1. Overview of Effect Pigments [7]

Pigment type	Examples
metallic platelets	Al, Zn/Cu, Cu, Ni, Au, Ag, Fe (steel), C (graphite)
oxide coated metallic platelets	surface oxidized Cu-, Zn/Cu-platelets, Fe_2O_3 coated Al-platelets
coated mica platelets	non absorbing coating: TiO_2 (rutile), TiO_2 (anatase), ZrO_2, SnO_2, SiO_2 selectively absorbing coating: FeOOH, Fe_2O_3, Cr_2O_3, TiO_{2-x}, TiO_xN_y, $CrPO_4$, $KFe[Fe(CN)_6]$, colorants totally absorbing coating: Fe_3O_4, TiO, TiN, $FeTiO_3$, C, Ag, Au, Fe, Mo, Cr, W
platelet-like monocrystals	BiOCl, $Pb(OH)_2 \cdot 2PbCO_3$, $\alpha\text{-}Fe_2O_3$, $\alpha\text{-}Fe_2O_3 \cdot n\ SiO_2$, $Al_xFe_{2-x}O_3$, $Mn_yFe_{2-y}O_3$, $Al_xMn_yFe_{2-x-y}O_3$, Fe_3O_4, reduced mixed phases, Cu-phthalocyanine
comminuted thin PVD-films	Al, Al (semitransparent) / SiO_2 / Al / SiO_2 / Al (semitransparent)

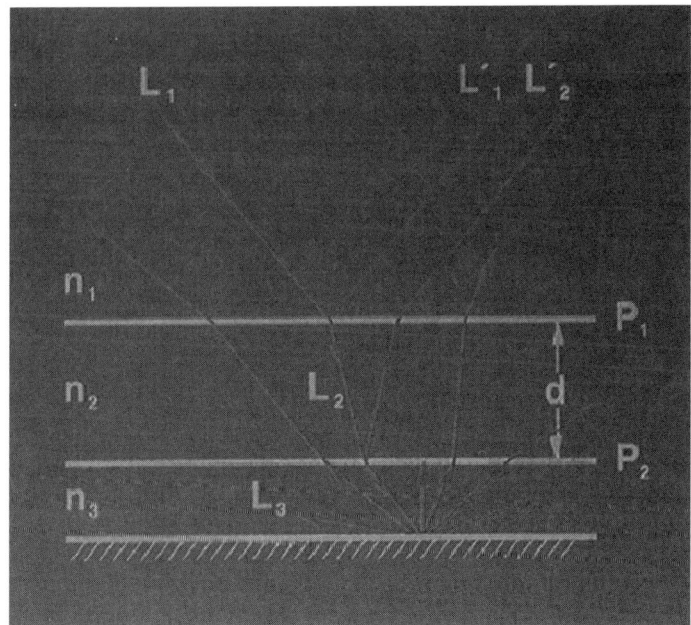

Figure 15.3. Simplified diagram showing nearly normal incidence of a beam of light (L_1) from an optical medium with refractive index n_1 through a thin solid film of thickness d with refractive index n_2 (L_1' and L_2': regular reflections from interfaces P_1 and P_2. L_3: diffuse scattered reflections from the transmitted light).

produces partial reflection [5, 6, 8, 9]. After penetration through several layers, depending on the thickness and difference between n_1 and n_2, virtually complete reflection can be observed for the case that the materials are sufficiently transparent.

Pigments that simulate natural pearl effects consist in the simplest case of platelet-shaped particles with two phase boundaries P_1 and P_2 at the upper and lower surfaces of the particles. An example of this is a single, thin, transparent layer of a material with a higher refractive index than the surrounding medium. The physical laws of thin, solid, optical films can be applied for small flakes with a thickness of ca. 100 nm [5, 6]. Multiple reflection of light on a thin solid film with high refractive index (Fig. 15.3) causes interference effects in the reflected light and in the complementary transmitted light. The intensity of the reflectance depends on the refractive indices n_1 and n_2, the layer thickness d, and the wavelength λ for the simple case of nearly perpendicular incidence in the following manner [10]:

$$I = \frac{A^2 + B^2 + 2AB\cos\theta}{1 + A^2 B^2 + 2AB\cos\theta}$$

$$A = \frac{n_1 - n_2}{n_2 + n_1} \qquad B = \frac{n_2 - n_1}{n_2 + n_1}$$

$$\theta = 4\pi \frac{n_2 d}{\lambda}$$

For known refractive indices, the maximum and minimum intensities of the reflected light, which are seen as interference colors, can so be calculated. The results are in good agreement with experimental data. Refractive indices of materials playing an important role for pearlescent pigments are listed below:

Vacuum / air	1.0
Water	1.33
Proteins	1.4
Plastics, lacquers	1.4–1.7
Mica	1.5
$CaCO_3$ (aragonite)	1.68
Natural pearl (guanine, hypoxanthine)	1.85
$Pb(OH)_2 \cdot 2PbCO_3$	2.0
BiOCl	2.15
TiO_2 (anatase)	2.5
TiO_2 (rutile)	2.7
α-Fe_2O_3 (hematite)	2.9

Practically applied platelet-like crystalline materials are produced with layer thickness (d) which are necessary to obtain the desired interference colors (iridescence) [11]. Most of the pearlescent pigments synthesized today consist of at least three layers of two materials with different refractive indices. A simplified structure of such pigments and their interaction with light is shown in Fig. 15.4. For the case of metal oxide-mica pigments, thin flakes of mica (thickness ca. 500 nm, compara-

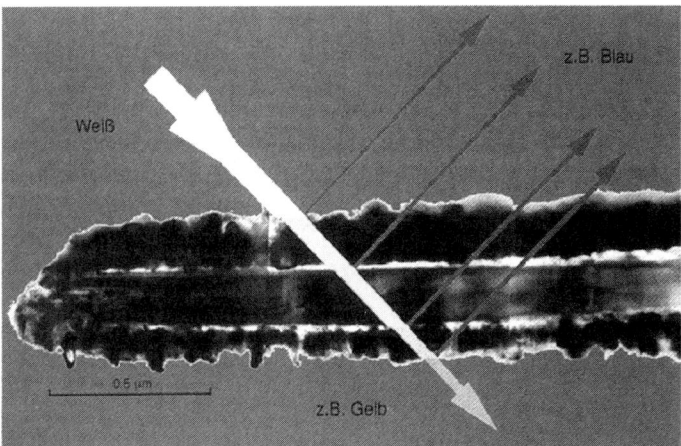

Figure 15.4. Transmission electron micrograph of a TiO$_2$-mica pigment with the schematic way of light through layers (blue interference effect, yellow transmission).

tively low refractive index of 1.5) are coated with a highly refractive metal oxide (e.g. titanium dioxide or hematite, layer thickness between 50 and 250 nm). The pigment particles have four interfaces and are therefore more complicated, but still predictable thin film systems. The optical behavior of more complex multilayer pigments containing additional, thin, light absorbing layers can also be calculated if appropriate parameters are known [12].

Color effects of pearlescent pigments depend on the viewing angle. The pigment particles split white light into two complementary colors depending on the thickness of the platelets in respect to the optically effective layer. The reflected (interference) color dominates under regular reflection, i.e., during observation the object at the angle of regular reflection. The transmitted color dominates at other viewing angles using diffuse viewing conditions for the case of a nonabsorbing (white) or reflecting background. Variation of the viewing angle therefore produces a sharp gloss (reflectance) peak. Luster and color resulting by this complex interaction in application systems can be measured goniophotometrically in reflection and at different angles [11, 13]. There is not a general standard for the measurement geometries up to now. However, colorimetric measurement always takes place under regular and diffuse conditions. One possibility for this is tilting of a standard pigmented film on a drawdown card. The colometric data are interpreted according to CIE as L*a*b* values. Pearlescent pigments are characterized by a minimum of three L*a*b* data sets measured under different conditions (e.g., 0°/45° black background, 22.5°/22.5° black background, 0°/45° white background). The analysis of these data yields the characteristic coloristic properties of pearlescent pigments, e.g., hiding power, luster, and hue [14].

There are many possibilities for blending pearlescent pigments with other colorants. In a blend with carbon black, the transmitted light is absorbed and the reflected interference color is seen as the mass tone of the material. It is necessary for blends of pearlescent pigments with absorbing colorants to hold the particle size of the latter below the scattering limit to realize transparency. This means that the pearlescent effect can be quenched by hiding pigments. This fact is also important for blends with the strongly reflecting metal effect pigments. Blends of pigments with different interference colors obey the law of additive mixing (e.g. blue + yellow → white), in contrast with the subtractive color mixing of pure absorption pigments (e.g., blue + yellow → green). In the case of combinations of interference pigments with absorption pigments, the pearlescent effects generally dominate under regular (gloss) viewing conditions and the absorbing effects under diffuse viewing conditions [11].

15.2. CLASSES OF PEARLESCENT PIGMENTS

15.2.1. Natural Pearl Essence

Natural pearl essence (Essence d'Oriente, Natural Fish Silver) is a pigment suspension mostly derived from fish scales. It consists mainly of a mixture of the purines guanine and hypoxanthine. The ratio of the purines depends on the fish species (e.g., herrings, sardines) and their geographical origin (e.g., Japan, Norway, northeast of the USA). Natural fish silver crystallizes in needles or longish platelets with a breadth of 1 to 10 µm, a length of 20 to 50 µm, and a thickness of only 0.025 to 0.075 µm. For the production, an aqueous suspension of fish scales is extracted with organic solvents to dissolve and remove the proteins. The remaining dispersion contains purine crystals and scales which are separated from one another by a complicated washing and phase-transfer process. Natural pearl essence is extremely expensive and is therefore used almost exclusively in cosmetic applications. It shows a strong agglomeration tendency in dry form. Some of the advantages over synthetic pearlescent pigments are the less fragility, low density (1.6 g cm^{-3}) and the very high, but soft luster (n = 1.8–1.9).

15.2.2. Metal Oxide-Mica Pigments

The most common class of pearlescent pigments today is based on thin platelets of mica (see Fig. 15.5). Mica itself is a natural mineral and belongs to the sheet layer silicates. Nacreous pigments are usually based on natural, transparent muscovite and only in some cases on synthetic phlogopite. Muscovite occurs worldwide, but only few deposits are suitable for pigment production. Mica is biologically inert and approved for use as a filler and colorant.

Selection and workup of the mica to produce thin platelets is one of the key factors which are responsible for the quality and appearance of metal oxide-mica pigments. The aspect ratio of the final pigments depends on the particle size distribution of the prepared mica platelets which have thickness of 300–600 nm and various diameter ranges (e.g., 1–15 µm, 5–20 µm, 10–50 µm, 10–130 µm, 40–200 µm; Fig. 15.5). Light is regularly reflected from the planes of the metal oxide coated mica and scattered from the edges. Therefore, brilliance and hiding power of the pigments are inversely related to each other.

Mica particles coated with a metal oxide film have three layers with different refractive indices (layer 1 and 3 are identical, layer 2 is mica) and four interfaces (see chapter 16.2). Interference of light is generated by reflections of all six possible combinations of the four interfaces. Some of them lead to equal effects. The thickness of the mica platelets varies according to a statistical distribution. As a con-

Figure 15.5. Mica platelets with different particle size distribution

sequence, interference effects involving the interfaces between the mica platelet and the oxide layers add together to give a white background reflectance. That is the reason why the interference color of a large number of pigment particles depends only on the thickness of the upper and lower metal oxide coating layers [4, 8]. The development of the metal oxide-mica pigments started with TiO_2 containing types. The next generation were brilliant, mass-tone-colored combinations, consisting of mica, TiO_2 and another metal oxides. There are combinations with one color (interference color same as mass tone) or two colors (interference color and mass tone are different) depending on composition and viewing angle. Further developments lead to mica particles coated with transparent layers of Fe_2O_3 or Cr_2O_3. For outside applications (e.g., in automotive coatings), an additional surface modification of the pigments is used to improve the long term weather resistance. Fig. 15.6 shows a summary of the main pearlescent pigment types based on the metal oxide-mica principle.

15.2.3. Titanium Dioxide-Mica Pigments

TiO_2-coated mica platelets were the first multilayer pigments which came into the market in the 1960's. The precipitate on the mica surface is hydrated titanium dioxide. The pigments are dried and calcined at temperatures between 700 and 900°C. The chloride process is preferred for the production of interference pigments with relatively thick TiO_2 layers because it is easier to control. It has been found that small amounts of SnO_2 can assist the conversion of anatase to the rutile structure. Rutile has the advantage of the higher refractive index, more brilliance and color intensity and superior weather resistance. A scanning electron micrograph of a TiO_2-mica pigment is shown in Fig. 15.7. The sequence of interference colors obtained with increasing TiO_2 layer thickness is in accordance with theoretical calculations in the L*a*b*-color space [5, 6]. Titanium dioxide-mica pigments are commonly used in color formulations with conventional pigments to produce brilliance

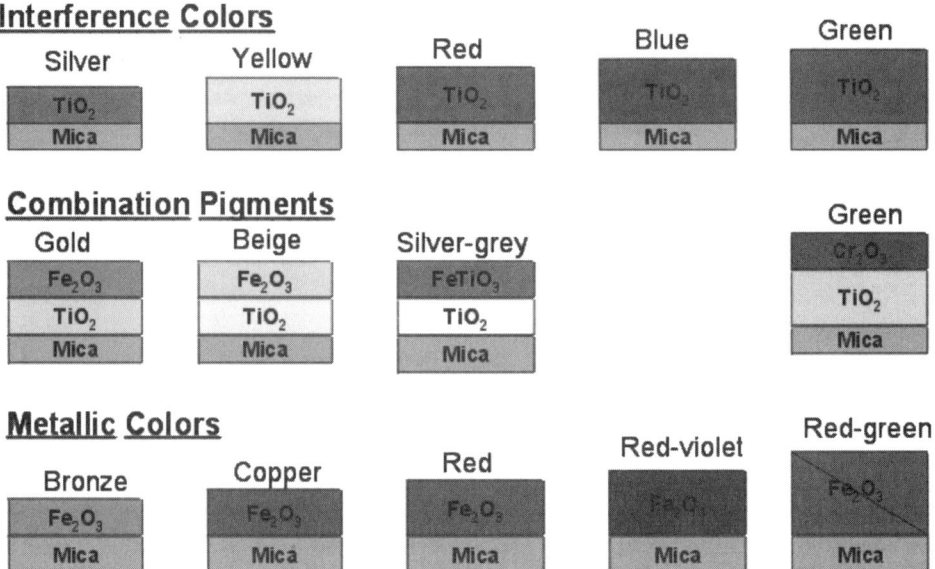

Figure 15.6. Schematic illustration of different metal oxide-mica pigments.

Figure 15.7. Scanning electron micrographs of a TiO_2-mica pigment prepared by the wet chemical process.

and luster in addition to color. Main application fields are plastics, coatings, printing inks, and cosmetic formulations. The major market for silver white pigments (pearl pigments, "white metallic") is the plastics industry.

15.2.4. Iron Oxide-Mica Pigments

Iron(III)oxide is also suitable for coating mica platelets. It combines a high refractive index (metallic luster) with good hiding power and weather resistance. Fe_2O_3-mica pigments are produced by precipitation of iron(II)- or iron(III)-ions in aqueous mica suspensions and then calcined at 700–900°C [5, 6, 8, 9]:

$$2FeCl_3 + mica + 3H_2O \rightarrow Fe_2O_3/mica + 6HCl$$

Brilliant, intense colors are obtained with 50–250 nm layers of Fe_2O_3. Absorption and interference colors are formed simultaneously and vary with layer thickness of iron oxide. The red shades are extremely intensive because the interference and absorption colors enhance each other. It is possible to produce an intense green-red flop with different viewing angles at a layer thickness according green interference [16]. Fig. 15.8 represents a scanning electron micrograph of a Fe_2O_3-mica pigment synthesized by the wet chemical process.

15.2.5. Combination Pigments on the Basis of Mica

Simple blending of transparent absorption pigments with pearlescent pigments is only one way to attain new coloristic effects. It is also possible to produce pearlescent pigments coated with a layer of transparent absorption colorants. This produces a more brilliant color with a sharper color flop, which allows the change of color with the change of viewing angle. An additional advantage of such pigments is the elimination of dispersion problems associated with transparent absorption pigments due to their small particle size and high surface area.

The thickness of the TiO_2 layer is decisive for the brilliance or interference effect under regular viewing conditions whereas the transparent colorant dominates at all other viewing angles. A deep, rich color with a luster flop at all angles can be attained. If interference color and masstone of the colorant are different, a color flop (two-tone pigments) is seen in addition to the luster flop.

Iron(III)oxide is the commonly used for combinations with titanium dioxide on mica flakes to produce brilliant, golden pearlescent pigments. Two routes are used to synthesize these pigments, and different structures are formed [5, 6, 8, 9]. In the first case, a thin layer of Fe_2O_3 is coated on the surface of a TiO_2-mica pigment. The overall interference color is the result of both metal oxide layers. The masstone is determined by the Fe_2O_3 layer, and gold pigments (e.g., reddish gold) are possible. In the second case, coprecipitation of iron and titanium oxide hydroxides on mica particles and calcination leads to greenish gold pigments. The masstone in this case is further modified due to the additional formation of the highly refractive yellowish iron titanate phase Fe_2TiO_5 (pseudobrookite) [17].

Coating of TiO_2-mica pigments with an organic colorant for a masstone or two-tone pigment is done by precipitation or deposition on the pigment surface in

236 Pearlescent Pigments/Flakes

Figure 15.8. Scanning electron micrograph of a Fe_2O_3-mica pigment prepared by the wet chemical process.

aqueous suspension, assisted by complexation or surfactants. A second route is fixing the colorants as a mechanically stable layer using suitable additives.

15.2.6. Basic Lead Carbonate

Basic lead carbonate was one of the first commercially introduced synthetic pearlescent pigment. In general, synthetic pearlescent pigments must fulfill the following demands: High refractive index, good transparency without own color, platelet structure with even surface, optimal thickness of the platelets. Basic lead carbonate with the chemical composition near to $Pb(OH)_2 \cdot 2PbCO_3$ is an ideal pearlescent pigment. It is synthesized in form of monocrystals by precipitation from aqueous lead acetate or lead propionate solutions with carbon dioxide. Under appropriate reaction conditions, regular hexagonal platelets (ca. 50 nm thick, 20 μm in diameter) can be prepared. Basic lead carbonate is optically equivalent or better compared to natural pearl essence due to the high refractive index (n = 2.0), high aspect ratio (>200), and has an extremely even surface.

The thickness of the platelets can be adjusted by controlling the reaction conditions to produce interference colors. The pigment crystals are mechanically sensitive and show a fast sedimentation behavior because of the high density (6.4 g cm^{-3}). Agglomeration tendency and occupational health (toxicity) risks are the reason why Pb(OH)$_2 \cdot$2PbCO$_3$ is not produced in powder form, but is flushed from the aqueous phase into suitable organic solvents or resins and handled as stabilized dispersions. Today, the application of basic lead carbonate is limited to artificial pearls, buttons, and bijouterie.

15.2.7. Bismuth Oxychloride

Bismuth oxychloride (bismuth oxide chloride) with the formula BiOCl was the first synthetic nontoxic nacreous pigment. Its production in form of monocrystals takes place by hydrolysis of acidic bismuth salt solutions in the presence of chloride ions. The desired pigment crystal quality can be controlled by varying the precipitation conditions (concentration, temperature, pH, pressure, addition of surfactants). The virtually tetragonal bipyramidal crystal structure is thereby modified into a flat platelet.

BiOCl pigments are available in three grades with different pearlescent effects depending on the aspect ratio and the crystal size:

- Low—or medium—luster powder (aspect ratio 10–15), mainly applied as a highly compressible, white, lustrous filler with excellent skin feel.
- Dispersion of high luster quality (aspect 20–40), consisting of square or octagonal platelets in nitrocellulose lacquers (nail polish) or castor oil (lipsticks).
- Dispersion of very high luster quality (aspect ratio > 50), consisting of lens-shaped platelets in nitrocellulose lacquer, castor oil, or butylacetate.

Pigments consisting of mica or talc coated with bismuth oxychloride and blends of BiOCl with organic or inorganic colorants are also available.

Bismuth oxychloride pigments are mainly used in cosmetic formulations, but the market for buttons, bijouterie, printing, and X-ray contrast in catheters is growing. A limit for some applications is the low light stability (color changes from silver white to metallic gray in sunlight), fast settling (high density of 7.73 g cm^{-3}), mechanical sensitivity, and the relatively high price. However, some low-luster BiOCl grades with improved light stability are now available.

Table 15.2 contains a comparative overview of TiO$_2$-mica pigments, basic lead carbonate, bismuth oxychloride, and natural fish silver. Some further physical data are summarized in Table 15.3.

15.3. WHAT MAKES PEARLESCENT PIGMENTS ATTRACTIVE FOR PLASTICS

The use of pearlescent pigments vastly increases the possibilities of color design. In comparison to other pigments, organic or inorganic, pearl pigments possess luster as an additional property. In all the pearlescent pigment series (Silver White, Interference, Gold and Earthtone), effects ranging from satin to glitter can be obtained

Table 15.2. Properties and Application of Pearlescent Pigments [15]

Pearlescent pigment	Advantages	Disadvantages	Main application field
natural fish silver	very low density high luster nontoxic light stable	high price low hiding power limited availability	nail lacquers
basic lead carbonate	very high luster good hiding power low price light stable	high density chemically and thermically only limited stable toxic	buttons bijouterie
bismuth oxychloride	very high luster good hiding power nontoxic	low light stability high density	decorative cosmetics buttons bijouterie
titanium dioxide-mica	high luster good hiding power (depending on the particle size) highest thermic, chemical, and mechanical stability nontoxic low price low density	inferior luster in comparison with top qualities of basic lead carbonate and bismuth oxychloride	plastics lacquers cosmetics printing inks ceramic products

Table 15.3. Technical Data of Pearlescent Pigments

Pearlescent pigment	Shape	Particle size (μm)	Thickness (nm)	Density (g cm^{-3})
natural fish silver	needles, longish platelets	10–40	40–50	1.6
basic lead carbonate	hexagonal crystals	4–20	40–70	6.4
bismuth oxychloride	flat tetragonal bipyramidal crystals	5–30	100–700	7.7
titanium dioxidemica	platelets	1–200	200–500	3

by modifying the particle sizes of the mica. Pearlescent pigments are often used without other added colors, but when combined with various transparent pigments, the possibilities of coloring plastics is practically limitless. Metallic effects can be simulated with the addition of small amounts of carbon black to a silver white pearlescent pigment. Pearlescent pigments are non-toxic, and do not contain heavy metals. The pigments are extremely heat and light stable and will not bleed or migrate. Pearlescent pigments can be used with all thermoplastic applications, and are appropriate for some thermoset uses. Pearl luster colored thermoplastics are commonly used in packaging applications (bottles, caps) and in household articles. The pigments can be easily processed through typical extrusion/compounding, injection molding or blow molding processes. Pearlescent pigments are also finding their way into rotational molding, thermoforming and blown film applications.

Since pearlescent pigments differ from conventional pigments due to their special physical and geometrical properties, important conditions must be observed to obtain the optimum effects. The pearl luster appearance is an optical phenomenon of light transmittance and reflection through a resin medium, the more transparent the medium or resin, the more striking the results. The pigmenting of less transparent and opaque plastics reduces the luster and the iridescence of the pigments. The homogeneous distribution of the dispersed pigments throughout the plastic is an important process in pigmenting plastics. If the pigment particles are not distributed evenly, color shadows or stripes may be visible. In order to show optimum pearl luster, the pigment platelets must be aligned parallel to each other and should also be parallel to the surface of the product. Pearlescent pigments can safely be incorporated into a solid or liquid form of color concentrate with pigmentation loadings ranging from 10–40%. For most plastic applications, the typical usage of pearlescent pigments in the final part should be between 0.5 and 2.0% loading by weight of the resin.

Pearlescent pigments can also be used in plastic formulations to produce marbleized and frosted effects. Marble effects using pearlescent pigments show the typical crystalline lattice structure of natural marble and make the effect appear more real in plastics. The effect known as the frost effect is easily produced while using the pearlescent pigments at very low loading levels (0.2%) in the transparent resin system.

15.3.1. Functional Pearlescent Pigments

Metal oxide-mica pigments were developed at first only because of their excellent coloristical properties. Meanwhile, they also have interesting functional uses. In coatings with a high content of platelet fillers, an advantageous overlapping rooftile type arrangement is possible that provides close interparticle contact, barrier effects, and dense covering. The composition of the oxide layer on the mica surface and its thickness are always responsible for the physical properties such as electric conductivity, magnetism, IR-reflectivity, or laser markebility.

A quality laser mark can only be achieved with very few plastic resins in their virgin state (e.g. PVC). The majority of resins used in packaging, including polyolefins and styrenics, have up to now proved difficult or impossible to mark adequately using the laser technology. Plastics with a low absorption level of laser light show practically no reaction to laser bombardment. The absorption of the plastic material is too low for the laser beam, thus the beam passes through the plastic material without creating a visible mark. The incorporation of pearlescent pigments to these low absorption plastics allows the resin to become receptive to laser light, consequently a high contrast, visible mark is achievable at a relatively low laser intensity.

Although pearlescent pigments are commonly known for their decorative enhancement to plastic products, the pigments can now be used in a functional

Table 15.4. Functional Metal Oxide—Mica Pigments

Pigment composition	Property	Application	References
$(Sn,Sb)O_2$/mica $Sn(O,F)_2$/mica	electrically conductive	conductive flooring, antistatic packaging materials, light colored primed plastic surfaces which can be electrostatically painted in further coating process, light colored conducting surfaces in clean room conditions which attract little dust	[18]
Fe_3O_4/mica	magnetic	magnetic surfaces	[9]
TiO_2/mica	IR-reflective	IR-reflecting plastic sheets, e.g. for domed and continuous rooflights	[19]
TiO_2/mica $(Sn,Sb)O_2$/mica	laser sensitive	laser marking of plastics, coatings	[20]

capacity. Today, due to further development of pearl luster pigments and close cooperation with the manufacturers of laser instrumentation, it has become possible to laser mark several types of plastic materials while retaining the flexibility in color design. The quality of the mark depends upon the plastic resin involved, the type of laser instrument and the pearlescent pigment used as well as its concentration. Recommended concentrations of 0.5% to 1.0% pearlescent pigments are adequate for the laser marking of plastics. Increasing the pigment concentration results in a darker mark, while the penetration of the laser beam into the plastic is reduced. The optimum pigment concentration for a well contrasted, sharply defined mark depends upon the type of plastic and thickness of the plastic product. The addition of these additives can easily be incorporated into existing or new color concentrate formulations.

Table 15.4 contains data about the functional properties of some metal oxide-mica pigments.

15.4. NEW DEVELOPMENTS BASED ON NON MICACEOUS SYSTEMS

Several developments are concentrated on the search for substrate containing systems, where mica is replaced by other platelet-like materials. Therefore it is possible to replace mica with kaolin or talc to produce bright, conductive pigments by coating with $(Sn,Sb)O_2$ [21]. New pigments based on transparent silica flakes show excellent optical effects which are different to mica pigments [22]. Angle dependent colors and other effects, achieved by the combination of these SiO_2-flakes with tin titania and/or iron oxide coating layers, lead to a new generation of pearl luster pigments. They contribute a whole new dimension to the existing possibilities of colorstyling with luster pigments.

REFERENCES

1. H. Simon, "The Splendor of Iridiscence", Dodd, Mead and Co., New York, 1971.
2. H. F. Taylor, Drugs Oil Paints, **3** (1937) 106.
3. H. R. Linton (DuPont), US 3 087 828 (1963) and US 3 087 829 (1963).
4. L. M. Greenstein, "Pigment Handbook", 2nd ed., Vol. 1, Wiley, New York, 1988, p. 829–858.
5. G. Pfaff and P. Reynders, Chem. Rev. **99** (1999) 1963.
6. G. Pfaff, K. D. Franz, R. Emmert and K. Nitta, in "Ullmann's Encyclopedia of Industrial Chemistry": Pigments, Inorganic, Section 4.3, 6th ed. (electronic release), VCH-Verlagsgesellschaft, Weinheim, Germany, 1998.
7. W. Ostertag, Nachr. Chem. Techn. Lab., **42** (1994) 849.
8. W. Bäumer, Farbe + Lack **79** (1973) 747.
9. R. Glausch, M. Kieser, R. Maisch, G. Pfaff and J. Weitzel, "Special Effect Pigments", Vincentz Verlag, Hannover, Germany, 1998.
10. C. Schmidt and M. Friz, Kontakte (Darmstadt), (2) (1992) 15.
11. F. Hofmeister, Eur. Coat J., **1** (1987) 400.
12. A. M. Gaudin, J. Phys. Chem., **41** (1937) 811.
13. J. A. Dobrowolski, E. C. Ho, and A. Waldorf, Appl. Opt., **28** (1989) 2702.
14. R. Emmert, Cosmet. Toiletries, **104** (1989) 57.
15. G. Pfaff and R. Maisch, Farbe + Lack, **101** (1995) 89.
16. G. Gehrenkemper, F. Hofmeister, and R. Maisch, Eur. Coat. J., **3** (1990) 80.
17. R. Emmert and M. Weigand (Merck KGaA), EP 307 747 (1989).
18. D. Brückner, R. Glausch, and R. Maisch, Farbe + Lack, **96** (1990) 411.
19. T. Daponte, P. Verschaeren, M. Kieser, and G. Edler, (Merck KGaA/Hyplast), WO 94/05727.
20. G. Pfaff and P. Reynders, Chem. Rundschau Jahrbuch, (1993) 31.
21. R. Glausch, G. Pfaff, and R. Maisch, XXIst FATIPEC Congr., Amsterdam, Vol. 2, 1992, 33.
22. S. Teaney, G. Pfaff and K. Nitta, Eur. Coat. J., 1999 (4) 90.

CHAPTER 16

Fluorescence

Christopher Newbacher and Apparao Jatla

DAYGLO Color Corporation
4515 St. Clair Avenue
Cleveland, OH 44103

16.1. THEORY AND MECHANISM

Fluorescence may be described as the unique property of a substance to absorb light of varying frequencies, and re-emit this energy as light of a longer wavelength causing the material to produce a brilliant "glow" of light. A fluorescent body can accomplish this because it has the capacity to convert short wavelength (High Energy) photons of light into a predominantly longer and more intense wavelength of energy. This is in contrast to a conventional colorant, which can only *reflect* a small portion of light from the visible spectrum (i.e., less than 100%, See Fig. 16.1).

For example, a conventional colorant such as orange, can absorb white light and reflect *only* the orange band (approx. 600 nm) of the visible spectrum. The rest of the spectrum (red, yellow, green, blue, indigo, and violet) will be dissipated as heat. A fluorescent orange color *however*, absorbs white light and converts the lower wavelength colors (yellow, green, blue, indigo, violet) along with longer uv energy into a single, more intense band of orange light (vs. losing these wavelengths as heat). This conversion of ultraviolet and visible radiation is why they are described as daylight fluorescent colorants.

Coloring of Plastics, Fundamentals, 2nd edition. Edited by Robert A. Charvat
ISBN 0-471-13906-8 Copyright © 2004 by John Wiley and Sons, Inc.

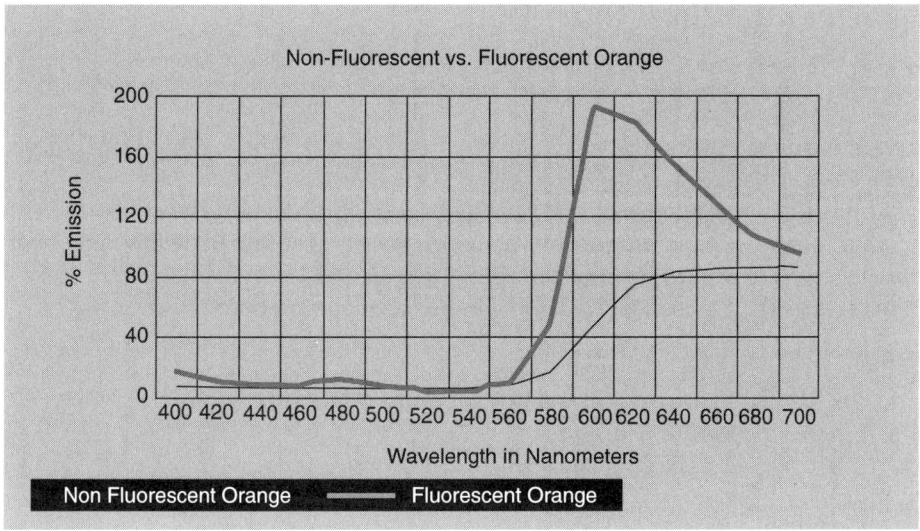

Figure 16.1. Non-Fluorescent vs. Fluorescent Orange.

16.1.1. Theory of Fluorescence

In order for a body to fluoresce, it must obey the *Three Laws of Fluorescence.*

The first law was discovered in 1942, by J. De Ment [1] who was among the first who attempted to describe processes of a luminescent nature. It states quite simply that, *before emission can occur from a luminescent system, absorption must first take place.*

The second Law of Fluorescence determined in 1852 is known as Stoke's Law [2]. It states that *the energy released as luminescence from a body is always less than the energy absorbed for initial excitation.* In other words, the wavelengths emitted from a fluorescing substance are always longer and therefore lower in energy than that of the initially absorbed light. The first law of fluorescence must be obeyed, before the second law can take place. Any wavelength of longer light will therefore appear much brighter to the human eye.

The reason for the lowering of energy may be explained by the following mathematical relationships.

Both visible and ultraviolet radiation arrive in discreet packets called quanta. Each quanta has a value of energy (E) equal to:

$$E = hv$$

Where h = Planck's constant (6.6208×10^{-34} Js) and v = frequency in inverse seconds. Next, the relationship between frequency and wavelength is described as:

$$vx\lambda = c$$

Where λ = wavelength

c = speed of light in a vacuum

Substituting $v = c/\lambda$ yields:

$$E = h(c/\lambda)$$

where it is implicit that if the wavelength were to increase, the energy of the system would have to decrease.

The third law of fluorescence [3] is concerned with luminescent efficiency, also known as quantum yield. In order to obey this law, it must first obey the first law of fluorescence (De Ment's absorption) and if the second law is also obeyed (Stoke's emission law) then the quantum yield can be calculated in the following way:

$$Eem/Eab = 1$$

Where Eem = energy emitted from a substance

Eab = energy absorbed by a substance

If we had 100% conversion, that is every quanta absorbed was re-emitted, unity would be achieved. Since some energy will be lost through thermal agitation, the thermal processes may be calculated according to the following equation:

$$E = h(vab - vem)$$

keeping all of these laws within the law of Conservation of energy.

From the above laws, we have learned that a molecule must first absorb, and then re-emit energy of discreet packets called quanta. We may determine the quantum efficiency of a molecule if we can measure the absorption and emission. In order to thoroughly understand this process, we must now look at the different energy levels within a molecule.

The state of a molecule *before* it absorbs radiation of any type is referred to as the ground singlet state. A singlet state refers to an electronic state where the spins of all the electrons in the molecule are paired. The electronic state of an electron in constant orbit around the nucleus of an atom in a dye molecule for example, can be represented graphically. It is in this region of space where there are various probabilities of locating an electron which are referred to as orbitals. All electrons will modify their bonding structures to obtain the lowest energy levels possible. This level is shown in Figure 16.2 as S_0.

Associated with the ground state of every molecule are different vibrational energy levels. If a molecule absorbs radiation from ultraviolet and visible regions of the electromagnetic spectrum, the electron from its ground state would be nearly instantaneously promoted to a next higher energy level. Instantaneously in this case, has an order of magnitude of 10^{-15} seconds. This new region of space occupied by the electron is referred to as the first excited state. Depending upon the molecular species, only quanta of certain frequencies will be absorbed. The frequency is dependent upon several variables, notably the type and length of bonds, symmetry, and multiplicity. It should also be noted that this process may be repeated, and that the electron may be promoted to higher excited states depending upon the frequency of the light and the absorbing species.

A **singlet** state can be graphically represented by the following diagram:

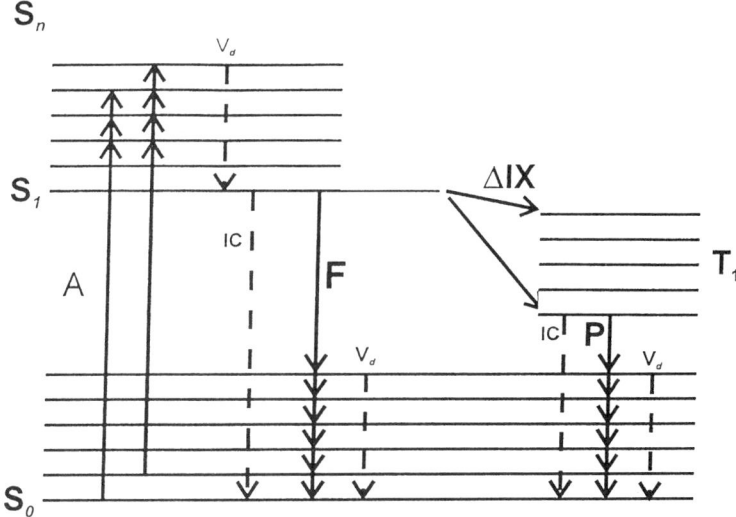

Figure 16.2. Fluorescent Energy Level Diagram.

Electronic States	Transitions	Lifetime, s
S_o = Ground Singlet State	A = Absorption of Energy	10^{-15}
S_1 = First Excited Singlet State	IC = Internal Conversion	10^{-13} to 10^{-11}
S_n = Other Excited States Possible	V_d = Vibrational Deactivation	10^{-13} to 10^{-11}
T_1 = First Excited Triplet State	F = Fluorescence	10^{-8}
	IX = Intersystem Crossing	10^{-8} to 10^{-7}
	P = Phosphorescence	10^{-4} to several sec.

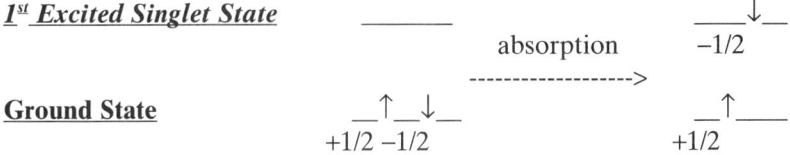

In order for two electrons to occupy the same orbital, they must have oppositely paired spin quantum numbers (denoted as +1/2 and –1/2). Any deviation from this rule violates the Pauli exclusion principle, which states that if two electrons share the same orbital, their spins must be opposed. There is also, another possibility upon promotion of the electron.

If upon absorption, the spin state of the electron changes, the new excited state is referred to as a **Triplet state** (see diagram below). The change from the lowest excited singlet state to an excited triplet state, is termed intersystem crossing (denoted IX in Fig. 16.2). It is dependent upon vibrational coupling between these two states, with a lifetime of 10^{-8} to 10^{-7} sec. This is the only means of obtaining a significant population of electrons in their triplet states.

246   Fluorescence

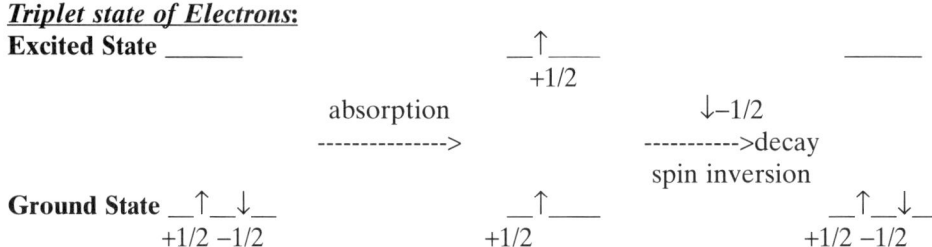

The only way an electron can return to the ground state from the triplet state is by first inverting its spin. This process takes longer than the singlet state (10^{-4} seconds) takes to return to the ground state and is termed **phosphorescence**. This is *not* a nearly instantaneous process like fluorescence. The longer time period for spin inversion is why a substance continues to "glow in the dark" long after the light source has been shut off.

Once promoted to the singlet or triplet states, there are a variety of mechanisms which may deduct or transfer energy from the system. In Figure 16.2, the A represents absorption of a quantum of light energy. This raises the electron from the ground singlet state (S_0) to the first excited singlet state (S_1). In a typical fluorescent dye, it is the pi electron that is raised in energy which is denoted as a pi $\rightarrow$ pi* transition. Upon this excitation, expansion between adjacent atoms occurs, with the energy for this promotion deducted from the pi* orbital. It is also possible for the incident photon to have enough energy to promote the pi electron to even higher states than the first excited state (S_1). These other energy levels may be represented by Sn, Sn + 1, Sn + 2.... There are now different ways to step down this excited state energy. Energy may be transferred to other atoms in a highly conjugated molecule through a combination of vibrations and rotations. If for example, a quantum with enough energy to promote the electron to the Sn level was absorbed by a molecule, it could also cause coupling to occur between vibrational levels of different states of the same multiplicity. This is known as Internal Conversion or (IC) Fig. 16.2. There is also another way for the energy to step back down after absorption. Transfer of energy lowers the electron's energy level to the near lowest or lowest first excited singlet state which is known as vibrational deactivation denoted (V_d). Both internal conversion and vibrational deactivation occurs within the range of 10^{-13} to 10^{-11} seconds. The electron has now returned to the very important lowest first excited singlet state (S_1).

It is from *only* this lowest first excited singlet state, that the decay phenomenon known as **fluorescence** may occur. This is denoted in Figure 16.2 as F. In order for these processes to occur, it takes very unique organic structures.

16.2. FLUORESCENT DYES

16.2.1. Fluorescein

The synthesis of fluorescing substances began in the late 1800s. Early experimentalist, such as Baeyer (1871), condensed resorcinol with phthalic anhydride in the presence of sulfuric acid to produce fluorescein (1).

Fluorescein is naturally colorless, until placed under UV radiation—which then causes it to fluoresce an intense green color. Other discoveries involving color soon followed, enhancing the daylight fluorescent properties of dyes.

16.2.2. Rhodamine B

In 1887, Ceresol produced a bright reddish violet (C.I. Basic Violet 10) which is now known as Rhodamine B (2a).

16.2.3. Rhodamine A

Then in 1891, Monnet esterified Rhodamine B with ethyl chloride to produce Rhodamine A (C.I. Basic Violet 11) (2b).

Virtually all important fluorescent dyes like those illustrated above, contain highly conjugated aromatic rings in their structures. According to the theory of chromophores and auxochromes, a number of groups such —N=N—, >C=C<, >C=O, —N=O or —NO$_2$, so called chromophores have to be present on benzenoid rings in order for compounds to have appreciable light absorption or color. Certain basic groups, so-called auxochromes, are necessary in addition, to bring out or intensify the color.

The benzene ring with its six *pi* electrons can act in conjugation with electron-donating groups (auxochromes) and electron-accepting (unsaturated) groups to produce strong absorption in the ultraviolet or visible regions which may give rise to fluorescence. Such a system of atoms, responsible for significant absorption of photons in the uv or visible regions, is referred to as a chromogen.

Organic dyes can be divided into four classes depending on the type of chromogen or unsaturated system present: (1) $n - p^*$ chromogens, (2) donor-acceptor chromogens, (3) cynanine-type chromogens, and (4) acyclic and cyclic polyene chromogens. The chromogens of class (1) are detrimental to fluorescence.

Donor Acceptor: In dyes in which a particular benzene ring carries a donor and an acceptor group, these groups are introduced in positions ortho or para to each other. In condensed-ring systems such as naphthalene, conjugated-bond paths between donors and acceptors are necessary for interaction.

For example, the amino group in 7-amino-4-methylcoumarin (3a) has a tendency to share its lone pair of electrons with the benzene ring, whereas the methacrylic ester group has a tendency to accept electrons from the benzene ring. Since these groups are para to each other, a partial electron displacement toward the carbonyl oxygen atom takes place, which on absorption of uv light can transform into a high energy polarized form. Fluorescence occurs if the molecule emits a photon of light in a transition from its first excited singlet state.

Alkylation of the amino group to a mono or dialkyl form strengthens the uv absorption and increases the wavelength of the fluorescent light. Hydroxy groups are very weak donors to the benzene ring. Greater depth of color in a chromogen can be obtained by providing more complex acceptor group or linear addition of acceptor groups. Thus, the 7-diethylamino coumarin (3b) is colorless with a blue fluorescence, whereas Coumarin 7 (4) is yellow with a green fluorescence. Dyes of type (3b) are used as optical brighteners whereas the dyes of type (4) are used as daylight-fluorescent dyes.

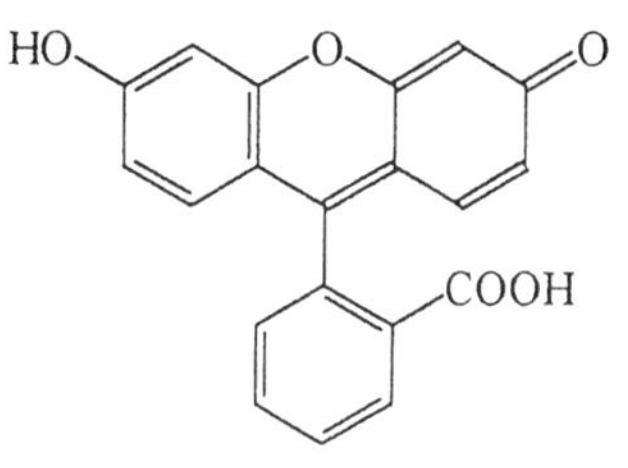

(1)

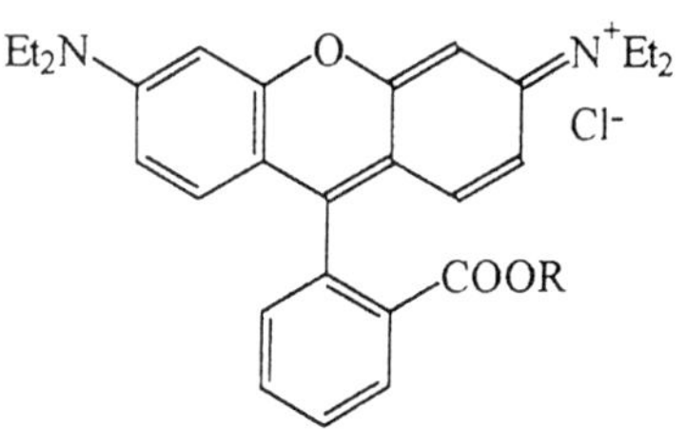

(2a) R = H
(2b) R = Et

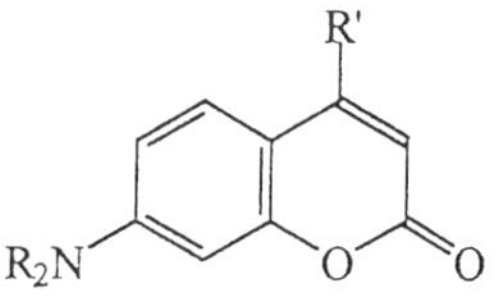

(3a) R= H, R'=Me
(3b) R= Et, R'=H

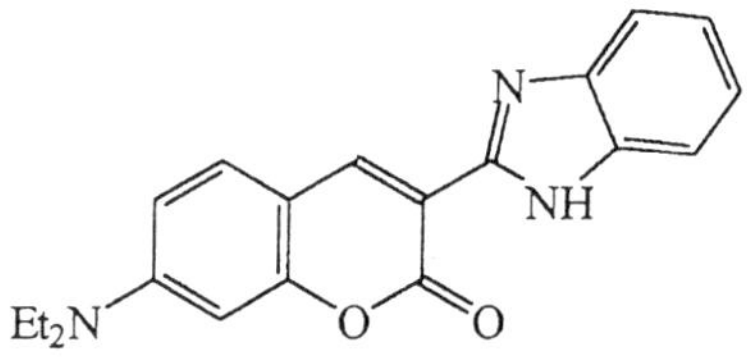

(4)

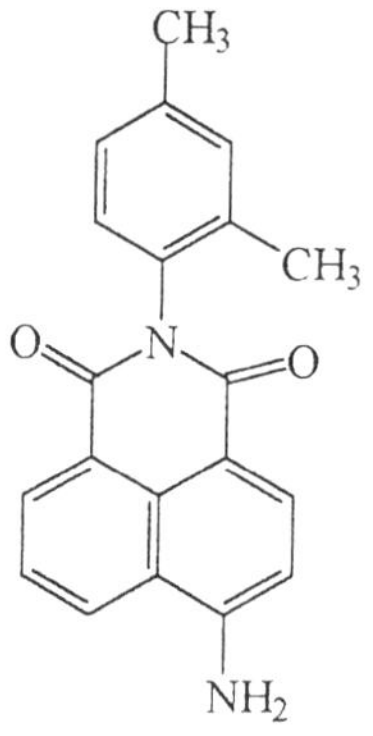

(5)

$$R_2\ddot{N}\text{-}(CH=CH)_n\text{-}CH=\overset{+}{N}R_2 \; X^-$$

$$\updownarrow$$

$$R_2\overset{+}{N}=CH\text{-}(CH=CH)_n\text{-}\ddot{N}R_2 \; X^-$$

(6)

Another example of Donor Acceptor type dye is CI Solvent Yellow 44 (5). Here, the two carbonyl groups are acceptors, the naphthalene is the chromogen and the amino group at 4 position of naphthalene nucleus acts as the donor.

Cyanine Types: Cyanine-type chromogens are odd-alternate systems that have at least two equivalent resonance forms as represented by (6).

In most cyanines the nitrogen atoms are part of heterocyclic rings. Astraphloxine FF (7) is an example, which is strongly fluorescent in the red region.

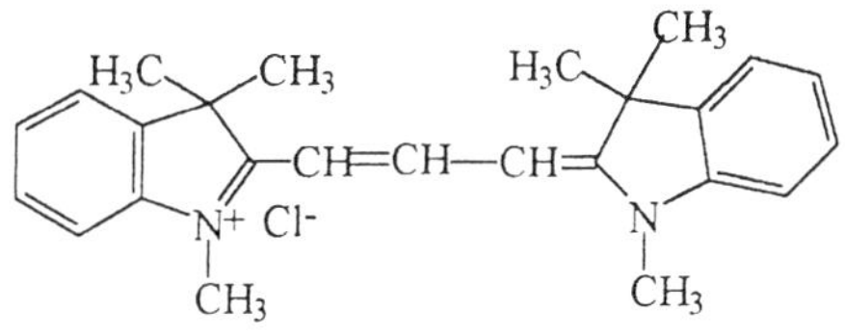

(7)

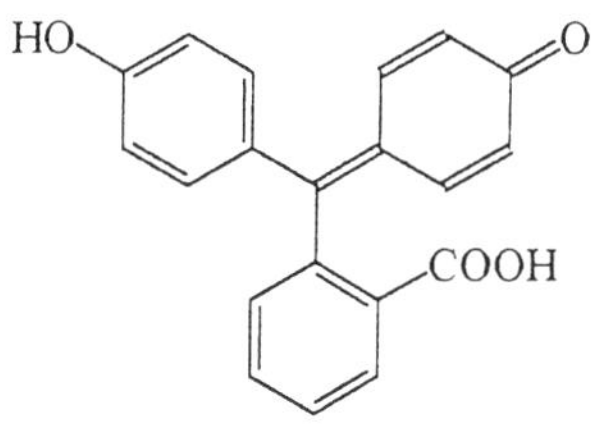

(8)

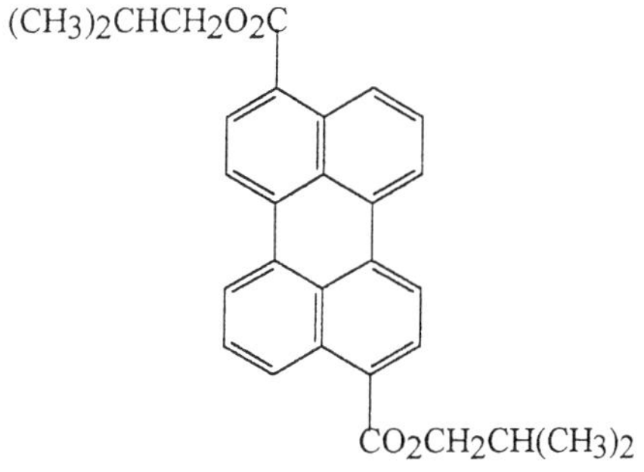

(9)

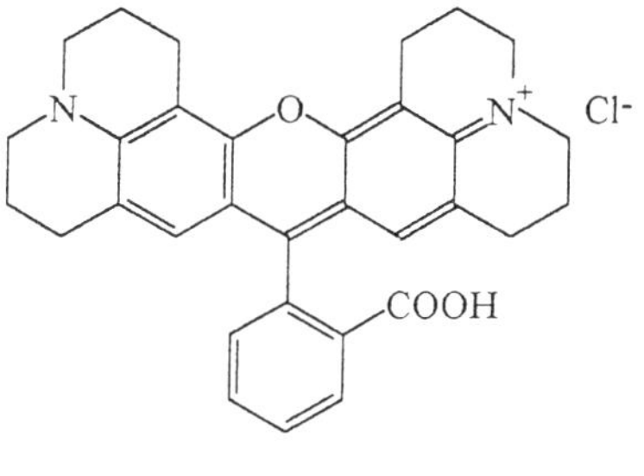

(10)

The well-known indicator phenolphthalein (8) is an example of tri-arylmethane dyes which represents a cyanine-type chromogen and does not fluoresce. If an oxygen bridge is introduced between the upper benzene rings, the intensely green fluorescing flourescein (1) results.

Cyclic Polyenes: This group of chromogens are represented by polycyclic aromatic and heteroaromatic hydrocarbons. One such example is Thermoplast Brilliant Yellow 10G (9) which is a green fluorescent dye.

Rigidity and Florescence: The more rigid the molecule, the less likely that low energy vibrations are initiated by transfer of energy from the first excited state before fluorescence can take place. The effect of a more rigid structure of the rhodamine series has been clearly demonstrated in Rhodamine 101 (10) which has its

Table 16.1. Physical Properties of Fluorescent Dyes Heat Stability*

C.I. Dye Type	% CONC.	GPPS	PC	PMMA	NYLON	PET
S.Y. 160 Lactone	.05	305°C	370°C	305°C	290°C	285°C
S.Y. — Thioxanthene	.05	305°C	315°C	305°C	290°C	285°C
S.Y. 131 Napthalimide	.05	265°C	NR	NR	NR	NR
S.O. — Thioxanthene	.05	305°C	370°C	305°C	290°C	285°C

*No color change after 10 minutes holding time at the given temperatures.

terminal nitrogen atoms each held in two rings and has a fluroescent quantum yield of virtually 100%.

16.2.4. Uses of Fluorescent Dyes

Dyes have broad applications in many fields beyond those of the plastics industry. Four classes of dyes are commonly used in textiles.

Textiles: Acid dyes are characterized by the presence of one or more water soluble anionic groups which is typically a sulphonic acid. These dyes are applied to nitrogenous materials such as wool, silk, nylon or modified acrylics. They are not fast to cotton.

Disperse dyes are relatively small molecules, with very low water solubility, which possess a high affinity for hydrophobic fibres such as cellulose acetate, polyester or blends thereof. The dyes are applied by transfer printing or high temperature steam fixation. Azo and anthraquinone dyes constitute the major portion of disperse dyes.

Reactive dyes typically contain the chlorotriazinyl group. The chloro substituent can be displaced by hydroxy functional groups from cellulosic fibres thereby incorporating the dye molecule into the fabric. Bright and lightfast colors are obtained by this approach. Reactive dyes for cotton generally contain a number of sulfonic groups to provide the water solubility which is required to apply the dyes from aqueous solutions.

Basic dyes are characterized by the presence of some cationic groups. These dyes are used for wool, silk, acrylic, polyamide and polyester fibre. These earliest synthetic dyes for textiles are exceptionally brilliant and have high tinctorial strength. However, their lightfastness is generally poor.

Biochemical applications: Dyes are finding new applications in biochemistry for affinity chromatography, affinity labeling, enzyme assay and in histology.

Plastics: Dyes are used in engineering plastics to make them visually attractive. More familiar engineering resins are GPPS, PC, PMMA, Nylon, and PET. Basically, there are two methods of coloring plastics. Surface coloration, e.g. painting, printing and dyeing; and mass coloration. The vast majority of plastics are colored by mass coloration. A pigment or dye is added at some stage during the manufacturing process to produce a homogeneously colored melt which can then be molded into the finished article. Compared to fluorescent pigments, some fluorescent dyes have far superior heat and lightfastness properties. Tables 16.1 and 16.2 indicate the maximum processing range and carbon arc lightfastness hours.

Table 16.2. Lightfastness Fadeometer Hours (Carbon Arc)

C.I. Dye Type	% CONC.	GPPS	PC	PMMA	NYLON	PET
S.Y. 131 Napthalimide	0.5	510	NR	NR	NR	NR
S.Y. 160	.05	500	500	500	16^d	1500^d
S.Y. —Thioxanthene	.05	1000^2	1000^2	1000^2	16^d	1500^d
S.O. — Thioxanthene	.05	1000^2	1000^2	1000^2	1400^d	1500^2

NR—not recommended.
2—no fading was observed.
d—slight darkening.

In addition to eye-catching products, several uses may be found for liquid oligomeric diol colorants. They can be used to color code polyurethane to identify specific treatments; for example where a flame retardant has been added. Another important use is to monitor the water content, particularly in the production of low density polyurethane foam. Another application is the use of water soluble oligomeric colorants as color coding dyes called VERSATINS. Due to their brilliance, fluorescent dyes have been used extensively in safety applications such as safety cones, barrels, signs, clothing, construction, and industrial purposes where safety is of primary concern. Another important use of fluorescent dyes is in children's toys to make them more visually appealing.

16.3. FLUORESCENT PIGMENTS

16.3.1. Formaldehyde Based Pigments

The fluorescent products used to color plastics fall into two broad classes of resin carriers, formaldehyde and non-formaldehyde chemistry. For the formaldehyde family, a fluorescent pigment may be produced by "dispersing a coloring material, such as a dye, in a resin which is a co-condensation product of an aromatic monosulfonamide, a cyclic triazine or benzoguanamine and an aldehyde such as formaldehyde [4]." These resins are prepared much in the same way other melamine or urea formaldehyde products are produced. Generally, N-methylol substitution of the acyclic triazine and monosulfonamide is first carried out in an aqueous solution utilizing either formalin or paraformaldehyde. Conditions such as temperature and pH of the solutions are chosen to optimize the reaction. Basic catalysis is usually employed during this process. The process then involves the removal of water and acidification of the reaction mixture to facilitate the condensation of methylol derivative to form methylene structures. The reaction is stopped by cooling the resin completely to a solid state. The colored resin is then ground to the desired particle size by means of mechanical grinding.

In order to produce a pigment with increased chemical resistance, increasing the amount of melamine (functionality = 6) proportionately increases the amount of cross-linking upon heating. The result is a thermoset resin with increased solvent resistance.

Fluorescent pigments based on formaldehyde chemistry produce the brightest and cleanest colors possible, but do have limitations in the plastics industry. One is

Table 16.3. Domestic Fluorescent Pigments—Formaldehyde Based

Pigment Type	Manufacturer	
	Day-Glo	Radiant
standard strength	A	R-105, R-103G
high strength	AX	R-106
std. strength (thermoset)	T	P-1600
high strength (thermoset)	GT	P-1700
std. strength (high fastness)	D	R-203-G

the potential for degradation and another is the potential release of formaldehyde odor upon processing at high temperatures. They also tend to build up when run through processes like extrusion, injection molding and blow-molding. For these reasons, non-formaldehyde, low plating out resin carriers were sought.

16.3.2. Non-Formaldehyde Fluorescent Pigments

One of the first non-formaldehyde fluorescent dye carrier systems for polyolefins was based on the reaction of polyfunctional amines with polyfunctional carboxylic acids to form relatively short chain polyamides [5]. These linear thermoplastic resins showed good solubility and friability making them suitable for the incorporation of dyes, which offered increased protection from thermal and UV degradation. This fluorescent resin showed a dramatic increase in color retention upwards of 288°C, even after a 10 minute hold period at this temperature. Due to its "non-idealized" polymeric nature, the polyamide chemistry suffered from preferential "plating out' or migration of polar oligomeric species not incorporated into the polymer chains.

The term "plateout" is one encountered specifically in the plastics industry. It refers to the example given above, where short-chained, highly mobile or polar oligomeric species tend to form or build-up on metal surfaces such as the mold of an injection molder, or the blow-pin of a blow molder. Given enough time, these build-ups will mottle the appearance of finished products and even close off air lines causing part failure to occur. Overcoming this obstacle was the next logical step in the advancement of resin systems.

A relatively lower plating out, polyester/polyamide pigment was the second generation of non-formaldehyde fluorescent pigments. It was shown that a coarser particle size of 20–60 microns vs. 10–12 micron finer grinds, produced higher loadings in color concentrate and decreased the amount of dusting when producing masterbatches. This colorant has sufficient heat stability to 300°C. Although the amount of plateout decreased, it was not sufficient to perform over long periods of time for the sensitive blow molding industry. In order to produce highly colored, attractive bottles for this "plateout sensitive" market, it was necessary to create a more stoichiometrically balanced polyamide type resin [6]. This produced a substantial decrease in the amount of oligomeric species plating out, making it acceptable to run for extended periods of time. A listing of non-formaldehyde products from various suppliers is listed in Table 16.4.

Table 16.4. Domestic Fluorescent Pigments—Non-formaldehyde

Pigment Characteristics	Day-Glo	Radiant	Application	Applicable Resins
Std. strength	Z	K-600	Injection Molding	Polyolefins, Vinyl Plastisol
Std. strength	ZQ	K-7000	Injection/Blow Molding	Polyolefins, Engineering resins
High Strength	NX	—	Blow Molding	Polyolefins
Std. strength	VC	—	Calendared, Extruded and Inj. Molding	PVC only
Std. strength	122-9168, 122-9046	—	Injection Molding	Ionomers

A typical masterbatch for extrusion, injection, or blow molding would consist of an appropriate carrier resin for the polymer being processed, a fluorescent pigment, and a wax additive. The following example illustrates a 25% color concentrate formula for injection molding which should be let down at 4% to yield a final color loading of 1%:

70% Low Density Polyethylene Carrier Resin
25% Fluorescent Pigment (ZQ, K-7000)
5% Wax Additive (See Table 7 for plateout results)
100% Total

16.3.3. Additives to Avoid in concentrates or with dry blends

Metal Ions: In addition to plateout concerns, it has also been shown that certain metal ions cause color shifting with fluorescent colorants. In a study conducted by Day-Glo Color Corp. [7], commonly used metal stearates were added at varying levels with fluorescent pigments. Two of the more common metal-based lubricants are calcium and zinc stearate, often used to aid in the dispersion of conventional pigments. When increasing amounts of calcium and zinc stearate were added to a polyamide based fluorescent pigment and injection molded in HDPE @ 400°F, a graph of delta E values vs. % Stearate were plotted. See Figures 16.3 and 16.4.

These Results demonstrated that as the amount of stearate increased, so did the color deviation from the standard (0% Stearate). It was also concluded that zinc produced more of a color shift than calcium for yellow pigments, with the opposite trend for pinks. In general, increasing stearate, temperature, or dwell time will all proportionately increase the color shifting phenomenon. Also, zinc causes colors to go blue and dirty, with decreasing fluorescent response, while calcium results in a yellow, less fluorescent response. Other common metal ions that may shift color include: free metal not 100% bound within conventional pigments (i.e., Fe in Fe_2O_3),

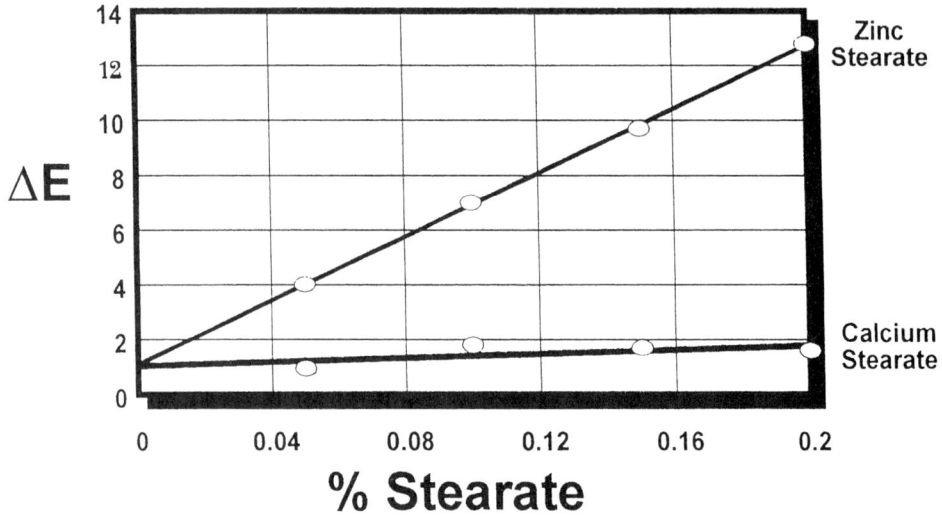

Figure 16.3. Plot of Metal Stearate vs. ΔE Values for Z-17 Saturn Yellow®.

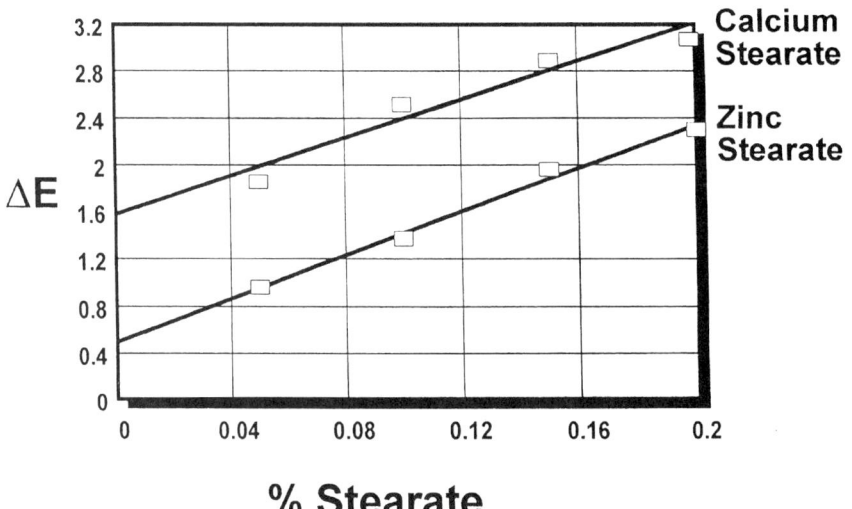

Figure 16.4. Plot of Metal Stearate vs. ΔE Values for Z-11 Aurora Pink®.

Na ions used in nucleating agents such as sodium benzoate, and residual catalysts in polymers (ie., Mg in ABS).

16.3.4. Additives

Plateout Reduction: A wide variety of external additives have been evaluated in an attempt to improve the compatibility of fluorescent pigments with the broad range of plastics that are currently available. In one study by Heyl [8], it was shown that fumed silicas added to masterbatches in small amounts (ie 4–6%) decreased the

Table 16.5. Quantitative Blowpin Plateout Wax Results (given in mg.)

Sample #	Product	Test #1	Test #2	Average	% Improvement
1	Control	2.3	2.0	2.2	0%
2	Allied 573	1.2	1.6	1.4	35%
3	Eastman G-3003	1.7	1.5	1.6	25.6%
4	Allied 597	1.8	1.8	1.8	16%

amount of plateout material deposited on both injection and blow-molding equipment. One drawback to the silicas however, is that due to their extremely low density, they are hard to feed into the throats of processing equipment like extruders. In another evaluation Hyche [9] rated various graft polymers of polyethylene and polypropylene with maleic anhydride to improve the compatibility and dispersibility of fluorescent pigments in polyolefins. This evaluation was carried one step further, to determine how these grafted copolymers affected the net plateout results of polyester/amide based products. In order to make this determination, the test procedure listed below was used.

16.3.5. Blowpin Plateout Test Procedure

A color concentrate is prepared according to the following percentages:

30.0% polyester/amide fluorescent pigment

5.0% maleic-anhydride PE copolymer

65.0% LDPE

and then letdown using a ratio of 1:10 in blow molding grade high density polyethylene. A Rocheleau blow molding machine Model SPB-2 was utilized in this testing procedure. The blow pin was modified to include a removable stainless steel insert on which plateout can be quantitatively determined. The amount of plateout is determined by weighing the insert before and after the blow molding of the specified resin. A total of 20 pounds of resin and color concentrate were blow molded, producing approximately 100 bottles. After the test is complete, the blow pin insert plus the plateout is weighed and the amount of plateout can thus be determined by subtracting the original weight of the insert, giving the amount of plateout in milligrams.

Each wax additive was tested twice for blowpin plateout according to the above procedure. Table 16.5 shows the % improvement over the Control (#1 no wax additives), ranked where 0 = no plateout, and the higher the number in milligrams, the higher the plateout. Results are reported in milligrams.

These results indicate an average of 35% in reduction of plateout using Allied ACX-573, and a 25.6% reduction using Eastman G-3003.

16.3.6. Toxicity

Daylight fluorescent pigments listed here are considered to be non-toxic. They are not however, FDA approved, due to the numerous variations of extractability in

Table 16.6. Results of Laboratory Animal Toxicity Tests

	Day-Glo A Series	Day-Glo Z Series
acute oral toxicity LD_{50}, g/kg	16.0	10.25
acute dermal toxicity LD_{50} g/kg	23.0	10.25
acute dust inhalation LC_{50}, mg/L air	4.4 (4 hours)	1.1
eye irritation	no significant irritation	mildy irritating

Table 16.7. Typical Heavy Metal Analysis for Some Fluorescent Pigments (in ppm)

Elements Determined	A and AX Series	T and GT Series	Z Series	NX Series	ZQ Series
Sb	<4.0	<4.0	<4.0	<4.0	<4.0
As	<4.0	<4.0	<4.0	<4.0	<4.0
Ba	<.50	<.50	<.50	<.50	<.50
Cd*	<.25	<.25	<.25	<.25	<.25
Cr*	<.50	<.50	<.50	<.50	<.50
Cu	1.0	.75	1.0	.75	1.3
Pb*	<1.0	<1.0	<1.0	<1.0	<1.0
Hg*	<0.05	<0.05	<0.05	<0.05	<0.05
Ni	<0.75	<0.75	<0.75	<0.75	<0.75
Se	<4.0	<4.0	<4.0	<4.0	<4.0
Ag	<0.50	<0.50	1.0	<.50	<.50
Zn	136.0	273.0	3,980.0	3,500.0	26,800.0

different resin systems. Table 16.6 gives results of laboratory animal toxicity tests of standard modified melamine-formaldehyde type pigments, Day-Glo A-series and also for the polyamide type Day-Glo Z-series.

The Radiant Color Company has also conducted toxicity tests on products similar to the A Series and found them to be non-toxic.

In heavy metal analysis of all types of Day-Glo fluorescent pigments for plastics, metals were found to be present in only trace amounts (Table 16.7). The values are given in ppm of total metal present, and may vary between colors within a series.

These pigments are in compliance with CONEG (Conference of Northeast Governors) requirements through January 1, 1994 (i.e., total Cadmium, Chromium, Lead and Mercury are less than 100 ppm).

16.3.7. Lightfastness

Fluorescent pigments are stable when used indoors, provided they are not directly exposed to uv containing daylight. Compared to conventional colorants however, fluorescent pigments have relatively limited outdoor lightfastness. In order to maximize the outdoor durability of a fluorescent article, the pigment concentration should be increased. This is true for final pigment concentrations up to the point at which quenching occurs. Quenching is the result of too many dye molecules competing for the same incident light. One dye body's emission is absorbed by an adjacent molecule which dissipates the initial excitation energy. The end result is a lowering of overall fluorescent emission or brightness. Another way to improve the lightfastness of a fluorescent pigment is through the use of a uv absorber or hindered amine light stabilizer. Specific recommendations are given by Ciba-Geigy or Cyanamid depending on the resin system being used. It should be noted that since their purpose is to block out or absorb uv radiation, this will diminish the black light response of a fluorescent article. Another method is to combine a small amount of conventional pigment (as clean and close in shade as possible) with the fluorescent colorant. When the fluorescence eventually does fade away, the conventional pigment will be there as a "back up" so that the article remains colored. Lastly, the use of opacifiers such as TiO_2 create clean, pastel shades and diminish the fade of the fluorescent article. This occurs because light is not only absorbed, but reflected back through the colored layer which results in surface fading.

16.3.8. Uses

Fluorescent dyes and colorants are outstanding for their extremely high visibility and uniqueness in their ability to attract attention. In a study by Xavier University [10], it was shown that children prefer fluorescent color **4:1** over conventional color, when given otherwise identical toy choices. It has also been demonstrated that fluorescent coated signs gain viewer's attention earlier, and hold it twice as long. These properties make fluorescent colors ideal for use in plastic packaging, safety, toys and many other consumer items aimed to heighten awareness or customer appeal.

BIBLIOGRAPHY

1. De Ment, "Fluorochemistry", Chemical Publishing Co. Inc., Brooklyn, New York, 1945 P. 2.
2. Calvert and Pitts, "Photochemistry", John Wiley & Sons, Inc., New York, 1966, P. 192,
3. De Ment, "Fluorochemistry", Chemical Publishing Co., Brooklyn, New York, 1945 PP. 6–9.
4. U.S. Pat. 2,938,873 (May 31, 1960), Zenon Kazenas (to Switzer Brothers. Inc.).
5. U.S. Pat. 3,915,884 (Oct. 28, 1975), Kazenas (to Day-Glo Color Corp.).
6. U.S. Pat. 5,326,621 (Aug. 17, 1993), DiPietro (to Day-Glo Color Corp.).
7. J. Cook, Day-Glo, Plastics Compounding, Vol. 11, No. 7, P. 36 (1988).
8. D. Heyl, Day-Glo, "Improving the Processability of Fluorescent Pigments". Technical Conference of the Society of Plastics Engineers, Oak Brook, IL. (1994) P. 87.
9. K. Hyche, R. Hollis, Eastman Chemical Company, "Preliminary Studies of Improved Dispersing Aids for Fluorescent Pigments in Polyolefin Plastics". Technical Conference of the Society of Plastics Engineers, Oak Brook, IL., P. 148.
10 W. M. Nelson III Ph. D., Department of Psychology, Xavier University.

GENERAL REFERENCES

P. F. Gordon and P. Gregory, "Developments in the Chemistry and Technology of Organic Dyes". Ed J. Griffithe, Published for the Society of Chemical Industry, London, Blackwell Scientific Publications. Oxford, 1984.

Odian, "Principles of Polymerization", John Wiley and Sons Inc., New York 1991.

CHAPTER 17

Introduction to Colorant Selection and Application Technology

Dennis I. Meade

Accel Corporation
38620 Chester Road
Avon, OH 44011

Color matching represents an investment made by colorant suppliers as a means to maintain and grow their business. It is also a service provided to their customers. Molders and extruders of plastic parts often need color costing information in order for them to cost their jobs out and to provide a more attractive, functional, and saleable plastic part to the marketplace. As an investment and as a service, it is essential that color matching be done as accurately and as quickly as possible.

The goal of this chapter is to provide the plastics color formulator with a basic system of gathering the information and materials necessary to produce colorant systems that meet customer requirements. The chapter will not be a discussion of specific colorant properties, as this subject is covered in other chapters. Instead, it will provide a method to aid in the correct selection of colorants and to create an acceptable match quickly. Getting the colorant system approved on the first submittal is always quicker and better than having to resubmit.

First, we start with a few basic definitions:

Appearance: The overall look of the plastic part. This encompasses the combined effect of color and surface texture of the finished molded or extruded part. Matte or textured surfaces make color "appear" lighter or less chromatic. Glossy surfaces make color appear more chromatic.

Binder: The polymer, wax, or liquid used to "bind" or hold colorants and additives together to form the colorant system.

Coloring of Plastics, Fundamentals, 2nd edition. Edited by Robert A. Charvat
ISBN 0-471-13906-8   Copyright © 2004 by John Wiley and Sons, Inc.

Colorant: A pigment, dye, or other chemical that imparts color (including black and white) or any special effect appearance to a polymer. These special appearances could be metallic, pearlescent, stone and fleck, flourescent, phosphorescent, thermochromatic, or photochromatic.

Colorant System: A mixture of colorants, additives, and/or polymers that when added to polymers deliver and disperse the colorants and additives into the finished molded part.

Color Concentrate: A colorant system where colorants and additives are compounded into a binder material. These can be in the form of pellets, coarse powders, granules, liquids, or pastes.

Metamerism: The condition that exists when two colored articles appear to match under one light source but not another. The three common light sources used in color matching are simulated daylight, incandescent, and cool white flourescent.

Processor: The manufacturer of a molded plastic part such as an extruder, an injection molder, rotational molder, blow molder, or compression molder.

17.1. POLYMER TO BE COLORED

The most basic question is to determine the polymer for which a color system needs to be developed? Some basic polymers are polystyrene, polyethylene, polypropylene, acrylonitrile-butadiene-styrene copolymer, flexible or rigid polyvinyl chloride (PVC), acrylic, polycarbonate, acetals, and many others. These plastic resins color differently. Some are clear, white, or pale yellow to varying degrees. Even resins within the same generic family have different colorability requirements. A good example is ABS. This resin varies from clear to pale yellow depending upon grade and manufacturer. Brighter and cleaner colors can be obtained easily in transparent grades while those that are yellow and more opaque require more colorant to mask the resin. Brighter, cleaner shades are more difficult to achieve in resins that have inherent color and opacity. These differences will determine the choice of colorants and the amount of colorant to be used.

The resin manufacturer and the specific resin product code should be established. Processors are very knowledgeable in molding and extrusion techniques but are not well versed in colorant use and resin compatibility. That is the job of the colorist. It may be helpful to contact the resin manufacturer directly for additional information regarding colorant and additive compatibility, especially when designing colorant systems for some of the more exotic engineering resins and alloys. The resin manufacturers may have done research or had experience that may help in selecting the colorant or avoiding certain problematic colorants or additives.

Some of the physical properties of the polymer, such as melt temperature and melt index, need to be determined. Design the colorant system to incorporate easily and completely into the base polymer without detracting from its physical properties.

It is necessary to determine whether the polymer to be colored contains any filler. Fillers have a definite effect on the colorability of a base polymer. They affect surface gloss and may behave like weak pigments. Some more common fillers used in plastic are talc, calcium carbonate, mica, glass fibers and beads, graphite, rubber, alumina trihydrate, and antimony trioxide. Also, the type of filler present and the filler loading level need to be determine. Certain pigments may have a negative effect on the reinforcing qualities of glass fiber, for instance. One colorant system will rarely produce identical colors in filled versus nonfilled polymers of the same generic type. Talc-filled, calcium carbonate–filled, glass-filled, and mica-filled polypropylene do not color the same as natural polypropylene. Thirty percent talc-filled polypropylene does not color the same as 40% talc-filled polypropylene. Supplier A's 30% mica-filled polypropylene may not color the same as supplier B's 30% mica-filled polypropylene. Fillers do make a difference.

Certain polymers need to be dried prior to molding or extruding. Pellet color concentrates are sometimes dried along with the base polymer. Drying temperatures are generally low; however, drying times can be as long as several hours. This could be an extensive heat history that may slightly darken the resin or may even affect the colorant system itself if heat-stable colorants are not selected.

17.2. APPLICATION

Knowing the end application for a plastic part when designing a colorant system is important. The end application plays an important role in colorant and additive selection. Some examples follow:

- Automotive applications may require colorants with superior lightfastness.
- Electronic applications may require colorants that have resistive or conductive properties.
- Thin cast, blown, or extruded films may require extremely small particle size pigments that will not cause tears in the film.
- Excellent dispersion and fine particle size are also required for fiber applications.
- Medical applications often require colorant systems to provide X-ray opacity.

Thus, it is necessary to determine the end application for the colorant system and select colorants and additives that will meet those requirements.

Also necessary is to know what the plastic part will be exposed to during its normal service life. Will the colorant system require highly lightfast or weatherable colorants for full-time outdoor exposure such as with vinyl house siding? Will colorants require chemical resistance such as in swimming pool applications. Will colorants require long-term heat stability such as in automotive underhood parts.

Colorant systems to be used in food contact, cosmetics, toys, pharmaceutical, and medical applications fall under various governmental regulations. The colorist needs to be familiar with these regulations and select colorants accordingly. Also, some states regulate the use of "heavy metal" pigments in packaging applications.

For some applications, the processor may want certain additives included in the colorant system to aid in the performance of the polymer or improve the durability of the colorant system during the expected service life of the plastic part. These additives may include light stabilizers, heat stabilizers, antioxidant, antistatic agents, mold releases and slip agents, fungicides, and antimicrobial agents. These additives should be considered when formulating. They may affect the shade and performance of the colorant system.

17.3. PROCESS

Establishing the process may seem like an obvious objective. However, its importance is often overlooked. Colorant systems designed for conventional injection molding will change appearance dramatically when used in injection-molded structural foam. Formulas designed for injection molding may have too much lubricant or dispersion aids present to allow for satisfactory part line bonding in blow molding applications. Colorant systems for blow molding may require colorants with higher heat stability because of long heat residence times for melted material held in the accumulator. Colorant systems for rotocast polyethylene applications may require higher heat stability due to the long heat exposure times in the rotocast oven compared with colorant systems for injection-molded polyethylene parts. Thin extruded or blown film applications may require more colorant loading for opacity than for the same polymer in an injection-molded part or thicker walled extruded profile. These are just a few examples of how the various part-forming processes can cause color variations even when dealing with the same polymer.

17.4. PROCESS CONDITIONS

The color formulator must learn as much as possible about the processing conditions the colorant system will encounter. Important conditions to determine are cycle times and process temperatures. Hot runner systems commonly used in today's injection molding processes may add temperature to the polymer melt and add to the time the polymer is under heat. Remember that the heat stability of any colorant and additive system is a function of both time and temperature. An excellent color match may be produced, but if the formulation cannot withstand the stress of processing without breaking down or changing shade, then effort and time have been wasted.

Other process conditions that may stress the polymer and colorant system would be high shear or high length/diameter extrusion screws. Extremely small gates on injection molds or extremely high injection pressures can cause momentarily high shear during the injection molding process. High shear equates to high temperatures. Colorant systems without the heat durability to withstand these conditions will darken, streak, or change shade.

The color formulator should also find out if the processor intends to post-decorate or apply labeling to the finished part. Certain pigments make possible the process of laser marking. Excessive dispersion aids or lubricants in the colorant system may interfere with adhesion. Incorrectly selected colorants could bleed or migrate into or through the decoration or label. If any postdecorating will be done, the colorant system should be formulated accordingly.

17.5. COLOR TARGET

A color formulator will be given "color targets" to match of all shapes, sizes, materials, and descriptions as well as plaques, cards, swatches, parts, chips, stones, glass, fruit, and fabric. He or she will be presented with anything from animal, to vegetable, to mineral that illustrates a color the customer wants to reproduce in plastic. This is the challenge and fun of being a color matcher.

Establishing a proper color standard is one of the most critical aspects of color matching. It is essential that the color matcher and customer agree on the color target or color standard. Both need to look at and measure the exact same thing. More color rejections occur because of incorrect or improper color standards than for any other reason.

We start with a bad example of setting a color target. An injection molder requests a match and a price quotation on a particular color. The customer walks out onto the shop floor, grabs a part from one of the bins, and says, "Here. Match this." You question your customer diligently, almost to the point of annoyance, about the polymer to be colored, the application, the process, and the process conditions. You take the color target back to your laboratory and pour over your technical data, reference charts, and colorant information to select just the right pigments, dyes, and additives to produce the closest and most economic match. Two weeks later, you are back in your customer's office with the color match. Your customer walks back out onto the shop floor, grabs a part, brings it back into the office, and lays it down next to your color display. To your chagrin, the color sample does not match. What is wrong with this picture and how could you have avoided this waste of time, money, and opportunity to do business?

There is absolutely nothing wrong with having a molded part as a color target. The problem with this example is that the colorist matched to one standard but the customer compared the color submittal to a different standard. Slight color variations in production lots of color concentrate, a change in injection presses, or a change in process conditions from the time the first "color standard" was given to when the second color standard was pulled from production could have created enough color difference in the two standards to cause a color rejection.

There are two possible solutions to this situation. One is to bring the first color target back to the customer when submitting the match. Then the submittal can be reviewed versus the same standard—the one the colorist matched. Care should be taken to maintain the integrity of the color standard if it is going to be handled and transported. It needs to be kept clean and protected from light and abrasion.

Another solution would be to cut the original color target in half when the customer first gives it. Both the colorist and the customer should sign, date, and initial each half. The customer should keep half and protect it from light, dirt, and abrasion. The other half is taken to the laboratory for formulation. This ensures that when the customer receives the color submittal, the same color target is being compared.

Many industries do a good job of generating color standard plaques and displays. Automotive companies regularly generate multiple plaques of each standard color and certify the uniformity and traceability of each separate plaque. The appliance, electrical, electronics, and furniture industries all have similar color mastering programs. These programs ensure that the standards given to colorant system suppliers and those used by the people judging the color matches are uniform. A few companies such as Pantone, Inc. and Munsell Color Services produce color standards and color catalogues that are commonly used in the plastics industry for color masters. The federal government also has a set of color standards listed in their Federal 595 catalogue.

On occasion, a colorist may be asked to match to a set of reflectance numbers, or L, A, B coordinates, generated by spectrophotometric measurement of a color standard from a remote location. In other words, he or she will be asked to match a color that they have not seen. While instrumentation technology has improved over recent years, there still may be enough variance between individual spectrophotometer to cause a color rejection. Instruments are extremely helpful quality control tools to measure color and numerically define color differences. However, their use as a device to reduce a color target to a set of numbers and then send that set of numbers to a color formulator as a target for matching is seldom successful. Physical color standards that can be viewed under various light sources are preferred.

17.6. COLOR APPROVAL CRITERIA

The primary function of a commercial color formulator is to develop a colorant system that meets a customer's specifications. Occasionally it may not be possible to make the "perfect" match. In these cases the color formulator should try to find out what criterion—quality (closeness of match) or cost—is the customer's priority.

Let us consider color match quality. The color formulator should ask the customer if the closeness of a match to the color target is the highest priority. Since there are hundreds of colorants available for the color formulator to select from, more than one possible combination of colorants may produce an acceptable match. Is the best, nonmetameric match of most importance to the customer and application? If a perfect, nonmetameric match is not possible, which is often the case when trying to match plastic colorant systems to color targets in paint, ink, ceramic systems, or plastic media, will the customer accept some metamerism? Also what light source will be the customer's preference when judging a slightly metameric match?

In some cases cost may be the dominant approval criterial. In applications where color is only an identifying attribute, the customer may be interested in just a close match to target, preferring that the colorist formulate for lowest cost. Examples of such applications could be wire and cable coatings, automotive dunnage applications, pallets, bread trays, and beverage crates.

17.7. INTRODUCING COLOR INTO PROCESS

There are several methods of delivering color into the polymer and each of these methods will play a role in the color formulator's selection of additives and binders:

Central Blending: Mixing the colorant system with polymers in a low-intensity blender at a central plant location, then conveying that blend pneumatically or by bulk container to the molding machine or extruder.

Metering Systems: Feeding devices that deliver colorant systems into the polymer stream just above the feed throat of the molding machine or extruder. Metering systems can be volumetric or gravimetric. When supplying colorant systems for volumetric metering systems, the bulk density of the colorant system must be controlled on a lot-to-lot basis. This is necessary to maintain a constant colorant percentage and therefore a constant color in the molded part.

This is also the time to establish with the customer the desired "let-down-ratio": the ratio of the colorant system to the polymer (e.g., 25:1, 50:1, 100:1). The first number in the ratio represents the amount of uncolored polymer, the second number the amount of colorant. Let-down-ratios are also expressed in parts per hundred. For example, 1 pph means 1 part of colorant system to 100 parts of uncolored polymer.

17.8. FINISHED COLORANT SYSTEM

Colorant systems are available in several forms, and it is a good idea at this juncture to establish with the customer the method they prefer to use to introduce color into their molded or extruded part. Colorant system form may have an influence on colorant selection and will influence additives and binder selection. Colorant systems are available in the following forms:

Dry Color: A high-intensity blend of pigments and/or dyes with additives provided to the user in the form of dry, free-flowing powders.

Powder Masterbatch: A blend of pigments and/or dyes with additives incorporated into a wax or polymer binder, then pulverized or granulated back into a coarse powder form.

Pellet Color Concentrate: A blend of pigments and/or dyes with additives incorporated into a wax or polymer binder, then formed by compression or extrusion into a round or cylindrical pellet.

Liquid or Paste Color Concentrate: A blend of pigments and/or dyes with additives incorporated into a liquid binder, then provided in a pourable liquid or viscous paste.

Precolored Compounds: Polymers that contain colorant and additives and are ready to mold or extrude without further processing. Precolored compounds can be in a pellet or dry powder form.

An example of a generic color match request form used as a tool to gather the above data is shown in Figure 17.1.

17.9. COLORANT SELECTION

Colorant selection is the most difficult step in the color-matching process. There are hundreds of pigments and dyes to select from and the trained colorist should be

CUSTOMER CODE	SALESMAN	REQUEST BY	DATE

CUSTOMER	SHIP TO

TELEPHONE NO	CONTACT	

PRODUCT APPLICATION	NEW APPLICATION	REFORMULATION #

FORMULATION NO _____
COLOR NAME _____
ADDITIVE NAME_____

PRODUCT FORM:	POLYMER:	PROCESS TYPE:
CONCENTRATE PELLETS @_____#PH		EXTRUSION [ ] OTHER [ ]
DRY COLOR_____	RESIN	BLOW MOLDING [ ]
COMPOUND_____	SUPPLIED	ROTO CAST [ ]
LIQUID CONCENTRATE_____	YES / NO	INJECTION [ ]

OPACITY:	PART THICKNESS	EXPOSURE:	HEAT STABILITY:
OPAQUE [ ]		INDOOR_____	HIGH (ABOVE 500)F [ ]
TRANSLUCENT [ ]	_____		MODERATE (400-475 [ ]
TRANSPARENT [ ]	(MILLS)	OUTDOOR_____	LOW (380-400) [ ]

COLOR TOLERANCE:	ILLUMINANTS: COOL WHITE.	HORIZON, DAYLIGHT, INCA, UV,

ADDITIVES REQUIRED:		SPECIAL APPLICATION:	
UV [ ]	SLIP [ ]	FDA [ ]	NONTOXIC [ ]
ANTIOX [ ]	FLAME RET [ ]	USDA [ ]	TOY [ ]
ANTISTAT [ ]	OTHER [ ]	CHEMICAL RESISTANT [ ]	

DATE STARTED_____	TIME STARTED_____
DATE COMPLETED_____	TIME COMPLETED_____
TECHNICIAN_____	

Figure 17.1 Color Match Request

familiar with as many as possible. It is helpful to build a library of information on colorants, additives, and resins. The best sources of technical information are the sales representatives from the raw materials suppliers. They can supply product data sheets on their materials as well as request technical support from their company's plastics laboratories. Based on information previously gathered, the colorist can select colorants and additives that are (1) compatible with the polymer, (2) suitable for the application, (3) suitable for the process, (4) durable enough for the process conditions, (5) of suitable hues to match the target color, and (6) able to meet the key approval criteria.

- ☐ **Establish the Polymer to be colored.**

- ☐ **Establish the Application.**

- ☐ **Establish the Process**

- ☐ **Establish the process conditions**

- ☐ **Establish the color target**

- ☐ **Establish the key color approval criteria**

- ☐ **Establish the processor's method of introducing color into the process**

- ☐ **Establish the finished colorant system form**

- ☐ **Select Colorants**

- ☐ **Blend the selected colorants to match the color target**

- ☐ **Compound the formula into the desired colorant system form**

Figure 17.2 Color Matching Checklist

17.10. MATCHING THE COLOR TARGET

The easiest part of the color-matching process is blending the various colors and additives to make the match. The selected colorants and additives should (1) match the hue of the target (watching for metamerism), (2) match the color saturation (chroma value), (3) match the lightness/darkness value, and (4) match the desired level of transparency or opacity. The colorist should consider and allow for the effect of any required additives on the color and evaluate the finished formulation for cost.

17.11. COMPOUNDING FORMULA INTO DESIRED COLORANT SYSTEM

The colorist should do the following:

   Blend the formulated colorants with the appropriate additives to produce a dry color.

- Compound the formulated colorants with the appropriate additives and polymer to produce a masterbatch or pellet concentrate.
- Mix the formulated colorants with the appropriate additives and liquid binder to produce a liquid color or paste.
- Compound the colorants and additives directly into the polymer to make a precolored compound.

The outline in Figure 17.2 can be used as a checklist to help gather all the needed information.

17.12. SUMMARY

The reader should learn as much as possible about the properties of colorants and polymers as well as about plastics processing and plastics applications. Colorant selection is the hard part. Color matching is easy.

CHAPTER 18

Color Compounding

Scott Russell

Polyone
33587 Walker Road
Avon Luke, OH 44012

18.1. INTRODUCTION

Among the technological innovations of the twentieth century, the compounding of polymeric materials stands out for its flexibility and diversity. Literally thousands of material combinations suitable for a myriad of end-use applications are the result of "compounding." For purposes of this chapter, compounding can be defined as "the combination and melt-mixing of polymers and 'value-adding' components to create more useful materials."

The attributes and properties of compounded plastics are applicable to many diverse applications. Although the focus of this chapter is the compounding of thermoplastic materials, many of the principles and comments also apply to thermoset polymers such as phenolics. The chapter will discuss compounding methods, the roles and challenges of additives, and techniques for introducing color to compounded plastics.

Blending and compounding polymers and additives in an optimum manner are the challenges of all plastics compounding processes. Although "added-value" enhancements of compounded plastics may often involve physical properties such as stiffness, impact strength, thermal resistance, and hardness, other performance characteristics also are often involved. These additional characteristics can include UV resistance and antistatic or frictional (tribological) properties, which are often key to successful end-use performance. To meet specific application needs, the compounding process must achieve a "balance" of properties, acquired through

---

Coloring of Plastics, Fundamentals, 2nd edition. Edited by Robert A. Charvat
ISBN 0-471-13906-8 Copyright © 2004 by John Wiley and Sons, Inc.

customized formulations that have integrated a broad group of components and characteristics into a single synergistic formula. A typical formulation involves 4–10 ingredients properly blended to assure the product's final performance.

The polymeric base of the blended mixture dictates many of the physical properties. There are at least 50 different polymers in multiple grades and hundreds of additives available to modify key property sets. Both polymers and additives come in many physical forms (e.g., pellets, powders, and liquids) with a wide range of melting point and thermal stability attributes. This variation in materials and properties is the reason there are five major compounding processes, with still more variations within these processes.

18.2. COMPOUNDING BASICS

All compounding processes share several basic scientific principles and parameters:

- *dispersive mixing*, that is, reducing agglomerates to their finest form;
- *distributive mixing*, that is, achieving a high degree of homogeneity throughout the polymeric matrix; and
- *thermal control*, that is, assuring the time–temperature profile is within the rheological and stability parameters of the polymer–additive system.

At times, low- or high-intensity blending alone can produce a suitable product for use by the fabricator. An example of this would be a polyvinyl chloride (PVC) blend used for several large-volume extrusion applications. More frequently, however, a compounding process is required to obtain the desired physical property. The five primary compounding processes used in the industry for controlling the above parameters are single-screw extrusion (SSE), twin-screw extrusion (TSE), continuous mixers, batch mixers, and kneaders. Table 18.1 summarizes key aspects of each process.

Table 18.1 Compounding Process Comparison

Process	Typical Feed	Dispersive Mixing	Distributive Mixing	Typical Upper Temperature Limit	
Single-screw extrusion	Flood	Poor	Good	600–650°F	315–345°C
Twin screw extrusion	Starve	Excellent	Excellent	600–650°F	315–345°C
Batch mixer	Not applicable	Excellent	Excellent	450–500°F	230–260°C
Continuous mixer	Starve	Excellent	Excellent	450–525°F	230–275°C
Kneader	Starve	Good	Excellent	475–525°F (oil) 600–650°F (electric)	245–275°C (oil) 315–345°C (electric)

Each process has a place in the compounding picture because of how it relates to the three key compounding parameters of dispersive mixing, distributive mixing, and thermal control. There is an overlap in the capabilities of these processes, but there are formulations that are best produced by only a single process.

A fourth and key parameter is cost. Compounding costs include initial capital, operating costs, and process clean-out ease. For example, SSE generally has a poor dispersive mixer but a good distributive mixer and offers a relatively low initial capital cost. On the other hand, TSE offers a broad and flexible approach with both excellent dispersive and distributive mixing capability but at a higher initial acquisition cost. Continuous mixers offer excellent high-shear mixing but are limited in their ability to process high-temperature polymers.

All five processes share several of the following steps: materials handling, pre-blending and/or drying, melting, mixing, degassing/venting, final form development (usually pelletizing), and packaging. Selection of the optimum compounding process and setup is heavily dependent on the specific formulation, the final quality needs, and the cost requirements of the application. Reactive compounding is also required for some formulations and will further define the details of the chosen process.

18.3. ROLE OF ADDITIVES/COMPONENTS IN COMPOUNDING

The starting product formulation defines the mixing and thermal challenges of the compounding process. For example, using PVC as a starting base polymer means that the compounding process must provide a low stock temperature capability and tight process control parameters are required to prevent decomposition by way of dehydrohalogenation of the compound. This selection of PVC as a base polymer reduces the compounding options available. It also tightly defines the processes suitable to meet the low-residence-time (shorter machines) and low-temperature requirements typical of this compound.

On the other extreme, a very high temperature polymer system such as polyethersulfone (PES) or polyphenylene sulfide (PPS) requires very high temperature machines to provide sufficient heat to properly control melt viscosity and rheology during processing. This selection of a high-temperature base polymer will often necessitate the use of relatively simple mixing formulations. Table 18.2 illustrate typical processing temperatures for a wide range of polymers.

Although thermoplastics are generally supplied in either powder or pellet form, additives come in forms ranging from liquids to solids. The thermal stability and form of the additive help define the materials handling, drying, feeding, melting, and mixing challenges of the total compounding process system. Table 18.3 lists typical types of additives used in formulating compounded plastics.

The various forms, particle sizes, melting points, and relative thermal stabilities of additive formulations require not only a range of compounding processes but also technical advances in materials handling, feeders, and all other aspects of the total compounding process. Many lubricant additives, especially at higher usage levels, create slippage of the screw in the early phase of the compounding process, reducing the ability of the process to provide shear for proper mixing. As a result, the selected compounding process must maintain acceptable production rates while assuring adequate dispersive and distributive mixing for the final product. Liquid

Table 18.2. Typical Processing Temperature Ranges for Selected Thermoplastics

Polymer Type	°F	°C
High-density polyethylene (HDPE)	400–500	205–260
Low-density polyethylene (LDPE)	350–425	175–220
Polypropylene (PP)	400–500	205–260
Polyvinyl chloride (PVC)	350–390	175–200
Acrylonitrile butadiene styrene (ABS)	390–500	200–260
Polyacetal, copolymer (POM)	375–425	190–220
Polybutylene terephthalate (PBT)	450–500	230–260
Polyethylene terephthalate (PET)	495–570	255–300
Polyamide (Nylon) 6 (PA 6)	480–580	250–305
Polyamide (Nylon) 6,6 (PA 6,6)	510–600	265–315
Polycarbonate (PC)	530–640	275–340

Table 18.3. Additive Types for Plastics

Type	Consumed During Year 2000 ($\times 10^6$)	
	kg	lb
Fillers	3795	8366
Plasticizers	929	2050
Reinforcements	727	1600
Colorants	302	665
Impact modifiers	91	200
Lubricants	68	150
Heat stabilizers	46	100
UV stabilizers	8	17
Antioxidants	27	60
Organic peroxides	32	70
Chemical blowing agents	8	17
Antimicrobial agents	8	17
Antistatic agents	8	17
Flame retardants	439	968
Other	78	172

additives, in particular, offer compounders the challenge of assuring that the proper composition is maintained while still obtaining the correct degree of mixing.

18.4. COLOR IN PLASTICS COMPOUNDING

The most visually obvious "attribute" developed in plastics formulation and compounding is the aesthetic appearance observed in the final fabricated part. Although the visual acceptability of the component may involve characteristics such as gloss and the absence of surface imperfections, color is one of the most evident and measurable final product attributes. The balance of this chapter addresses issues presented by adding color to the plastics compounding process.

18.5. COMPOUNDING WITH COLORANTS

All the considerations discussed in the formulating and compounding of plastic products also apply when colorants are added to the picture. Like other additives, colorants come in many forms and particle sizes, with a wide range of melting point and thermal stability characteristics. The average price of colorants, however, is often far higher than that of other types of additives, requiring good formulation and conservation practices by compounders to control costs.

Generally, colorants are classified as either pigments or dyes. A pigment is a colorant that is insoluble in the polymer base being used, whereas a dye is soluble in the base resin. Pigments, dyes, and other colorants can affect other key physical and mechanical properties of the base resin and can, in turn, be affected by the polymer's relative pH (alkalinity or acidity). As chemicals, color additives can synergize or interfere with needed reactions during the compounding process. For example, virtually all formulas involve the use of chemical reactions to stabilize the polymer during the various thermal histories it will be subjected to during processing, and certain colorants can interfere with these reactions. Further, colorants often influence the final thermal and UV stability of the material blend. Mechanical properties, particularly impact strength, also can be altered based on the colorant formula used. For example, the impact strength of many products can be severely affected by the addition of TiO_2. If the base polymer requires very high fabrication temperatures, the colorant also must be stable at those temperatures, which can exceed 600°F (315°C).

Pigments are either inorganic or organic chemicals. Examples of inorganic pigments include ultramarine, titanium dioxide, iron oxides, cobalt blue, and chrome oxides, whereas common classes of organic pigments are phthalocyanine, azo, polycyclic, and carbon black. Pigments are insoluble in the polymer matrix; therefore, any agglomerates must be reduced to the minimum particle sizes (dispersive mixing) to optimize color development, reduce load levels and thus cost, and minimize negative influence on other properties. Selecting the correct pigment system and maximizing dispersion (including distributive) efficiency are the greatest technical challenges confronting the compounding processor. Inorganic pigments offer similar challenges as those that are experienced when working with small-particle-size reinforcing fillers such as calcium carbonate, talc, barium sulfate, or mica.

With pigments, assuming thermal stability is adequate, the issues of dispersive and distributive mixing are dominant concerns in the compounding process. If the pigments are not reduced to their smallest particle size through proper compounding, the color will not be fully developed. As a result, higher pigment levels will be required. This leads to a negative effect on both cost and, potentially, resulting properties. For example, inorganic violets are particularly weak and difficult to disperse and therefore require higher additive levels. In contrast, luminescent pigments often require minimum shear to avoid breakdown of the relatively larger particles needed to obtain this desired special effect.

Material handling problems with very fine inorganic pigments can create housekeeping problems. Pigment suppliers are constantly developing improved handling characteristics for their products.

Organic pigments have grown in popularity and use during the 1990s, especially as some classes of inorganic pigments like cadmium have come under environmental scrutiny. Typically, inorganic pigments are limited to the coloring of opaque materials. Organic pigments, however, can be used in either opaque or transparent systems. The most significant disadvantages of organic pigments are that they are more thermally unstable and more difficult to disperse. They tend to act more like additives than fillers during compounding. Compounders must recognize these disadvantages in their process selection and setup and make adjustments to allow for the relative differences. Aside from thermal stability and dispersion differences, organic and inorganic pigments are treated in similar manners during the compounding process.

Dyes are generally used with transparent materials and are classified by their Colour Index generic name and constitution number. The latter is a five-digit number relating to a dye's chemical structure, such as xanthenes, quinolene, and azine. Dyes are usually added at low levels for tinting purposes. Since they are primarily used in transparent materials, their formulation systems are usually very simple. As with other pigments, dyes must be selected to match the polymer and additives using the processing temperature parameters required. The compounder must then select the process to meet these requirements and provide the needed mixing and control to achieve a quality product.

The need for cleanliness to prevent contamination is absolutely essential to managing the processing needs of these compounds. Mixing requirements are often low based on the simplicity of the formulation and the low percentage of dye used. Table 18.4 compares the three major types of colorants.

At times, a concentrate or masterbatch will be used instead of raw organic or inorganic pigments or dyes during the compounding process. Benefits of a masterbatch can include improved color development, higher productivity, ease of housekeeping, and faster process cleanout. These concentrates may be purchased from a specialty manufacturer or produced internally. Using this approach also can permit the selection of a simpler, more economical compounding process.

One disadvantage of the masterbatch approach is the inability to adjust individual colorants if color adjustments are required to meet quality standards. The use of several individual single-pigment concentrates can overcome this problem.

Table 18.4. Types of Colorant and Their Relative Characteristics

Type of Colorant	Loading Level	Behavior in Polymer Matrix	Coloring Effects	Challenges
Pigment				
Inorganic	High	Insoluble	Opaque	Dispersion
Organic	Medium	Insoluble	Opaque transparent	Dispersion, thermal stability, cost
Dye	Low	Soluble	Transparent	Compatibility, thermal stability

Table 18.5. Forms of Colorant and Relative Performance Comparison

Form of Colorant	Materials Handling	House Keeping	Color Match/ Adjustment	Raw Material Cost
Raw/dry colorant	− (time consuming)	−	− (time consuming)	Low
Single colorant concentrate	+	+	+	Medium–high
Precolor concentrate	++	++	++	High

Although a masterbatch can be more expensive because of the cost of raw materials, the advantages of advanced color development, higher productivity, easier housekeeping, and process cleanout and the option of a less complicated compounding process can help offset higher raw material costs.

The fabricator has several methods to choose from when blending pigments, dyes, or a masterbatch into the polymer compound. The choice must ensure that the final part meets the color needs of the end-use application.

One method is adding color via concentrates or a masterbatch at the fabrication step (molding or extrusion). The process involves a two-component approach: a natural polymer and a concentrated (solid or liquid) additive masterbatch metered and blended at the press. In this option, the fabricator must take responsibility for preblending the product, properly mixing and dispersing the two phases, and performing checks and adjustments to assure final color quality.

A second method for introducing color is known in the plastics industry as "salt and pepper." This process requires that a supplier blend the base polymer and color concentrate together before shipping the mixture to the fabricator. Care must be taken to prevent any segregation of the two phases, but this is a very effective route for some applications.

A third method is to use precolored product, where the needed pigments and additives are precompounded into the base resin through a melting process and then sold as a single-component product. The precolored compound approach requires that the supplier provide the fabricator with product that is already formulated and color matched to meet the needs of the end-use application. This is accomplished by using the coloring additives in a melt-mixing process to assure final appearance, property development, and consistency. The technique also requires that the fabricator maintain an acceptable and consistent processing window to protect final color quality. Table 18.5 compares the three approaches discussed.

18.6. QUALITY CONTROL

Color adds a subjective aspects to the quality of compounded products. Clear targets must be established for the desired color when the initial match is performed. When the match is approved, a specification needs to be established that reflects the target and an allowable variation for the compounded color.

As discussed in other chapters, the instrumental methods for measuring color quality have improved and gained a great deal of acceptance within the industry. Using an approved light source, ΔE values are often placed in internal specifications as the color control quality target. Upper and lower limits are established on key values, such as lightness/darkness (L value) or blue/yellow (B value) to ensure that unacceptable color excursions do not occur. Injection or compression molded chips of acceptable thickness are typically used as samples for testing. Color chips may be sent to the customer to confirm or document color as well as instrument values. If there are concerns that colorants may affect key physical properties, a minimum property value, such as impact, is usually placed on the internal specification. Blown film or other tests are sometimes used to confirm acceptable dispersion levels. Sampling frequency is dependent on the quality systems and statistical methods being used by the compounder. Customers often require certificates of analysis (COAs) to prove that specifications are met. However, visual conformance using trained personnel, combined with instrumented methods, is considered the best method to assure the quality of the product.

18.7. SAFETY, HEALTH, AND ENVIRONMENTAL

The materials handling, blending, feeding, and compounding of colors generally have the same safety, health, and environmental issues as all additives used during the process. Chapter covering summarizes these issues for this segment of the industry. The principles outlined there also apply to precolor compounding. Topics for consideration in this area include raw material Material Safety Data Sheet (MSDS) considerations and air and water quality. Once encapsulation in the compounding process via melting occurs, solid waste is evaluated by requirements outlined in the Federal Government Publication 40 CFR 262.11.

CHAPTER 19

Dry Color Concentrate Manufacture

Joseph M. Cameron
RTP Company
580 East Front Street
Winona, Minnesota 55987

19.1. TERMS USED IN THIS CHAPTER

In their manufacture and finishing process, pigments come close enough together to form clusters of primary particles called agglomerates. In order to gain the full color potential of the pigment, these agglomerates must be broken apart and the individual particles scattered in all directions. The term *dispersion* deals with how well the process of breaking apart the individual particles has been accomplished. *Distribution* deals with how evenly the dispersed particles are scattered or spread throughout the material.

The term *color release* is used to describe how well a concentrate melts and distributes in the final product. A concentrate with poor color release will give a weak or streaked finished part appearance.

Processing refers to the methods used to convert plastic pellets into finished parts and can include injection molding, extrusion, and roto-molding.

The terms *color at the machine* or *dry coloring* are used to describe coloring product during processing (i.e., injection molding) using concentrates. *Dry color* is also used to describe compounding from neat dry pigments as opposed to using concentrates. Are plastics people consistently inconsistent?

Delamination describes how two incompatible resins molded together can form

---

Coloring of Plastics, Fundamentals, 2nd edition. Edited by Robert A. Charvat
ISBN 0-471-13906-8 Copyright © 2004 by John Wiley and Sons, Inc.

layers that peel apart after cooling. They may only appear to have small blisters, but when scraping at the spot, layer after layer of polymer peel away.

Cube blends, also called *salt and pepper*, employ a technique of blending concentrate and natural pellets prior to processing rather than metering concentrate in during processing.

Precolor compound is material containing colorant throughout and ready to mold as is.

Single-pigment dispersions (SPDs) are concentrates that have only one colorant ingredient.

Letdown (L/D) is a term plastics people use in different ways depending on the circumstances. In concentrate color parlance, letdown refers to the ratio at which a concentrate is designed to be used with natural polymer in order to obtain proper color. In color compounding, the term is used to describe a diluted pigment as opposed to a neat dry colorant. Some call this a "cut" pigment. The dilution in this context is usually done to reduce the strength so dosage is more accurate. Another use in plastics of L/D refers to screw length in relationship to diameter for extruders and injection molding machines.

Mixing and *compounding* are used synonymously to describe the action of physically combining a number of different materials into one.

19.2. TYPES AND FORMS OF CONCENTRATES

There are many types of color concentrates (also called masterbatches) to choose from, including liquid, paste, resinated blends, so-called freeze dried, universal concentrate, resin specific, and encapsulated. All types have strengths and weaknesses that have to be considered when choosing what is right for an application. This chapter will attempt to clarify some of those differences. The liquid and paste types, however, will be left to another chapter since they are somewhat different from the rest, in both how they are made and how they are used.

Color concentrates are supplied in a wide range of physical forms:

1. Liquid concentrates are much like latex paint but range in consistency from liquids to thixotropic gels and pastes. The major advantage of a liquid form is distribution efficiency.
2. Resinated blends are dry mixtures of pigment plus fine solid additives and in some cases ground polymer resin into a powdered mixture. The resulting mixtures are either trumble blended with natural pellets prior to processing or metered in during processing. Very little in the way of special equipment is needed to make resinated blends. They are made in typical high- or low-shear blenders and packaged without melt compounding.
3. Freeze-dried concentrate is called that because the first material made looked like freeze-dried coffee crystals. It is actually pigment dispersed in a waxy additive that is then used either by grating material off a bar like grating cheese or by flaking the material into irregular granules that can be metered in with natural resin during processing.
4. Universal concentrate is more a description of the carrier system than it is a form of concentrate. Universal concentrates use a carrier system such as EVA

or EMA that is compatible with a wide array of polymers. Some use low-melting-point waxes or solid stabilizers as the binder. This form is usually converted into pellets but is also made into small beadlike particles called prills. Another universal concentrate system is a blend of polystyrene and colophony ester, which is a rosin derivative that functions as a compatibilizer. This product is sold in irregular granule form as well as pellets.

5. Resin-specific concentrates are colorants dispersed in the same polymeric material as the one in which they are intended to be used: acrylonitrile butadiene styrene (ABS) for ABS, Polycarbonate (PC) for PC, Polyphenylene sulfide (PPS) for PPS. These are almost always sold in pellet form; however, they are occasionally ground back to a powder form for use when color release problems occur. The latter approach is generally employed with single-pigment concentrates that are used in color compounding as opposed to coloring at the machine. Another use of this technology is in roto-molding, where the polymer compound as well as the colorant are ground to a fine powder, blended, and tumbled in a hot mold to create a hollow precolored part. The reason the materials have to be finely ground for roto-molding is because there is little if any flow during this process to provide distribution of the colorant. By grinding the ingredients to a powder, the colorant can be physically distributed throughout the polymer prior to melting, thus giving more uniform part coloration.

6. Encapsulated concentrates are fine granules of pigment that have been prilled into waxy additives or coated with polymeric material. The coating keeps the colorant from contaminating or staining the work environment yet allows enough friability to break down and distribute during mixing or processing. Holcobatch and Holcomax, trademarks of Holland Suco and Holland America, respectively, are examples of this technology.

19.3. CARRIER CONSIDERATIONS

Selection of a carrier resin is much more complex than one would naturally suspect. Some of the considerations seem obvious; for instance one cannot use a carrier that is incompatible with the letdown resin. For example, low-density polyethylene (LDPE) is incompatible with ABS. It leads to delimitation in parts, poor impact, and low tensile strength. However, small amounts of LDPE are not a problem in ABS. It simply disappears in the matrix and has no measurable effect. Would the writer recommend using LDPE as a carrier for ABS concentrates? Certainly not. However, pigments weighed in polyethylene bags are thrown bag and all into batch mixers with ABS everyday. The point is that questionable contaminants in polymers can be acceptable as long as the amounts are small and they are well distributed in the matrix.

Waxes are another excellent example. They can be very good carriers for color concentrates but, if not properly used, can cause serious problems in finished parts. Ethylene bis-stearamide (EBS) wax is widely used as a carrier medium. If EBS loading levels in the final product are too high, a number of problems can result, including excessive screw slippage during processing, plate-out on molds and dye lips, splay, latent blooming on parts, and poor adhesion of pad printing or other decorating problems. The EBS wax also reacts with PC, causing polymer chain breakage, leading to high flow and low impact strength.

Waxes and waxy additives like UV stabilizers make good carriers because they melt early in the extrusion or molding process, leading to good distribution. They can also accept very high pigment loading. The resulting letdown ratios are very favorable with regard to contamination of the base polymer. These carrier systems cannot be used for cube blends with polymers that must be hopper dried. The carrier will melt in the drier and can fuse together into one large mass that has to be chopped out of the drier.

To give good color release, a concentrate needs to melt at or below the temperature of the finished compound and have viscosity similar to or lower than the resin into which it is being let down. In typical precolored polymer the level of pigment is low enough that the effect on the polymer's viscosity is negligible. As pigment levels are increased and dispersed in a carrier, the viscosity of the system increases until viscous and shear forces get so high that friction heating can result in total decomposition of a polymer during compounding. This is because the polymer is being filled with pigment to the point that there are not enough polymer chains between the pigment particles to slip and slide. Oil absorption is an indirect measure of surface area of the pigment and can give a good clue as to how much one can fill a polymer without having color release problems. Carbon black pigment has a surface area of 20–1000 m^2/g depending on type. Most organic pigments run 10–100 m^2/g while inorganic pigments have 5–50 m^2/g of surface area. To put this in more visual terms, a pound of phthalocyanine pigment ("phthalo") has a surface area of about 5 acres! This means that the carrier in a 50% concentrate of phthalo blue would have to be spread 5 acres to the pound in order to wet out all pigment surfaces. Pick up a pound of polymer, approximately 1 pint of pellets. Imagine how thin it would have to be made to cover 5 acres and one will understand why 30–35% loading of phthalo is about the highest level for a phthalo blue concentrate.

In order to keep resin-specific carriers from getting too viscous when loaded with pigment, three options are available to the compounder:

1. using lower molecular weight grades of polymer that have higher melt flow than the letdown polymer,
2. using plasticizers and other additives to lube the resin thereby reducing the melt viscosity, and
3. keeping the pigment loading as low as possible in the concentrate.

In practice, all three approaches are employed. In cases where lower molecular weight grades are not available, the resins are sometimes treated with additives to break down the polymer backbone, yielding a higher flow substrate to serve as a carrier. In the case of PC, it is possible to use recycled Compact Disc (CD) grade material as a concentrate carrier. The CD grades have higher flow in the first place, and the reprocessing further reduces molecular weight and increases flow. In addition to making a good high-flow carrier, the recycled CD PC material is low cost, a primary prerequisite for making concentrates. Use of plasticizers is widely practiced in making concentrates and can be very effective. All of these options have positive benefits. At the same time they can cause all sorts of headaches. The lower molecular weight carriers may reduce impact and other toughness properties below acceptable levels. Plasticizers can erupt at the surface during molding and cause blemishes called splay marks or condense on mold surfaces and cause haze on parts. They also can solubilize organic pigments to a dye state, causing them to fluoresce or overdevelop and make the concentrate off shade. If they migrate to the surface of the concentrate, as they are prone to do, plasticizers can actually interfere with color release by causing what the writer calls the watermelon seed syndrome. If you pinch a watermelon seed between your fingers, instead of squashing the seed, the pressure will cause the seed to slide out of your grip. Inside processing equipment the same thing can occur. Instead of the concentrate particle melting, squashing, and distributing among the natural pellets, the plasticizer can make the outside slippery and the concentrate just slides around when it gets squeezed in the bed of natural pellets rather than breaking down into smaller domains and melting thoroughly. The result is uneven distribution of the concentrate and poor color development. Probably the worst problem caused by plasticizers is fuming and odor during processing. Plasticizers can come off at the vent or at the dye in extrusion operations and fill an entire processing shop with a hazy cloud that is not at all pleasant for the operators. Plasticizers are more soluble in heated polymer than when the polymer is cool. If too much is used in a concentrate, when the colored parts cool, the plasticizer can bloom to the surface. The blooming may not be noticeable in all colors. It may only show up on dark colors or colors that were formulated with a colorant that can migrate with the plasticizer.

Plasticizers and waxy additives can give good initial color and physical property results yet cause the finished product to have poor heat deflection properties that can lead to latent part warpage. Much of this delayed defect phenomenon is associated with crystalline or semicrystalline products that must recrystallize after melting in order to reach their full property potential. Waxes and plasticizers can hinder recrystallization, and the effects can be different in thick sections that cool slower than thin sections.

A key reason concentrates are used is because they are clean and less contaminating than neat pigments and dyes.

19.4. COMPOUNDING EQUIPMENT

All common mixing techniques for plastics and some used for paints are used to manufacture color concentrates:

- batch mixers,
- continuous mixers,
- single-screw extruders (SSEs),
- kneaders,
- twin-screw extruders (corotating and counterrotating),
- cone and drum plasticators and mills (two- and three-Roll), and
- Z mixers.

The SSE is the most common type of mixer used for producing concentrates. While not necessarily the best tool for the job, they probably are the least expensive mixers to purchase and operate. The SSEs are quite versatile and capable of compounding practically all types of polymers. Longer L/D SSEs are capable of the high-shear dispersive mixing needed to make high-quality concentrates. Screen packs are often used to remove undispersed pigment agglomerates from the melt. Equipped with gear pumps to increase pressure, tortuous path melt filters can be used to assure the exceptional particle size reduction needed of concentrates for use in blown film and fiber applications.

Batch mixers are ideally suited for concentrate mixing because of the ability to sequentially add low-bulk-density ingredients. When making concentrates at high colorant loadings, the dry ingredients are frequently so bulky that everything will not fit in the mixer in one pass. This situation causes double-pass compounding on extruders and adds cost to the manufacturing process. With batch mixers, once the batch fluxes and densified, the mixer can be opened, allowing additional pigment or extenders to be added. A major drawback of batch mixers is their lack of pressure output. To compensate for this, the batch is usually dropped to a hot-melt pump for heat removal and pelletizing.

Twin-screw extruders (TSEs) and continuous mixers are probably the most efficient mixers for making concentrates. They are particularly well suited for polymers that have very low melt viscosity (a phenomenon called shear thinning) such as acetals and nylons. Both machines can be configured such that all material must pass through a high-shear dispersive mixing section before exiting the mixer compared to other mixer types that may allow some material to short circuit around the high-shear sections. Thus they are efficient for making low agglomerate concentrates needed for film and fiber applications. Corotating intermeshing TSEs are very versatile mixers for concentrates but cannot develop the high pressure needed to filter out undispersed agglomerates. Where melt filtering is necessary, crosshead single screws or gear pumps are used to force molten polymer through screen packs or torturous path filters. The biggest drawback for TSEs is high investment cost. They typically run two to three times the cost of SSEs, and small TSEs cost nearly as much to produce as large ones. Since production runs of concentrate are small in relationship to precolor runs, the return on investment for a TSE making concentrates can be both long and small.

Kneaders are effective mixers for making color concentrates because they are excellent for both dispersive and distributive mixing. The main attribute of a kneader is uniform control of heat and dwell since there is no localized high-shear zone to overmix or overheat sensitive polymer like polyvinyl chloride (PVC).

Cone and drum plasticators as well as two- and three-roll mills are other ways of making concentrates. These along with Z mixers are probably best suited for low-viscosity polymer carriers, plastisols, or paste concentrates. Some concentrates based on wax and stabilizer carriers are produced on this type of mixer.

One "universal" type of concentrate was manufactured in Pug Mills of the type typically used to mix clay in the pottery industry. No one manufacturing method works for all polymers and generally the more versatile the process the less optimized it is for any one product.

19.5. STRANDING AND DICING PROBLEMS (OPPORTUNITIES?)

One of the crucial elements of concentrate manufacture is dealing with the output from the mixer. Carriers that are highly filled with pigment frequently will not have very good integrity as a hot strand. When it is warm and flexible enough to run through rollers of a water bath, it may gum up and stick in the cutter blades of the pelletizer. Chilled enough to cut properly, the strands might be so brittle they break, coming through the water bath, or the pellets might shatter when struck by the cutter and give excessive dust and fines. Softer carriers may strand out nicely and cut well only to stick together in the package as one large clump instead of being the free-flowing pellets desired. How these situations are addressed is as plentiful as the ways concentrates are compounded. In some cases they are side stepped altogether, as in the case of prilling, beading, and flaking methods used on many of the waxy carrier systems. Typical dicers and die-faced pelletizers are most frequently used, especially for polymer-specific carrier systems. Granulators are also used, particularly for very small lot runs. Dicers usually have cube-shaped pellet output. Die-faced cutters generate lenses, ovoid or round particles that are smooth and that easily slip and slide between the natural pellets. It is advantageous for a concentrate to have the same size and shape as the natural it is to color. The slippery nature of a die-face cut makes the size factor doubly important since it is size more than any thing that prevents the concentrate from segregating from the natural during transportation processing. So, one of the best means of avoiding the problems of pelletizing concentrates introduces a new problem. Nothing is ever simple is it?

19.6. PARTICLE SIZE CONSIDERATIONS

19.6.1. Making Concentrate

A wide variety of dispersants or wetting agents are used in the manufacture of concentrates. These ingredients and specifics of the compounding techniques employed represent the proprietary technology of the concentrate manufacturers and cannot be covered in detail. Some of the important considerations that must be taken into account include the following. The wetting process is affected by the relative surface tensions of the pigment and the polymer. Solids such as pigments have very high surface tension and are not easily wetted by the polymer medium. To aid the wetting process, pigments for plastics are usually ground to a finer particle size, thus increasing the surface area. Pigments made for nonplastics applications tend to have less surface area and give much lower tinctoral strength when used in plastic applications as compared to equivalent plastics grade pigments of the same chemical type; this is particularly true of organic pigments. This creates a fine balance (no pun intended) for ease of dispersion versus performance. To reduce surface tension (and also to allow the polymer to be formed), the polymer is heated to high temperature to give a fluid molten product. The colorants must have adequate heat stability to withstand the resulting processing temperature. Many organic pigments can become soluble at elevated temperatures, particularly at low concentration. This dissolution can be because of pure solubility in the polymer or solubility in the decomposition products formed by heat and shear. In either case, this solubilization can lead to migration-related problems, color change, or fluorescence. Dyestuffs for plastic applications also need to be heated above the solvation temperature to give good distribution of the dye. The wetting agents selected can contribute to the problems associated with solubilization and may force the formulator to use different dispersant ingredients for one color than are used for another color.

19.6.2. Coloring Plastics

When a processor uses concentrates, the means of addition of the concentrate can affect the result. If the concentrate is going to be tumble blended into the natural, then directly fed into the processing machine, it is important to have the pellet size of both the natural and the concentrate reasonably similar. Otherwise the concentrate will segregate due to vibration, and flow effects give variation in dosage part to part, resulting in color drift, part rejection, and delays for the processor. This is particularly critical for cube blends that are mixed to the proper letdown ratio and packaged as salt-and-pepper blends prior to shipment. Just the free fallout of a feed hopper is enough to separate disparate particle size blends, but equal-size particles will retain their distribution. For long glass fiber material with pellet cuts up to $\frac{3}{4}$ in. in length, the concentrate is made with the same $\frac{3}{4}$ in. in length cut to assure consistency.

If the colorant is to be fed separate from the natural pellets, such as by a slave feeder, the only concern is that the concentrate size and shape are uniform and consistent from lot to lot. Ideally, to increase the physical distribution of color and reduce the amount of distributive mixing needed from the processor screw, the coloring agent needs to be as small and finely divided as possible. Some concentrate producers have even patented the size and shape of concentrate, which are purported to perform with more consistency at lower dosage rates.

19.7. DISTRIBUTION VERSUS DISPERSION

When making concentrates, it is the producer's job to achieve good dispersion of the colorants and design the concentrate to give good color release. If the job is done well, the concentrate will provide parts with a minimum of defects such as lumps, bumps, and undispersed agglomerates or color streaks. Distribution of the concentrate is a joint responsibility of the maker and the user. When a concentrate has adequate color release, the processor should be able to run it at conditions that give good distribution. Efforts to reduce costs by the processor can introduce distribution problems. Sometimes a change as seemingly insignificant as a different natural pellet cut can upset the process. An independent change by the processor to increase part output rates or reduce dosage can lead to problems of poor color distribution that are not the fault of the concentrate or its maker. Both parties should jointly review changes such as these.

19.8. "COLOR RELEASE" ISSUES

Theoretically, the higher the concentration of colorant in a concentrate the lower the dosage or higher letdown ratio needed to obtain properly colored parts. With lower dosage, however, the need for distribution of the concentrate (equivalent to distributive mixing by the processor) becomes more critical. The law of diminishing returns is alive and well in concentrates as in most other pursuits. Concentrate letdown levels as high as 300:1 can be achieved in some color applications. At this level, 1 pellet of concentrate has to be melted and distributed thoroughly enough to color about 300 natural pellets. This in and of itself is a tall order. Assuming it is possible, that is, you get 100% color release and the concentrate is 95% colorant, this translates to 0.316% pigment. For dyes and some of the stronger organic pigments, this can be an adequate loading; however, for most colorants at least 0.5% is required to obtain a stable formulation that will cover any polymer color as well as a potential color shift of the polymer when exposed to higher than the normal temperature range used in processing. In most cases, full distribution is not achieved. The formulation usually does not contain all strong dyes and pigments; therefore a higher dosage of concentrate is needed.

19.9. LOADING LEVELS IN FINAL PRODUCT AND CONCENTRATE

A typical colorant loading level in a finished product is 0.5–2.0% pigment. If there are opacity and/or light stability requirements or the polymer system is inherently difficult to color, as much as 5% pigment or more may be required to obtain stable colors. The higher the colorant level required in the finished product, the *lower* the letdown ratio number. Letdown ratios are expressed in two ways; for example, 4/100 or 25:1 both indicate a concentrate designed for use at nominally 4% loading. Some people refer to this as a 4% concentrate. A 50% colorant level concentrate used at 25:1 will deliver approximately 2% colorant to the final product. The same strength of concentrate used at 10:1 provides almost 5% colorant in the finished product.

19.10. QUALITY TESTING CONCENTRATES

As a general rule, parts that are concentrate colored do not have as good color consistency as those made from precolored products. Some of the blame for this can be placed on the molding conditions required. Those conditions most suited to give good part appearance rarely provide good distribution of the color concentrate, leading to improper shade or streaks. When conditions are optimized for concentrate distribution, one might find excellent colored parts, but they exhibit surface blush or splay. The lot-to-lot variation in color can be due to inadequate testing techniques during manufacture of the concentrate. If the color quality inspection method employed is simply a pressout or molded chip comparison of the current lot versus the standard in concentrate form, major variations in shade, strength, or color release can go undetected. Color variation must be checked in the polymer type for which the concentrate was developed and at the designed letdown ratio. Even though by this comparison the color may be within acceptable release tolerances, problems with concentrate strength could be missed. To check strength, the concentrate should be let down in tint if it is a dark color or in reduction if it is a light color. To do this test, white should be used for the tint test and black for the reduction test. The white should have just white pigment at a low loading; 0.5–1.0%

usually works well. For black reduction, material with as little as 0.02% carbon black is appropriate. The tint base or reduction base materials should be precolored instead of concentrate colored to make sure they are thoroughly consistent in the white or black level. Using the tint or reduction base in place of natural polymer, some test chips should be molded at the design letdown ratio. At least 12 chips should be molded before saving test samples to make sure the molding process is stable. More chips may be needed to get to steady state, depending on the material, machine, and shot size. An Repeatability & Reproducibility (R&R) study to test consecutive part output from the molding machine until color variation, chip to chip, reaches steady state must be run to determine the appropriate number to reject at startup. This may show that color cycles in lightness or chroma over a range of 5–15 chips. In that case one need to take enough multiple chips to cover the normal fluctuations. The two most important properties to measure in the tint or reduction letdown tests are *color consistency* and *tint strength*.

Color consistency testing can be as simple as visually comparing samples of a new concentrate lot to samples prepared the same way from an agreed-upon standard concentrate sample. Most likely, though, there will be instrument measurement data to eliminate subjective human judgment. A good way to do this is by reading multiple samples on the lot and standard using a color spectrophotometer and then comparing color difference values. It is important to average multiple measurements on each color chip because spectrophotometers respond to differences in gloss and opacity on a chip that human observers overlook. Also, both the standard and lot should be prepared side by side to make sure any variation is not due to preparation.

The most common means of determining colorant tint strength is by using the Kubelka–Munk theory, which deals with absorption (K) and scattering (S) for densely packed particulate mixtures. The basic assumption is that a chromatic pigment will absorb light, thereby preventing its reflection from the white pigment in the tint base. The stronger the chromatic pigment, the lower the reflectance (R as a decimal) of the material and vice versa. In this technique the K/S of the lot is divided by the K/S of the standard and then multiplied by 100 to give the percentage of tint strength:

$$K/S_{\text{lot}} = (1-2R)^2/2R$$
$$K/S_{\text{std}} = (1-2R)^2/2R$$
$$\frac{K/S_{\text{lot}}}{K/S_{\text{std}}} \times 100 = \text{Percent Tint Strength (White Tint)}$$

Typically a fixed wavelength is selected in the region of maximum absorption and that same wavelength is used for that color every time the test is run. For example, one might select 520 nm for Solvent Red Color Index 135 (*SR*-135) and 540 nm for Solvent Red Color Index 52 (*SR*-52).

When dealing with a white concentrate, the black reduction base is used instead of the tint base because in this case the measurement is how strongly the white reflects out of the constant black substrate. This method is also used when testing pale yellows. When using a black reduction base to test the color strength of a white or yellow, the inverse of the function is used to calculate tint strength:

$$K/S_{\text{lot}} = (1-2R)^2/2R$$
$$K/S_{\text{std}} = (1-2R)^2/2R$$
$$\frac{K/S_{\text{std}}}{K/S_{\text{lot}}} \times 100 = \text{Percent Black Reduction Tint Strength}$$

This is because a stronger white concentrate would give higher reflectance and a weaker one less reflectance.

19.11. SPECIAL EFFECT CONCENTRATES

Some special effects such as granite and mottles are best done from concentrates because the end result is so dependent on the molding process. The dose rate can be adjusted at the machine to obtain the desired appearance. Metallic flake and glow in the dark are exceptions. Aluminum flakes are sensitive to shear during mixing and do not do well because of the high-intensity mixing required to make the concentrate. Glow-in-the-dark pigments must be loaded at such a high level to be effective that letdown ratios become unreasonable. Clear and translucent colors can be made successfully via concentrates and are quite easy to manufacture. The dosage rate needs to be high enough to give good distribution prior to melt. Many of the clear resins are lightly colored to cover up the natural yellowness of the polymer, so it is very important to determine which natural tint is being used.

CHAPTER 20

Liquid Color Concentrates

Richard L. Abrams

Ferro Corporation
7500 East Pleasant Valley Road
Independence, Ohio 44133

A liquid color concentrate is a dispersion, in a liquid medium, of pigment and/or functional additives for a thermoplastic. Liquid color is usually added during processing, rather than being preblended with a resin. Liquid color can be used almost anywhere a solid concentrate of color or additive can be used but is particularly effective at extremely low levels. Letdown ratios as low as 0.1% have been used successfully in transparent polyethylene terephthalate (PET). Blowing agent concentrates are particularly suited to liquid, since the blowing agent is dispersed without partial degradation from process heat. Liquid concentrates can also be made of antioxidants, UV additives, or virtually any other kind of additive. No carrier resin needs to melt in the manufacture, so production of the liquid concentrate can be done near room temperature.

20.1. HISTORY OF LIQUID COLORS

Liquid color was developed in the late 1960s and early 1970s. Most attribute the development to Inmont Corporation. The product was probably oversold at the time. Many people saw its advantages, particularly economic advantages, and tried to use it everywhere. Unfortunately, the product was not fully developed and understood. In

---

Some of the material in this chapter was derived from McKinney and Wienckoski, "Effects of Various Liquids in Thermoplastics," paper presented at the 1988 SPE Color and Appearance Division/Polymer Modifier and Additives Division joint RETEC, Newport, RI.

Coloring of Plastics, Fundamentals, 2nd edition. Edited by Robert A. Charvat
ISBN 0-471-13906-8 Copyright © 2004 by John Wiley and Sons, Inc.

288 Liquid Color Concentrates

particular, controlled ways of adding liquids were not fully developed. People first encountered housekeeping problems. A spill of a solid is not a big problem and is soon forgotten. A large spill of a concentrated liquid color is hard to forget.

Other problems occurred. Specks of color appeared when mechanical dispersion was not good enough. Specks were also common when surfactants were of the wrong type or the wrong amount was used. If some liquid colors were stored for too long, a solid precipitate formed on the bottom of the vessel, which could not be redispersed. This is called "hard settling" and was a common source of specks. In addition, some liquids were very lubricating. In the worst case, screw slippage was so bad that little output came from the extruder die or short shots occurred in molding. For less severe cases of overlubrication, streaks occurred and distribution was poor. Liquid-rich areas and a natural polymer existed in distinct layers throughout the part.

Pumping of liquid colors was inefficient until the development of the microprocessor. Some liquid color was simply measured into batches with dippers. If a pump was used to deliver liquid color to an extruder, synchronizing the pump with the process was nearly impossible. If the speed of the extruder varied, color concentration in the product varied. When the extruder stopped, the liquid color could overflow and make a horrible mess.

For those of us who used liquid color in the early days, it is natural to not want to consider using it now. However, several developments in liquid color have been made in the last 25 years, some because of bitter experience. Developments in mills have improved grinding of the colorants. Computer-controlled pumps now allow for consistent delivery of liquids. Formulation has improved through better understanding of the mechanism of the coloring process, better carriers, and better surfactants. Even the effect of incremental development over that span of time is significant. Liquid color is now worth attention when starting a new operation or changing an existing one. It has advantages and disadvantages, as do all coloring methods. It is best to compare all available methods and choose the one that best fits the process.

20.2. SUITABILITY FOR USE

Liquid colors can be used in injection molding and extrusion. Thin films usually require large amounts of color and so are not usually made with liquid color. Thick films and sheet are suitable applications for Liquid. Usually, 1–3% is the practical limit for a letdown ratio, but liquids have been used up to 8% in special circumstances. For letdown ratios of 1% and less, liquid color is often the preferred method of coloring or introducing additives.

20.3. MANUFACTURING OF LIQUID COLORS

Production of liquid colors is done in various ways, similar to paint and ink production. Predispersion is often done using a mixer, called a "dissolver." The carrier is put in a container, and a mixer is lowered into the carrier. The mixer has a vertical shaft, with a drive at the top, similar to a drill press. Its blade often looks like a horizontal saw blade, a disk with offset teeth on the perimeter. Speeds at the blade tip may be in the 5000-ft/min range. The solid ingredients, pigments, and/or additives are then added under agitation. For less demanding applications, with "stir-in" pigments, it may not be processed further. Variants on the dissolver/mixer may contain slower moving elements, coaxial with the "saw" blade, multiple elements on

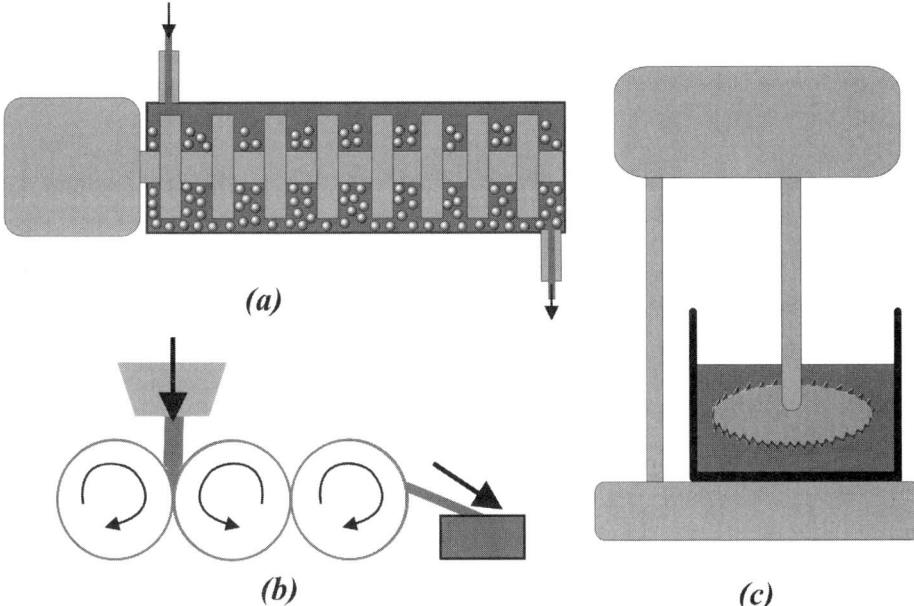

Figure 20.1. Common dispersing equipment for liquid color: (*a*) horizontal media mill; (*b*) three-roll mill; (*c*) dissolver mixer.

the shaft, or other designs. Common processes for fine dispersion include the three-roll mill and various media mills: horizontal or vertical. See Figure 20.1 for schematics of the different mills.

The choice of a mill may depend on the desired characteristics of the liquid color needed. If a heavy viscosity is acceptable, the three-roll mill is an excellent tool. The premixed liquid color is introduced at the first nip. After passing through that nip, about half the liquid clings to the second roll and is carried around to the second nip. About half that liquid clings to the third roll and is scraped off and collected by a doctor blade (also called the apron). That is the finished liquid. Since the liquid is split twice, about three-quarters of the liquid that passes through the first nip is returned to be reprocessed at the first nip.

Dispersion occurs at the two nips, between the roll at the feed end and the middle roll, and between the middle roll and the third roll. Gaps and roll speeds can be adjusted to give excellent dispersion. The nips may have some crushing force for agglomerates, but elongation shear is generally considered the main force for dispersion. In effect, the particle is torn apart by being trapped between fluid moving up and fluid moving down. Lubricating carriers reduce the quality of the dispersion. In effect, they let the particles "slip away." The property of "tack" is important to transmitting shear to agglomerate particles and breaking them up. Along with tack, viscosity of the carrier liquid is important to dispersion. The higher the viscosity of the fluid, the more force is transmitted to the agglomerate. If a liquid color is passed through a mill more than once, the viscosity is usually greater on the second pass, because of the effect of the greater surface area of pigment exposed by the dispersion in the first pass. A liquid carrier adsorbs on the surface of the pigment, and viscosity increase. Ultimate fineness from the three-roll mill can be better than 2 μm. A media mill is more commonly used with less viscous liquids than the three-roll mill, although a range of viscosities exists where both work well. The mill consists of a vessel with stirred media, ranging from sand to moderate-size beads of steel or ceramics. Its shaft can be oriented vertically or horizontally; both types are popular. Fluid is fed in one end and flows through the stirred media, and the suspended solids are ground to finer particle size. Media size, flow rate, and shaft speed can be adjusted to control dispersion.

Other types of mills are in use, but the above are the most common in liquid color.

20.4. DISPERSION AND DISTRIBUTION OF LIQUID COLOR

For this discussion, *dispersion* of a pigment or other material into a carrier is the breaking up of large particles of the solid into smaller particles and the wetting of those particles by the carrier. As much as possible, dispersion is intended to reduce pigments to single particles, each surrounded with a carrier. *Distribution* is the even mixing of the carrier and color into the end-use resin, free of streaks and other uneven coloration.

Dispersion of pigments in liquid colors can be extremely fine, good enough for fine denier fibers. In one project, liquid concentrates of pigment for Nylon were studied extensively for dispersion quality. In a wide range of colors—ranging from white to organic pigments, to inorganic pigments, to black—no agglomerates were detected greater than $2\,\mu m$. The same colors all passed the test of a polymer, with the liquid color being extruded for 24 h through a 1400×160-mesh screen with a minimal pressure rise.

Distribution is generally quite good with liquid colors. The liquid can be intimately coated onto the resin particles before melting, so only fine-scale mixing is necessary. A problem occurs if the carrier is too lubricating, called "screw slippage." Screw slippage is generally peculiar to liquid color and wax concentrates. It is a problem that occurs in the feed zone of an extruder or on the screw of an injection molder when the screw "loses traction." It cannot force the material forward or build up enough frictional heat to melt the polymer. In the severest form of screw slippage, an extruder screw can turn with no output at the die. If it is not severe, streaks can form or distribution may not be good, affecting color strength and reducing additive effectiveness. If distribution is poor, laminar flow results, causing streaks and the possibility of delamination. Screw slippage can usually be overcome by formulation. Another way of eliminating screw slippage is to inject the liquid color at a point as far downstream in an extrusion or molding process as possible. In an extruder, the injection point can be past the "melt seal," where the polymer melts. Various methods of downstream injection include putting the color in at a vent, forcing the liquid through the barrel of the extruder under pressure, or injecting the liquid into a groove in the barrel near the melt seal. One interesting example of the latter method is the "CC120" method developed by the Bemis Company (Sheboygan Falls, WI). The intent of the CC120 is to change colors in less than 2 min. This method uses tapered grooves in the barrel of the extruder or molding machine. Liquid color is injected into these grooves and enters the polymer near the melt seal.

Another injection point, which can be used with either extrusion of molding, is before the die or nozzle. A "static mixer" can be added that mixes the liquid color into the melt. The mixer has a series of channels that split and recombine the melt stream, giving good distributive mixing. Usually, the further downstream a liquid is added to a natural resin, the less chance there is of separation. This may require high pressure pumping to assure the liquid enters the melt stream which is under processing pressures. Downstream addition also makes cleanup easier, since less metal is actually wetted with color.

20.5. DELIVERY OF LIQUID COLOR

Liquid color is added to thermoplastics in extrusion by continuous slow pumping, either directly into the throat of the extruder or near to the throat. In injection molding, the liquid is injected in cycles by a computer-controlled pump. Small color adjustments in either process are easily made by increasing or decreasing pumping rates. The liquid carrier, along with the color and additive, becomes adsorbed onto the polymer through mixing. If the formulation of the liquid color is correctly formulated, the liquid absorbs well into the polymer during later melt mixing.

20.6. PUMPS

A common type of pump used in injecting liquid color is the peristaltic pump, which works by passing rollers over a resilient tube. Liquid is forced forward by the moving constriction in the tube. The force that fills the tube behind the rollers is the resilience of the tubing. The walls of the tube spring back, drawing the liquid in. Peristaltic pumps work best with lower viscosities and when the color reservoir is at the same level as the pump or higher. If the pump is too far above the reservoir, the pressure head decreases pumping efficiency. Its outlet side is less sensitive to height differences, because the rollers provide a positive force.

One reason for the popularity of peristaltic pumps is their versatility; they can be programmed for low or high speeds. Tubes vary in diameter to cover a variety of flow rates. The various sizes of tubes are color coded for diameter and output rate. Another advantage is ease of cleaning: After a run, just throw out the inexpensive tubes and put in clean ones. The tubes can also be capped and preserved with the container of color to be used the next time it is run. Peristaltic pumps are self-priming, and a cycle is included in the programming of the pump to fill the output tube up to the extruder or molder quickly.

We sometimes find a problem in linearity at high pumping rates. Linearity means that if the pump runs twice as fast, output doubles. Linearity is usually good over 10–20% changes in the rate. Calibrating the pump for each liquid is recommended, and periodic rechecks are good practice. The tubing has a life sufficient for most runs, but abrasive materials and excessive use can reduce pumping efficiency.

Another popular type of pump is the gear pump. Gear pumps are not self-priming, but they can lift liquids higher than a peristaltic pump. The liquid is forced forward by interlocking gears, which contact the liquid. This makes gear pumps much harder to clean than peristaltic pumps. The direct contact with the gear surface also causes wear, especially when potentially abrasive materials such as titanium dioxide white in liquid color are pumped. Gear pumps are less sensitive to viscosity of the liquid color and are more linear than peristaltic pumps.

A third type of pump is used largely for pumping at high rates, often with a white liquid color. This is a progressive cavity pump. Like the gear pump, it is very linear and can handle high viscosities. It has a convoluted tubular stator and a convoluted rodlike rotor, which intermesh, moving the liquid forward, much like an extruder screw.

Any of these pumps can be controlled by a microprocessor that is sensitive to the actions of an extruder or molder. The microprocessor can control the pump, precisely adjusting the liquid color flow. The pump electronics can interface with process machine controls, so pumping rates can vary in extrusion with screw speed, or with an injection molder cycle. See Figure 20.2 for photographs of typical liquid color pumps. Probably the most popular supplier of the peristaltic pumps is the Maguire Company.

Another device often used with liquid color is the intermixer (see Fig. 20.3.). An intermixer is installed between the resin hopper and throat of an injection molder or extruder. The main parts of an intermixer are a mixing chamber and a motor-driven impeller. One end of the chamber is usually transparent, so the mixing process is visible. The resin falls into the chamber of the intermixer from the top, and liquid color is pumped into the chamber from the side. The impeller turns at a rate of about 60 rpm and distributes the liquid color uniformly on the surfaces of the resin pellets.

20.7. PACKAGING

Liquid color are shipped in various sizes of containers. Small samples are sent in 1-qt to 1-gal collapsible containers (Hedpack). More common containers are 5-gal pails, 30-gal drums, and 55-gal drums. For really long runs, reusable totes are available. All-steel or all-aluminum totes are widely used, and totes that consist of a polyethylene tank supported by a cage of steel rod are becoming more popular. Totes are particularly efficient for large-volume colors, often black or white. The tote can be returned and refilled with the same color, reducing the chance of cross-contamination.

20.8. ADVANTAGES AND DISADVANTAGES OF LIQUID COLORS

Advantages

- Better distribution can make more effective use of color and additives.
- They are usually 10–40% less expensive than solid concentrates and considerably less expensive than precolor.
- They are effective for a wide range of polymers, often without formula change.
- Low-temperature processing is best for temperature-sensitive materials, such as chemical blowing agents.
- There is fast turnaround time from order.
- Increasing or decreasing letdown ratio can be done very quickly online by adjusting pump rate.
- There is a fast color change.
- Less material is lost on color change than in solid concentrates.
- The even coloring from liquid color allows less total color to be used, since overcompensating for areas of less color is not necessary.
- They are good for letdown ratios as low as 0.1%, possibly lower with care.
- They are multifunctional—color and additives are in one package.
- Less space and weight to inventory are needed for the same amount of a polymer to color.

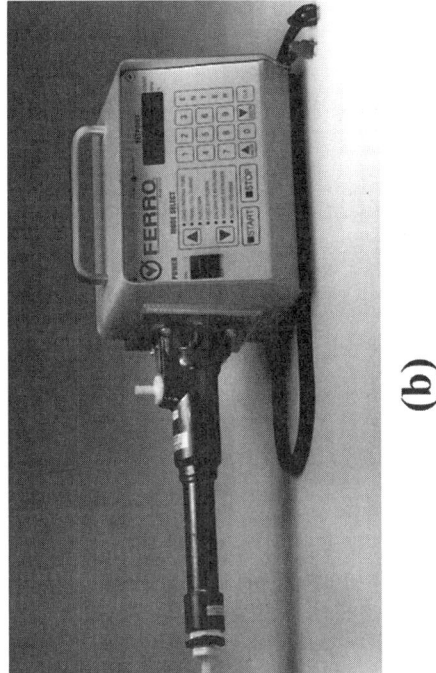

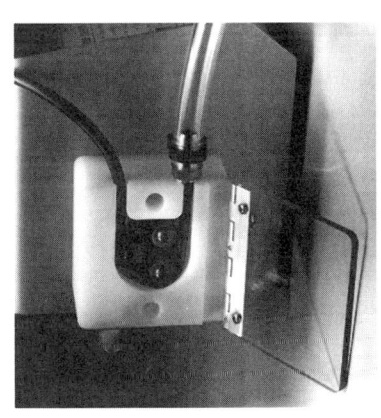

Figure 20.2. Typical liquid color pumps: (*a*) peristaltic pump; (*b*) detail of peristaltic pump head showing tubing; (*c*) high-volume progressive-cavity pump.

294 Liquid Color Concentrates

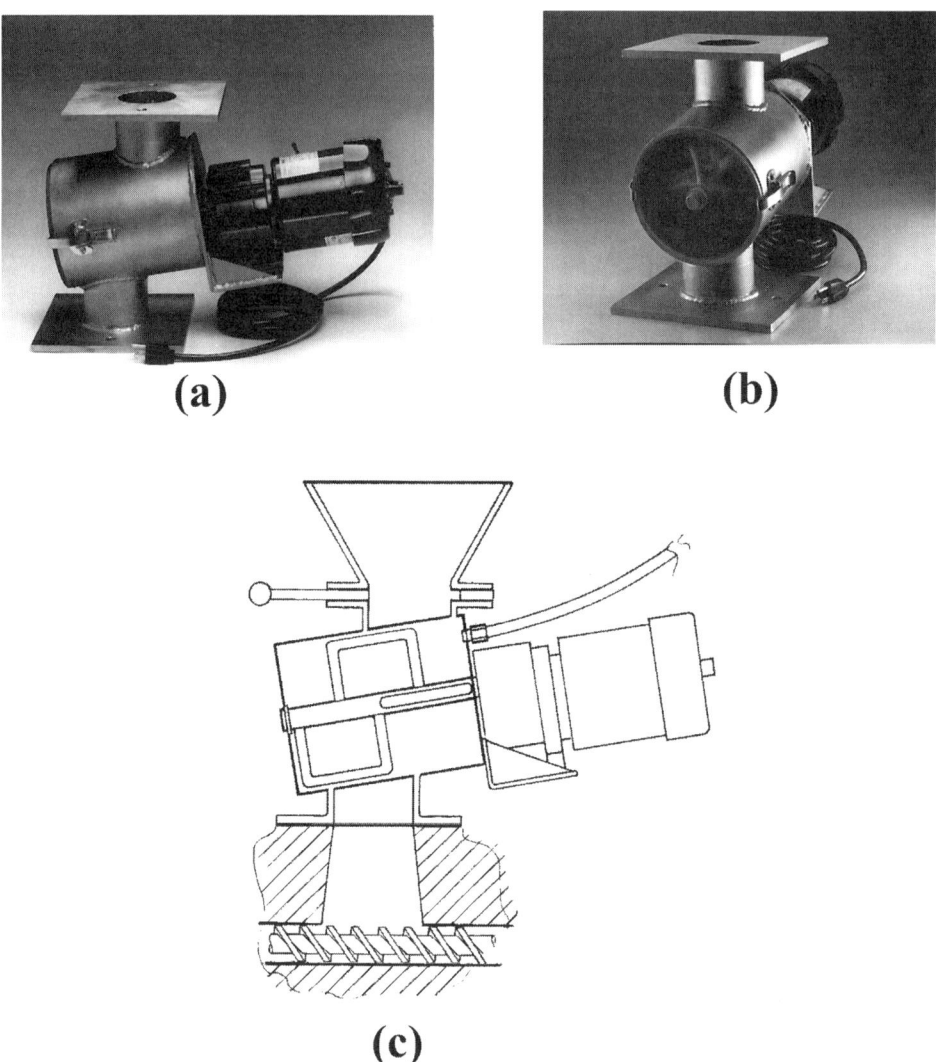

Figure 20.3. (*a*) Intermixer, side view; (*b*) intermixer, front view; (*c*) diagram of intermixer.

- They can be readily blended, so a large, consistent batch can be made by cross-blending.
- They are available in containers from 5-gal pails to large totes.
- A liquid purging material is available for cleaning out extruders, molders, or molds between colors.

Disadvantages

- Poorly formulated liquid colors can give screw slippage, particularly at greater than 5% letdowns.
- Shelf-life is shorter than solid concentrates, particularly if wide temperature fluctuations occur in storage. A one-year shelf-life is typical. (Although even settled materials can often be refreshed by vigorous stirring or reworked into a different color without experiencing any further heat history.)

- Liquid carriers may affect physical properties of polymers.
- Spills and housekeeping can be a problem, but good practice can reduce the difficulties.

20.9. RHEOLOGY

A suspension of heavy inorganic particles in a liquid will settle; the main question is how fast? The densities of the liquids and solids are usually set by other formulating considerations, which dictate material choice. Particle size and viscosity of the liquid can be affected by process and formulation.

Stokes's law describes the relation of settling to densities, viscosity of a carrier, and particle radius. Note that particle radius is a squared term and so has a large effect on settling. Particle size is dependent on dispersion quality and the fundamental particle size of pigments or fillers. Very fine dispersions show less settling than those with agglomerates present. Settling rate varies with the square of the particle diameter, so good dispersion is important in preventing settling. Stokes's law is given as

$$V = \frac{218 r^2 (\rho_s - \rho)}{\eta}$$

where V is the settling velocity (cm/s), r is particle radius (cm), ρ_s is the density of the solid particle (g/cm^3), ρ is the density of the liquid (g/cm^3), and η is the viscosity of the liquid in Poise (P).

We saw that the viscosity of the liquid is an important consideration in settling; the higher the viscosity, the lower the settling rate will be. Settling is, however, not the only consideration for viscosity. Low viscosity is good for pumping and flow. What we need is a fluid that is high viscosity at rest and low viscosity for pumping. Low viscosity is good for coating a liquid on a resin pellet; high viscosity is good for keeping it on the pellet. The seemingly impossible task of making a liquid that is both high and low in viscosity can be achieved with good formulation. To understand how this is possible, a little background in rheology, the science of flow, is necessary.

The variation of viscosity with shear can be of three types:

1. Newtonian. Viscosity is independent of shear rate. Pure liquids usually show this type of behavior. No matter how fast water is stirred, the viscosity stays the same.
2. Shear Thinning. Viscosity goes down with shear rate (sometimes called thixotropic). One example is paint. when you open latex paint, you can tilt the can, and the paint does not flow. When you stir, the paint thins out and will flow. This is the magic effect that is good both for settling and for flow.
3. Dilatant. Viscosity goes up with shear. If you pour sand, it pours like a liquid. If you punch sand hard with your fist, it hurts. The sand feels like a continuous solid.

One liquid color can display all three characteristics. In the container, it can have very high viscosity with shear thinning, because pigment particles are still attracted to one another. In the range of viscosity for normal pumping, it may act

Newtonian. If the dispersion is very highly loaded, it may be dilatant at very high shear rates. The dilatency is caused by pigment particles being jammed together, like the sand.

20.10. COMPARISON WITH OTHER COLORING METHODS

The following is a comparison of selected coloring methods:

Form	Cost	Warehousing	Dispersion	LDR Range (%)	Heat History	Physical Effect
Dry	Low	Small	Poor	1–5	None	Some
Wax	Medium	Small	Fair	1–5	Low	Little
Pellet	Medium	Medium	Good	4–20	Medium	Little
Salt and Pepper	High	High	Good	N/A	Medium	Little
Precolor	High	High	Good	N/A	High	None
Liquid	Medium	Small	Good	0.1–8	Low	Some

LDR, Let Down Ratio;   N/A, Not Applicable.

Colors and additives can be added to thermoplastics by several means, some discussed elsewhere in this book. The most common forms are as follows:

Dry Color. Raw pigments or additives are mixed with a lubricant in a high-intensity mixer without melting.

Wax Concentrates. Pigments and/or additives are dispersed in wax by a melting process.

Pelletized Concentrates. Resin, additives, and polymer are made by a melting and extrusion process to produce pellets of concentrated colors and additives, which can be blended into natural resin for coloration.

"Salt and Pepper" (S&P). A preblended mix of a pellet concentrate and natural resin, shipped to the molding or extrusion house for direct use.

Precolor. Preextruded colored resin, ready for final processing in molding or extrusion. Each pellet has a full coloration and additive package.

Each type of coloration has its own advantages and disadvantages. A discussion of these follows.

Dry color is a matched color made from raw pigment with some dry lubricant, usually a wax or metallic soap (commonly calcium or zinc stearate). It is often blended on an intensive dry mixer to achieve some dispersion. Dry color mixes well at low letdown ratios. The dry pigment powder clings to the pellets, sometimes by electrostatic force. Some users coat a small amount of liquid, such as mineral oil, on the pellets before mixing in the dry color. The limitation on letdown ratio is the surface area of the resin pellets. After the pellets are coated, the excess sifts to the bottom of any storage vessel. This causes a variation in pigment loading through the lot.

Wax concentrates may be in the form of prills, bars, flakes, or granules. Prills are very small spherical pellets, usually made from a pigment blend and wax. They may be blended, like dry colors, or dispensed by a metering device at the throat of the extruder. If they are blended, it is best to use a powder resin, such as polyvinyl chloride (PVC), so they do not sift to the bottom of the container. The other physical forms are applied in different ways; some are designed to be metered in at the hopper, the rest to be blended in before reaching the hopper. The carrier may lubricate, so letdown ratios are limited.

Pellet concentrates and S&P depend on the mixing of pellet concentrate with natural resin. To achieve good distribution, the pellet should melt and spread before the natural pellets melt, or at least spread more readily than the natural resin. The more particles of pellet concentrate in 100 pellets, the better the distribution can be. For that reason, pellet concentrates are preferred for extremely high pigment loadings, such as white film. The preferred method of mixing in pellet concentrates is premixing before putting into the hopper. Customers are starting to request small pellets for use in continuous dispensing systems. Conventional pellet sizes put too much color into each pellet for even dispensing.

Precolor shifts most of the responsibility for good coloration to its supplier. The pellets need only be put into the hopper of an extruder or molder and processed. Color is prematched, and all mixing has been taken care of. Precolor works best when a large amount of a product has been contracted for in advance. Precolor is more expensive per pound than other coloring methods, although capital expenditures for blending equipment are avoided. The shipping costs are high, warehouse storage requirements are very high, and cancellation of an order can leave a lot of useless inventory to dispose of.

Liquid colors can be loaded up to 80% with inorganic pigments and white, up to 50% with organics such as quinacridones, 30% with phthalo blue, and up to 30% with medium-color carbon blacks. Liquid concentrates can also be made of specialty pigments, such as aluminum flake and pearlescent, fluorescent, and phosphorescent pigments. Liquids are often used for concentrates of blowing agents—both exothermic and endothermic, flame retardants, antioxidants, antistatic agents, mold release agents, and generally a full line of functional additives. You can combine the color for a product and the full additive package in one liquid concentrate.

Liquid colors are compatible with a wide variety of polymers, including PVC, polyethylene vinyl acetate (EVA), nylons, PET, polycarbonate, alloys, high-temperature polymers, and the full range of polyolefins and styrenics. Many liquid colors are suitable to color food-contact polymers and comply with all government regulations worldwide for shipping, use, and disposal. Lead times for liquid color are typically less than those for pellet or other melted color concentrates, since the processing is quick and efficient.

Figure 20.4 gives an indication of why given coloring methods are effective at different letdown ratios. Liquid color coats each pellet, particularly if an intermixer has been used. No further time or energy is needed to distribute the material to the pellet level. The limitation in use of liquid color is at high letdown ratios. Too much carrier can lubricate too much. Also, if there is an excess of liquid color, it can separate out to some extent, particularly if mixed resin and color are allowed to sit in a stopped extruder.

298 Liquid Color Concentrates

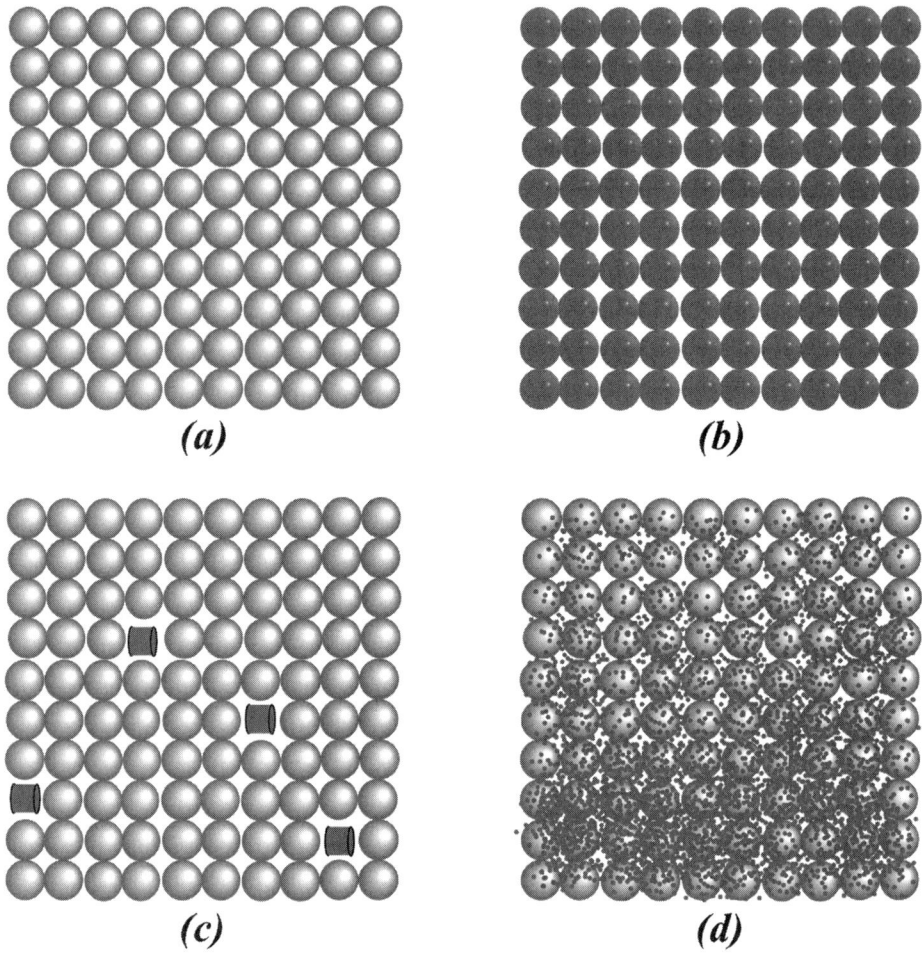

Figure 20.4. Common methods of coloring plastics: (*a*) 100 pellets of natural regin; (*b*) liquid color on pellets; (*c*) 4% pellet concentrate; (*d*) excess of dry color.

20.11. EFFECT ON PHYSICAL PROPERTIES

Table 20.1 shows the effect that a liquid flame-retardant concentrate at 4.5% in both homopolymer and copolymer polypropylene (PP) had on physical properties of PP. Similar tests have been carried out using liquid colors in PP, polyethylene (PE), polystyrene (PS), nylons, polycarbonate, and other resins, with similar results. There is a slight plasticizing effect, which improves impact properties, slightly reducing tensile yield strength while increasing tensile break strength. Tensile and flexural moduli are reduced, and flexural strength is slightly reduced. Rockwell hardness is slightly reduced, and heat distortion temperature is slightly reduced. The change in properties ranges from about 5 to about 10% at this level of pigmentation; effects are less with smaller letdowns.

The better dispersion of pigments in liquid colors may give improved flexural strength, because undispersed pigment agglomerates form sites for cracks to start. It cannot be said that liquids never affect physical properties, but the changes are slight and may be beneficial.

Table 20.1. Summary of Test Results

Product	Izod impact (/ft/lb/in.)	Tensile Yield Strength (psi)	Tensile Break Strength (psi)	Tensile Modulus (10^3 psi)	Flexural Strength (psi)	Flexural Modulus (10^3 psi)	Rockwell Hardness (R Scale)	Heat Distortion (°C)
PP homopolymer	0.39	4750	3020	2.17	5230	1.65	63	54.3
PP homopolymer 86-00751 W/4.75%	0.57	4550	3030	1.89	5010	1.66	66	53.1
PP copolymer	1.47	3730	2630	1.9	4330	1.44	39	51.2
PP copolymer 86-00751 W/4.75%	1.76	3540	2780	1.37	3830	1.28	30	48.3
Averaged data								
PP homopolymer + PP copolymer	0.93	4240	2825	2.04	4720	1.55	51	51.8
PP homopolymer 86-00751 W/4.75% + PP copolymer 86-00751 W/4.75%	1.17	4045	2905	1.63	4420	1.47	48	50.7

Source: From Carlson, "Liquid Flame Retardant Technology/Polypropylene/V-2 Systems," paper presented at the 1989 SPE PMAD RETEC, Toronto, Ontario.

Table 20.2. Liquid Color: Physical Properties in Various Resins

Physical Property	Shell 5A64 Homo PP	Dow 8354N HDPE	Huntsman 3037 HIPS	Zytel 1016 Nylon 6	Lexan 141 PC	Rohm & Haas V920 Acrylic	Dow Magnum 342EZ ABS
Tensile yield strength	96.2	96.4	91.8	98.0	110.0	100.0	98.0
Tensile yield strain	110.5	110.0	75.0	100.0	100.0	100.0	100.0
Tensile break strain	116.6	100.0	133.0	100.0	100.0	—	—
Flexural strength	95.2	91.0	88.8	100.0	110.0	97.0	100.0
Flexural modulus	92.0	90.0	100.0	100.0	110.0	100.0	100.0
Notched Izod	100.0	100.0	100.0	100.0	100.0	100.0	100.0
HDT at 264 psi	96.0	100.0	98.9	95.0	96.0	100.0	100.0

Note: Liquid color/resin (2.50% LDR of 70% TiO_2 dispersion) vs. resin: % retention of resin physical properties.

Table 20.2 shows the effect on physical properties of a 2.5% letdown of a 70% TiO_2 dispersion in liquid color. Tensile yield strength, Heat deflection temperature (HDT), and flexural strength and modulus are the only properties where there is some negative effect, essentially a plasticizing effect. Several properties in various resins are enhanced or unaffected. Consult this table before choosing a coloration system and decide what properties are important.

20.12. SUMMARY

Liquid color is not a universal coloring system, but nothing is. It can perform in a variety of processes, in nearly all resins. Liquid is at its best in small percentages, where its distribution in the resin is excellent; it does modify physical properties somewhat but enhances some properties. It is excellent for heat-sensitive additives, such as blowing agents. Look into using liquid color as a coloring system; the economics may work out favorably.

ACKNOWLEDGMENTS

The author would like to thank Ferro Corporation for its support and to acknowledge the contributions of Robert Opalko, Thomas Loschiavo, Allen Carlson, Robert Brokaw, and J. Michael McKinney, all employees of Ferro Corporation.

CHAPTER 21

Use of Color Concentrates

Basic Colorant and Additive Production Processes: A Regulatory Compliance Perspective

Ralph A. Helfer

Polyone Corporation
2900 Shawnee Industrial Way
Suwanee, Georgia 30024

Scott D. Russell

Polyone
33587 Walker Road
Avon Luke, OH 44012

21.1. INTRODUCTION

This chapter is designed to provide a basic understanding of the synthetic resin (plastics) processing that is necessary to improve or alter the properties of various synthetic resin materials so these resins may be turned into more desirable and useful consumer products. It also discusses the health, safety, and environmental issues associated with the industrial processes used to formulate colorant and additive products.

Coloring of Plastics, Fundamentals, 2nd edition. Edited by Robert A. Charvat
ISBN 0-471-13906-8 Copyright © 2004 by John Wiley and Sons, Inc.

The final product may take any one of a vast multitude of final forms. These include, but are not limited to, construction materials, toys, automotive parts, kitchen aids, industrial parts, sports and recreational equipment, containers of all sizes and description, and an unlimited variety of packaging products for foods, drugs, cosmetics, and other consumer products. The processes that are used to produce or "mold or fabricate" the final products are also varied and depend on the design and purpose of the product. Some of these processes include extrusion, compression molding, roto-molding, blow molding, sheeting, and films.

In general, the purpose of the final product and the manufacturing process by which it is to be produced determines what type of resin (plastic) must be used and specifically which additives must be incorporated into the resin to enable it to meet the design specifications of the final product. This in turn determines the colorant and/or additive production process or processes that should be used to produce the desired effect in the final product. Depending on the specific purpose of the final product, additives processed into the resin may include, but are not limited to, color pigments, dispersion agents, antioxidants, UV absorbers, light stabilizers, plasticizers, wax lubricants, fire retardants, and a myriad of other ingredients designed to alter the properties of the resin to meet the specific production process and final product needs.

The required additives may be incorporated into the final product by two general methods. First, the molder may add a concentrated formula product to the resin at the "point of manufacture." This procedure affords the manufacturer greater versatility without having to maintain an unreasonably large inventory of resin products. Second, molders producing large quantities of a single or a limited number of products may opt to purchase "compounded" resin products already containing the desired additives. "Compounding" is a process usually confined to relatively large volume production of a single resin product for a specific purpose. The compounder adds the desired additives, including color, to the resin before it is shipped to the final product or molding facility. Molders using the compounded resin are restricted in versatility and are thus usually large-quantity producers of a limited product line.

21.2. COLORANT AND ADDITIVE PRODUCTION PROCESS DESCRIPTIONS

Companies will custom formulate colorant and additive products designed to be used by plastic molders, who will, in turn, produce the final consumer products. The raw materials for colorant and additive products may be in powder, liquid, or solid form. The products formulated from them may also be in powder, liquid, or solid form. Dry color formulations (powder form) currently comprise less than 5% of the total colorant and additive products being produced today. Liquid formulations account for another 5%; however, this form of product is increasing in popularity and is expected to capture a larger share of the colorant and additives market in the near future. The solid form, known as "concentrates or masterbatch" products, are concentrated ingredients encapsulated in a carrier resin that is usually in pellet form. This type of product comprises the overwhelming majority of the formulated products used by molders and compounders today. A discussion of the basic production processes associated with the production of the various colorant and additive product types is presented below.

21.2.1. Dry Formula Products

Weigh-Out

Essentially all product formulations start out the same way. Formulation of dry color and dry additive products involves the use of powdered raw materials that must be weighed out with reasonable precision. Large-quantity formulations allow most of the raw materials to be handled in their original container, usually plastic-lined paper bags of known weight. Smaller quantities usually require manual weigh-out, which is performed by placing increasing amounts into small bags until the desired quantity is reached. After the required proportions of the specific formula raw materials have been weighed out, the "batch" is usually placed on a pallet and labeled in preparation for mixing. However, some plants are set up to directly place the ingredients into a special mobile mixing chamber that will contain the formula in a sealed environment through the mixing process and on to the packaging process.

Mixing

Once weigh-out has been completed, the formula is usually transported to a mixer where the ingredients must be mixed to uniformity. The ingredients are usually manually dumped into the mixer, after which the lid is sealed and the mixer is operated until the formula reaches the desired uniformity. After mixing, the formula is usually manually removed from the mixer and placed into a "hopper" or container with a controlled outlet and is ready for the next operation.

At the more modern plants where the formula is automatically weighed out and added to a sealed mixing chamber, the sealed mixing chamber is next moved to the mixer, where it is attached for simultaneous rotation and tumbling until formula uniformity is achieved. The mixing chamber is then moved to the next operation—packaging.

Packaging

In the case of dry color or dry additive formulations, the final step is packaging. Regardless of the type of package to be used, the dry formula is generally accessed via a control valve at the bottom of the formula container. The packaging may take many forms but is usually a manual weigh-out of product into the final package. These packages may be in the form of plastic-lined paper bags, fiber drums, or boxes but may also be unlined plastic buckets or plastic bottles of various size. In all cases, packaging meets Department of Transportation (DOT) requirements as outlined in 49 CFR.

In the more modern plants, bulk packaging is usually automated. The formula container is attached to packaging equipment, and product packaging is completed by computer-automated equipment.

21.2.2. Concentrate (Pellet) Products

Weigh-Out

Unlike dry color or additive formulations that require the use of powdered raw materials, formulation of pelletized color and additive concentrate products may also involve the use of pelletized raw materials. Although pelletized raw materials still require precision weigh-out, their handling does not create a significant dust control problem. Again, large-quantity formulations allow most of the raw materials to be handled in their original container, usually plastic-lined paper bags of known weight. Smaller quantities usually require manual weigh-out, which is performed by placing increasing amounts into small bags until the desired quantity is reached. After the required proportions of the specific formula raw materials have been weighed out, the "batch" is usually placed on a pallet and labeled until it is ready for mixing. However, some plants are set up to directly place the ingredients into a special mobile mixing chamber that will contain the formula in a sealed environment through the mixing process and on to the encapsulation process.

At a more modern plant, raw materials are usually procured in large bulk bags. Other than the time when a vacuum hose is inserted into the top of the bulk bag, the system for preparing specific formulations is a computer-operated totally enclosed system that performs the necessary weigh-out of raw materials and automatically places the formula ingredients into a sealed mixing chamber. The mixing chamber is then moved through the mixing process where it is attached directly to the feed mechanism of the equipment that will encapsulate it into resin.

Mixing

The mixing process for concentrate pellet product production via Leistritz-type extrusion equipment is essentially identical to that of dry color product mixing. Once weigh-out has been completed, the formula is usually transported to a mixer where the ingredients must be mixed to uniformity. The ingredients are usually manually dumped into the mixer, after which the lid is sealed and the mixer is operated until the formula reaches the desired uniformity. After mixing, the formula is manually removed from the mixer and placed into a "hopper" or container with a controlled outlet and is ready for the next operation.

Some types of thermal processing equipment, such as a Banbury, are designed to perform the preliminary mixing before the extrusion step. In this case, a weighed-out formula batch is directly added to the feed throat of the equipment.

At the newest plants where the automatically weighed formula is added to a sealed mixing chamber, the sealed chamber is moved to the mixer where it is attached for simultaneous rotation and tumbling until formula uniformity is achieved. The chamber is then moved to the next operation—encapsulation.

Encapsulation (Concentrate Pellets)

After mixing, the formula is transported to thermal processing equipment where the formula container or hopper is usually attached directly to the feed throat of the encapsulation process equipment. The ingredients and properties of the product formula will determine the exact process equipment that will be used in the encapsulation or extrusion step. All equipment performing this process uses controlled flow and thermal processing to perform the required encapsulation. The actual processing temperature range will depend on the type of resin and additive ingredients to be used.

When the equipment is within the desired temperature range, resin (in powdered or pelletized form) and the colorant/additive formula are fed into the encapsulating equipment. The resin is melted and forced to move through the equipment at a specific flow rate. This forced movement of the resin and additive ingredients also creates a mixing and dispersion action necessary to produce a uniform color concentrate. Pigment loading depends on the properties of the ingredients and encapsulating resin and the desired specifications of the final product; however, they typically range from 5 to 40%.

The formula mixture is finally forced (extruded) through a set of dies where it is drawn into long, thin strands. These strands are pulled through contact cooling water for a few seconds, which causes the resin to set before it enters a pelletizer. The pelletizer is a sharp fan or "chopper" that rotates at a specific speed coordinated with the strand feed rate to slice the strands into pellets of the desired size. The pellets are then automatically screened, and pellet product of the correct specification size range is dropped into the desired packing container. Rejected pellets and particle product are returned as feedstock to the front of the extrusion process for reprocessing.

Packaging

The final bulk packaging is via plastic-lined cardboard gaylords or fiber drums that are placed under the screen at the end of the pelletizer. Gaylords are large ($36\,\text{ft}^3$) and may contain 1000–1500 lb of final product. Each gaylord is individually weighed and labeled. Fiber drums are used for smaller quantity deliveries (300 lb or less) and are also individually weighed and labeled. Of course, smaller quantities may involve the use of 5-gal buckets as the desired package for the customer. In all cases, packaging meets DOT requirements as outlined in 49 CFR.

21.2.3. Liquid Colorants and Additives

Weigh-Out

Liquid colorant and additive product formulation may involve both powdered and liquid ingredients. All dry constituents are generally manually weighed out in the same manner as for the products described above. However, proportioning of liquid raw materials requires the use of volumetric measurement methods.

Mixing

After the precise formulation of the liquid product constituents has been prepared, the constituents are manually added to a liquid mixer. This type of mixer is simpler in design than dry product mixers and more closely resembles a large drum with impeller blades.

Packaging

After mixing, the liquid product is ready for packaging. The packaging used for liquid products may include, but is not limited to, 1-liter plastic bottles, 5-gal plastic buckets, and various size plastic or steel drums. In all cases, packaging meets DOT requirements as outlined in 49 CFR.

21.3. DISTRIBUTION CHANNELS

21.3.1. Direct Sales Versus Distributorships

Companies will normally provide their customers with a wide and diverse selection of custom-formulated products. In addition, companies also have the ability to formulate new products for new resins or new consumer product applications. Because customers may select from more than 2000 different products that have already been formulated, the diversity of product options requires that direct sales to the customer be practiced in response to specific product orders. It would be impractical, if not impossible, to maintain the immense inventory that would be required to offer a distributorship-type operation. Conversely, compounders who produce a limited number of bulk products may effectively use a distributorship system to provide customers with their product.

21.3.2. Shipping Information

Forms of Shipment

A variety of shipping methods are used, which may include both company-owned vehicles and commercial transportation modes. Small-quantity orders may be shipped via UPS or a similar commercial delivery service. Many use a limited number of company–owned trucks to deliver products to local consumers. However, the overwhelming majority of bulk shipments are made using commercial freight lines (heavy truck transport).

Manifest/Bill of Lading

A manifest and bill of lading meeting DOT requirements as outlined in 49 CFR is prepared for each product and accompanies the shipment to its final destination.

Material Safety Data Sheets

A product-specific material safety data sheet (MSDS) is prepared for each product that has been formulated for a customer. Any time a change in the formula or an ingredient substitution is made in that product, a new update of the MSDS information is performed. In addition, the customer is notified that the original product has changed and the new MSDS is the document that should be consulted for product health, safety, and environmental information. The most current product MSDS then accompanies each shipment of product from the manufacturing facility.

Labels

An evaluation of each formulated product is performed to determine labeling requirements. This activity requires evaluating product constituency as well as determining the customer's geographical location (or the product's destination) to determine the federal, state, and/or local labeling requirements. These labeling requirements may include but are not limited to DOT labels, emergency labels, right-to-know labeling, and any of several special state program information labels.

21.4. WORKPLACE EXPOSURES AND CONTROLS

21.4.1. General

Historical studies performed in association with corporate pollution prevention and waste reduction programs coupled with the analytical results obtained from routine ambient air monitoring performed to meet Occupational Safety and Health Administration (OSHA) workplace requirements have provided valuable information concerning the sources of emissions and exposure to the materials being handled at a colorant and additive production facility. In addition, these studies have provided a characterization of the emissions being monitored and waste streams that may be generated from the operations performed at these facilities. Three important facts of information stand out and pervade manufacturing operations. The first is raw material costs. These costs may range from $0.65 to $2.00 per pound for resins to $500 per pound or more for pigments or special additive products. Therefore, these products must be handled with the greatest of care. No operation of this type can afford to lose a significant amount of these relatively expensive raw materials and stay competitive in today's global marketplace.

The second fact is that the manufacture of colorant and additive products involves an insignificant amount of volatile organic compounds (VOCs). Other than the routine small-quantity use of these materials in maintenance shop repair activities, which are mainly confined to solvent parts washing stations, the widespread use of VOCs is not really required in facility operations. The third fact is that raw material and/or product losses at production facilities normally average less than 1% and have been documented to be at a level below 0.1% in better managed operations. In addition, more than 99% of all lost materials are usually accounted for in the form of dust recovered from the facilities' dust collection systems and solids produced from the equipment purge process that must be performed between every product run on each piece of thermal processing equipment at each facility. The dust recovered from the facility's air pollution control system is essentially a mixture of powdered product formulas that have been run in the production area. Although this mixture will no longer meet specific product specifications, the relative consistency of pigment and additives to resin in the recovered dust remains fairly constant. Therefore, some of larger production facilities generally use the recovered dust and the reground purge plastic materials as feedstock to produce a wide spec ification black colorant product. This product is generally offered as an alternative low-cost colorant product to the manufacturers of recycled plastic products or manufacturers of plastic products that do not require critical specification colorant and additive products for their final product (flower pots, nursery flats, automobile splash guards, etc.). However, several facilities vend the purge plastic (with color and additives) directly to recycling facilities as raw materials for a variety of specific plastic recycling processes or special uses.

There are, however, several potential health and environmental impacts associated with handling the various raw materials associated with the formulation and production of colorant and additive products destined for use in the plastics industry. The greatest potential health or environmental impact would be represented by potential exposure to dust created by handling and processing powdered formula ingredients and final products. In addition to the general nuisance dust particulates and Parts per Million (PM_{10}) exposure, the dust consistency may include heavy metal compounds and a wide variety of toxic inorganic and organic compounds.

Process evaluation and routine ambient air-monitoring studies indicate that, from a standpoint of significant air emissions, a facility may be divided into two general areas: nonproduction and production. Except for a potential disaster situation (tornado, hurricane, etc.), virtually no significant emissions are generated outside of the production processing area. A discussion of the two basic facility areas is presented below:

21.4.2. Materials Handling and Air Pollution Point Sources

Nonproduction Areas

Nonproduction areas include those areas normally identified as performing the following functions: loading dock/receiving, raw material storage (including exterior storage such as silos), finished product storage or warehouse, shipping dock, offices, parking lots, and the exterior property Random air monitoring performed in these areas from 1989 through 1992 has typically yielded 8-h sample results below a detection limit of $0.1\,mg/m^3$ for total dust.

Although random sampling indicated that no significant emissions are generated on a routine basis in the nonproduction areas, it does not mean there is no potential for emissions to be generated in these areas. Emissions may result when bulk resins in powder form are transferred from rail cars or trucks to facility silos; however, all silos are equipped with baghouses to capture the dust generated by this activity. Studies of the bulk powder resin transfer process completed by an independent environmental consultant indicate that large operations may generate less than 100 lb of dust emissions per year, and the evaluation of bulk pelletized resin-handling activities indicates this form of product generates insignificant dust emissions. Once the resin is in the silo, it is transferred to the production process via an enclosed system and dust generation is eliminated.

Other than bulk loading to a silo, the potential for any large emissions being generated in nonproduction areas is minimal. Although spills may occur during receiving or while the raw materials are being transferred from the raw material storage area to the weigh-out room in the production area, material losses from accidental spills in nonproduction areas always remain relatively small because the materials are always packaged in relatively small containers. The vast majority of raw materials are contained in 80-lb plastic-lined paper bags; however, some products may be contained in plastic-lined fiber drums and a few resins are available in gaylord containers. Regardless of the quantity of material involved in an accidental spill, company policy should require that all spills occurring outside must be cleaned up immediately. Significant spills that occur inside the facility are cleaned up immediately, and any powdered material recovered in this manner is added to the dust recovered from the facility's dust collection system and is ultimately returned to the production process as a wide specification product raw material. Other spills that occur inside the facility are cleaned up at the end of each work shift and usually take the form of routine floor scrubbing.

Production Area

The production area at each facility represents the greatest potential for generating dust emissions and thus represents the most significant potential for personnel exposure to the materials handled at a colorant and additive production facility. In addition to the requirement for use of appropriate personal protection equipment (PPE), such as protective gloves, uniforms/coveralls, safety shoes, and protective eyewear, in all production areas, written corporate policy and procedures require all personnel entering areas of potential significant dust generation to wear appropriate respiratory protection. Minimum respiratory protection requires the use of half-face respirators unless the individual has a physical problem or deformity that would preclude the use of this type of equipment or the individual is wearing facial hair that would preclude satisfactorily passing a thorough fit test. In these circumstances, full-face powered air respiratory protection (PARP) is required.

In addition to the PPE provided for all personnel who work in or near the production areas, each facility must maintain a variety of air pollution control equipment to reduce potential exposures and limit emissions experienced outside of the production facility. Potential exposures and pollution control measures associated with each specific operational function carried out in the production area are discussed below.

WEIGHT-OUT. Manual weigh-out of powdered raw material constituents for product formulas represents a significant potential for personnel exposure to dust emissions and the various chemicals these emissions contain. All weigh-out scale areas should be provided with intakes and/or hoods that are part of an air pollution control dust collection system. Written corporate policy and procedures require that a facility's air pollution control system must be in operation any time weigh-out activities are performed. In addition, all personnel entering the weigh-out area must wear appropriate PPE, including respiratory protection, for activities performed in this area.

Based on routine monitoring results, exposure levels experienced by personnel directly performing weigh-out activities may range from a low of $0.35\,mg/m^3$ to a high of $6.55\,mg/m^3$ (total dust). The typical average exposure level over a 5-year period might be $1.64\,mg/m^3$. However, approximately 10 ft from the point at which this activity is being performed, dust exposure levels usually drop below a detection limit of $0.1\,mg/m^3$.

In a new facility, a scalable mixing chamber can be connected directly to an automated computer-controlled weigh-out system. This, of course, represents an enclosed system that virtually eliminates dust generation and employee exposures associated with the weigh-out process.

MIXING ROOM. Mixing vessels are sealed and, thus, no emissions are generated during the actual mixing process. However, manual loading of powdered formula constituents into the mixer and unloading the mixed formula into a hopper when mixing is completed represent the greatest potential for generating dust emissions during the production process and personnel exposure to the dust and the various chemicals these emissions contain. Therefore, each facility must confine this activity to a segregated area or room specifically designed for that purpose.

All mixing room areas must be appropriately equipped with intakes and/or hoods that are part of the air pollution control dust collection system. In addition, air intake collars designed to fit around a mixer lid are used during loading and unloading operations. As stated above, written corporate policy and procedures require that a facility's air pollution control system be in operation any time mixing activities are performed. The policy and procedures also require all personnel entering the mixing room area to wear appropriate PPE, including respiratory protection.

Routine monitoring results of exposure levels experienced by personnel performing loading and unloading operations of mixing equipment may range from a low of $0.55\,mg/m^3$ to a high of $8.20\,mg/m^3$ (total dust). A typical average exposure level over a 5-year period may be about $2.12\,mg/m^3$. However, outside of the mixing room area, exposure levels drop to below detectable limits. This would tend to indicate that significant exposure levels are confined to within a reasonably close proximity to the area where these operations are being performed.

In newer facilities, the sealed mixing chamber that came from the formula weighout operation is connected directly to the mixer during the actual mixing process. Then, without being opened, it is moved to the next step, which for dry colorants and additives would be packaging. Thus, generation of dust emissions is eliminated from the mixing process.

THERMAL PROCESSING: CONCENTRATE PRODUCTION. In the case of concentrate production, the next step is thermal processing. The hopper that was loaded with the mixed formula is attached to the throat (or feed) of the equipment that will be used to encapsulate the powdered constituents. In the past, feeding powdered constituents into thermal processing equipment represented a very significant emission source. However, several equipment modifications have been and are continually being made at most production facilities to reduce or eliminate emissions from problem area. Some equipment has been amenable to modifications that enclose this operation and virtually eliminate the generation of dust emissions. Because they save a significant amount of very expensive raw materials, these modifications are very jealously guarded trade secrets.

Routine monitoring indicates that typical exposures at the point of thermal process equipment feed range from a low of $0.10\,mg/m^3$ to a high of $2.93\,mg/m^3$ (total dust). The average exposure level over a 5-year period may be about $0.71\,mg/m^3$.

In the newest plants, the sealed mixing chamber would be directly connected to the feed throat of the thermal processing equipment before it is opened. When the vessel is emptied, it is again closed and sent to a special equipment wash bay for cleaning. This enclosed system virtually eliminates dust emissions throughout the thermal encapsulation process.

The thermal process encapsulates all constituents into the desired resin. At the terminal end of the thermal encapsulation process, the resin/additive mixture is forced or extruded through a set of hot dies where it is drawn into long, thin strands. As the resin/additive mixture passes through the hot dies, minute quantities of the petroleum wax lubricants and emulsifiers contained in the product are volatilized. This emission, commonly called "die gas," is an opaque smoke comprised of paraffin and other volatilized petroleum waxes that are relatively heavy. Die gas emission is confined to within close proximity of the die and is controlled through the use of a filtered exhaust system that has a pickup hood at the die location. Thus die gas emissions do not represent a significant exposure potential. This condition primarily results from the effective use of the pollution control equipment and is aided by the fact that the die gas waxes rapidly condense once they are away from the heat source (die). In fact, the overwhelming majority of the wax emission condenses in the first 10 ft of the exhaust system ducting and never reaches the exhaust system filter. Although wax condensation means that air emissions exiting the stack are extremely low, the condition also requires that the ducting be changed at frequent regular intervals to prevent stack fires. A die gas stack emission study performed at a number of facilities in 1990 indicated the annual emission rate at the worst case facility to be 69 lb. Most of the emission was fugitive dust picked up from the vicinity of the die location that passed through the die gas exhaust system filter.

After the product is pelletized, the final concentrate product is gravity fed into the desired container. Again, there is no dust generation associated with this process step. Although after manufacture rough handling may at times damage the pelletized concentrate product, the form of damage is usually pellet deformation and, on rare occasions, chipping with the more brittle resin products.

An understanding of the following information is necessary to gain a full appreciation for the change in potential health and environmental exposures that result from the encapsulation process:

1. Virgin resin in pellet form contains emulsifiers that give the pellet surface a thin, "oily" coating that acts to collect dust, preventing it from becoming airborne. Colorant and additive products also contain emulsifiers and/or lubricating waxes (especially when powdered resins are used) that also impart this property to the final product pellet.
2. The pigment and additive ingredients of the formula have been evenly dispersed throughout the resin. Once the pigment and additive ingredients have been locked into the resin, they are not subject to significant leaching or environmental release.
3. When the encapsulated resin exits the thermal process, it no longer represents a significant potential for generating dust emissions. At the pelletizer, commonly called a chopper, the resin is still relatively pliant and is actually cut or sliced rather than chopped. This operation does not create dust emissions.

21.5. ENVIRONMENTAL EXPOSURES

21.5.1. Air Emissions

Of the two types of air emissions associated with the production of colorant and additive products (stack and fugitive emissions), fugitive emissions represent an insignificant potential for environmental impact. In 1996 facility fugitive emissions ranged from an estimated high of 69 lb to a low of less than 1 lb (mass balance calculation). Facility evaluation monitoring performed usually indicates fugitive emissions are confined to the immediate vicinity of the production processes that generate them. These emissions are seldom detected beyond 10 or so feet from the generating source in the production areas and are almost never found in the nonproduction areas of the facility's interior.

Stack emissions represent close to 100% of the air emissions associated with an environmental release. Of the stack emissions, the overwhelming majority of these are contained in the dust collection system exhaust. In 1996 stack emissions (total dust) averaged 452 lb per typical facilities. However, the actual total dust emissions experienced by individual facilities may range from a high of 1327 lb to a minimum of 63 lb.

21.5.2. Wastewater

Generation

Wastewater may be generated from three general sources within a facility. Sanitary wastewater is generated from the employee showers, industrial wastewater results from equipment and production floor cleanup, and an occasional discharge may result during clean-out of the contact cooling water system. Of these, only the contact cooling water is classified as a "categorical wastewater discharge." However, this water is contained in a recirculating system and is not discharged on a routine basis.

Pretreatment

Typical wastewater pretreatment systems at concentrate facility locations include hydroxide precipitation, filtration, resin bed/ion exchange, and clarifier sedimentation. Wastewater discharge at all facilities must meet the requirements of Environmental Protection Agency (EPA), individual states, and the conditions outlined by local wastewater pretreatment ordinances.

21.5.3. Storm Water

All raw materials and products must be stored under cover. As a result, routine storm event monitoring will usually show no discharges are contained in storm water runoff.

21.5.4. Solid Waste (Process)

All solid waste materials generated by operations must be evaluated in accordance with the requirements outlined in 40 CFR 262.11. Waste classified as hazardous is shipped to a qualified disposal facility. However, the better facilities located in the United States do not generate any hazardous waste. The only facilities classified as a large-quantity generator is manufacturing products for the wire and cable industry and is locked into meeting customer-dictated product specifications requiring the use of lead chromate pigments. A discussion of waste types and disposal practices is presented below.

Dust

Nonhazardous waste dust may be generated from two sources, the facility dust collection system or from spills in the storage or production areas. Dust that is not used to produce a wide specification colorant product is usually encapsulated with virgin and/or purge resin and sent to a qualified solid waste landfill for final disposal. The average annual waste dust landfilled per facility is approximately 4.5 tons. However, the facility generating the highest annual quantity of dust (20.6 tons) ships this material to an industrial waste incineration facility.

Purge Resin

Purge resin generated by thermal processing equipment clean-out is a nonhazardous material that is either reentered into the production process or sold to plastics recyclers. However, small quantities of this material do end up in a facility's routine trash and ultimately ends up in a solid waste landfill.

Wastewater Pretreatment Residues

Toxic Characteristic Leaching Protocol (TCLP) testing of this material indicates it to be nonhazardous. After drying, this material is sent to a solid waste landfill. The average annual quantity of wastewater residue landfilled per facility is approximately 1.1 tons.

Waste Oil

The average waste oil generation from the colorant and additive process is approximately 200 gal (4 drums) per year per facility. This material is sent to a qualified recycling facility where it is blended into fuel for energy recovery.

Other Waste Materials

Usable scrap metals, pallets, paper, and cardboard are segregated from other waste and are sent to commercial recycling facilities.

21.6. REGULATORY CONTROLS AND STANDARDS

21.6.1. Consumer Product Safety Commission (16 CFR Parts 1000–1799)

The requirements in 16 CFR Parts 1000–1799 contain the provisions of the Consumer Product Safety Act and the Hazardous Substances Act that are designed to ensure consumer safety. Although these regulations primarily concern the final plastic products to customers, colorant and additive formulators are required to ensure these products, when used in the customer's final commercial product, will meet these regulatory requirements. Among these requirements are constituency limits for certain hazardous ingredients (such as toxic metals) and product flammability criteria.

21.6.2. Food and Drugs (21 CFR)

Since there are a multitude of plastic or synthetic resin products regulated under the provisions of 21 CFR that use colorant and additive products, any section of this code may apply to any colorant or additive product under any number of final product applications or uses. Although, as with the Consumer Product Safety Act regulations, these regulations primarily concern the final plastic products of customers, colorant and additive formulators are required to ensure their products, when used in the customer's final commercial product, will meet these regulatory requirements. Among the more important parts of this set of regulations are the sections listing approved ingredients and specifications for colorants and additives used in indirect food contact, drug, and cosmetic applications.

21.6.3. Occupational Safety and Health Administration (29 CFR)

In addition to the multitude of OSHA requirements listed in 29 CFR, Sections 1903, 1904, and 1910 (Subparts A–Z), that apply to all industries in general, the following sections of 29 CFR 1910 have particular impact on colorant and additive manufacturing processes.

Personal Protective Equipment (29 CFR 1910 Subpart I)

The standard 29 CFR 1910.132–1910.138 outlines the PPE for personnel required to perform activities in and around colorant and additive production operations. This PPE includes, but is not limited to, respiratory protection, safety glasses, hard hat, hearing protection, safety shoes, coveralls, and gloves.

Toxic and Hazardous Substances (29 CFR 1910 Subpart Z)

AIR CONTAMINANTS (29 CFR 1910.1000). OSHA requires that employees who must work in or near industrial operational areas with an ambient air exposure limit above a set level wear appropriate respiratory protection. The permissible exposure levels (PELs) of major concern for colorant and additive formulators are as follows:

Pollutant	PEL	Remarks
Total dust	$15 \, mg/m^3$	
Lead	$50 \, \mu g/m^3$	
Cadmium	$5 \, \mu g/m^3$	
Cadmium	$200 \, \mu g/m^3$	Dry color formulators
Chromium	$0.5 \, \mu g/m^3$	Proposed

LEAD STANDARD (29 CFR 1910.1025). Since colorant and additive production involves the use of lead compounds, the lead standard (29 CFR 1910.1025) becomes important. This standard applies to all occupational exposures to lead and lead compounds and requires respiratory protection when workplace exposure levels for lead reaches or exceeds $5 \, \mu g/m^3$.

CADMIUM STANDARD (29 CFR 1910.1027). Colorant and additive production also involves the use of cadmium compounds, thus, the cadmium standard (29 CFR 1910.1027) also becomes important. This standard applies to all occupational exposures to cadmium and cadmium compounds and requires respiratory protection when workplace exposure levels for lead reaches or exceeds $5\,\mu g/m^3$ ($200\,\mu g/m^3$ for dry color formulators).

HAZARD COMMUNICATION STANDARD (29 CFR 1910.1200). The hazard communication standard requires that all personnel receive training concerning the types of materials handled in the workplace and the potential hazards associated with handling and use of these materials. In addition, the standard requires that a MSDS for each hazardous material be made available for individual employee reference. The MSDS outlines specific material chemical and physical properties, exposure information, emergency response information, regulatory information, and any other information of significance concerning the material.

21.6.4. Protection of the Environment

Numerous sections of the Federal Code (40 CFR) apply to the operations associated with formulation and production of colorant and additive products. Most of these regulations are augmented by supporting state regulatory programs that, for the most part, are more stringent than the requirements contained in 40 CFR. Most state regulatory programs lack consistency with other state programs and tend to be complex beyond practical enforcement. However, all these regulatory programs add significant cost to the final product, and this cost is ultimately borne by the consumer.

In addition to the sections of 40 CFR applying to all industrial operations in general, the following regulations have specific direct impact on colorant and additive manufacturing processes.

Air Quality (40 CFR Parts 58–72)

Sections 58–72 of the environmental protection regulations address the air quality pollutant standards that must be met by stationary sources and the requirements for state air quality programs. Of these, air quality permitting performed at the state level has the most immediate impact to colorant and additive formulation operations. Consistency between state programs is nonexistent, and often these programs result in major delays in upgrading operational equipment or adding newer equipment that is designed to increase productivity and reduce overall air emissions. Tighter air quality standards for specific regional air quality management zones usually result in a direct impact on the conditions contained in the permits issued to all industries in that district; thus, the time required to evaluate a new permit application or a modification to an existing permit creates greater delays.

Without a doubt, the main reason for major delays experienced with obtaining permits and maintaining an effective air emissions management program is a general lack of qualified personnel at the regulating agencies. There are not enough qualified engineers to perform permit application evaluations, and delays may exceed 12 months. In addition, many air quality agency representatives are not familiar with plastics industrial machinery and associated air pollution control equipment. In fact, too many state and local agency representatives just do not understand the regulations and equipment system conditions they are required to enforce.

EPA-Administered Permit Programs: NPDES (40 CFR Parts 121–129)

This set of regulations outlines the purpose of the National Pollution Discharge Elimination System (NPDES) special programs, particularly for control of storm water discharges. Unless the best management practices (BMPs) for eliminating or controlling and monitoring storm water discharges are specifically included in an NPDES permit issued for a direct discharger or the industrial wastewater discharge permit issued for an indirect discharger, a separate storm water discharge permit is required. Colorant and additive formulation operations are defined by a Standard Industrial Classification (SIC) code that is included in the EPA's list of processes subject to the requirements of this program.

The raw materials associated with the production of colorant and additive products are relatively costly and are ruined if they contact storm water. Final colorant and additive products are even more expensive. Therefore, except in the case of accidental spill while shipping or receiving, these materials are never stored in exposed positions. However, many states define significant storm water contact to include potential air emissions and require colorant and additive production facilities to participate in storm water monitoring programs.

Although the monitoring requirements may sound good when one sits at a dry desk and thinks about what conditions should be met by an effective storm water monitoring program, serious evaluation of the required monitoring conditions will quickly indicate these conditions to be highly impractical and very difficult to meet. Results of all storm water monitoring performed by color facilities should indicate operations should have no impact on the composition or quality of the storm runoff water from those facilities.

Hazardous Waste Management (40 CFR Parts 260–278)

Obviously, the hazardous waste management regulations promulgated under the provisions of the Resource Conservation and Recovery Act and certain provisions of the Comprehensive Emergency Response, Compensation and Liability Act, represent a significant compliance responsibility and fiscal impact upon colorant and additive formulation operations. General activities complying with the requirements of this set of regulations were discussed above. Managing this liability is a full-time responsibility and represents a very significant added cost to the overall colorant and additive production process. Minimization of this liability and its associated costs require a continuing pollution prevention and waste reduction effort that was initiated before the EPA began its current P_2 programs and will continue to be a factor as long as there is commercial competition in the marketplace. Consumer demand, production costs, and overall profit are by far a greater incentive for pollution prevention and waste reduction than any program devised by the EPA.

Emergency Planning and Community Right to Know (40 CFR Parts 350, 355, and 372)

The requirements of the Superfund Amendment and Reauthorization Act (SARA) apply directly to colorant and additive formulation operations. In addition to the time and resources necessary for meeting the emergency planning and reporting required under these regulations, proposed new requirements pose a special threat for inadvertent disclosure of confidential business information. Although these regulations were designed to make certain potential hazardous chemical information available to surrounding communities, relatively few of these communities (or the people living in them) even bother to use it.

Federal Water Pollution Control Act (40 CFR Parts 401–471)

In addition to the Federal Water Pollution Control Act regulations discussed above, 40 CFR Parts 401–471 outline additional requirements for local municipal wastewater treatment authorities and the pretreatment program requirements that apply to industrial facilities that use municipal sewage treatment systems. Two sections of these regulations are of particular importance to colorant and additive formulating operations. These are Part 403, General Pretreatment Regulations for Existing and New Sources of Pollution, and Part 463, Plastics Molding and Forming Point Source Category.

The standard 40 CFR Part 403 outlines the requirements of the National Pollution Discharge Elimination System (NPDES) program for local district management of industrial wastewater discharges and their treatment. Part 463 identifies plastics molding and forming operations, directly pertaining to colorant and additive production operations, as a Federal Categorical Point Source and outlines the requirements for wastewater discharge from these specific industrial operations.

Toxic Substances Control Act (40 CFR Parts 700–723)

The regulations promulgated under the Toxic Substances Control Act (TSCA) apply to all manufacturers, importers, and processors, including distribution operations. In addition to compliance with the health and safety recordkeeping requirements of TSCA, company policy should require that all materials associated with all operational activities be TSCA listed. This requirement is designed to ensure positive identification of the materials being used so proper evaluation may determine health and safety requirements for handling these materials as well as the requirements for meeting acceptable criteria concerning regulatory compliance issues.

State and Local Requirements

In addition to the many federal regulatory programs directly affecting colorant and additive formulating processes, there is a myriad of state programs and local ordinances that must be considered. These range in complexity from the various state programs for primacy management of EPA programs (which are often more stringent than the federal programs) to special and more specific regulatory programs that tend to duplicate and unnecessarily expand similar federal regulatory requirements. Some examples of these programs are as follows:

- **CONEG.** The Model Toxics in Packaging Legislation, developed by the Coalition of Northeastern Governors (CONEG) and adopted in one form or another by 18 states, directly affects the colorant and additive industry.
- **Proposition 65.** "Proposition 65," promulgated as part of the California Safe Drinking Water and Toxic Enforcement Act of 1986, requires special product labeling for any product containing any ingredient "known to the State to cause cancer or reproductive toxicity."
- **New Jersey Worker and Community Right to Know.** The product labeling requirements of the New Jersey Worker and Community Right To Know regulations (Title 8, Chapter 59) are complex. Pennsylvania also has a very similar program with Title 34, Part XIII, Chapter 309.

In addition to the examples cited above, there is a vast multitude of well meaning state regulations affecting plastics packaging and plastics recycling that are grossly ineffective and fail to meet designed objectives. All of these regulations impact the colorant and additive industry and add costs to the final product.

21.7. EMPLOYEE EDUCATION

Employee training is a continual process. Monthly safety meetings (25 hours annually) and specific training necessary to fulfill the requirements of the various special OSHA and EPA programs (approximately 24 hours annually) are mandatory. Employees required to operate or work in close proximity to production equipment at any color compounding facility should be required to, at a minimum, complete 40 hours of specialized on-the-job training under the direct supervision of a qualified operator (for the equipment of concern).

Special training sessions include, but are not limited, to the following topics:

OSHA

- Lead standard
- Cadmium standard
- Hazard communication
- Lock-out tag-out
- Confined space entry
- Fork lift operation
- Incident and/or accident reporting
- Safety equipment use and maintenance

EPA

- Emergency preparedness
- Emergency response and reporting
- Hazardous materials management
- Pollution prevention and waste reduction
- General regulatory compliance

21.8. SUMMARY

Customer demand has the greatest single impact on the selection of ingredient raw materials used to formulate colorant and additive products. Cost is, without a doubt, also among the customer's highest priorities. Even the most overwhelming demand for special products that exhibit certain desirable characteristics is tempered by the cost of the product. Companies should promote and/or develop alternative or substitute products that eliminate the requirement to use several of the more stringently regulated hazardous or toxic constituents associated with coloration and manufacture of plastic products. A customer must be willing to pay the additional (and sometimes significant) costs associated with use of these products or find a more cost-effective production process. Companies will do whatever is necessary to safely produce the product of demand and meet regulatory requirements.

Color compounding and color concentrate companies must be responsible industrial partners that place a very high priority on and a substantial resource commitment to personnel safety and environmental protection. It is important to note that material handling and operational practices have always been consistent with sound pollution prevention and waste reduction principles. These activities have generally preceded regulatory programs that mandated these actions because they are very cost effective and directly translate to the bottom-line profitability and, in general, are plain common sense good business. In addition, we all must live in this ecosystem.

Note: All references to government rules and regulations by EPA and other Federal organizations are available through those organizations and are Listed in the Federal Register.

CHAPTER 22

Product testing

22.1. INTRODUCTION

Why test? We have been coloring thermoplastics for half a century and have a knowledge base that could fill the New York City Public Library. Why then does it seem that each new coloring project turns into an adventure that does not come to a close until an exhaustive (and exhausting) battery of property tests are run? The reasons behind this problem are those very things that attract most of us to the color industry. First, there is the public's insatiable thirst for variety. Then, there is the sheer number of different materials to be colored. But most of all, the materials we color were not necessarily designed with color in mind. We may have replaced "traditional" materials such as wood, metal, and glass with our polymeric inventions, but we maintain a traditional mind set regarding the separation of properties from appearance. Thermoplastic materials are designed by one group of specialists and colored by another. Consequently, we need to retest the material properties after the color is added. To compound the issue, there is more than one way to introduce color to a fabricated thermoplastic part:

1. Start with precolored plastic granules.
2. Blend the natural (uncolored) material with "dry color."
3. Blend the natural with a masterbatch or color concentrate.
4. Pump in liquid color at the machine.

Deliver the same color formulation by these four different means, and you are guaranteed four different colors (slight differences, to be sure) and four different resulting sets of material properties.

In this chapter we describe many of the tests commonly used in our industry to evaluate the characteristics of colored thermoplastics parts, starting with colorant dispersion. The focus will be on the objectives of these tests, not specific ASTM or ISO procedures. References to specific test procedures will be given in the citations.

Coloring of Plastics, Fundamentals, 2nd edition. Edited by Robert A. Charvat
ISBN 0-471-13906-8   Copyright © 2004 by John Wiley and Sons, Inc.

22.1.1. Terminology

In this chapter, the term *colorant* is used to refer to any chemical substance we add to a thermoplastic with the deliberate intent of imparting color. There are two fundamental classes of colorants, pigments and dyes. *Dyes* are defined operationally as colorants that dissolve when compounded into the plastic matrix. A substance may act as a dye in one thermoplastic system (i.e., dissolve into the matrix) but not another. All of the dyes we use in thermoplastics are synthetic organic compounds. Likewise, *pigments* are defined operationally as colorants that do *not* dissolve when compounded into the plastic matrix. Many of the pigments we use in thermoplastics are minerals (inorganic chemicals). Examples include oxides, chromates, sulfides, and silicates. The majority, however, are synthetic organic chemicals. In practice, the distinction between dyes and pigments is not always crisp. At high concentrations, many of the "dyes" are only partially soluble. Conversely, most organic pigments are partially soluble in thermoplastic materials at the high temperatures and pressures characteristic of melt processing, and sometimes this solubility manifests itself in undesired color shifts.

The term *dispersion* refers to how well the pigments are "wetted out" in the plastic matrix (see below), whereas *distribution* refers to the uniformity of color in the part. Finally, in referring to the parts fabrication process, the term "molder" is used below to represent any the following pieces of equipment: injection molding machine, injection or extrusion blow molding machine, or rotational molding machine.

22.1.2. Types of Tests

Dispersion Testing

Since dyes dissolve in the resin matrix, our concern with dispersion focuses on pigments, which are suspended in the matrix in a colloidal state. The quality of the dispersion affects the following:

1. the color strength realized from the pigments in the formulation,
2. the surface qualities of the part, and
3. the physical properties of the part.

Given the potential consequences of inadequately dispersed pigments, pigment manufacturers apply proprietary surface treatments to make their products easier to disperse, and colorists who formulate with these pigments devote at least as much effort to dispersion technology as they do to color formulation technology.

Color Durability Testing

In practice, durability refers to how well the formulation holds up to environmental stresses. The properties of interest are the formulation's heat stability toward processing temperatures, its fade and weather resistance, its resistance toward migration into various solvents, and its chemical resistance toward alkali, acid, and bleach (strong oxidizing agents).

Physical Properties Testing

We take a broad view here. In addition to strength properties, we consider rheological properties and the dimensional changes that can take place due to shrinkage. The influence of color on rheology in turn affects how the thermoplastic processes and can be an indication of polymer degradation. Colorant-induced breakdown of the polymer chains, a chemical reaction, leads to a reduction in viscosity. On the other hand, the mere presence of pigments in the matrix can inhibit the flow of the polymer chains, resulting in higher viscosities.

22.2. PIGMENT DISPERSION TESTING

The color formulator's job is not just to combine the pigments and dyes in the correct ratio to match the customer's target color, but to assure that the color strength of each component is fully developed and that all of the pigment particles is wetted out. On the other hand, the customer's job is to distribute the color uniformly (except if a marble appearance is desired) in their molder or extruder. The dispersion process is illustrated in Figure 22.1, which shows that dispersion is actually a sequence of processes involving (1) particle size reduction and (2) displacement of any air and moisture surrounding the pigment particles by the thermoplastic matrix. The objective is the formation of a uniform emulsion. Color formulators may seek to aid emulsification by employing surface-active chemicals (surfactants) or by coating the pigment particles with a wax after size reduction.

The most obvious undesirable effect of poor dispersion is color strength reduction, which translates immediately into higher cost. For a single-pigment dispersion, the color strength is determined by measuring the absorption of light at the wavelength of maximum absorption, relative to some standard, as shown in Figure 22.2. The strength of a formulated color can be similarly evaluated by looking at its reflectance curve. A poor dispersion yields higher reflectance values at all wavelengths and appears visibly "washed out" (less chromatic) compared to a well-dispersed standard. Why should poor dispersion lead to reduced color strength? After all, the correct weight percent of pigment is in the formula. A simple model provides the answer. The color we see from the pigment is a combination of absorption and scattering. Each particle of a given pigment in the dispersion has about the same ability to scatter the incident light back to the observer. The more particles per pound of pigment in the dispersion, the more color. By breaking down the aggregates and agglomerates, you are, in effect, creating more color-scattering sites and a stronger color.

The effects of poor dispersion on processing (molding and extruding) are no more well documented than in the case of fiber spinning. Pigment agglomerates can cause fiber breaks. To prevent this, the fiber producers use tight screen packs in the heads of their extrusion lines to filter out agglomerates and any other large particles. A poor dispersion will clog the screen in hours or even minutes. Even with automatic screen changers, the lines will suffer production losses.

Undesirable appearance blemishes due to poor dispersion show up immediately in sheet or film extrusion and in extrusion blow molding. Agglomerates near the surface get caught up in the die, and as the plastic matrix flows around the large particles, flow lines appear that are often described as "fish eyes"(see Fig. 22.3). Within each fish eye is a pit, usually visible with the unaided eye, which leads to a loss in surface gloss. Under magnification, you will find a pigment agglomerate in the center of each pit. If you hold a transparent or translucent section of the part up to the light, the agglomerates reveal themselves as unsightly specks.

The presence of a pigment particles with a high aspect ratio, such a pearlescent or metallic pigment (plate structure), accentuates the appearance problems due to undispersed pigment. The author has "seen" Colour Index (CI) pigment green 7 particles as small as 10 µm cause visible flow lines to appear in a green pearlescent formulation. As the pearl particles flow around the agglomerate, they flip on their side leaving a dark line. Visually, no pitting was observed, nor were any specks visible. The microscopic agglomerates that produced the flow lines were observed under 60× magnification, however.

Pigment agglomerates can also have an effect on tensile elongation and strength and impact strength, especially at temperatures well below the resin's T_g. Polyamide cable ties, injection molded in lengths up to 4 ft, are used in a variety of home and industrial applications, including bundling over head communication cables. The installer pulls on the cable with great force, sometimes while standing in a cherry picker 30 ft above ground in −20°F weather. If, due to loss of tensile strength, the cable tie breaks, the results are disastrous. In this application, pigment agglomerates as small as 10 µm can cause the ties to fail under these extreme conditions of use Polypropylene copolymer automotive interior parts must pass a cold dropping dart impact test carried out at −40°F. Again, pigment agglomerates serve as crack initiation sites that lead to part failure in this application as well. Similar specifications must be met by exterior parts.

22.2.1. Dispersion Test Methods

The most widely used test methods are optical, including computerized image analysis, and pressure rise testing. In most cases, examination of the formulated color or a colored finished part under an optical microscope at 100× will reveal a problem. At this magnification, agglomerates as small as 5–10 µm are easily seen. A more dramatic test is the use of blown film. Typically, the colorant is let down to yield about 0.5% pigment in film that is about 5 mils thick. Film provides high surface area per weight, and all agglomerates in the field of view will be visible. The subjectivity in these tests can be reduced by establishing a dispersion quality scale based on the number of defects counted per unit area viewed. This tedious process can be automated through use of image analysis, a technique involving computer-assisted automated surface scanning and data collection. Figure 22.4 shows a typical print-out in the form of a particle size distribution.

Pressure rise testing is widely used in the fibers industry because it simulates, in part, the fiber spinning process. Figure 22.5 provides a schematic view of the test. For single pigment dispersions, the colorant is let down in the test resin at 8.33% final pigment loading. As the blend is extruded through a screen pack (typically 60/100/325/50 mesh), the pressure in the head is monitored. The test is run until a prescribed weight of pigment has been filtered. The difference in pressure rise between the test sample and that obtained with the natural resin is then recorded. For most fiber applications the desired result is <100 psi. At first glance it would appear to be the ideal test since it simulates in part the fiber spinning process. In practice, however, correlation between laboratories, or even between test machines within a given laboratory, often is not good, and the test can give false-positive ("good") readings. Nonetheless, within a given laboratory, sufficient data can be acquired over time with a particular pressure rise test procedure to give the formulator a high enough degree of confidence to make pass–fail judgments on the product.

22.3. COLOR DURABILITY TESTING

22.3.1. Heat Stability

Sometimes, a color formulation that appears to be on target when tested in the quality assurance (QA) laboratory triggers a complaint when run on the customer's equipment. A cursory examination reveals it has "burned out." Given that the value we bring to our customers is our formulation knowledge, this type of complaint is embarrassing, let alone costly to ail. To avoid complaints of this nature, get all of the required processing information up front. In addition to the nominal heat settings in the molder or extruder barrel, you also need to know the following:

1. Is there a hot-runner manifold and, if so, what are the temperatures?
2. What is the gating like? Pin gates generate shear heat that can cause local temperature increases of 50°F or so near the gate.
3. What is the average dwell time in the barrel? Large parts often have molding cycle time in excess of a minute. If there are three shots in the barrel at any time, that means your formulation needs to hold up for at least 3 min at the maximum processing temperature.

For any application, you need to know the heat stabilities of the pigments and dyes you are using. Most often this information is provided by the colorant manufacturer, and most of the time you can believe it. But, they test under one set of conditions in two or three resins, usually low-density polyethylene (LDPE), polystyrene (PS), and polyvingl chloride (PVC). A successful test in PS at 450°F does not mean the pigment or dye will hold up in styrene acrylonitrile copolymer (SAN) at this (or any other) temperature. There may be unexpected chemical reactions. The following straightforward plaque injection molding test can be used to evaluate the heat stability of a pigment, dye, or color formulation in a particular resin:

1. Set the barrel heats to the recommended temperature profile for the resin.
2. Using the recommended molding cycle for that resin, mold about 20 plaques of the natural resin to achieve thermal equilibrium.
3. Add the resin/color blend to the hopper and then mold plaques until the color first appears.
4. Throw away the next four plaques and collect the next 10.
5. raise the barrel temperatures in 25°F increments and repeat steps 1–4.

6. Repeat step 5 until there a visibly obvious color changes or until the resin begins to degrade.
7. Repeat steps 1–6 holding each shot in the barrel a given dwell time (e.g. 3 min).

If you suspect *shear heating* will be a problem, you can try to simulate your customer's aggressive molding conditions by molding at various injection rates through a pin gated mold. The effects of a hot runner manifold can be simulated by a modification of the plaque molding test cited able. Extend the hold time between shots up to 5 min and add 25–50°F to the expected barrel temperature settings.

It is important in this or any other heat stability test to mold the natural resin as a control. Some resins, especially acrylonitrile butadiene styrene (ABS), darken considerably when heated above a particular threshold temperature. In the case of ABS, the butadiene rubber component may begin to go yellow and dark over 450°F. Although the customer may set its machines below 450°F barrel temperatures, shear heating or heating in the hot runner manifold could yield local melt temperatures in excess of 500°F and burn the natural resin.

22.3.2. Fade and Weather Resistance

Thermoplastics have been replacing painted metal parts in a variety of durable goods, including appliances and automobiles, and they are expected to retain their color after long exposure to sunlight and the elements. *Fade resistance* refers to the ability of the part to resist color changes due to the effects of sunlight and artificial light sources alone. By *weather resistance*, we mean the ability of the part to hold its color in its intended outdoor environment, that is, when exposed to cycles of sunlight and darkness and associated temperature changes as well as to cycles of moisture and dryness. In the case of an interior application, such as an automobile door pillar or an indoor carpet, we are concerned more with fade resistance. With an outdoor application, such as a car bumper or a municipal trash container, we seek outstanding weather resistance. Reference 1 provides an excellent introduction to outdoor and accelerated weather testing.

Outdoor weather testing stations represent one approach to setting up a testing environment. The test samples are placed on racks oriented at a specific angle that expose them to sunlight from a specific direction; for example, the samples are given a westerly exposure at an angle of 45°. By placing the sample in a glass-covered box, the effects of sunlight (and temperature) can be assessed. At the test site weather instruments record the amount of sunlight and rain and continuously monitor the temperature and humidity. Commercial stations such as this, for example, have been set up in the United States in Florida (hot and humid), Arizona (hot and dry), and the Midwest (industrial emissions) to test for the effects of a variety of environments. Typically, you would want to test in each environment to which the part is likely to be exposed.

Fade and weather resistance can also be evaluated using laboratory test chambers, fadometers, and weatherometers, designed to simulate a variety of actual environments in an accelerated fashion. The types of tests are classified according to the nature of the light source:

(1) carbon arc lamp,
(2) xenon arc lamp, or
(3) fluorescent UV lamp.

Each of these sources represent various different portions of sunlight, with the xenon arc light source matching daylight spectra closest of all the artificial light sources used. Two widely used automotive accelerated test methods based on xenon arc radiation are Society of Automotive Engineers (SAE) J1960 for exterior parts and SAE J1885 for interior trim [2]. Since outdoor results depend on the actual weather conditions during the test period, there is no precise factor to convert from hours of weatherometer exposure to years of South Florida (or any other region) outdoor exposure. A rough rule of thumb is that 1300 h of exposure according to the SAE J1960 procedure corresponds to about one year of actual South Florida exposure. Figure 22.6 shows several commercial test apparatuses used for accelerated weather and fade testing.

The J1885 test developed for automotive *interior* trim enables the colorist to screen color/additive formulations quickly. On the one hand, the body of pigment, dye, and UV stabilizer data is so large for this application that "safe" formulations could readily be constructed from historical data with only an occasional "audit" required in the weatherometer. But the automotive customers seek the "optimal" formula in terms of price and performance. This requires testing to determine the minimal amount of added cost, in terms of UV stabilizers, to provide the requested protection. Failure in these tests is rarely due to the colorant system. Almost always the observed color shifts are due to chemical changes in the thermoplastic resin. (Reference 3 discusses the mechanisms of light-induced polymer degradation.) Oxidation of the resin can form chromophores and cause yellowing (increase in the b^* value, if you are measuring in CIE $L^*a^*b^*$ units). It can also lead to chain scission and produce a network of microscopic surface cracks (crazing), which in turn makes the part appear lighter over time (increase in the L^* value). A work of caution: Accelerated weather testing needs to be backed up with actual outdoor exposure. Run side-by-side testing, accelerated versus outdoor, and continue the outdoor testing even beyond product introduction. This will enable you to establish an acceleration factor for your application.

Our industry has been collecting outdoor weathering data for PVC building products applications such as vinyl siding for decades and has established correlations with several accelerated tests. The pigments of choice in these applications are the mixed metal oxides, which are among the most durable available. Vinyl siding testing is well covered in ref. 4. Despite all of its other good properties that recommend its use in exterior siding, PVC yellows rapidly when exposed to the UV component of sunlight. Suppliers have developed several competing stabilization approaches. One of the more popular ones is to use a high loading of TiO_2, 8–10%, in combination with chemical stabilizers that prevent oxidation of the PVC and other light-induced chemical reactions. The TiO_2 serves as a UV absorber. Each variation on this theme yields slightly different performance characteristics, and a particular color formulation needs to be tested in each of the PVC compounds in which it will be used.

In outdoor weathering, reactions of moisture are just as important as the direct interaction with light. Ultramarine pigments can react in typical acid rain conditions found in the northeast region of the United States to release H$_2$S, resulting in a reduction in chroma. In the case of PVC siding, the evidence from long-term weather studies suggests that reactions between water molecules and the TiO$_2$ in the PVC form free radicals that initiate the breakdown of the PVC polymer chains.

22.3.3. Color Migration

Some color formulation components, especially dispersion aids and dyes, are soluble in the plastic matrix during melt processing but may come out of solution as the melt cools. Over time, these components migrate to the surface of the part, a process we call *blooming*. If the component is a liquid at room temperature, an oily film will appear. If it is a solid, a haze appears. In the field, blooming may take days to manifest itself. For example, the molder may pack what appears to be good parts and ship them to the customer. Three days later, when the customer goes to put the part on the shelf or on an assembly line, the bloom is evident and a complaint is registered. This is an expensive scenario, and good will evaporates quickly.

Three properties that contribute the most to migration are (1) incompatibility with the resin, (2) low melt point, and (3) low molecular weight. Examples of noncolorant additives that are prone to migrate are (1) lubricants/dispersants such as metal stearates and bis-stearamides, (2) mold release and slip agents, and (3) antistatic agents. All of these ingredients serve useful purposes in color formulations. If you exceed their solubility limit in the end-use resin, however, you risk blooming and its consequences.

Colorants can also exude to the surface and rub off. If the concentration of a dye exceeds its solubility in the plastic matrix, it will be squeezed out and migrate to the surface. Also, some organic pigments are slightly soluble in various organic solvents, including some plastics additives. For example, CI pigment violet 19 (a quinacridone pigment) dissolves to some extent in ethylene bis-stearamide wax. It can then bloom to the surface along with the wax and rub off and stain anything it comes into contact with, for example, merchandise and people's hands.

To avoid problems in the field, run a simple *accelerated test* in which samples are placed in a laboratory oven at an elevated temperature for various times. For example, set the over for 120°F and check for blooming after 24, 48, and 72 h. A visual check against parts stored at room temperature reveals any problems. This simulates the worst storing and shipping temperature conditions and provides a definitive answer in about three days.

Leaching is used migration into an external solvent and is a risk whenever the color formulation is used in packaging containing a liquid, cream, or moist solid. Again, leaching is not limited do dyes. Organic pigments may have slight solubility in one of the solvents and leach into the contents. Special care should be taken in developing color formulations for packaging of personal care products (liquids, lotions, and creams) and food products. Even if the color represents no harm, the customer will perceive it as a contaminant.

Unfortunately, there is no universal test for leaching. One very aggressive test is the *white vinyl test*, in which a plaque of the colored plastic is pressed against a sheet of white plasticized vinyl using a 500-g weight. You run this test in an over set at 150°F and check after a predetermined time (typically 6, 12, or 24 h) for staining of the white vinyl. While this test can give a false positive due to the aggressive solvent (plasticizer) and test conditions, it tells you what pigments to avoid. See ref. 5 for other published test methods.

In the case of food packaging, most applications are covered by the regulation cited in the Code of Federal Regulations, Volume 21, paragraph 178.3297 (21CFR178.3297), which deals with indirect food contact. These regulations do not cover all cases, however. For example, the customer demands a color that cannot be achieved with the colorants cited in the regulation, or the conditions of use are not covered by the regulation. Regulation 21CFR175.3 provides testing guidelines for a variety of food types and use temperatures that you can use for evaluating colorants not cited in 21CFR178.3297. In such cases, seek an expert opinion from your legal department before submitting an interpretation of the test results to your customer or the Federal Drug Administration (FDA).

22.3.4. Chemical Resistance

Solvents and other environmental chemicals may undergo a variety of chemical reactions with the pigments and dyes in a color formulation. The two types of reactions most frequently encountered are acid–base and oxidation–reduction reactions. Reference 6 a identifies source table of most of the colorants for plastics and their properties, including reactivity. If, on the basis of this table or other available information, you suspect an acid–base or redox reaction is likely, then test the color formulation under the expected conditions of use. One pigment class that is particularly susceptible to reactions with acid is CI pigment blue 29, ultramarine blue. The sulfide ions in these pigments can react with acid to form H_2S, resulting in an unpleasant odor and possible loss of color. Color formulations intended for synthetic fibers need to hold up to alkaline laundry solutions and be resistant to bleach (oxidation). The American Association of Textile Chemists and Colorists (AATCC) has published standard methods for testing these properties [7].

22.4. PHYSICAL PROPERTIES TESTING

22.4.1. Strength Properties

The strength properties more often specified for plastics materials are (1) tensile strength and elongation, (2) flexural strength, (3) Izod and Gardner impact, and (4) heat deflection temperature under load. Our purpose here is not to describe each test in detail but to point out some of the known effects that colorants and other formulation ingredients can have on these properties. Table 22.1 lists the ISO and ASTM test methods for most of the physical properties, and ref. 1 (pp. 7–112) describes each of the methods in detail. Table 22.2 lists typical values of the above cite four properties for selected thermoplastics.

Pigment agglomerates serve as fracture initiation sites for thermoplastics. Similar effects are seen with large-particle-size special effects pigments (e.g., mica). If the application specifies retention of strength properties, testing is in order. At the very least, check the tensile and Gardner impact values relative to the natural resin. Pigment agglomerates have less of an effect on the properties of elastomeric thermoplastics, such as high impact grades of copolymer polypropylene, at least at room

Table 22.1. ASTM and ISO Designations for Physical Properties Tests

Type of Test	ASTM No.	Units	ISO No.	Units
Tensile strength	D 638	psi	R 527	MPa
Tensile modulus	D 638	psi	R 527	MPa
Flexural strength	D 790	psi	178	MPa
Flexural modulus	D 790	psi	178	MPa
Compressive strength	D 695	psi	604	MPa
Izod impact	D 256	ft-lbf/in.	R 180	kJ/m
Charpy impact	D 256	ft-lbf/in.	R 179	kJ/m
Heat distortion	D 648	°C	75	°C
Gardner impact	D 3029	in.-lb	6603	J

Table 22.2. Typical Values of Strength Properties for Selected Thermoplastics

Property	HDPE	PPRO	HIPS	ABS	PA 6/6	PC
Tensile strength at yield (psi; D 638)	4,000	5,000	3,700	6,000	12,000	9,000
Elongation at break (%; D 638)	>400	10	47	40	60	125
Flexural modulus (psi; D 790)	150,000	180,000	300,000	350,000	410,000	340,000
Notched Izod impact (ft-lbf/in; D 256)	1.6	0.5	1.6	5.0	1.0	15.0
Heat deflection temperature (°C; D 648)						
66 psi	80	100	—	110	235	137
264 psi	—	80	100	100	90	132

Abbreviations; HDPE, high-density polyethylene; HIPS, high impact polystyrene; ABS, acrylonitrile butadiene styrene; PA 6/6, polyamide 6/6; PC, polycarbonate.

temperature. But they readily degrade the highly valued low-temperature impact properties of these materials. In general, thermoplastics are not very forgiving of poor pigment dispersion.

Take special care in formulating color for glass fiber-reinforced thermoplastics. Pigments with high moh hardness values should be avoided. These include TiO_2 and certain complet inorganic color pigments. Such hard pigments can etch the glass fibers and promote breakage during processing. If the fiber length is reduced below as critical value (length-to-diameter ratio), the reinforcing effect will be lost. The properties most likely to be affected are tensile and impact strength. Again, if these pigments are required from a coloristic standpoint, testing should be done.

Dyes can also have an undesirable effect on properties. Because they dissolve in the matrix, they can sometimes have a plasticizing effect. This will reduce the material's tensile and flexural strength as well as the HDT (heat deflection temperature under load). The plasticizing effect of dyes can also influence the way they process

330 Product testing

on the customer's equipment. Screw slippage, for example, can slow down injection and blow molding cycles and extrusion rates.

The color formulation binder system can also have undesirable effects on strength properties. If the binder in a liquid color concentrate, for example, plasticizes the let-down resin, you can expect the flexural properties and HDT to decrease. Some "universal" carrier systems have only limited compatibility with many resins and separate out as a separate phase. The unwanted effects include delamination and loss in tensile and impact strength.

22.4.2. Other Special Tests

Engineers have devised tests for a variety of applications that place unusual stresses on thermoplastic parts. In the consumer products packaging industry, for example, the filled plastics containers have to survive crating and shipping and the odd chance that they will be accidently knocked from the top shelf of the display rack in the store. An obvious, somewhat messy test to simulate use conditions is to fill the containers with water and drop them from successively greater heights. Crystallizable polyethylene terephthalate (CPET) trays intended for frozen foods are subjected to numerous freeze–thaw cycles and tested for Gardner impact at freezer temperatures. Automotive manufacturers require instrumented impact testing down to −40°F for thermoplastics used in exterior applications and for interior trim.

22.5. RHEOLOGICAL TESTING

Rheology is a complex subject, and in the formulated colorant industry two simplified approaches have served well in the majority of cases. These are the single-point method, *melt flow*, and the molding simulation test, *spiral flow*.

22.5.1. Melt Flow

The melt plastometer (Fig. 22.7) is an instrument designed to measure the flow of a thermoplastic material at a single temperature and (low) shear rate. A given weight of sample is dispensed into its heated barrel at a specified temperature. At the base of the barrel is a plug bored with a standard orifice (length an diameter). A piston is then inserted. After a prescribed conditioning time with minimal weight (50 g) on the piston, a larger weight is placed on the piston and the molten material is forced through the orifice. The flow is timed, usually for 1 min, and the material collected is weighed. The results are reported as weight in grams per 10 min of flow. Table 22.3 gives the ASTM D 1238 standard conditions for running the melt flow test for a number of important resins. The conditions are selected to yield a value of about 10 g/10 min for a molding grade resin.

The melt flow test is routinely used as an indicator of whether or not a color concentrate will be adequately distributed throughout the natural in injection molding applications. A simple rule of thumb is to design the melt flow of the concentrate to be about 1.5–2 times higher than that of the natural resin. The assumption is that the colorant will flow slightly faster than the natural and readily blend in with it. At traditional let-down ratios of 20:1 and 25:1, this rule of thumb has served the indus-

Table 22.3. Sandard Melt Flow Test Conditions for Selected Resins

Resin	Reference	Stand Designation	Temperature, °C	Total Load, kG
Acetal, PE	E	190/2.16	190	2.16
Acrylic	H	230/1.2	230	1.2
Polypropylene	L	230/2.16	230	2.16
ABS, PS	G	200/5.0	200	5
PS	I	230/3.8	230	3.8
Polycarbonate	O	300/1.2	300	1.2

Note: PE, polyethylene; PS, polystyrene. Based on ASTM D 1238.

try well. Considering that injection molding shear rates are typically 100 times higher than that used in the melt flow test, the success is surprising but welcome. Requirements are changing, however. In the quest for greater economics customers are demanding 50:1 and higher let-down ratios. Each pellet of concentrate now has to spread out and color 50 or more pellets of natural resin, and the assumptions inherent in the melt flow test are strained. Nevertheless, once a color formulation is established and stable, the melt flow test can be used as a simple quality control (QC) measurement to predict performance of a color concentrate or precolored material. A typical specification will be target ± 25%.

Typically, the small amounts of pigment or dye used to color a thermoplastic part will not have a measurable impact on the resin's melt flow. Exceptions arise in cases where strong interactions between the colorants and the resin are possible. For example, polycarbonate (PC) is readily degraded by chloride ion. Pigments such as TiO_2, which have residual chloride ions from their manufacturing process, can degrade PC through chain scission. Chloride ion can also be a contaminant in some dyes that are used in PC, CI solvent yellow 163, for example. The resulting loss in molecular weight greatly reduces the resin's highly valued strength properties. It also drives up the melt flow. Testing the melt flow then serves as a quick and dirty test to determine if a particular colorant is chemically compatible with PC.

22.5.2. Spiral Flow

A material's mold flow properties are critical. They influence the cycle time, the surface appearance, and the strength of the part. The *spiral flow test* can be used to assess a color formulation's effect on the material's ability to fill and pack the mold at a given set of molding conditions. It is an injection molding test that measures the length of flow in a specially designed spiral channel. The channel has embossed rulings, and the length of flow can be read directly from the molded piece.

22.6. DIMENSIONAL STABILITY TESTING

22.6.1. Shrinkage

All thermoplastic materials shrink as they cool and set up in the mold, and they continue to shrink slightly after ejection from the mold, especially crystalline mate-

Table 22.4. Shrinkage Values for Selected Thermoplastics

Resin	Shrinkage (%)	Comments
Noncrystalline		
ABS	0.6	Unfilled
	0.1	30% glass fiber reinforced
Acrylic (PMMA)	0.4	Unfilled
Polycarbonate	0.6	Unfilled
	0.2	30% glass fiber reinforced
Polystyrene	0.6	Clear; unfilled
	0.2	30% glass fiber reinforced
Crystalline		
Acetal (homo- or copolymer)	2.2	Unfilled
	0.4	25% glass fiber reinforced
Polyamide 6	1.2	Unfilled
	0.3	33% glass fiber reinforced
HDPE	2.0	Unfilled
Polypropylene	1.8	Unfilled homopolymer
	1.1	Unfilled copolymer
	1.2	40% talc-filled homopolymer
	0.5	40% glass fiber-reinforced homopolymer

Source: From the *International Plastics Selector*, 13th ed. D.A.T.A. Business Publishing, Engelwood, CO, 1992.

rials. The amount of shrinkage is a property of the material and is measured in percent. Consequently, mold cavities are designed to allow for the material's expected shrinkage. Table 22.4 lists the published shrinkage values for a variety of thermoplastic materials. Note that, as a class, crystalline thermoplastics have greater shrinkage rates than noncrystalline thermoplastics. Pigments are known to affect shrinkage of crystalline thermoplastics, and if a material does not shrink as expected due to its color formulation, then the resulting part may not function properly. For example, a cap may not fit properly on a bottle. The effects of a wide variety of pigments on the shrinkage of high-density polyethylene (HDPE) and LDPE are given in ref. 1 (pp. 7–112). We can measure shrinkage with the aid of metal fixtures that have precision gages attached. Table 22.5, based on ref. 8, shows the effects that some pigment classes have on shrinkage.

22.6.2. Warpage

Warpage may be viewed as "differential shrinkage." As the part cures, in the mold and afterwards, it may not shrink at the same rate throughout the part. Many organic pigments are known to promote this phenomenon. (See Parker's well-known paper [8] on this subject.) Warpage is associated with crystalline rather than amorphous resins and is believed to be related to the recrystallization process. In practice, warpage is often seen in the case of large, thin-walled polyolefin parts such as lids. It is also observed in the necks of blow-molded HDPE bottles, yielding an out-of-round part. The usual approach to solving the problem is replacing the suspected

Table 22.5. Effects of Some Pigment Classes on Shrinkage

Pigment Class	Colour Index Generic Name	Effect
Disazocondensation yellow	Pigment yellow 93	Very slight
	Pigment yellow 95	Very slight
Quinophthalone yellow	Pigment yellow 138	Slight
Naphthol red	Pigment red 170	Moderate
Phthalocyanine blue	Pigment blue 15:3	Slight
Phthalocyanine green	Pigment green 36	High
Carbon black (channel process)	Pigment black 7	Moderate
Titanium dioxide[a]	Pigment white 6	Very slight
Synthetic iron oxide	Pigment red 101	Very slight

Note: All tests performed using concentrates of 95/5 white tints let-down to yield 0.5% total pigment in HDPE. Adapted from ref. 3.
[a] Tested at 0.5% in HDPE.

organic pigments with inorganic pigments or special low-warpage grades of organic pigments. Almost always, we compromise on the color and pay a cost pealty.

One hypothesis is that many organic pigments have a mild nucleating effect on polyolefins, polyamides, polyesters, and other crystalline resins. This hypothesis has gained some support from reports that occasionally warpage problems have been solved, or at least reduced, by the addition of strong nucleating agents, such as sodium benzoate, that promote uniform crystallization throughout the part. Another hypothesis [8] relates to the relative polarity of the pigment and the resin. Inorganic pigments, according to this study, are too polar to interact strongly with most thermoplastics, and their presence is ignored by the plastic as it recrystallizes. The hypothesis suggests that a low-warpage grade can be made by applying a highly polar coating to an organic pigment so that it will act more like an inorganic. A successful application of this hypothesis to CI pigment red 254 is also given [8].

One test for measuring the warpage potential of pigments is to mold thin-walled plaques [9] and measure the extent of curling. Parker [8] describes related tests that make use of molds in the form of a cylindrical container and a large-diameter thin-walled lid. Molding conditions can influence the test results. Specifically, molded in stress can lead to warpage. Before evaluating the effects of a particular color formulation on warpage, first run a designed experiment on the natural resin to determine the optimal set of molding parameters to minimize warpage of the natural resin. Then test the color formulation under these optimized conditions.

22.7. SUMMARY

Color match accuracy is just one of many requirements that have to be met in any color formulation project. If pigments are used, the quality of the pigment dispersion is critical to the formulation's cost and performance. Light stability and physical strength properties deteriorate if the pigments are not fully dispersed. Our customers are continually challenging us with new applications, some of which have no standard testing procedures associated with them. We have to work harder and smarter to keep up. On the one hand, testing is unavoidable. On the other, there are two practices that can help minimize the quantity of tests needed to run on a daily basis. First, maintain a data base of the properties of the pigments and dyes you use. The data base should include heat and light stability, chemical resistance (toward acids, bases, oxidation and reduction), and resin compatibilities, including tendencies to promote shrinkage. More importantly, get your processes under control. You cannot afford surprises.

REFERENCES

1. V. Shah, *Plastics Testing Technology*, Wileys, New York, 1984, pp. 126–145.
2. The Society for Automotive Engineers, 400 commonwealth Drive Warrendale, PA 15096.
3. F. Gugumus, "Light Stabilizers," in R. Gaechter and H. Mueller, Eds., *Plastics Additives Handbook* 3rd ed. Oxford University Press, New York, 1990, pp. 129–270.
4. "Standard Specification for Rigid Poly(Vinyl Chloride) (PVC) Siding," ASTM Designation D 3679-95, American Society for Testing and Materials, West Conshohocken, PA.
5. "Standard Test Methods for Bleeding of Pigments," ASTM Designation D 279-73 (Reapproved 1979), and "Standard Test Methods for Rubber Property—Staining of Surfaces (Contact, Migration, and Diffusion)," ASTM Designation D 925-83, American Society for Testing and Materials, West Conshohocken, PA.
6. "Colorants," in *Modern Plastics Encyclopedia '97*, McGraw-Hill, New York, 1996, pp. C48–C61.
7. "Test Methods for Resistance to Laundry Detergent," AATCC 8-1966; "Test Method for Resistance to Chlorine in Pools," AATCC 162–1991; and "Test Method for Resistance to Non-Chlorine Bleach," AATCC 172-1996, American Association of Textile Chemists and Colorists, Research Triangle Park, NC.
8. J. Parker, "Organic Pigments and Their Relationship to Distortion in Linear Low Density and High Density Polyethlenes," SPE RETEC, Color and Appearance Division, 1989, pp. 185–193.
9. P. Bugnon, J. Boechet, V. Dudler, and D. Merian, Huron, Ohio *Chemia*, **48**, 436 (1994).

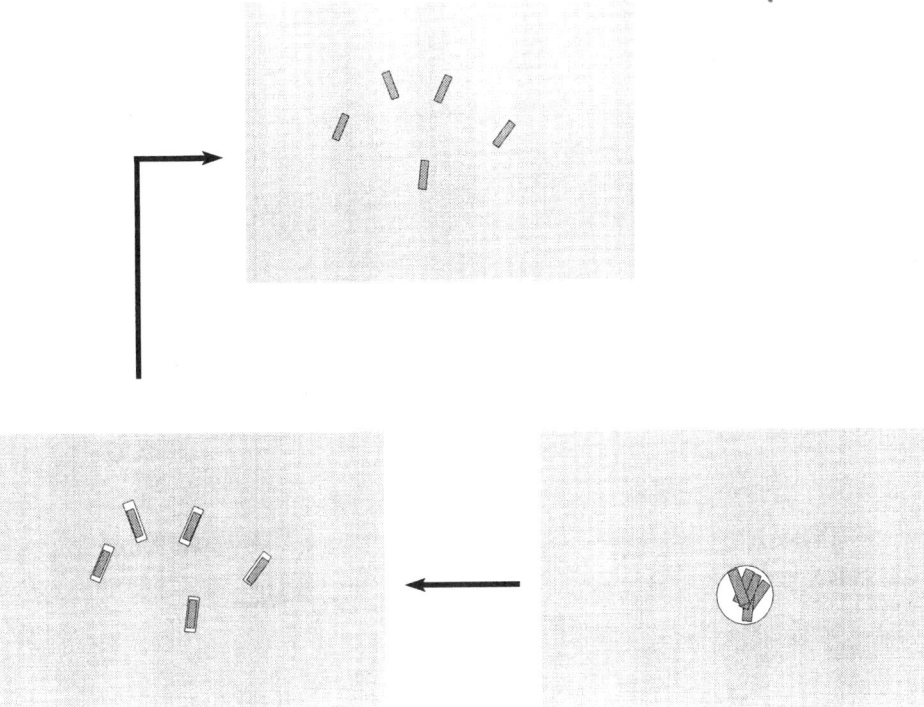

Figure 22.1. Dispersion can be viewed as a two-step process. An agglomerate, composed of numerous smaller pigment particles and surrounded by entrapped air and moisture, is first broken down into its component particles. The resin matrix then displaces the air and moisture, wetting out each particle.

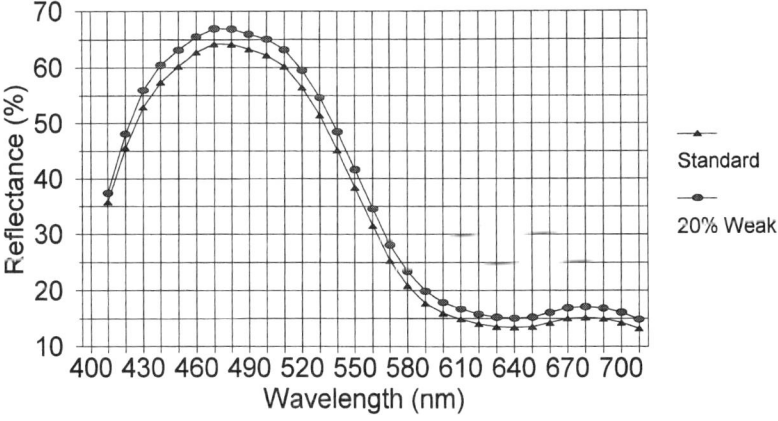

Figure 22.2. Evaluating pigment strength using reflectance curves of pigment blue 15:3. The weaker batch of pigment dispersion has a lower absorption at each wavelength, and the color appears visually to be washed out relative to the standard.

336   Product testing

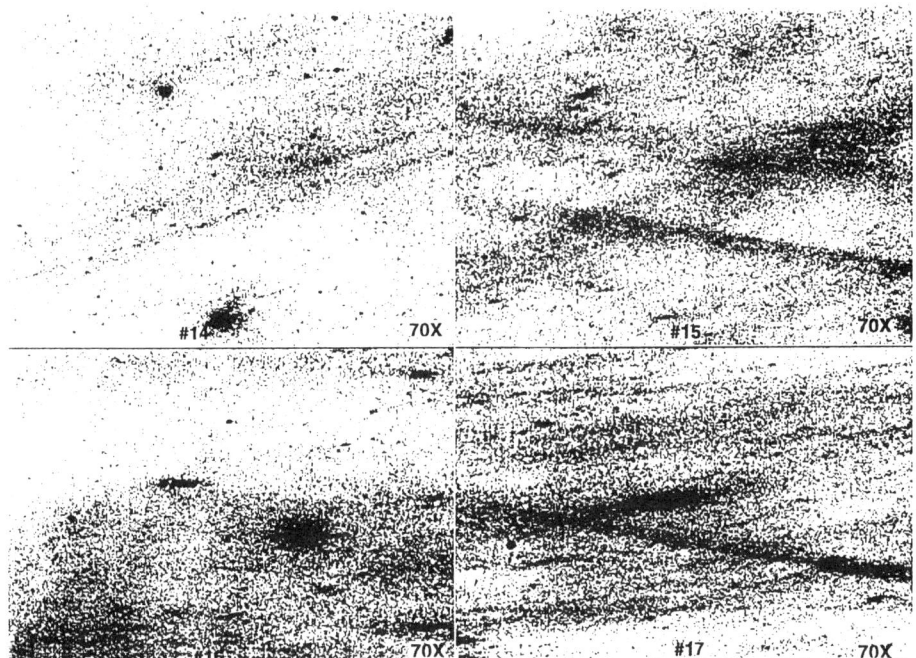

Figure 22.3. Evaluation of pigment dispersion using optical microscopy. The photograph shows four fields of view of LDPE film colored with 40% masterbatch of carbon black letdown to yield 1% pigment in the film. Fields 14 and 15 show agglomerates of various sizes. "Fish-eye" defects are evident in fields 16 and 17. (Courtesy of Ferro Corporation.)

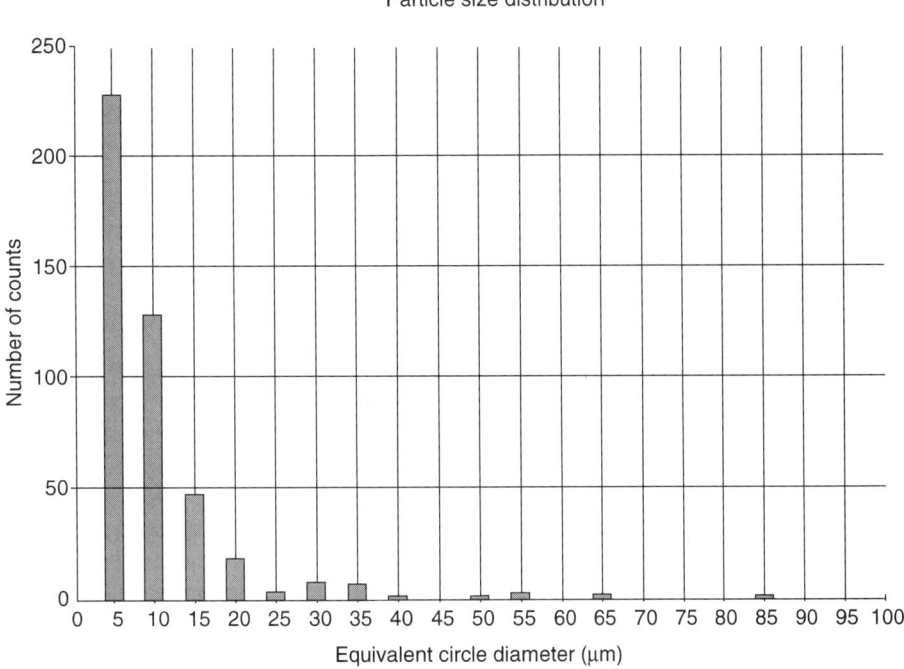

Figure 22.4. Image analysis. The graph represents a scan of one of the fields of view in the black LDPE film shown in Figure 22.3.

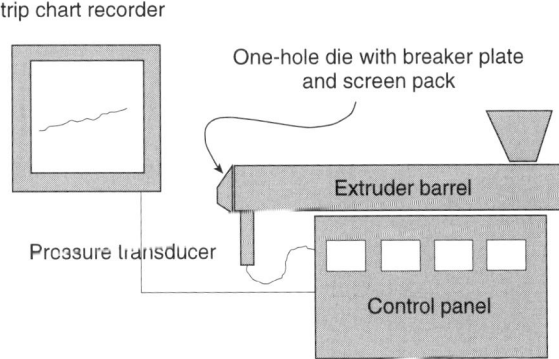

Figure 22.5. Schematic representation of a pressure rise test for dispersion quality.

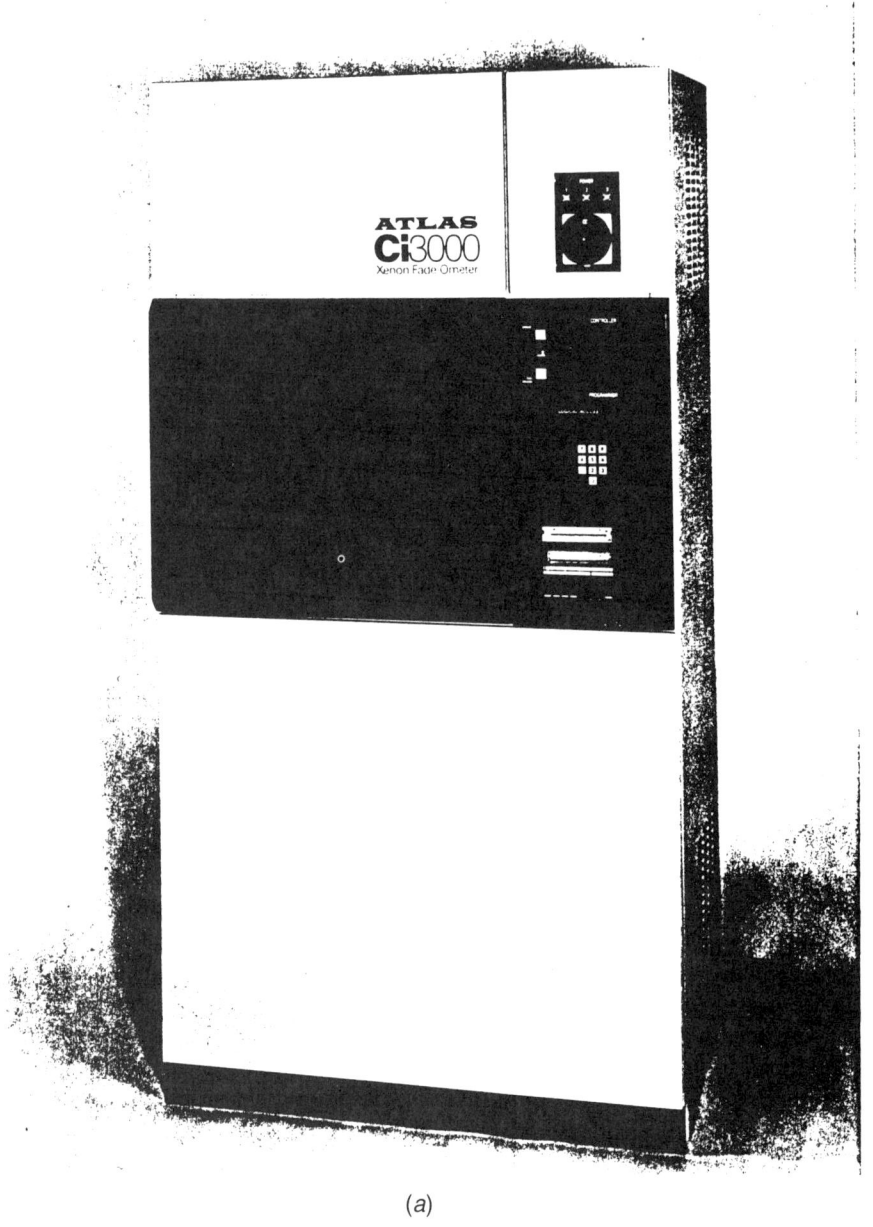

(a)

Figure 22.6. Accelerated weather testing. (*a*) Xenon arc Fade-Ometer. (Courtesy of Atlas Electric Devices Company.) (*b*) Fluorescent lamp "QUV" accelerated weathering tester. (Courtesy of Q-Panel Company.) (*c*) Xenon arc Weather-Ometer. (Courtesy of Atlas Electric Devices Company.)

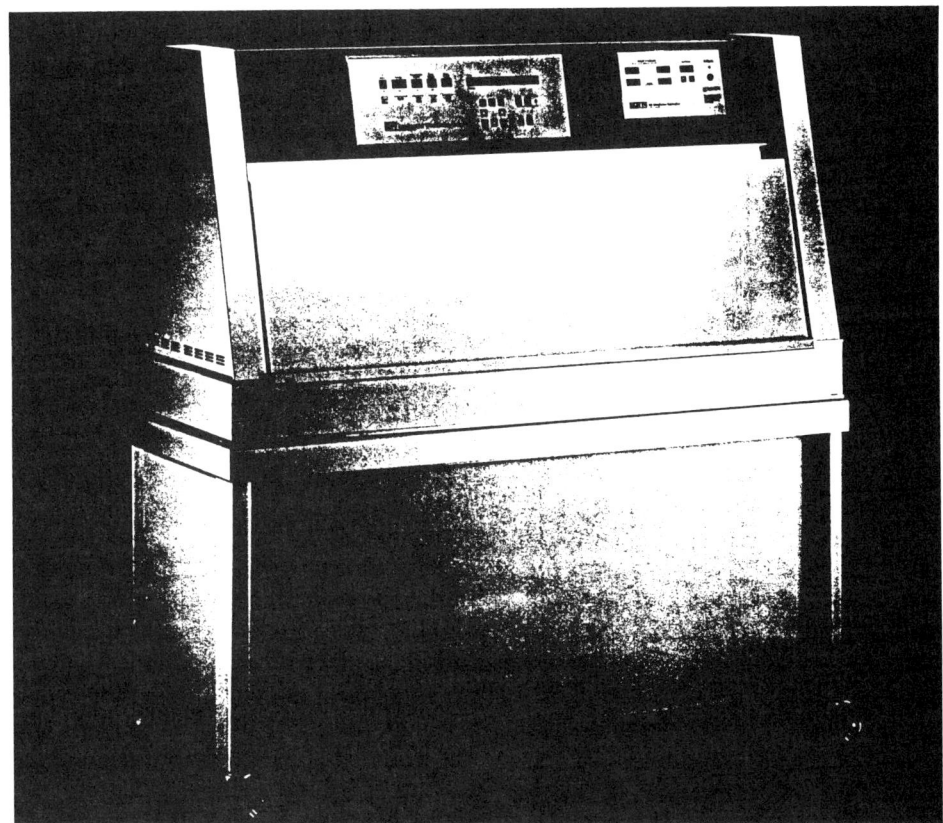

(b)

Figure 22.6b. *Continued*

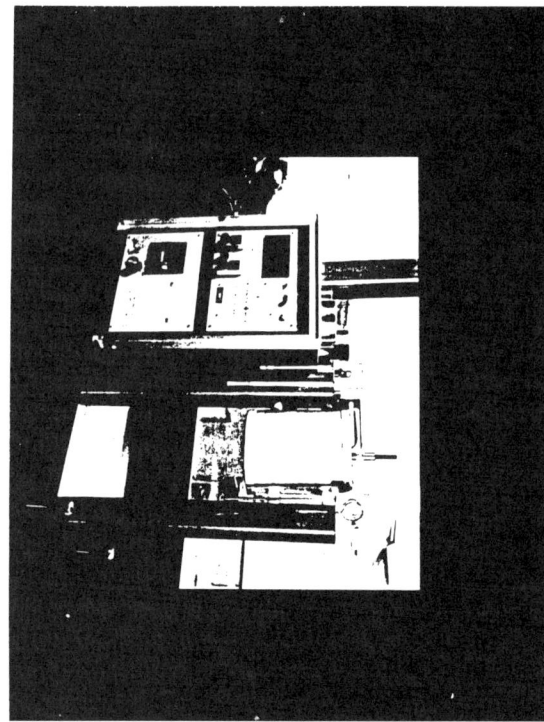

Figure 22.7. A melt plastometer. (Courtesy of Ferro Corporation.)

CHAPTER 23

Effect of Additives on Coloring Plastics

Bruce M. Mulholland

Ticona
8040 Dixie Highway
Florence, Kentucky 41042

23.1. INTRODUCTION

Additives play an important role in the total resin system used by processors today. While there are only several dozen truly different polymers available worldwide (polyethylene, polypropylene, nylon, etc.), there exists tens of thousands of different grades of plastic resin. Each grade is unique, not necessarily by the polymer type, but by the additive system incorporated into the total system to modify one or more properties of the base resin. In general, additive systems can be formulated to modify viscosity, increase mechanical properties, improve thermal stability, increase lubricity and/or mold release, reduce shrinkage, impart ultraviolet light stability, make the polymer flame retardant, or improve wear performance, just to name a few benefits of additive systems. Additive systems can be as simple as a single antioxidant incorporated into a polymer to reduce oxidation during processing or as complex as a system containing a primary and secondary antioxidant, a three-component flame-retardant package, fiberglass, and an impact modifier all in one grade of resin. In fact, some formulations contain as much as two-thirds additives to one-third resin by weight, with the resin seemingly just serving as a binder to hold all of the additives together!

So it is easy to see the importance of the addition of additives to a resin system to help tailor and maximize specific properties. And it is these properties that make

Coloring of Plastics, Fundamentals, 2nd edition. Edited by Robert A. Charvat
ISBN 0-471-13906-8   Copyright © 2004 by John Wiley and Sons, Inc.

it possible to use the plastic in applications where the unmodified resin is deficient. It is equally important then to realize that these modified resins must be colored to fulfill these applications. The coloring process must take into account not only the effect of the colorants on the polymer but also their effect on the additives and the properties being optimized. This chapter has been written not to teach the reader how to modify polymers to achieve certain properties but rather to understand the effect of additives on the coloration of polymers. Furthermore, when coloring problems do occur, this chapter will serve to remind the reader to consider all components of the resin system—the polymer, colorants, and additives when finding a solution.

Before we can discuss the effect of additives on the coloration of polymers, it is important that the reader has an understanding of the visual color perception process. This is discussed in great detail elsewhere in the book. We include it here in brevity as a review. In order to see color, three things must exist: a light source, an object, and an observer. In this chapter, the light source and observer are not discussed other than to say we assume the light source contains all wavelengths (i.e., white light) and the observer possesses normal color vision. The object in our case is an article molded from a colored resin. A beam of light reaches the surface of our object. A portion of the light is reflected due to the surface interface and is called the specular reflection or gloss component. The remainder of the light penetrates the surface of the object, where it is modified through selective absorption, reflection, and scattering by the colorants, polymers, and additives. Selective absorption and reflection by wavelength create color. For example, if an object absorbs all wavelengths of light other than blue, blue light will be reflected or transmitted and the object will appear blue.

Scattering occurs when the light beam contacts particles or regions within the polymer system that have an index of refraction that is different from that of the base polymer. The change in refractive index that the light beam encounters causes the light to be redirected and, if the index of refraction is increasing, to slow down. The index of refraction is a physical property of the substance. It is determined by the equation

$$\frac{\sin \theta_1}{\sin \theta_2} = \eta$$

where θ_1 and θ_2 are the angles formed by the incident ray and the refracted ray versus the normal, as shown in Figure 23.1.

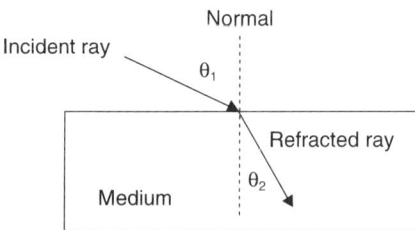

Figure 23.1. Light direction change due to refractive index differences.

Table 23.1. Index of Refraction for Various Substances

Medium	Index of Refraction (Na 5893 Å)
Air (1 atm and 20°C)	1.0003
Water	1.33
Glass	1.52
Most polymers	1.50–1.54
Diatomaceous earth	1.43
Barium sulfate	1.64
Iron spinel	1.85
Zinc oxide	2.01
Antimony oxide	2.09
Zinc sulfide	2.37
Titanium dioxide (Anatase)	2.56
Titanium dioxide (Rutile)	2.61

The index of refraction for a substance depends on the wavelength of incident light. This relationship is most often observed by refracting light with a prism. In a prism, the blue wavelengths of light are refracted more than the red region of the spectrum because of this function. The result is the obvious separation of wavelengths and our ability to see the individual colors of the spectrum. For anisotropic crystals, the index of refraction also depends on the crystal axis on which the light is incident. For these materials, the index of refraction as a number is typically reported as an average value of the various axes. The index of refraction for various substances is shown in Table 23.1.

If scattering occurs nearly equally at all wavelengths with no absorption, the object will look white. That is how titanium dioxide appears to impart its white color to objects, by scattering virtually all of the incident light. In mixtures of titanium dioxide and dark pigments such as carbon black, the index of refraction as a function of wavelength becomes apparent. Because the index of refraction for the titanium dioxide is higher for shorter wavelengths of light (blue light), the net result is that more blue light will be reflected than red light in these dark colors. Blue wavelengths of light see greater light scatter from the titanium dioxide and do not penetrate as deeply into the medium as longer wavelengths do (red light). The resulting shorter path length for the blue portion of the incident light cause less absorption by the carbon black so that more blue light emerges and is reflected toward the observer compared to the red light. This scattering causes the blue flop observed when titanium dioxide is present in black color formulations. This discussion was included here because this phenomenon can also hold true for any additive that has an index of refraction significantly different from the base polymer. In the case of the titanium dioxide example, if the blue flop was problematic in color matching, one can most likely reduce or remove the pigment from the formulation, minimizing this problem. When this blue flop is caused by the additive system, the effect can only be minimized by increased pigment loading. This phenomenon is also a characteristic of the base polymer as well. Most crystalline polymers will exhibit more bluish flop compared to amorphous or semicrystalline resins due to this dependence of index of refraction on the wavelength of incident light.

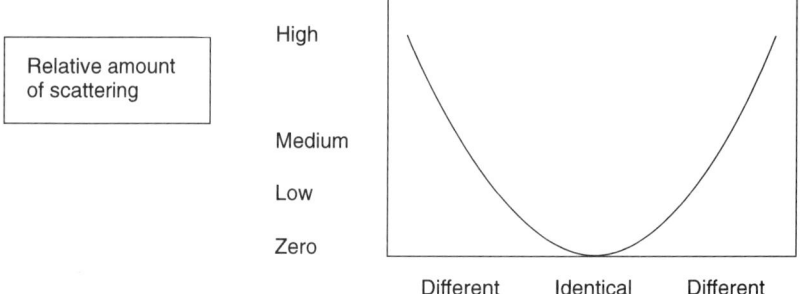

Figure 23.2. Scattering as a function of refractive index difference.

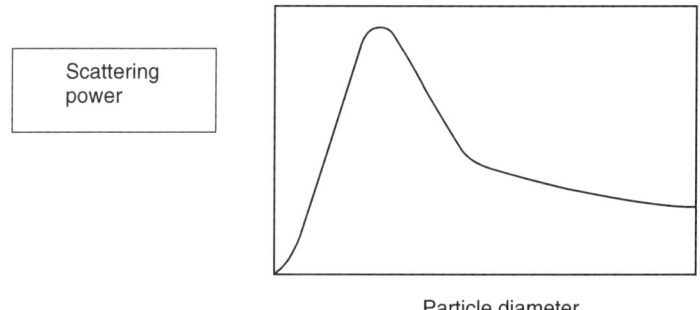

Figure 23.3. Scattering as a function of constituent particle size.

As expected, the amount of scattering depends on the magnitude of difference in refractive index between the polymer and the scattering substance. The direction of difference does not matter, only the magnitude, as depicted in Figure 23.2.

Finally, there is an optimum particle size for scattering to occur that is dependent on the index of refraction of the substance and the medium, and the wavelength of light. This optimum particle size is generally in the range of half the wavelength of the incident light. Smaller or large particle sizes will scatter less light, as shown in Figure 23.3.

It is important to understand the mechanism of scattering because the majority of the problems in coloring polymer systems containing additive packages can be related to the intrinsic whiteness or scattering of the polymer system itself. Increasing the amount of scattering in the total resin system will increase the amount of diffuse reflection (white light) that is mixed with the reflected colored light generated by our pigment interactions. This mixing will dilute the color strength, and the color of our object will appear lighter and less bright to the observer. An analogy can be percolated from everyday life: coffee. Coffee without cream is said to be black (actually more like dark brown) due to light absorption by the coffee extract dispersed in water. Adding cream to the coffee increases the light scattering and the color is transformed from dark brown to light tan. And it is obvious that once the cream is added, it is impossible to color the mix with more coffee extract to achieve the original black color.

So is the case in polymers containing additive systems. As will be shown, some additive systems can impart so much light scattering to the base resin that certain colors can no longer be achieved. Or if they can be achieved, other properties may be adversely affected, such as impact strength and cost. In either case, the practical color gamut or palette that is obtainable with this particular resin system is reduced. The discussion below presents the effects that the polymer and its additives can have on colorability. Color data presented in the following tables have been calculated under illuminant D-65, 10° observer, specular included, expressed in CIELAB units, unless otherwise noted.

23.2. POLYMER TYPES

We will begin our discussion with the base polymer itself. While it is not an additive, it is obviously a main component of our total system and must be discussed in coloristic terms. From a color standpoint, there are three classifications of resins: transparent, translucent, and opaque resins.

23.2.1. Transparent Resins

From the discussion above, it should be intuitive that transparent resins such as polystyrene or polycarbonate do not scatter light and therefore can achieve the most brilliant colors. Colors are generated through transmitted or reflected light without having to mix with white light from diffuse reflection since none exists from the polymer. The difficulty here is making the system opaque (no light transmission). If opacity is required, a scattering opacifier such as titanium dioxide is needed and apparent color strength will be reduced.

23.2.2. Translucent and Opaque Resins

Most polymers fall in the class of translucent resins. These include acetal, polyamide, polybutylene terephthalate (PBT), polyethylene, and polypropylene as examples. There are very few neat polymers that are truly opaque (this depends on thickness as well). Liquid crystal polymer (LCP) is an example of a typically opaque polymer. It is theorized that these semicrystalline and crystalline resins will scatter some portion of incident light due to spherulitic crystal structure and the amorphous–crystalline region interfaces themselves.

The degree of translucency can be measured by calculating a contrast ratio. This number is the ratio of the L value in color measurement obtained when backing the natural sample with a black tile divided by the L value of the sample backed with a white tile. A completely opaque sample will have a contrast ratio of 1.00. Contrast ratio measurements are specific to the thickness of the sample and the intensity of the light source. Contrast ratios for several resins are presented in Table 23.2.

The data in Table 23.2 show that for these crystalline resins, the contrast ratio generally increases with increased degree of crystallinity. Nylon 6,6 has the lowest contrast ratio (less opaque) compared to liquid crystal polymer, which has an

Table 23.2. Contrast Ratios of Selected Resins

Resin	L Value[a]		Contrast Ratio at 3.2 mm
	White	Black	
Nylon 6,6	85.65	66.75	0.78
Acetal	93.01	82.05	0.88
PBT	92.70	87.14	0.94
PBT (annealed)	92.80	90.94	0.98
LCP	84.10	84.10	1.00

[a] Specular excluded, expressed in Hunter Lab units

extremely high crystallinity and is completely opaque. Furthermore, the two PBT samples show that the annealed sample (higher crystallinity) is also more opaque. The higher opacity indicates that more light scattering is occurring by the polymer, and typically a more restricted color gamut will be achieved. Polymers with a high degree of intrinsic whiteness will exclude deep, dark colors and bright, high-chroma colors from the achievable color gamut.

To illustrate this effect on colors, three single-pigment colors (blue, red, and yellow) were developed in three different resin systems. Blue and yellow were produced in acetal (POM), polyphenylene sulfide (PPS), and LCP. The red color was prepared in nylon 6,6, PPS, and LCP. The acetal and nylon resins are translucent while the PPS and LCP are opaque at the 3.2-mm sample thickness. In all three colors, the more translucent resins produced visually more brilliant, higher chroma colors than the more opaque resins with increased diffuse scattering.

Table 23.3 presents color data generated from these same samples. Numerically, PPS and LCP produced lighter colors (higher L^* values) and/or lower chroma (C^* values) compared to the acetal and nylon controls, which is consistent with what is seen visually. The base resin light scattering will increase the L^* value and decrease chroma in these colors. In addition, the PPS and LCP were evaluated from 2 times to 15 times the pigment loading (relative concentration) of the acetal and nylon, and this still holds true. Another indication of the increased scattering from the PPS and LCP is the relative strength data of Table 23.3. As shown, even at the higher relative pigment concentrations, these resins produce colors that appear only 16–50% as strong.

This discussion illustrates the effect light scattering of the polymer has on the resulting color. In real life, if LCP is required for a particular application, the examples in Table 23.3 show that deep, dark colors or bright, high-chroma ones cannot be achieved. A customer currently using these higher chroma colors in other resins such as polycarbonate needs to be educated that these same colors cannot be achieved in LCP.

The above discussion deals with the base polymer only. One can see that if the additive system imparts light scattering similar to these examples above, the effect on coloring the total resin system can be dramatic. We begin our discussion of additives by looking at polymer blends, where the secondary polymer can be considered an additive to the continuous base polymer.

Table 23.3. Relative Color Strength in Various Resins

Color	Resin	Relative Concentration	L^*	C^*	Relative Strength (%)
Blue	POM	X	37.05	55.01	
	PPS	$3X$	47.78	34.34	42
	LCP	$3X$	61.60	27.93	16
	LCP	$6X$	56.14	35.09	26
Red	Nylon 6,6	X	35.91	41.61	
	PPS	$3X$	44.43	33.67	48
	LCP	$3X$	56.87	34.13	23
Yellow	POM	X	77.52	77.94	
	PPS	$2X$	69.35	52.45	50
	LCP	$15X$	78.13	51.03	31

23.3. POLYMER BLENDS

One can assume that blends of polymers will be more difficult to color than any single component by itself. Diffuse reflection can increase due to internal light reflection or scattering at phase interfaces if the polymers are at least partially immiscible or their refractive indices are significantly different. Blends of translucent polymers are typically more opaque than either resin alone. Furthermore, colorant stability (thermal or chemical) can be adversely affected by the presence of the other polymer(s). As in the case of neat polymers, both circumstances will result in a restricted achievable color gamut for the polymer blend. An example of a prominent polymer blend is GE's Noryl (PS/PPO), which certainly colors much differently than the polystyrene component by itself.

Blends of polymers can pose their own unique problems as well. An example is where colorants exhibit preferential dispersion to one of the polymer phases. The other polymer phase remains virtually uncolored. Macroscopically, this may not be a problem as the molded part appears uniformly colored. But even at this level, if wall thickness is very thin, color striations may become apparent. Other performance measures may be adversely affected as well. At the microscopic level, since all of the colorant is dispersed in one phase, impact strength and other properties may be reduced at pigment concentrations that are much lower than expected. This would primarily occur in blends where the colorant prefers the resin phase that provides the toughness to the blend.

23.4. ANTIOXIDANTS

Virtually all polymeric resins undergo oxidation in the presence of oxygen. To retard this degradation, antioxidants are typically added. These additives are usually hindered phenols, amines, hydroxylamines, phosphites, or thioesters. In general, antioxidants will have little effect on colorability since they are typically used at low levels. At higher levels, they may increase light scattering and impact colorability depend-

Table 23.4. Effect of Flame Retardants on Colorability[a]

Color	$DL*$	$DC*$	Relative Strength (%)
Bright yellow	2.3	−3.5	56
Bright red	4.9	−7.1	53
Medium tan	7.0	−2.4	57
Bright organe	4.0	−1.5	74
Royal blue	6.3	−3.2	78
Dark green	9.3	n/a	52
Medium gray	5.0	−0.5	n/a
Black	3.4	−0.9	72

[a] Flame-retardant resin versus neat polymer.

ing on the polymer type. There are remote instances where antioxidants have been linked to problems with graying bright colors or "pinking" whites. But these are very polymer specific and usually result from a chemical instability within the system.

23.5. ANTISTATICS

Antistatic additives are designed to be present on the surface of the molded part to achieve the full antistatic benefit. The types of additives used to enhance antistatic properties include quaternary ammonium salts, alkyl sulfonates or phosphate plus alkali metals, ethoxylated amines, or gylcerol esters. Antistatics are typically used at higher levels than other additives such as antioxidants. Therefore, antistatics are likely to increase light scattering, making it more difficult to achieve the higher chroma colors.

23.6. COUPLING AGENTS

Coupling agents such as silanes and titanates will increase light scattering. Both types can impact colorability if they are incorporated at high levels.

23.7. FLAME RETARDANTS

Typical flame-retardant formulations will include one or more of the following compounds: phosphorus-based compounds, metal hydrates such as aluminum trihydrate, halogen compounds (bromine or chlorine) such as decabromodiphenyloxide or tetrabromo-bisphenol A, antimony trioxide, zinc borate, and zinc molybdate. All of these additives will significantly increase light scattering and reduce the color gamut of the resin. Table 23.4 presents color data to show this effect. Eight flame-retardant polyester colors are listed, with lightness, chroma, and strength differences calculated versus the same colors in neat polyester resin. All colors are lighter (positive $DL*$) and duller with lower chroma (negative $DC*$) with the presence of the flame-retardant system. Relative color strength data also snows that the colors are weaker and less intense. It would be difficult, if not impossible, to exactly match the

Table 23.5. Effect of Impact Modifier on Colorability

Color	DL*	DC*
Gray	4.5	−1.8
Bright red	2.8	−2.4
Bright yellow	3.6	−4.0

neat polymer color in the flame-retardant system due to the light scattering of the additives.

23.8. FOAMING AGENTS

The author has little experience with foamed resins. These resins contain a chemical foaming or blowing agent (CFAs or CBAs) that produces a cellular structure in the plastic part. One would speculate based on all of the above information that foaming would increase light scattering since the molded parts would contain a complex cell structure with all of its polymer–gas interfaces. Knowing that the indexes of refraction for the polymer and air are different (see Table 23.1) substantiates this thinking.

23.9. HEAT STABILIZERS

Antioxidants can be referred to as heat stabilizers and were previously discussed. Other heat stabilizers are metal complexes based on tin, lead, cadmium, or copper–halogen systems used in polyvinyl chloride (PVC) and nylon. The largest impact on colorability is with the copper–halogen systems used to increase the continuous-use temperature rating in nylon. In that resin these systems can significantly reduce colorability due to discoloration via reaction with the colorant or by thermal degradation. Light colors in heat-stabilized nylon are virtually impossible to control since a slight change in residence time or temperature will significantly drive the color to tan or brown. The presence of oxygen will accelerate this. Darker colors are less affected and are preferred in heat-stabilized nylon.

23.10. IMPACT MODIFIERS

This class of additives covers a broad range from butadiene to acrylic polymers. Since these additives are polymeric in nature, diffuse reflection will occur at the polymer–modifier interfaces similar to polymer blends. Again, this will result in colors that appear lighter and duller. Table 23.5 contains three examples of impact-modified colors again in polyester compared to the neat resin without modifier. As expected, the impact-modified colors are lighter and have lower chroma. In practice, acrylonitrile butadiene styrene (ABS) or HIPS would have a more restricted color gamut compared to their transparent SAN and polystyrene (PS) base polymers.

Table 23.6. Effect of Lubricant on Colorability

Color	$DL*$	Relative Strength (%)
Bright red	1.6	85
Dark blue	2.7	80

Table 23.7. Effect of Fiberglass on Colorability

Color	$DL*$	$DC*$
Gray	−1.8	−0.2
Dark blue	−1.5	−0.7
Bright yellow	−5.9	−8.2
Dark green	−3.5	−2.9
Bright red	−1.7	−0.6

23.11. LUBRICANTS/MOLD RELEASES

Like antistatics, lubricants and mold releases are designed to reside on the molded part surface. Compounds used as lubricants or mold releases represent a broad range of chemistry. Some common examples include metallic stearates (calcium or zinc), hydrocarbons (polyethylene waxes, mineral oil, paraffin wax), fatty acids and fatty alcohols (gylcerol monostearate, bis-stearamides), complex esters, and silicone. Light scattering can be increased depending on the chemistry and concentration of additive. Generally, this effect is not a problem unless these additives are used at high levels or they have a significantly different index of refraction compared to the base polymer. Table 23.6 shows color difference data for two colors containing a high level of lubricant versus a polymer with no lubricant prepared in acetal copolymer. Increased light scattering is evident by the lighter color (positive $DL*$) and lower relative color strength.

23.12. REINFORCING AGENTS

Reinforcing agents are typically either a mineral such as talc and calcium carbonate or fiberglass. As expected, mineral types used at their high levels can scatter a large portion of the incident light depending on the refractive index and particle size. In mineral-filled resins, it is usually impossible to achieve deep, dark colors such as chocolate brown or forest green. Bright colors are also made duller. Because a high amount of mineral is generally used, bright white colors such as appliance white can also be difficult to achieve.

Fiberglass poses a different problem in that colors are typically darker and more dingy looking in glass-reinforced resins versus their unfilled counterparts. Table 23.7 lists color difference data for several 30% glass-reinforced PBT colors versus the same color in unfilled PBT. The darker, dingier look is evident by the negative lightness and chroma difference values. Brighter, high-chroma colors in glass-reinforced resins either require significantly more colorant to achieve (at higher cost) or cannot be exactly matched due to the resin. Overcoming the darkness makes bright white

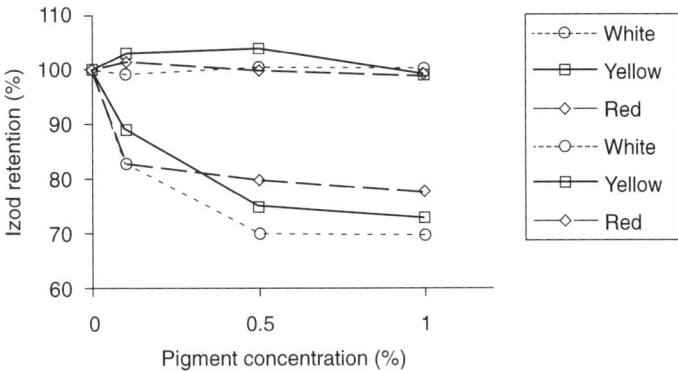

Figure 23.4. Effect of colorants on Izod impact retention.

colors difficult to achieve as well. Contributing factors to the dark, dingy appearance are most likely the glass-sizing agent and the elevated processing temperatures for glass-reinforced resin versus the unfilled product.

Fiberglass adds another problem to coloring: mechanical property retention. A number of widely used colorants will abrade glass fiber length or can affect fiber wetting, significantly reducing properties. For example, in 30% glass-reinforced PBT, certain colorants can reduce tensile strength up to 20% and notched Izod impact by as much as 30%. An example of this is shown in Figure 23.4. Here three colors (white, yellow, and red) are formulated to preserve Izod impact strength, and a corresponding three colors are shown that reduce Izod strength. As is shown, this reduction can occur using colorant concentrations that would be termed typical and not excessive. Mechanical property retention may further limit the types and numbers of colorants that can be used in the glass-reinforced resin, further limiting its achievable color gamut.

23.13. UV STABILIZERS

A large number of colored polymers and polymer blends are used in applications where UV stability is important. First and foremost, many common colorants do not possess satisfactory lightfastness to be used in UV applications. Therefore, the color gamut can be reduced due to limited availability of colorants with acceptable lightfastness in certain polymer systems. Ultraviolet stabilizers themselves can also impact colorability. These additives generally include benzophenones, benzotriazoles, and hindered amine light stabilizers. Many of these additives impart a yellowish tint to the polymer that must be overcome. Also, some do increase light scattering, making the more popular automotive maroon and dark blue colors difficult to achieve. Finally nickel complexes can also be used as UV stabilizers, although they are green in color, which limits their use.

23.14. CONCLUSION

Hopefully the discussion and examples presented here have clearly shown that the colorability of a resin is determined by the total system and not just the base polymer. Most of the discussion has been centered around just the neat polymer or the polymer plus a single modifier or additive. Coloring issues further escalate when polymer blends are used or when multiple additives are incorporated. Would it be difficult, and would the color gamut be somewhat restricted, if one was to color an impact-modified, flame-retardant, glass-reinforced PBT–polycarbonate alloy? The answer now better be a resounding YES!

CHAPTER 24

Recycling

Jack W. Blakeman

505 Sackett Street
Maumee, Ohio 43537

24.1. HISTORY OF RECYCLING

Recycling has been going on for many centuries. Analysis of early American native stone tools shows that some have been reworked, such as a broken spear point reworked into a scraper for cleaning animal skins. In the early part of this century people made clothes from grain and flour sacks. Clothes handed down to younger brothers and sisters is another form of recycling. Metals such as automobiles, broken farm implements, tin cans, copper wire, and lead batteries are recycled for their valuable metal content. At one time "junk collectors" had regular routes with their horse and wagons or hand carts in cities and rural areas collecting metals, glass, papers, rags, and bones. Plastics are the most recent materials to be recycled. Due to the numerous different and incompatible plastic resins, they pose a more difficult situation.

24.2. COLLECTION

Currently, the most visible recycling is the curbside collection, although many other recycling efforts are proceeding unnoticed by the average person. Plastics are being reclaimed from automobiles, appliances, computer cases, printer cases, and carpeting. Automobiles are currently a challange due to the use of many different types of plastic.

Coloring of Plastics, Fundamentals, 2nd edition. Edited by Robert A. Charvat
ISBN 0-471-13906-8   Copyright © 2004 by John Wiley and Sons, Inc.

At curbside many communities are collecting aluminum and tin cans, glass bottles, newspaper and plastic bottles. High-density polyethylene (HDPE) and polyethylene terephthalate (PET) are the most commonly recycled plastic bottles. Some areas that have a market for other plastics also pick them up for recycling. Sorting is done to some degree by the resident, the collector, and the reprocessor.

24.3. REPROCESSING

The reprocessor receives baled bottles from a collection company. The bales are opened and conveyed to a grinder. Workers stand by the conveyer and remove undesirable items such as cans, pumps, sprayers, and closures. They can also sort various fractions from the conveyor, such as black, white, natural, or various color sorts.

The desired bottles then proceed to the grinder and the washing machine. The ground plastic is washed to remove product and labels. The paper labels sink and the HDPE is floated off the top and conveyed to a drier. Contaminants such as glass and metals also sink and are discarded with the paper labels. After the material is dry, it can be boxed or extruded and pelletized. Color or additives can be added to the resin prior to extrusion if necessary.

Automatic electronic sorting equipment is being used to some extent and soon will be the major method, as faster and better electronic sorting equipment is developed and put to use.

24.4. USE OF RECYCLED PLASTICS

To efficiently use recycled plastics in molded parts, it is beneficial to work with several recycling suppliers to obtain a resin that is as close to the color of the product as possible. The resin that most closely matches the product may cost more per pound due to special sorting necessary to obtain a special color. This resin will probably cost less to use since it will require less corrective colorant to adjust the recycled resin to match the target. Recycled resins that are farther off color will require much more pigment to overpower the colorants in the recycled resin.

Other properties that are important to the product or process include stress crack resistance, melt flow, and stiffness. Milk bottle resin has less stress crack resistance, higher flow, and slightly higher stiffness. If these properties would adversely affect the product/process, you should specify that you do not want significant quanities of milk bottle resin in the recycled resin. Other HDPE bottles are made from a lower density resin that has higher stress crack resistance and lower melt flow.

After obtaining the best possible samples from several suppliers, you should have several custom color concentrate suppliers make special color concentrates that will convert the recycled resin to match the product color. After receiving the samples of correcting colorants, they should be evaluated to determine which one works best and most economically in your situation. This is accomplished by making a plaque of 100% postconsumer recycle (PCR) or using a molded chip from the supplier. Also samples using PCR are prepared with about three-fourths of the recommended mix ratio and a sample with about one and a half times the recommended mix ratio. Samples like this are made using each correcting colorant in all the different manufacturers PCR samples. Also a sample of each correcting colorant is prepared using the recommended mix ratio in natural resin. It would be acceptable to use color chips from the color supplier if they are molded in natural resin. These samples can be processed on a laboratory roll mill, extruder, or injection molder. You will need flat parts or plaques on which color readings can be taken. If you do not get flat parts from the process, then pieces cut from extrusions or moldings can be pressed on a platten press to make flat plaques that are similar in thickness to the end product.

The next step is to measure the color of all these samples on a spectrophotometer or colorimeter so they can be plotted in L,a,b or L,c,h color space. (See Fig. 24.1 for an example of a $\Delta L,a,b$ plot and see Fig. 24.2 for an example of a $\Delta L,C,H$ plot.) Most spectrophotometers will have a program to plot these samples in either color space. If your instrument does not plot them, plot the sample readings on graph paper. Label the graph with the "a" or "c" scale horizontally and the "b" or "h" scale vertically on the paper. Make a separate graph for each recycled resin and include the target and all correcting colorants that are to be evaluated. Also include the samples made with PCR and various mixes of the correcting colorants. Choose the range of values that will separate the samples but be sure that all samples plot on the graph. Now draw a line from the recycled resin to each correcting colorant in the virgin sample. The line that passes through or nearest to the target will be the best colorant to correct this recycled resin. A line drawn perpendicular from this line to the target will intersect the line at the approximate mix ratio needed to correct the color of the recycled resin (Figs. 24.1 and 24.2).

Now that the colorant and recycled resin mix ratio have been determined in the laboratory, you are ready to evaluate the mix in a production machine. You need to mix a sufficient quantity of recycled resin with the appropriate amount of correcting colorant. Then try adding 10% of this mixture to the production material and mold parts. Measure the color of these parts, and if the color is acceptable, try using 20% of the recycled mix, and so on, until you reach the amount you want to use or the color becomes unacceptable. If the color is not acceptable at the desired percentage, add more production colorant to the mix. If you are molding multilayer parts, add more production colorant to the outer layer. It is much more efficient to add colorant to the outer layer whenever you have that option since the outer layer is normally made of 100% virgin and colorant.

24.5. POLYETHYLENE TEREPHTHALATE RECYCLING

In most cases, before attempts are made to recycle plastic resins, the used parts must first go through a series of sorting processes. The purpose of these sorting processes is to separate the desired recycle plastic from possible contaminants in the incoming stream. These contaminants can take many forms. They include dirt, paper, glue, and metal as well as other recyclable plastics and even the same desired plastic but of a different color. The most postconsumer recycled plastic today is PET. This is the plastic from which all plastic soft drink beverage bottles are made.

The 1995 volumes for the use of thermoplastic PET was 3.8 billion pounds. The recycling rate for PET in 1995 was approximately 50%. In the 10 states that have deposit laws, the rate is greater than 75%. The recycling process begins with col-

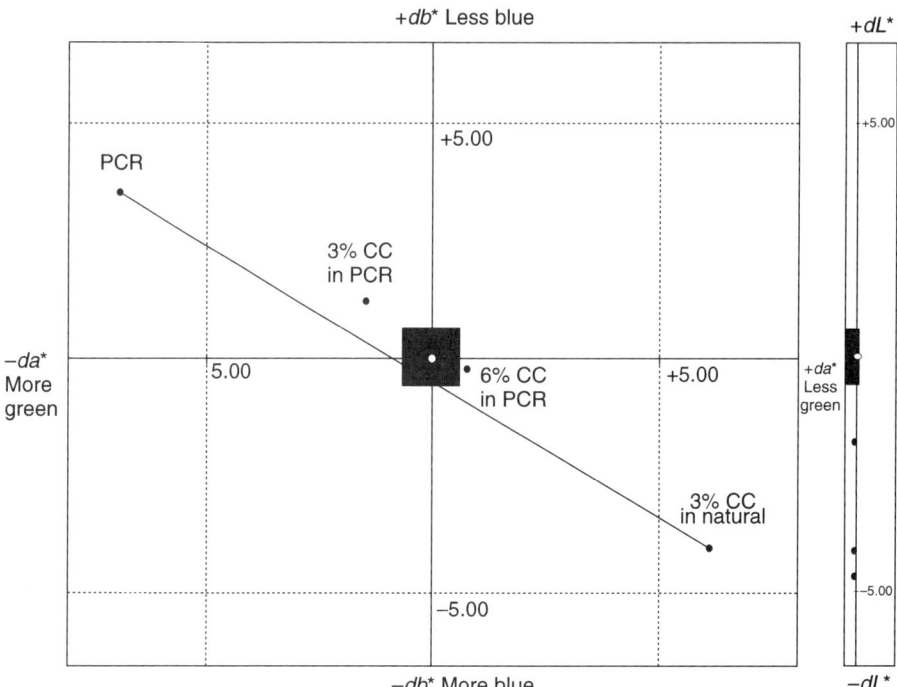

Figure 24.1. CIELAB $L^* a^* b^*$ color difference plot on samples of PCR, correcting color let down in natural PCR with 3 and 6% correcting colorant. The plot shows that the best mix ratio falls between 3 and 6%.

lecting the bottles at either a drop-off site, a curbside collection system locale, or retail outlets that sell beverages in plastic bottles in the deposit states. Regardless of the source, this raw material stream must first go through a basic sort to sort the metal cans, polyethylene milk jugs and soap containers from the PET bottles. This can be done manually or with automatic sorting equipment that uses special cameras and light sources to look at bottle shape, material type, and/or color to sort the stream. The PET bottles are then pierced and baled to densify the stream for transportation.

Once delivered to the recycling plant, a much more thorough sort takes place. The crushed bottles are mechanically debaled and washed and the labels are removed. The bottles are dewatered and sent to an automatic sorting machine that separates the green bottles from the clear bottles. There are several places in the process where metal, if detected by sensors, is automatically removed. The process flow is now split into a green stream and a clear stream, each with its own grinders, wash tanks, and flotation tanks. The sorting of the polyolefins from the PET takes place via a float sink tank. The olefins (i.e., HDPE basecups and polypropylene closures) have a density less than water while PET has a density of 1.35, greater than water (1.00). The olefins float off the top of the tanks and the PET is dragged off the bottom of the tanks with an auger. Soaps and agitation are used in the wash tanks and surfactants are used in the float/sink tanks. The flake is dried, tested for color and contamination, and stored for either sale or further cleaning. In flake form, the clear flake can be used for fiberfill or carpet fiber. The green flake can be used for pallet strapping and some thermoformed sheet applications. The clear PET is

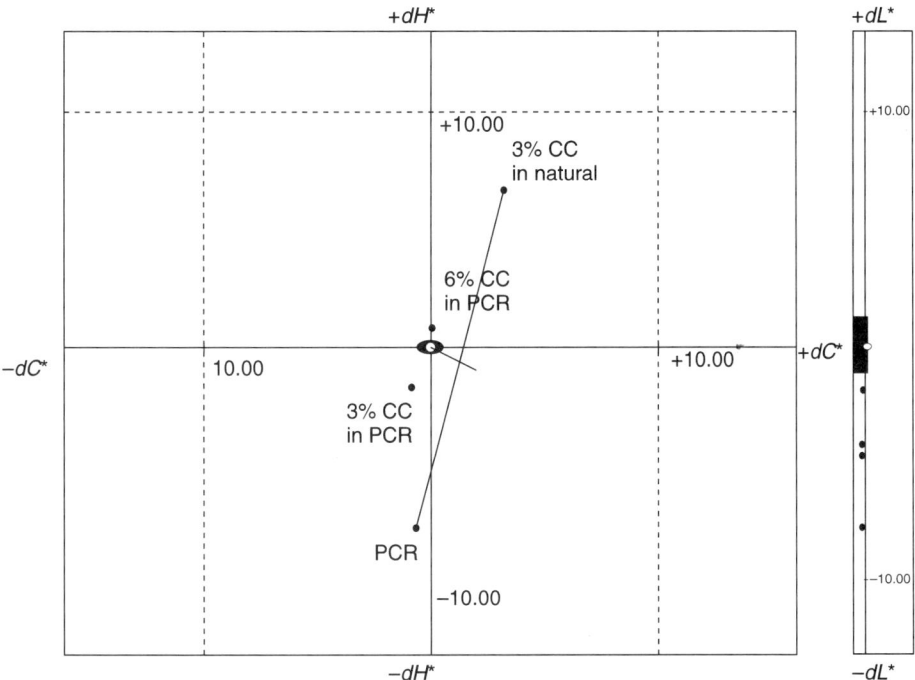

Figure 24.2. Plot of dL^*, dC^*, and dH^* on samples of PCR, the correcting colorant let down in natural PCR with 3 and 6% correcting colorant. The plot shows that the best mix ratio falls between 3 and 6%.

typically pelletized through a melt filtration unit that will further remove fine contaminants. The resin can then be used in injection molding and blow molding processes for nonfood applications such as detergent and chemical containers. Postconsumer recycle PET is often compounded with fillers that improve its impact properties for automotive end uses.

In order to bring the quality of the resin back to virgin grade for food contact applications, for instance, other proprietary processes can be used. The flake can be depolymerized through a glycolysis process into monomer units or methanolysis can be used to get the resin back into its original raw materials of terephthalic acid and glycol. These raw materials can then be distilled to remove all foreign contamination. Distillation after methanolysis can even take the colorants out of the resin. It can then be blended with virgin raw materials and repolymerized into PET resin with PCR content. Recent developments have provided a proprietary mechanical washing process that is capable of generating a resin clean enough to be used in food contact, according to the U.S. Food and Drug Administration.

The recycling processes that are used to reprocess the PET resin often involve heat and chemicals. At times these may have a degradative effect on the resin, turning it yellow. The color is constantly monitored throughout the processes. A blue tint is sometimes added to the resin in the pelletizing step to offset the yellow cast. While a sacrifice in light transmittance is made by doing this, it provides a more appealing shade for clear packaging applications.

CHAPTER 25

The Environment and Government Regulations

Patrick Surgeon

323 Dayielson Way
Chandler, AZ 85225

This chapter presents the issues involved in environmental and governmental regulations from the position of how a company and/or individual must approach these issues to be in compliance and be a successful and viable company.

Today, it is imperative to understand the environmental aspects of your business. Closely related to environmental issues are those of worker health and safety, which is why so many companies have combined these tasks into an HSE (health, safety, and environmental) function. This chapter will discuss environmental, health, and safety regulations from the perspective of how a company and/or individual must approach these issues to be in compliance and be a successful and viable operation. I will present several alternatives for assessing your current situation, determining whether change is warranted, and, if so, how to go about making the needed changes. Finally, I will discuss recent developments such as the International Organization for Standardization (ISO) standard 14000 and the Occupational Safety and Health Administration's (OSHA) Voluntary Protection Program (VPP) and how they may fit into your future plans.

Like most journeys, the route to environmental compliance and beyond can be looked at in terms of knowing where you are, deciding where you want to go, determining how you want to get there, and developing and implementing a strategy to make it happen.

Coloring of Plastics, Fundamentals, 2nd edition. Edited by Robert A. Charvat
ISBN 0-471-13906-8   Copyright © 2004 by John Wiley and Sons, Inc.

Step 1: Where are you now? To begin the process, it is necessary to understand the issues and how the existing regulations affect your company and its current business.

Step 2: Determine where you want your company to go. You may decide that a "compliance-only" strategy is sufficient or you may choose to pursue higher goals. What about ISO 14000?

Step 3: How do you want to get there? Will you be customer/market driven? If you choose to exceed minimum standards, will you do so unilaterally or wait until competitors make a move? Will you take an adversarial approach or choose to become a stakeholder in the regulatory process?

Step 4: Examine the costs and benefits of making changes in your environmental, health, and safety programs. Will you develop the expertise internally or seek outside assistance to develop and implement your plan?

25.1. ENVIRONMENTAL AND WORKER SAFETY REGULATIONS: A BRIEF HISTORY

Although laws and regulations covering these areas have been in place at various governmental levels (local, state, and federal) for decades, it was not until 1970 that both the U.S. Environmental Protection Agency (EPA) and the U.S. Department of Labor's OSHA were created. It is often said that EPA owes its origins to the first Earth Day, April 22, 1970. Not only is there a U.S. EPA, but all states also have an equivalent, whether it is known as a state EPA or a Department of Environmental Quality or by some other designation. Similarly, states with approved plans are permitted to run their own "state OSHA" rather than submit to federal control of their worker safety and health programs. Any state may enact an environmental or worker safety law that is more strict than federal laws, but the federal laws may not be weakened in any case.

Although the EPA may have begun as a result of Earth Day, there were certainly many environmental laws in existence before then. There was a Clean Water Act of a sort and limited clean air regulations (e.g., automobile exhaust controls), especially in California. But famous sights in the late 1960s such as the Cuyahoga River "burning" in Cleveland, smoggy skies in Los Angeles, and areas of the Great Lakes being nearly "dead" combined to create a public outcry for tighter controls on sources of pollution, both industrial and individual.

OSHA began on December 29, 1970, as Public Law 91-596. [1] This law was described as "An Act...to assure safe and healthful working conditions for working men and women; by authorizing enforcement of the standards developed under the Act, by assisting and encouraging the States in their efforts to assure safe and healthful working conditions; by providing for research, information, education, and training in the field of occupational safety and health; and for other purposes." This act is also known as the Occupational Safety and Health Act of 1970.

According to Congress, it was necessary to create OSHA because "(the) Congress finds that personal injuries and illnesses arising out of work situations impose a substantial burden upon, and are a hindrance to, interstate commerce in terms of lost production, wage loss, medical expenses, and disability compensation payments." [1]

Congress outlined the following strategy to achieve its goal of "assuring so far as possible every working man and woman in the Nation safe and healthful working conditions and to preserve our human resources": [1]

1. by encouraging employers and employees in their efforts to reduce the number of occupational safety and health hazards at their places of employment and to stimulate employers and employees to institute new and to perfect existing programs for providing safe and healthful working conditions;
2. by providing that employers and employees have separate but dependent responsibilities and rights with respect to achieving safe and healthful working conditions;
3. by authorizing the Secretary of Labor to set mandatory occupational safety and health standards applicable to businesses affecting interstate commerce and by creating an Occupational Safety and Health Review Commission for carrying out adjudicatory functions under the act;
4. by building upon advances already made through employer and employee initiative for providing safe and healthful working conditions;
5. by providing for research in the field of occupational safety and health, including the psychological factors involved, and by developing innovative methods, techniques, and approaches for dealing with occupational safety and health problems;
6. by exploring ways to discover latent diseases, establishing causal connections between diseases and work in environmental conditions, and conducting other research relating to health problems, in recognition of the fact that occupational health standards present problems often different from those involved in occupational safety;
7. by providing medical criteria that assure insofar as practicable that no employee will suffer diminished health, functional capacity, or life expectancy as a result of his or her work experience;
8. by providing for training programs to increase the number and competence of personnel engaged in the field of occupational safety and health;
9. by providing for the development and promulgation of occupational safety and health standards;
10. by providing an effective enforcement program that shall include a prohibition against giving advance notice of any inspection and sanctions for any individual violating this prohibition;
11. by encouraging the states to assume the fullest responsibility for the administration and enforcement of their occupational safety and health laws by providing grants to the states to assist in identifying their needs and responsibilities in the area of occupational safety and health, to develop plans in accordance with the provisions of this act, to improve the administration and enforcement of state occupational safety and health laws, and to conduct experimental and demonstration projects in connection therewith;
12. by providing for appropriate reporting procedures with respect to occupational safety and health, which procedures will help achieve the objectives of this act and accurately describe the nature of the occupational safety and health problem; and

13. by encouraging joint labor–management efforts to reduce injuries and disease arising out of employment.

Much of OSHA's early work centered around abating mechanical and electrical hazards in the workplace. Beginning with the Hazard Communication Standard in 1986, however, a larger emphasis was placed on eliminating chronic hazards associated with workers' exposure to chemicals, including colorants.

REGULATORY STRATEGY. In the beginning, the EPA felt that the fastest and best way to get the nation and its industry into compliance with the new laws was to use a "command-and-control" strategy. This involved, for example, establishing laws and legal limits for effluent concentrations, then controlling compliance through inspections, enforcement, and penalties.

In the 1980s this strategy began to change. The EPA realized that the command-and-control method, while successful in achieving early compliance and effluent reductions, would not be effective in meeting future goals. An early example of the EPA's new strategy was the disclosure of the data from the "Form R" reports. Known as the Toxic Release Inventory, or TRI, these reports of industry's release of certain toxic materials to all environmental media (air, water, or land) are made available to the public.

Through the TRI, the EPA hoped that it could shift the burden of applying pressure for compliance from itself to the public and the news media. So far, this seems to have been effective. Newspapers annually publish compilations of a state's biggest "polluters." The normal response of a named company is "How do I remove my company's name from this list?" Clearly, to eliminate itself from the roster, it must reduce its effluent significantly, which is, of course, the EPA's goal. But the EPA does not have to go to the time, trouble, and expense of inspecting the company and enforcing the rest of the command and control procedures; all it does is require the company to submit an annual report of its releases, publish the report, and let the public and news media do the rest.

Just as environmental laws existed before the EPA, there were many worker safety laws in place prior to OSHA's creation, particularly in notoriously dangerous occupations such as mining. OSHA was an effort to bring good safety practices to all aspects of private sector employment. OSHA has the responsibility of creating regulations to protect workers from injury in the workplace. In recent years, this effort has expanded beyond protection from dangerous work rules and injury-producing machines and tools to include exposure to hazardous chemicals in the workplace. Most of the physical, mechanical, and electrical hazards OSHA works to abate are of an acute nature, meaning their effects are immediately felt and recognized. In contrast, the effects of exposure to chronic hazards, such as inhalation of colorant dusts, are not immediately recognized, and their effects may not be realized for years or decades. This lack of recognition is what can make chronic hazards more serious than acute hazards.

OSHA at first considered extending the proposed regulations to create a community right-to-know law but decided that its true mission was to provide a mechanism to protect only exposed employees. Legislation to give the community surrounding a facility the right to gain information concerning chemicals used inside that facility was covered in subsequent legislation, which turned out to be the part of EPA regulations known as the Superfund Amendments and Reauthorization Act (SARA).

Probably the most significant worker health and safety development affecting the coloring of plastics is OSHA's Hazard Communication Standard (HCS). This regulation brings many colorants under regulation due to possible chronic effects of exposure.

The HCS, also known as the Employee Right-to-Know Law, is basically concerned with the exposure of workers to hazardous chemicals in the workplace. Under this standard, since 1986, all manufacturers, importers, and distributors of hazardous chemicals have had to label the containers they ship, listing the ingredients and hazards of the material in the containers, and provide Material Safety Data Sheets (MSDSs) to all downstream users, handlers, and storers of these chemicals.

Under regulations set forth by the HCS, an employer must

1. develop a written hazard communication plan and make it available to all employees who are exposed to hazardous chemicals in their work;
2. ensure that the proper labels and/or warning statements appear on all containers of hazardous chemicals;
3. train and inform all exposed employees so that they will be aware of the physical and health hazards of each hazardous chemical to which they are exposed, how to recognize and react to an accidental release of hazardous materials, and how to protect themselves from exposure;
4. obtain or develop an MSDS for each hazardous chemical in the workplace and make it available to all exposed employees; and
5. develop and make available to all exposed employees a list that includes all hazardous chemicals known to be present in the workplace.

WHAT IS AN "OSHA HAZARDOUS CHEMICAL?" Under HCS, each manufacturer or importer had to determine the hazards of the chemicals it manufactured or imported. These were broken down into many different physical and health hazards. If a substance possessed any of these properties, it was, by definition, a hazardous chemical and therefore had to have an MSDS created for it, listing certain properties and hazards in a loosely specified form. Even if, after testing, it was determined that the chemical did not fall into any of these 23 categories but was listed in certain other references, the standard said that it was also hazardous, by definition.

OSHA recognized four references as establishing that certain chemicals are hazardous. Any materials listed in these references were also, by definition, hazardous chemicals. Those sources are

1. Code of Federal Regulations (CFR) 29 Part 1910, Subpart Z-Toxic and Hazardous Substances;
2. International Agency for Research on Cancer (IARC), Monographs on the Evaluation of Carcinogenic Risk of Chemicals to Humans;

3. *Threshold Limit Values for Chemical Substances and Physical Agents in the Work Environment*, American Conference of Governmental Industrial Hygienists (ACGIH); and
4. National Toxicology Program (NTP), *Annual Report on Carcinogens*.

A great deal of frustration has arisen over the way hazardous chemicals are defined by OSHA. Most people feel more comfortable if there is a specific list of chemicals to which they can refer, but there is no such list provided under the Hazard Communication Standard. In addition, chemicals that were not previously considered hazardous because they are virtually inert or are not harmful to humans by contact with skin are now considered a dust hazard (particularly inorganic, nontoxic, noncombustible dusts that can enter the body via the respiratory system). This is where the regulation of colorants under OSHA occurs via the Hazard Communication Standard.

Many manufacturers felt that it was easier to provide an MSDS for each product they made, rather than sort through the list of chemicals used at their facility to determine whether or not an MSDS was really necessary for each substance. In this way, materials that were nonhazardous ended up with an accompanying MSDS.

EXPOSURE LIMITS: PELs AND TLVs. Threshold limit values (TLVs) were developed in the 1940s, long before the existence of OSHA. The TLV is defined as the airborne concentration of a contaminant to which it is believed that most workers may be repeatedly exposed, day after day, without developing adverse health effects. Developed and maintained by the American Conference of Governmental Industrial Hygienists (ACGIH), these values have been determined based on industrial experience and animal and human studies.

The reason for setting a TLV varies depending on the nature of the effect it may cause. For this reason, they cannot be used to compare the relative hazards of materials. For instance, one substance may cause an acute effect such as eye and respiratory irritation while another may cause chronic carcinogenic effects. Clearly, these effects are very different, and thus the substances' TLVs are likely to be dissimilar as well.

The permissible exposure limit (PEL), a term used by OSHA, is essentially the same as the TLV, except that PELs carry the force of law through OSHA enforcement, whereas TLVs are recommendations.

25.2. CURRENT STATUS: EPA

1990 Clean Air Act Reauthorization

In 1990 the Clean Air Act was reauthorized by Congress. This was another wide-ranging environmental law that will continue to be phased in through at least the middle of the next decade. Senator John Chaffee, a sponsor of the legislation, claimed that the Clean Air Act reauthorization was "primarily a health law" referring to its intended impact on the improvement of public health (rather than just the quality of the environment itself). EPA studies indicate that exposure to "air toxics," which are specifically defined in the regulations, may result in 1000–3000 cancer deaths annually in the United States.

What are the targets of the Clean Air Act? There are several, usually referred to by their "title" names in the regulations. There is a title for urban air pollution problems of ozone (smog), carbon monoxide, and particulate matter; another for motor vehicle pollution, including ozone precursors (volatile organic compounds, nitrogen oxides), fuel quality (volatility, sulfur content, gasohol), and evaporation during refueling; one to control acid deposition (acid rain); and one for the infamous problem of stratospheric ozone depletion [chlorofluorocarbons (CFCs)].

You have undoubtedly seen and felt the effects of some of these titles already. The air pollution and motor vehicle titles have been addressed by new laws requiring oxygenated motor fuels and low-sulfur coal for electric power plants. The Clean Air Act has affected refrigerants in automotive and stationary air conditioning and refrigeration equipment and the manufacture of some types of foamed plastics.

Of the various titles, the only direct concern for plastics coloring is known as Title III. An inventory of 189 toxic air pollutants has been generated for this title. Included on this list of substances are many commonly found in colorants, including compounds of antimony, arsenic, cadmium, chromium, cobalt, lead, manganese, mercury, nickel, and selenium. To avoid being a regulated facility under this law, you must confirm that your operation cannot emit more than 10 tons per year of any one of these or 25 tons annually of all of them combined. Based on these limits, Title III of the Clean Air Act has little effect on the coloring of plastics, though it could affect colorant manufacturers and high-volume compounders. Other titles do, however, have a significant impact on resin manufacturers and a possible effect on plastics processors in general, depending on where in the United States their facility is located.

There may even be some new opportunities under this law, as plastics processors who now paint or otherwise use a solvent-based coating on their products may be forced to stop coating and use molded-in color.

Clean Water Act

The Clean Water Act is an example of a federal law that is frequently delegated to the states for enforcement. Often the responsibility is passed down to the local level for setting water quality standards. The local authority, such as the publicly owned treatment works (POTW), then regulates industries that discharge their effluent to it.

This act does not have a major effect on the coloring of plastics, with the following caveat: Water quality testing for metals content is done via digestion in strong acid and subsequent atomic absorption analysis for elemental metal concentration. Most permit levels for the metals commonly found in colorants are in the low-parts-per-million range. Even though most colorants are nearly insoluble in water, it is possible to have significant concentrations of metals in your effluent because of the test method employed. Though this manner of test does not necessarily simulate the likely fate of colorants in the environment, it is nonetheless the standard method. Thus, it is imperative to prevent colorants that contain metals for which your facility has a regulated permit level from entering the water discharge.

Toxic Substances Control Act

OVERVIEW. The Toxic Substances Control Act (TSCA) became effective January 1, 1977. Its major goal is to prevent "unreasonable risk of injury to health or the environment from exposure to chemical substances." It has been said that TSCA is the most powerful, wide-ranging environmental law in the world. Prior to its enactment, there had been extensive public discussion about the possible harmful effects to workers and the general public from the manufacture, processing, distribution, use, and disposal of chemicals.

Through enactment of the TSCA, Congress established a number of new requirements and authorities for identifying and controlling toxic chemical hazards to human health and the environment. Programs now exist under this law to gather information about the toxicity of particular chemicals and the extent to which people and the environment are exposed to them, to assess whether they cause unreasonable risks to humans and the environment, and to institute appropriate controls after weighing their potential risks against their benefits to the nation's economic and social well-being.

The law also enables the EPA to require companies to test selected existing chemicals for toxic effects and requires the EPA to review most new chemicals before they are manufactured in commercial quantities. To prevent unreasonable risks, the EPA may select from a broad range of control options under the TSCA, ranging from requiring hazard-warning labels to outright bans on the manufacture or use of certain hazardous chemicals. The EPA may regulate a chemical's unreasonable risks at any point in its life cycle, from manufacturing, processing, distribution in commerce, use, or disposal.

Under the TSCA, the EPA has authority to regulate chemical substances and mixtures via the following mechanisms:

- premanufacture review of new chemicals before they are manufactured commercially or imported into the United States;
- requiring manufacturers, importers, processors, and users of chemicals to maintain appropriate records and to report to the EPA information about their adverse health or environmental effects;
- specific prohibitions on the manufacture, use, and disposal of certain chemicals (e.g., asbestos, PCBs, and CFCs); and
- requiring manufacturers, importers, and processors to conduct tests on chemicals that may present a significant risk or present a substantial human or environmental exposure.

Since its inception, the EPA has proceeded to implement the TSCA via the issue of many additional regulations. Enforcement actions by the EPA to ensure compliance have continued, with recent emphasis on the imposition of fines where appropriate.

The TSCA regulations are found in Title 40 of the Code of Federal Regulations (40 CFR), Parts 700–789. Probably the most important sections of TSCA to the plastics industry in general and the coloring of plastics in particular are Section 5: Premanufacture notification (PMN) and SNUR (Significant New Use Rules); Section

8: recordkeeping and reporting; and Sections 12 and 13, export and import rules, respectively.

Under Section 5, the EPA has the authority to require notification before any person manufactures or imports a chemical not on the TSCA inventory or manufactures or processes a chemical for a use the administrator has determined to be a significant new use. The former falls under the PMN requirement and the latter under SNUR. There are R&D exemptions under this section for small quantities of material to be used under controlled conditions by technically competent individuals.

Section 8 covers several topics including mandatory reporting on chemical production, use, and processing, the TSCA inventory itself, and how to handle allegations of adverse health and environmental effects of chemicals.

Section 12 covers export regulations. There is an exemption from most provisions of TSCA for chemicals that are exported, provided that it can be shown that there is no unreasonable risk of injury to health or the environment of the United States.

In addition, Section 12(b) requires exporters to notify the EPA before exporting certain hazardous chemicals. The intent of this provision is to enable the EPA to notify the foreign nation, through diplomatic channels, of the fact that these chemicals will be entering their territory. Notification to the EPA must be made of the first shipment of a given calendar year of a given chemical to each country to which it is exported and must be postmarked by the date of export or within 7 days of forming the intent to export, whichever is earlier. The notification document should be sent to TSCA Document Processing Center (TS-790), Room L 100, Office of Toxic Substances, U.S. Environmental Protection Agency, 401 M Street SW, Washington, DC 20460.

There is a comprehensive list of TSCA-regulated chemicals available on the Internet. This list, known as the Chemicals on Reporting Rules (CORR) list, is available through the EPA's World Wide Web site, which is http://www.epa.gov. To access the list from the EPA home page, choose Offices, Labs, and Regions; Office of Prevention, Pesticides, and Toxics; and Office of Toxic Substances. The Internet CORR list, which is updated quarterly, should be found there. You can also get the CORR list by writing to the Environmental Assistance Division (7408), U.S. Environmental Protection Agency, TSCA Assistance Information Service, 401 M Street, SW, Washington, DC 20460, or faxing your request to (202) 554-5603.

CONEG

The CONEG legislation, which began to take effect in the early 1990s, has to do specifically with packaging materials. An acronym for Coalition of Northeastern Governors, CONEG prohibits the intentional addition of four heavy metals (mercury, lead, cadmium, and hexavalent chromium) into these materials at any level and sets the maximum incidental combined quantities for these metals as follows:

600 parts per million by weight effective two years after adoption of the legislation, 250 parts per million after three years, and
100 parts per million four years after and beyond.

The stated purpose of the legislation was to keep these heavy metals out of landfills and municipal incinerators, thereby supposedly protecting and improving public health and environmental quality.

Though their origins were in the northeastern United States, similar laws are now in effect not only throughout that region but also in several midwestern states and elsewhere. Despite the fact that it is not a federal law, CONEG has effectively eliminated the use of materials containing these metals in packaging components throughout the United States as manufacturers were not willing to produce "CONEG-compliant" and "non-CONEG-compliant" grades of products.

The colorants affected by this legislation include the lead chromate and lead molybdate yellows and oranges, cadmium pigments, and mercury cadmium pigments (which were little used by the time this legislation began to take effect).

Tier I and Tier II Reports and Toxic Release Inventory

There are two major annual environmental reports, both mandated by SARA Title III, that can have an impact on colorants. Both of these reports cover activities for the previous calendar year. The first, due by March 1, concerns inventories of materials. The second, due July 1, details releases of certain materials to the environment.

TIER I AND TIER II. Title III's Section 311/312, "Inventory Reporting," requires reporting any substance for which you have an MSDS and have had a minimum of 10,000 pounds at your facility on any given day in the previous calendar year. These reports are intended to furnish information to the community and the local emergency responders and are provided to the state, local, or county Emergency Planning Commission and the local fire department. There is a slight difference in the amount of information provided in the Tier I and Tier II reports; most states now require the Tier II report, which is more comprehensive.

Among the data provided are the identity of the material(s), amount of material stored (average and maximum), conditions of storage (temperature, pressure, type of container), and location within the facility where the material is kept. This report is intended to provide the necessary information to enable the community to respond to a "chemical emergency" in the most effective manner. A chemical emergency could take the form of a fire or explosion, a weather-related situation, or an "act of God" such as an earthquake or flood.

TOXIC RELEASE INVENTORY. Section 313 of Title III is "Emission Reporting." Under this program, the EPA wants to know how much material goes into the air, the water, and the land. Chemicals on the Section 313 list can be individual materials such as ethylene glycol, methyl isocyanate, and ammonia or chemical categories defined as compounds of certain metals or organic compounds. These chemical categories include almost all the metals whose compounds produce colors: antimony, barium, cadmium, chromium, cobalt, copper, lead, manganese, nickel, selenium, and zinc. The standard threshold reporting quantity for manufactured or processed chemicals is annual usage of 25,000 lb.

The purpose of this report seems to be to twofold: to force industry to take a hard look at its processes and the materials it uses with an eye toward improving efficiency and reducing wastes and other emissions and to provide data to the public and increase awareness of industrial releases of toxic substances. Of course, the EPA also uses the information to understand the sources and annual amounts of toxic materials released to the environment.

Among the data collected are the identity of the material(s), amount of material released and to which medium (media) it is released, the name and location of the POTW and/or other disposal facility that treats the substance or its ultimate disposal location, and estimated future releases. There is one important distinction between the Tier I/II and the TRI reports. The inventory reports are concerned with the total amount of material stored. The TRI report details the amount of substance released, but in the case of metal-containing colorants, it is necessary to track the amount of compound (colorant) used to determine whether you have exceeded the reporting threshold. You must then convert the compound to the actual parent metal(s) for purposes of reporting.

An example is lead chromate yellow, which is reportable as both a lead and a chromium compound. It may contain 60% lead and 20% chromium by weight. If your facility used 25,000 lb or more of lead chromate (or other lead and chromium compounds), you must report any releases of these metals. If you found that you released 100 lb of lead chromate, you would report 60 lb of lead released and 20 lb of chromium released.

The EPA continues to refine and improve the TRI reports. The first year's report, for example, required the reporting of releases of titanium dioxide as a "toxic" substance. In later years, phthalocyanine blue and green pigments and barium sulfate have been delisted, as data have been submitted to the EPA proving their lack of toxicity.

How did these compounds end up on such a reporting list in the first place? The phthalocyanines are, of course, copper compounds, and, initially, all copper compounds were regulated as "toxic" materials. Similarly, all barium compounds were regulated, capturing barium sulfate. This has been a sore point among pigment manufacturers and users since the inception of the TRI reports. It is well understood that the mixed metal oxide pigments, while comprised of antimony, chromium, manganese, zinc, and other reportable metals, are not toxic in the truest sense, yet their release must be reported under this program.

Another recent modification to the TRI program is that an optional threshold has been established for facilities with low annual reporting amounts. Under this exemption, a facility that releases a combined total of 500 lb or less of a TRI chemical to all media is eligible to use an alternate "manufacture, process, or otherwise use" threshold of one million pounds for that chemical. It is still necessary to substantiate your status under this exemption through maintenance of records and documentation.

25.3. 33/50 INITIATIVE

The EPA is also investigating the use of voluntary means to reduce effluent. One such voluntary program is the Industrial Toxics Project, aimed at reducing environmental releases of 17 high-risk chemicals. This initiative is more frequently referred to as the "33/50" program, named for its goal of reducing emissions of the 17 chemicals by 33% by 1992 and 50% by 1995. In selecting these chemicals, the EPA considered several factors, in order to make the most significant improvements in environmental quality. The criteria include quantities produced and released, the seriousness of the chemical's known health and environmental effects, and the potential for reducing emissions or discharges.

The program began in February 1991 when the head of the EPA contacted more than 600 large companies to encourage their participation in the project. The EPA used TRI data to identify these companies, which account for 80% of the annual chemical releases reported to the EPA. The agency ultimately asked nearly 6000 companies to join the program.

In March 1992, the EPA announced that 734 companies, including almost half of the 600 largest waste generators, had joined the program and committed to overall emission reductions of over 300 million pounds by 1995. The EPA has touted the 33/50 program as a success, but environmental groups have pointed to the low participation rate as evidence of the weakness of the voluntary approach.

However one views the success of the 33/50 program, the initiative illustrates several characteristics of the EPA's pollution prevention strategy. The program rejects the traditional command-and-control, single-media approach to regulation, instead leaving companies free both to set their own reduction goals and devise their own methods to reach those goals. The EPA believes that companies that have joined the program are discovering that pollution prevention actually saves them money. The public relations value of participation also may be a factor in companies choosing to join. Finally, wise companies realize that if voluntary programs prove ineffective, the government may enact more restrictive alternatives.

25.4. FUTURE DIRECTION OF EPA

The EPA's preference for today and the foreseeable future is for businesses to engage in pollution prevention. By their definition, this is defined as "the use of materials, processes, or practices that reduce or eliminate the creation of pollutants or wastes at the source. It includes practices that reduce the use of hazardous and nonhazardous materials, energy, water, or other resources as well as those that protect natural resources through conservation or more efficient use." Of course, most businesses today must reduce or eliminate their wastes to remain competitive. No one can afford to discard valuable raw materials as wastes.

There seems to be a new sense of cooperation between the regulated community and the EPA. Perhaps some of this is due to the realization that industry has made great progress in reducing (or eliminating) pollution from its processes and also that many of the voluntary programs have been successful. Today, in many cases, nearly all of the possible gains have been squeezed from industry, and many of the major sources of pollution are now at the individual level (e.g., passenger cars). There have already been, and will continue to be, howls of protest as individual sources are controlled more closely through such means as auto exhaust testing, more costly reformulated gasoline, mandatory carpooling, and possible restrictions on such mundane activities as lawn mowing and barbecuing.

25.5. ISO 14000

The ISO 14000 series of standards, known informally as "Environmental ISO," is a set of generic standards developed by the ISO to provide organizations with a framework for managing the environmental impact of their activities. Environmental disciplines including the basic environmental management system, product labeling, criteria for auditing, and life-cycle assessment are included.

Many companies have registered their facilities to the ISO 9000 quality standards. Similarly, under ISO 14001 your organization will establish objectives and targets and commit to effective and reliable processes, including continual improvement of your environmental management system (EMS). The EMS is defined as the organizational structure, responsibility, practices, procedures, processes, and resources for developing, implementing, achieving, reviewing, and maintaining the environmental policy.

What is ISO 14001? Among other things, it is a framework for managing the environmental aspects of your business. It is a voluntary consensus, private-sector standard that can be used by any company, any size, anywhere in the world. ISO 14001 is not required; it is strictly voluntary. It does not establish performance standards or require you to reach "zero pollution" in your operation. But it does require a company to commit to continuous improvement of its EMS.

REASONS TO INVESTIGATE ISO 14000. Benefits expected to accrue from registration under ISO 14000 include customer endorsement, possibly including government bodies in the United States and elsewhere. It is believed that customers may reduce the frequency of compliance audits or consider ISO 14001 companies as preferred suppliers.

Other reasons include the possible avoidance of major costs from environmental problems that may be eliminated through an effective EMS. Greater acceptance of ISO 14001 may remove some of the redundancy of multiple international, national, and regional environmental standards, making it easier for your company to meet one global standard. It may level the playing field between you and some of your less regulated international competitors. Finally, it may become part of the "ante" of doing business, especially with large, multinational concerns. Even small, local or national firms may want to consider implementing or conforming to the ISO 14001 standard without certifying to it, because it is a good environmental management tool. Finally, a number of U.S. state and federal regulatory agencies are introducing initiatives to allow companies a greater degree of latitude in self-inspection, reporting, and correction if they develop formal environmental management programs along the lines of ISO 14001.

In the end, however, the decision to embrace ISO 14001 is an economic one, not a technical one. Organizations are beginning to realize that some of the greatest bottom-line enhancements in their operations can come from improvements in the way they address environmental management. The key challenge will be for organizations to create programs that meet the requirements of ISO 14001 yet are still not so unmanageable that they are useless in generating effective results on the plant floor.

25.6. FUTURE DIRECTION OF OSHA

Recently, OSHA developed the Voluntary Protection Programs (VPP), designed to recognize and promote effective safety and health management. This is an example of OSHA's stated goal of being more of a partner with industry than solely an enforcement agency. In the VPP, labor, management, and OSHA establish a cooperative relationship at a workplace. Management agrees to operate an effective program that meets an established set of criteria. Employees agree to participate in the program and work with management to assure a safe and healthful workplace. And OSHA initially verifies that the program meets the VPP criteria. It then publicly recognizes the site's program and removes the site from routine scheduled inspection lists. OSHA may still investigate major accidents, valid formal employee complaints, and chemical spills. The VPP concept is a recognition by OSHA that compliance enforcement alone can never fully achieve the objectives of the Occupational Safety and Health Act. Good safety management programs that go beyond OSHA standards can protect workers more effectively than simple compliance.

Many benefits have been cited by current VPP participants, including improved employee motivation to work safely, better quality and productivity, and reduced workers' compensation costs. In addition, participants are not subject to routine OSHA inspections, because OSHA's VPP onsite reviews ensure that safety and health programs provide superior protection.

Knowing the Regulations and How They Affect You and Your Company

If you are in charge of developing or maintaining your company's HSE programs, you probably have several questions:

How and where do I get the information I need?
How do I know which regulations apply?
How do I keep current?
What about any personal liability?

Sources of Information

In the HSE field, it is important to keep up to date with changes as they occur. There are many available sources of information, each with distinct advantages and disadvantages.

Attorneys: Lawyers can interpret regulations from legalese into layman's terms, but usually at a high cost.
Consultants: It is an important part of a consultant's job to remain well informed. They will, of course, want to sell their services to you.
Internet: Available at low cost, with immediate access and 24-hour-a-day availability. The information found there may not be the most current, however. In the case of U.S. government Internet sites, the date the information was last updated is listed.

- **Local Libraries:** Another source of low-cost information. There can be problems with access; also, if you are seeking specific regulatory knowledge, the public library may not have what you need on their shelves.
- **Material Safety Data Sheets (MSDSs):** These are furnished at no cost, and the information they provide must, by law, be accurate. The inconsistent formats currently used can make interpretation difficult; also, these may contain minimal data, requiring further searching to find what is required.
- **Networks of Other Professionals in Your Business:** Low (or no) cost. Others in your field may have the same, or a very similar, situation as you. You may not get complete information or may get an incorrect interpretation of regulations.
- **Seminars:** Attendance at seminars can help keep your information current and provide a chance to "network" with your peers. Drawbacks are the costs, which can sometimes be high, and many of these may not be focused on your business.
- **Trade Groups:** Information received from your trade association is tailored to your business or industry. To gain access to the information, usually membership in the organization is required.
- **U.S. Government Printing Office:** Federal regulations are available from this source at a relatively low cost. All necessary information is present, but the information often must be interpreted to your specific situation. This is where the attorneys can be helpful.
- **Waste Disposal Vendors:** These people can be an excellent source of information on regulations specific to the generation, transportation, and disposal of wastes. Since it is their business, they must know the applicable regulations. Almost always, the cost is paid as part of their services.

How Do I Know Which Regulations Apply?

First, it is critical to know the amount and composition of your raw materials and other substances you have on site. Do not forget your facility's maintenance area, where toxic and reportable materials are often used and stored. Most environmental laws have thresholds of regulation; if you use less than the threshold amount of material, you do not fall under the regulation.

The next step is to find out which of the materials you have are regulated. There are many lists on which these materials may appear, as described above. Access the OSHA and EPA lists to begin, then others as appropriate. Check your MSDSs; they vary in quality and content but often contain valuable information. Finally, get professional help such as attorneys and consultants.

Keeping Current

California Proposition 65 is the State's Safe Drinking Water and Toxic Enforcement Act of 1986. The purpose of the regulation is to prevent contamination of drinking water and to advise the public of any chemicals in products that are known to the state to cause cancer or reproductive toxicity. The governor of California is required to revise and republish the list of chemicals at least once a year. A Scientific Advisory Panel recommends additions to the list periodically. These additions are reviewed at a public hearing, may be incorporated in the official list, and go into effect at a specific date. The official list has grown from the original 29 chemicals to several hundred. The large majority of substances listed are organic chemicals such as ethyl alcohol; however, metals listed are compounds of arsenic, cadmium, and hexavalent chromium. Lead is listed as a reproductive toxin.

Note that Proposition 65 is a labeling law. There is no prohibition on the use of these materials; there is only the requirement that their presence must be noted on a warning sign and/or label. Which is why when you go into an establishment in California that serves adult beverages there are signs posted to warn you that alcohol is a known reproductive toxin. The only colorants affected by this law are the lead- and cadmium-containing pigments.

The list of chemicals is updated at least annually, effective October 1. You should compare your materials with those on the list to determine whether your product will require Proposition 65 labeling. To receive the updated list, write to the State of California in care of the California Environmental Protection Agency, Office of Health Hazard Assessment, 301 Capitol Mall, Sacramento, CA 95814; call to order the list at (916) 455-6900; or access it via the Internet at http://www.calepa.cahwnet.gov/oehha.

SOURCES OF OSHA INFORMATION. OSHA standards, interpretations, directives, and additional information are available on the World Wide Web at http://www.osha.gov and http://www.oshaslc.gov.

The U.S. Government Printing Office produces a CD-ROM that includes OSHA standards and more. It can be ordered by writing to the Superintendent of Documents, P.O. Box 371954, Pittsburgh, PA 15250-7954, or phoning (202) 512-1800. Specify OSHA Regulations, Documents, and Technical Information on CD-ROM (ORDT), GPO order number S/N 729-013-00000-5. An annual subscription (four quarterly releases) costs $38.

Personal Liability

Some environmental and worker safety laws carry not only corporate or civil but also personal criminal liability. Usually this applies only in extreme cases of willful neglect or intentional falsification of records, but it is there to "put some teeth into the laws." These penalties are usually reserved for top-management levels but could apply to a rank-and-file worker if it was determined that he or she deliberately did not follow a clear procedure and that there was an intent to enforce penalties for not following the procedure. In other words, if it is management's intent to "look the other way" in the case of employee misconduct, then management, not the employee, is at fault and is therefore liable.

Specific "Problem" Colorants

"HEAVY METAL"–CONTAINING COLORANTS. There is a great deal of uncertainty surrounding the subject of exactly what constitutes a heavy metal pigment. As our friends in the legal profession might say, "it all depends." There are several definitions; each is correct in its own right but may not be satisfactory for a specific situation.

For example, one dictionary defines a heavy metal as one whose specific gravity is 5.0 or higher. Under Resource Conservation and Recovery Act (RCRA) regulations, metals analyzed for with the Toxic Characteristic Leaching Protocol (TCLP) extraction test to determine whether a waste is hazardous are arsenic, barium, cadmium, chromium, lead, mercury, selenium, and silver. EPA Form R release reporting covers the following metals: antimony, barium, cadmium, chromium, cobalt, copper, lead, manganese, nickel, selenium, and zinc. The Food and Drug Administration (FDA) defines heavy metals as those precipitated as sulfides: silver, arsenic, bismuth, cadmium, copper, mercury, lead, antimony, and tin. And, for packaging applications, the CONEG metals are cadmium, lead, mercury, and hexavalent chromium.

With all these interpretations of what a heavy metal is, the question then becomes, "according to whom?" When many people refer to heavy metals, they mean the "Big Two"—lead and cadmium. Packaging users are, of course, concerned with the four CONEG metals. Beyond these, some people have their own preferences. The best suggestion is to determine the preferences prior to choosing any metal-containing colorants for use.

Regardless of the heavy metal, the real issue involved with human health and the safe handling and use of colorants containing these metals is bioavailability. That is, is the metal present in such a form as to make it possible for it to enter the bloodstream and move throughout the circulatory system to the place(s) in the body where it can do harm? For almost all colorants, including lead chromate and cadmium sulfide, the answer is "no." These colorants are so sparingly soluble in bodily fluids as to be little more than nuisance dusts. Why, then, has the use of lead- and cadmium-containing pigments been practically eliminated in plastics applications? Part of it has to do with "guilt by association." Lead chromate in particular has been lumped together in the mind of the public with "lead-based" paint, which contains the highly soluble (i.e., bioavailable) lead carbonate.

There's no question about it—lead isn't good for you, and when you disclose on a label or an MSDS that a product contains lead, the general public is not sophisticated enough to make a distinction. This is where the good people who believed that bioavailability was the concept that would "save" heavy metal pigments were wrong. Public perceptions about these colorants had already been established and would be difficult, if not impossible, to sway. Moreover, laws had been written restricting or banning their use, reminding us again that laws are written by lawyers, not scientists. The net result is that lead- and cadmium-containing pigments are nearly obsolete in plastics applications in the Western world, and their future is, at best, dim.

LEAD CHROMATE YELLOW AND LEAD CHROMATE/MOLYBDATE ORANGE PIGMENTS. These are among the oldest and most thoroughly studied colorants. OSHA has such great concern about workers exposed to lead that it has created a separate standard (29 CFR 1910.1025) for the element. The atmospheric exposure limits for lead are quite low, established at $50\,\mu g/m^3$. The long-term health effects of exposure to lead are well understood, involving the blood, gastrointestinal, nervous, and reproductive systems. The OSHA lead standard gives details on the physical and biological monitoring required in the workplace when using these pigments. The chromium component of lead chromate is hexavalent chromium, which is regulated as a suspected carcinogen.

Lead chromate is almost insoluble and has very low bioavailability. Studies on lead chromate pigment have shown that it does not have the carcinogenic potential of soluble chromates, such as zinc chromate. As with other heavy metals, solubility is the key to toxicity. With proper safeguards such as adequate ventilation and personal protective equipment, chrome yellow and molybdenum orange can be safely used in the workplace.

The EPA regulates both lead and hexavalent chromium in its hazardous waste regulation. Solid waste containing lead chromate or lead molybdate should be tested for toxicity prior to disposal via the TCLP test.

CADMIUM SULFIDE AND SULFOSELENIDE. Like lead, OSHA has created a specific standard for cadmium. Cadmium pigments are alleged to be carcinogenic and are regulated as such. The EPA regulates both cadmium and selenium in its hazardous waste regulation, and these pigments are forbidden for use in packaging under CONEG. These pigments, like lead chromate, are of course still legal to use but should be used only in areas where adequate ventilation is present and suitable personal protective equipment is provided. Solid waste containing cadmium pigments should be tested for toxicity prior to disposal via the TCLP test.

CARBON BLACK. The International Agency for Research on Cancer (IARC), an agency of the United Nations World Health Organization, recently conducted a scientific review of carbon black. Based on its survey of available information, the IARC made a change in the classification of carbon black from category 3 (insufficient evidence to make a determination) to category 2B (known animal carcinogen, possible human carcinogen). This, despite the fact that nearly 60 years of human epidemiological evidence indicates lower than expected deaths from cancer among U.S. carbon black workers. This has resulted in the requirement that MSDSs for carbon black and products that contain it above minimum levels be updated to include the new IARC classification information.

GERMAN AZO DYES. A regulation was adopted in Germany in July 1994 that bans, in consumer textiles that come into contact with the skin, the use of certain azo dyes. Upon reductive cleavage of the azo groups, these dyes could form any of 20 aromatic amines. Many acid dyes and direct dyes were affected by this legislation. If your application involves coloring of textiles for skin contact, you should investigate whether this regulation affects your products and raw materials.

CARBAZOLE VIOLET. These pigments are "chloranil-derived" compounds. In 1991, the EPA investigated the manufacture and use of these pigments as they and their precursors were suspected of being carcinogenic. No further action has been taken to date as a result of this investigation.

DIARYLIDE PIGMENTS. Studies published in the early 1990s showed that these pigments can decompose at temperatures in excess of 200°C, liberating hazardous compounds such as 3, 3'-dichlorobenzidine. This has restricted their use primarily to inks, coatings, and polymers processed below this temperature.

NICKEL-CONTAINING PIGMENTS. Recently the IARC classified nickel and nickel compounds as Group 1 carcinogens. (Group 1 designates sufficient evidence for causing human cancers.) This includes both organic and inorganic pigments that contain nickel as a part of their structure. However, the IARC states that the evaluation applies to the group of chemicals and not necessarily to all individual chemicals in the group. Furthermore, the IARC says evidence points to nickel refining as having a carcinogenic risk to workers. There is no known evidence that nickel titanate pigments have caused adverse health effects. Therefore, though the Hazard Communication Standard mandates that MSDSs include the IARC findings, the pigment may avoid being labeled as a suspect carcinogen.

COBALT-CONTAINING PIGMENTS. Even though some of these pigments are approved by the FDA for coloring polymers that are used in direct food contact applications, all cobalt compounds are also suspected carcinogens, according to the IARC. FDA evaluation determined that while there is "inadequate evidence" for the carcinogenicity of cobalt and cobalt compounds in humans, these materials are "possibly carcinogenic to humans."

CHROMIUM-CONTAINING ORGANIC AND INORGANIC COLORANTS (OTHER THAN LEAD CHROMATE). The chromium present in these colorants is the "good," trivalent variety. Trivalent chromium compounds have PEL and TLV values of $0.5 \, mg/m^3$ and are recognized as noncarcinogenic. In fact, this type of chromium is an essential trace element in humans. The EPA also differentiates in the disposal of trivalent chromium compounds with a regulation that specifically exempts these compounds from listing as hazardous waste. Therefore, waste from pigments containing only trivalent chromium can be disposed of as nonhazardous waste.

REGULATED COLORANTS THAT ARE FDA APPROVED. It often seems a contradiction that many of the colorants that are suitable for use in food contact applications must also appear on an MSDS as a "reportable" or "hazardous" material. Examples include zinc oxide and zinc sulfide, both zinc compounds; chromium oxide green (pigment green 17), a chromium compound; carbon black; and cobalt aluminate blue (pigment blue 28), which is not only reportable because of its cobalt content but is also a suspected carcinogen.

How, you may ask, can the FDA be so irresponsible as to approve for use in food contact applications a substance that is a suspected carcinogen? A bit of common sense, often a rarity in environmental matters, prevailed in this situation. The FDA recognized that the only way cobalt compounds are carcinogenic, if indeed they are at all, is via the respiratory route. Since it is not reasonable to anticipate that food contact plastics will be inhaled, and since there is no other discernible hazard, the FDA has approved the use of these colorants.

Recently, the IARC has pronounced carbon black to be a Group 2B carcinogen. The same situation exists in this case; the FDA realizes that the hazard (if any) of carbon black is from respiration, not ingestion, so carbon blacks have retained their approval for use in certain food contact uses.

Step 2: Where do you want your company to go? Why? There are essentially three options for this: try to get by with the bare minimum, perhaps not even reaching compliance; achieving compliance with the law, but nothing more; and going beyond compliance to some extent. I will assume that you are planning to at least achieve compliance and may be unsure as to whether you should go further in this area.

There are strong arguments on both sides of the issue of whether to go beyond compliance with environmental and worker safety laws. Among the benefits of exceeding the requirements in these areas are being proactive and staying ahead of competitors; your employees may appreciate your concern for their safety and well-being, resulting in improved morale and productivity; possible reductions in costs associated with waste disposal and possible environmental mishaps; savings associated with fewer safety and health problems (e.g., reduced Workers' Compensation insurance costs); and public and industry perception as a leader and a good corporate citizen. Dangers of this strategy include the very real possibility of intentionally walking away from business opportunities; extra costs associated with training, engineering controls, and medical surveillance; additional costs for material substitution and reengineering of products and processes; and additional costs for new equipment, redesign of work areas, training, and possible lower productivity.

For most processes involving plastics coloring, achieving compliance with environmental and worker safety standards can be done with readily available equipment and easily understood procedures. Most exposure standards can be met through the use of low-dusting grades of pigments and a good dust collection system. From an environmental perspective, most pigments would be classified as nonhazardous wastes. The largest environmental issue in this case is the metal content of the colorant. Most colorants will pass the TCLP extraction test as received. When encapsulated in polymer as a concentrate or in a finished part, any metal extraction is almost certainly below regulatory levels. Of course, it is the responsibility of the waste generator to determine whether or not the waste is hazardous.

Nevertheless, for the reasons cited above, your company may wish to move beyond compliance. This path, if chosen, will have a significant impact on the future operations of your company and so should not be entered into without full consideration.

Step 3: How do you want to get there? Achieving compliance is not only relatively easy, it is also the law, so it is assumed you will choose to operate legally. If you want to move beyond compliance, there are many possible routes to take and timetables to follow.

Will you take your lead from your customers? In many cases the largest multinational corporations set the tone for the marketplace because they must reach the highest global standards in at least one of the markets in which they operate. If one of these businesses is your customer, they may demand that you exceed regulatory standards.

You may choose to take the lead and exceed minimum standards unilaterally, rather than waiting until competitors make a move. This can be both a business and an emotional decision. There maybe tangible economic benefits and a discernible value added to your business through your customers' (and the public's) perception of environmental excellence. Moreover, it may also be an individual decision of those in top management.

Today, there seems to be an increased willingness to have those who will be affected by proposed regulations provide input to and participate in the regulatory process. This is possible either as an individual company or via the route of a trade association. The Society of the Plastics Industry (SPI) is an example of an industry group that has been a participant in the regulatory process.

Step 4: As more business is done globally, many people believe that certain standards (e.g., wages, concern for worker safety) are being driven to the lowest level. If this is true, it certainly does not apply to the concern people have for their environment, and environmental laws in general. If anything, these seem to become ever more strict. Even third world nations are copying the regulations of developed countries, and enforcement of these new laws, if lagging today, will eventually come.

For this reason, it makes sense to examine the costs and benefits of making improvements in your environmental, health, and safety programs. Make the HSE area part of your company's planning process, so that change can be proactive rather than reactive. Develop a schedule of implementation, as it is difficult to tackle all of these areas at once. You will need to decide whether to develop the expertise internally or seek outside assistance to devise and accomplish your plan. Training is readily available if you want to do it yourself, and there are many capable consulting firms ready to assist.

Whether or not you believe that many of the existing safety and environmental laws are sensible, the reality is that they exist and must be followed. Developing an excellent environmental management system is a good way to avoid present and future risks and is a sound business strategy.

25.7. GLOSSARY

ACGIH: American Conference of Governmental Industrial Hygienists, an organization devoted to the administrative and technical aspects of occupational and environmental health. ACGIH is a professional society, not a governmental agency.

CONEG: Coalition of Northeastern Governors, the group that devised the regulations on the "heavy metal" content of certain packaging materials.

FDA: Food and Drug Administration, a branch of the U.S. Department of Agriculture.

IARC: International Agency for Research on Cancer, an agency of the United Nations World Health Organization.

Life Cycle: Consecutive and interrelated stages of a product's existence, from raw material acquisition or generation to final disposal.

OSHA: Occupational Safety and Health Administration, part of the U.S. Department of Labor.

PEL: Permissible exposure limit. Similar to TLV, these are established by OSHA and are legal mandates.

POTW: Publicly owned (or operated) treatment works, another name for a municipal sewage treatment plant.

TCLP: Toxicity Characteristic Leaching Protocol (Procedure?), an EPA-required laboratory test to determine whether harmful amounts of material leach from wastes in simulated landfill disposal conditions.

TLV: A trademark of the ACGIH, it refers to the airborne concentration of a contaminant to which it is believed that most workers may be repeatedly exposed, day after day, without developing adverse health effects.

TSCA: Toxic Substances Control Act, a set of federal regulations governing the commercial use of chemicals.

REFERENCE

1. OSHA Act of 1970—Public Law 91-594 http://www.osha-slc.gov/oshAct/OSHACT.html#1

CHAPTER 26

Use of Statistics in the Coloring of Plastics

Thomas V. Edwards

489 Windsor Drive
Elyria, OH 44035

26.1. OVERVIEW

The intent in this chapter is not to present in great detail the mathematics behind the statistical methods discussed. An excellent reference manual assembled by the Automotive Industry Action Group (AIAG), *Fundamental Statistical Process Control*, details process control systems, variation, action on special or common causes, process control and capability, process improvement, control charting, and benefits derived from using each of these tools. "Reprinted with permission from the Fundamental Statistacal Process Control Reference Manual (Chrysler, Ford, General Motors Supplier uality Requirements Task Force", Measurement Systems Analysis, MSA Second Edition, 1995, ASQC Press.

This chapter presents an in-depth understanding of the what, where, why, when, and how of applying statistics and statistical tools specifically for use in the coloring of plastics. First we need to present some terms:

Statistics refers to the scientific methods applied to the collection, organization, interpretation, and presentation of information—numerical *data*. For statistical process control (SPC), data types are divided into attributes or variables.

Data is the plural of *datum*, which is Latin for "fact."

Coloring of Plastics, Fundamentals, 2nd edition. Edited by Robert A. Charvat
ISBN 0-471-13906-8   Copyright © 2004 by John Wiley and Sons, Inc.

Now that the "facts" are straight, the goal of statistics is to make sense of the "matter of direct observation" in a way that helps us expose properties of those observations that might otherwise be hidden. For the colorist, direct observation could refer to visual or spectrophotometric data. Once these statistical inferences are laid bare, we then develop intelligent decisions based on our findings—search for facts that support the human observations that lead us to collect the statistical data in the first place.

All statistical analysis is dependent on the data collection methods employed—the sampling, testing, and measurement techniques used to gather data. Typically batch color data, generated from a given set of samples, is accepted as fact. Pass–fail or go–no go decisions are made based solely on these facts. More often than not, color determinations are made with little or no regard for sampling, testing, or measurement errors associated with the process that generated those numbers. Delta color data is most likely plotted on a two- or three-dimensional Delta Lab (*DLab*) or Delta LCH (*DLCH*) chart.

Further, color specifications are set with the same limited attention toward error and predilection in the data collection system. To complicate matters more, the color language used to produce those numbers contributes its own *biased error* when applied to batch-to-batch control or "linear" color tolerancing:

Color is a biased human observation. A complex response to light, an object, our eyes, and subsequent analog signals that is processed by our emotional brains to form a mental "interpretation" called color, that is, no numbers—just simple subjective human observations or responses to a stimulation.

Spectrophotometric color data are generated to fulfill the colorist's desire to have a nonbiased, common scale by which to compare different observations, that is, numbers interpreted as objective facts.

Coding or scaling of color data provides a basis for representing visual observations in three-dimensional, mathematical terms: lightness, chroma, and hue

Color Specifications are *digital* representations of what we humans might accept or reject—our "tolerance" or "intolerance" to color differences. Human color observations are *analog*. Linear specifications are only as good as the analog-to-digital conversion factors used to generate the color data. Again, the principal difficulty in making the rules of observed color and color language match up with the facts originating from the same scientific attempts to convert our analog color observations into digital representations.

Therein lies the great paradox: Biased, human, visual observations tell the color scientist if these digital color representations are true (facts) or false (biased).

So how do we get at the "truth" and determine which statistical methods may be applied to color of plastics? Johann Wolfgang von Goethe said, "It is easier to perceive error than to find *truth*, for the former lies on the surface and is easily seen, while the latter lies in the depth, where few are willing to search for it." Several years ago, a process engineer returned from a Genichi Taguchi, Taguchi Methods: Design of Experiments, ASI Press, 3/1993 (DOE) course and immediately applied this breakthrough problem-solving technique to a chronic color manufacturing problem. Approximately forty trials and over $100,000 in off-grade material later, all testing was abruptly stopped by manufacturing. The engineer soon realized not all quests for truth yield cost-effective solutions to everyday batch-to-batch quality problems in the real world.

Elaborate DOEs have their place on the research and development (R&D) work bench (very controlled, small beakers) and as tools to direct "scaleup" changes, but in manufacturing environments (not as well controlled, big beakers), "right the first time" is mandatory. There is often little or no room for costly experimentation on the plant floor—the main problem with the DOE involved replication of previous tests.

My father taught me one of my first valuable lessons in statistics: "Spit in one hand and wish in the other. Now, tell me which is the fullest?" (An absolute gage is needed to judge absolute differences!) My father-in-law taught me the second: "If a frog had wings, it wouldn't bump it's b__!"(How do I "gage" the unknown or "unknowable"?)

College taught me fewer practical lessons about statistics:

- As taught in a classroom setting, probability and statistics were uneventful subjects. Only accounting comes to mind when I try to recall my least desirable academic subject. There were moments filled with visions of jelly beans orbiting overhead, pondering the probability of one dropping from the sky to land squarely in the center of the room.
- Much of what I learned from "formal" statistics—as with other higher academic subjects—had more to do with stimulating thought than with the day-to-day problems in the real world. Academic statistics constituted little more than professors who came straight from high school, only to graduate from college and in turn teach students what they were previously taught by similar academic types who preceded them. As I remember it, theoretical statistics yielded approximately the same antediluvian obsolescence as the "fact" that our world is the center of the universe.

As Henry Brooks Adams said, "Nothing in education is so astonishing as the amount of ignorance it accumulates in the form of inert facts."

College provided me with the "basics." My first job as a chemical technician involved a daily grind (pun intended) of sample preparation for production quality control, process control, and quality assurance. Day after day, I sat at an X-ray spectrophotometer, equipped with an 8K RAM Olivetti "computer" (this tells you how many years ago). Regression lines were programmed in, using machine language, to calculate percentage of metals based on goniometric peak-height response against "known, wet-chemisty standards."

One day, after two years of the same daily toil, the chief chemist approached me with my first real challenge. Apparently, he had gotten involved with some round-robin testing among several analytical laboratories to determine percentage of cerium (Ce) in steel. The samples ranged from 0.001 to 0.030% known concentrations.

The initial pass for known 0.001 and 0.030% concentrations revealed a significant difference in peak height response for the two samples. However, analysis of the additional eight "blind" samples yielded very poor results against the known %Ce concentrations. The chief reported our poor correlation back to the round-robin committee. Now most of my life I have found it difficult to accept "It just can't be done" without plausible explanations for why not.

Determined to find a why not, I returned to the X-ray unit. Installed to the side of the goniometric counter was an oscilloscope—never turned on. The chief had given strict instructions when I started the job that I was not to touch the thing, let alone make adjustments to the instrument "window" based on the oscilloscope. Like the kid in a candy store, recording ALL predetermined settings, I did the unthinkable—I switched on the oscilloscope to look at each of the 10 samples individually. Starting with those samples that deviated most from the regression line, I noted a bimodal peak-height response for each L-α measurement.

The "window" was set up wide enough initially to pick up both peaks. The samples that measured "high" compared to wet analytical methods exhibited a higher right peak response, while samples below the regression line contained a lower right response. The left-hand peak seemed to remain the same, regardless of concentration.

Doing the forbidden, I adjusted the window to exclude the left-hand peak and count only the right-hand peak-height response. Rerunning the regression analysis for all 10 samples, R^2 went from less than 0.6 to greater than 0.9. Excluding one outlier, $R^2 = 0.96$.

Reporting my new finding to the chief, including the changes made to the instrument settings, took a great deal of courage on my part, but the joy and excitement over this discovery far exceeded any retribution.

From this experience I learned the following:

- It is easier to ask forgiveness than permission when trying something new. Be brave! Be bold! Dare to try a novel approach (within reason of safety and practicality).
- Unless you try something new, you can never say "I failed," and you will not prevail,
- Often, when confronted with seemingly unsolvable problems, we just need to take additional time. Simply step back and take a *fresh look* at what you and others have accomplished or consider to be a fact.
- Use the tools you have at hand. When possible, create new ones.
- A picture is truly worth a thousand words. Picturing a problem (looking at the oscilloscope window) often sheds new light on a once insurmountable problem.

With new insight, we owe it to ourselves to explore alternatives, even though we know "they" or "we" may have been right all along: "There is no failure except in no longer trying. There is no defeat except from within, no really insurmountable barrier save our own inherent weakness of purpose" (Kin Hubbard). Moving on to pigment manufacturing and quality control a year later, I have spent the last 20 years applying these principles to everyday problems encountered in manufacturing and quality control.

When I started in a Color Quality Laboratory, I trained under a 40-year veteran to color manufacturing and control. Most of the instruction I received in my apprentice position was similar in nature to that earlier cerium-in-steel situation. More often than not, conventional wisdom and experience were right; however, once in a while they were wrong. To the scientist, that is exactly when the fun begins: "If your capacity to acquire has outstripped your capacity to enjoy, you are on the way to the scrap-heap" (Glen Buck).

26.2. ROOT OF BATCH-TO-BATCH VARIABILITY WITHIN COLOR CONTROL: TESTING FOR COLOR VARIATION

Reading the AIAG manual on statistical charting and process control, assembly lines with operators measuring widgets for changes in gap or thickness come to mind. Widgets seem to flow in an endless rhythm from the manufacturer to the consumer. If you have visited an assembly line lately, you begin to wonder what does widget making and SPC charting have to do with batch control, let alone statistical control for color?

In batch manufacturing, the closest thing to a line stop is when the batch color fails to meet specifications and either requires a formulation adjustment or must be scrapped. Decisions are more of a go–no go nature. Testing variation more often than not accounts for a large portion of the perceived batch-to-batch variability. Typically, generating batch color data occurs from a single "grab" sample, or better still a composite sample.

Rework time is the main problem that confronts a batch operations—not getting it right the first time. So how does one separate the test variability from actual batch process variations? Based on a single sample and a single test, how do I know the batch or lot sent to the customer is really OK for their end use?

Classical statistical control methods assume "assignable causes for product variability usually *signal* an 'out-of-control' condition in the manufacturing of the product." First, a signal is defined as a sign for giving notice, especially at a distance, something to attract the eye or ear. My first thoughts begin with smoke signals from a distant bluff or someone's car alarm. I have yet to experience a single quality control scheme that delivers this clear a signal. Second, a process, in simplest terms, is said to be out of control if three or more assignable cause events in 1000 result in deviation of a product quality characteristics to be outside $\pm 3\sigma$.

These definitions only raise more questions! How do I read and interpret these signals, and if I start up a new run and the first batch is "in control" but "out of specification," how do I know if or when a correction is needed? After all, the process is in control.

A starting point: Getting at the root of any batch-to-batch variability in any process begins with the test. Roughly translated to widget making, this refers to gage R&R (that is, repeatability and reliability, not rest and relaxation). Taking color measurements on a batch or lot of color is the same as taking physical measures on widgets like length, thickness, or gap. However, there are some fundamental differences not accounted for in classical widget R&R. First is the test.

Widget SPC generally provides a direct measure of the manufactured piece. Color SPC generally requires further sample preparation prior to measurement. The AIAG's translation of gage R&R falls short of addressing the full scope of color measurement R&R because of the multiplicity in the shear number of steps required to produce color readings.

So what's a colorist to do? It all depends on whether you want the short but sweet answer or the answer that might lead to real breakthrough quality improvements on the plastics manufacturing plant floor. If you require a more in-depth understanding of not only your process but also the numbers you rely on to gage that process, you must first look at all the components that make up the test. Strictly speaking, this means the physical properties, each step in the test, and the varying techniques and equipment used to produce the final "piece."

Once analyzed, the colorist can then determine the portion of batch-to-batch variability associated with test. Remember the window on the X-ray unit? After removing the left peak, only the true %Ce counts were left. What is left after removing test variation is true process variation. It is only then that you can begin comparing test data to process for "common" causes and long-term chronic sources of variation.

One further point: Everything discussed up until now has been statistical quality control (SQC), not SPC, quality data and not process data. Process controls should be used to control quality, not the quality to control the process. This point seems to be missed in the majority of literature. Again, old habits die hard.

To take this one step further in the evolution from SQC to SPC, the color scientist must grasp the concept of "linking" process color quality with critical process control paramaters. Generally, this is accomplished in a color manufacturing operation through such measures as statistical problem solving, regression analysis, analysis of variance, and simple graphing techniques.

Once these links and meaningful process control plans are made, operators can for the first time experience true statistical process control beginning with the critical few parameters that control color quality for the final product.

26.3. PROCESS MEASUREMENT SYSTEMS ANALYSIS

The term *process measurement systems analysis* comes directly from the AIAG reference manual on SPC. Previous references to analysis of measurement systems in relationship to actual process versus total variation were expressed in terms of the relation

$$\frac{\sigma^2}{\text{Total}} = \sqrt{\frac{\sigma^2}{\text{process}} + \frac{\sigma^2}{\text{measurement}}}$$

Solving for process variation,

$$\frac{\sigma^2}{\text{Process}} = \sqrt{\frac{\sigma^2}{\text{total}} - \frac{\sigma^2}{\text{measurement}}}$$

Again, the AIAG assumes only gage (colorimetric) measurement error. Expressed in everyday terms for the color industry, the "true" process color difference from standard for a given batch is made up of the total measured color variance minus any measurement error associated with sampling, test, and color measurement.

The AIAG did lay an excellent foundation for exploring true product or process variations in relationship to batch-to-batch color control. First they recommended less than 30% measurement error. If the combined sampling, test, and color measurement error exceeds 30%, then one needs to further examine the problems within their "gage system."

The second tool the AIAG provided was a basic blueprint for getting at percent gage R&R. They assigned measurement repeatability to equipment variability (EV) and reproducibility to appraiser variability (AV). The AIAG R&R further defines PV as the actual part-to-part variation. Part variation can represent the process, the product, or the sample variation depending on how the measurement analysis is set up.

Equipment variability assumes an appraiser (color technician) takes a part (final test piece) and measures that part two to three times using a gage (spectrophotometer). Appraiser variability is comprised of two to three appraisers doing the same thing for the same parts. This process is repeated for an additional 1–9 parts. Color values for the 2–10 parts are then entered into a measurement unit analysis form (Tables 26.1 and 26.2).

From Table 26.1

$$\text{Average range between technicians } (\overline{R}) = (0.12 + 0.09)/2 = 0.105$$

$$\text{Equipment Variation } (\mathbf{EV}) = \overline{R} K_1 = (0.105 \times 3.05) = 0.32$$

where K_1, K_2, and K_3 are derived from Table 26.2:

$$\text{Appraiser variation } (\mathbf{AV})$$
$$= \sqrt{(\overline{X}_{\text{DIFF}} K_2)^2 - (\text{EV}^2/nr)} = \sqrt{(0.03 \times 3.65)^2 - (0.10/30)} = 0.093$$

Where n = number of lots tested and r = number of trials per lot.

$$\mathbf{R \& R} = \sqrt{\text{EV}^2 + \text{AV}^2} \qquad = \sqrt{0.32^2 + 0.093^2} \qquad = 0.33$$

Part (batch or sample) variation (**PV**)
$$= \text{process range} \times K_3 \qquad = 0.23 \times 1.62 \qquad = 0.37$$

Total variation (**TV**)
$$= \sqrt{\text{R\&R}^2 + \text{PV}^2} \qquad = \sqrt{0.33^2 + 0.37^2} \qquad = 0.50$$

and

%EV = 0.32/0.50 × 100 = 64%   Estimated equipment σ   = 0.32/5.15 = 0.062
%AV = 0.093/0.50 × 100 = 19% Estimated appraisal σ   = 0.093/5.15 = 0.018
%R&R = 0.33/0.50 × 100 = 66%   Estimated measurement σ = 0.33/5.15 = 0.064
%PV = 0.37/0.50 × 100 = 74%   Estimated process σ   = 0.37/5.15 = 0.072

The 5.15 value used to estimate sigma for each of the variation components originates from the AIAG calculations for K_1, K_2, K_3, which uses 99.0% of the area under the normal distribution curve (5.15σ). Note also that the R&R and estimated PV are not additive to 100% of total variation. Using $\overline{X}$ and s for this product, estimated total variation is equal to 0.48(6 × 0.081σ), which compares well with the gage R&R value of 0.50.

To get at the sample variation (SV) component of total variation (TV) of the

Table 26.1. Repeatability and Reliability Data Sheet

Technician/Test Number	Batch or Lot Sample CIE Da										Average
	1	2	3	4	5	6	7	8	9	10	
Method A											
1	0.03	−0.04	−0.14	−0.03	0.14	0.05	0.05	0.04	0.20	−0.08	
2	0.01	−0.03	−0.09	−0.15	0.11	0.03	0.00	−0.09	0.15	0.14	
3	−0.11	−0.03	−0.10	−0.11	0.03	−0.06	−0.12	−0.08	0.11	0.05	
Average range	−0.02– 0.14	−0.03– 0.01	−0.11– 0.05	−0.10– 0.12	0.09– 0.11	0.01– 0.11	−0.02– 0.17	−0.04– 0.13	0.15– 0.09	0.04– 0.22	0.00– 0.12
Method B											
1	−0.03	−0.07	−0.08	−0.05	0.09	−0.07	−0.08	−0.01	0.21	0.24	
2	0.05	0.00	−0.05	−0.03	0.07	−0.07	−0.03	0.01	0.04	0.20	
3	0.09	0.04	−0.01	0.03	0.14	0.06	−0.04	−0.08	0.19	0.22	
Average range	0.04– 0.12	−0.01– 0.11	−0.05– 0.07	−0.02– 0.08	0.10– 0.07	−0.03– 0.13	−0.05– 0.05	−0.03– 0.09	0.15– 0.17	0.22– 0.04	0.03– 0.09
Process average	0.01	−0.02	−0.08	−0.06	0.10	−0.01	−0.04	−0.04	0.15	0.13	Range 0.23

Table 26.2. K, Constants Tables for R & R.

		2	3	4	5	6	7	8	9	10
Number of trials,	K_1	4.56	3.05							
Number of operators,	K_2	3.65	2.70							
Number of parts,	K_3	3.65	2.70	2.30	2.08	1.93	1.82	1.74	1.67	1.62

test, one simply samples a batch 2–10 times. Each sample is tested two to three times by two to three color technicians.

To find true batch-to-batch color variation (PV), exclusive of test variation, each batch is sampled and composited, then tested twice by two to three technicians. If the above data represented those 10 samples drawn from 1 batch or lot, then TV within this lot is 0.50.

The other component in this model is useful for determining the variation between two different test methods. If you have a Brabender and a two-roll mill or two different injection molders, then method A can be one unit and method B the other.

A color technician tests 3–10 batch samples on each piece of equipment two to three times each. The AV now becomes appraisal reproducibility for each EV piece of plastics dispersion equipment. This technique is also useful for comparing supplier Certificate of analysis (COA) data against your own incoming quality control (QC) testing to determine reliability between the two sets of data. To accomplish this, you must request duplicate sets of color data for each lot over 3–10 consecutive lots.

Once test measurement error (R&R) is determined for a given sample or test plan, then process variation can be calculated using the previous subtractive method. Assuming the same test variation of 0.33 and total variation of 0.081 were known from the previous example, then

$$PV = \sqrt{(\sigma_{total} \times 6) - (R \& R)^2} = \sqrt{(0.081 \times 6)^2 - 0.33^2} = 0.36$$

This subtractive value of 0.36 compares well to the 0.37 PV number obtained from the gage R&R study. The 0.33 R&R value is marginally acceptable under the AIAG guidelines for a ±0.5 *DLab* color tolerance (1.0 total variation with 33% R&R). Plotting cumulative portions of AV, PV, and EV, a total variation of 0.50 reaches 100% of specification at 0.5 tolerance (Fig. 26.1). This point would approximate 1.0 Cpk, or three standard deviation units, in classical capability analysis studies.

26.4. OTHER SOURCES OF TEST VARIABILITY

With today's R&R, spectophotometers make up only a small portion of total measurement error, assuming they are diffuse spherical geometry and halogen flash-lamp source units. If you have a good spectrophotometer, then the majority of test error originates from sampling and test preparation.

If the spectrophotometer is suspect, one would substitute each unit as A and B in Table 26.1, holding all other variables constant. Ten test pieces could be read three times each by the two color measurement units. Once color data are entered into

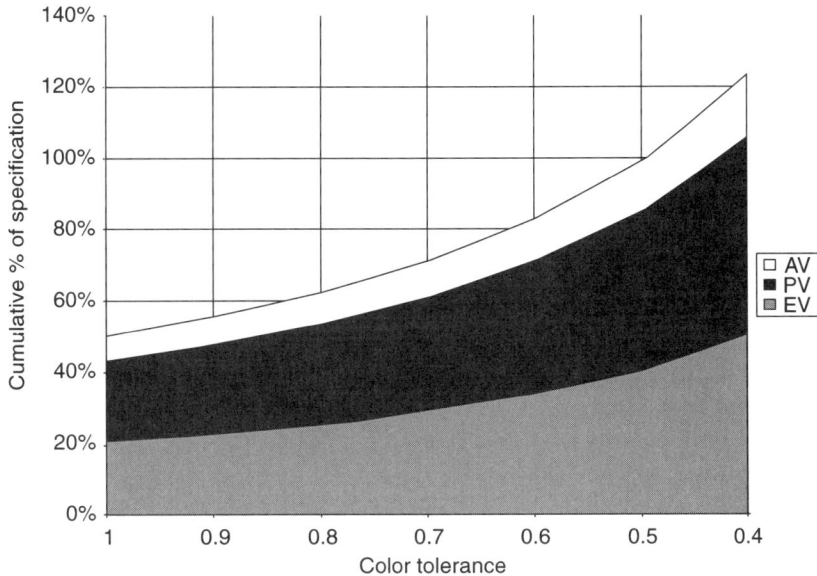

Figure 26.1. Percentage of specification.

the AIAG R&R model, then AV would report reproducibility between each unit, while EV would yield repeatability.

26.5. PROCESS COLOR CONTROL USING HUNTER, CIE. CMC *DE*

If you are knowledgeable about color space, then you know it is not uniform, it is nonlinear, and it represents a poor substitute for one of our human senses—sight. Humans associate emotions with color: green with envy, red hot anger, yellow bellied, and good guys wear white. So what good comes of statistical process control when it relates to subjective human perceptions. The first step might be to convert those unreliable, nonuniform color data measurements (and some of our subjective perceptions) into a form that will accommodate AIAG's recommendations for SPC.

Because color is three dimensional, the majority of pigment manufactures and their customers have settled on one of two color languages. When one describes batch color control, colorists generally use Hunter or CIE *Lab* values. Some variations might include CIE LCH: lightness, chroma, and hue.

More complex color systems have evolved through the years: the three-dimensional Friele, MacAclam Cheekering (FMCII) and now the newer single-dimension CMC *DE*. Conceived in response to color "nonlinear" characteristics, each system possesses its own unique characteristics. Of the systems implemented and used, CMC appears to be the most promising and is being embraced by the major automotive manufactures and their supporting plastics and coatings suppliers.

However, adopting a new color system for suppliers and customers may mean discarding years of archived data (and perceptions) from their old system (and thinking). Three-dimensional *Lab* values would be useless, unless there were a way to convert those old numbers into a newer CMC tolerance.

If your old batch color control system is in Hunter and you want to convert to CMC, the first step is to transform all old *DLab* values back into tristimulus *XYZ* values.

If both standard Hunter absolute *Lab* and lot *DLab* values—as well as the illuminant (A, C, D_{65}) and observer (2° or 10°)—are available, one can determine the standard tristimulus values from the Hunter *Lab* equations by solving for *Y*, *X*, and *Z*. According to E. I. Stearns ("Weights for Calculation of Tristimulus Values," in F. W. Billmeyer and M. Salzman, Eds., *Principles of Color Technology*, 2nd ed., p. 62 John Wiley & Sons New York 1981), equations for illuminant D_{65} and the 10° observer become:

$$Y_s = 100\left(\frac{L}{100}\right)^2 \quad (1)$$

$$X_s = 94.83 \frac{(a\sqrt{Y/100})}{(175\sqrt{0.0102 \times 94.83}) + (Y/100)} \quad (2)$$

$$Z_s = 107.38 \frac{((Y/100) - b\sqrt{Y/100})}{(70\sqrt{0.00847 \times 107.38})} \quad (3)$$

To calculate lot *XYZ* values, add the batch or lot *DLab* values from the historical data to the standard absolute *L*, *a*, *b* values in the equations

$$Y_L = 100\left(L + \frac{DL}{100}\right)^2 \quad (4)$$

$$X_L = 94.83 \frac{(a + Da)\sqrt{Y/100}}{(175\sqrt{0.0102 \times 94.83}) + (Y/100)} \quad (5)$$

$$Z_L = 107.38 \frac{(Y/100) - (b + Db)\sqrt{Y/100}}{(70\sqrt{0.00847 \times 107.38})} \quad (6)$$

Subsequent to determining standard and lot tristimulus *XYZ* values, one simply plugs those calculated values into the CIE *Lab* color equations. Why the CIE system? Because CMC uses the *DLab* values, generated in CIE color space, to determine *DC*, *DH*, and eventually CMC *DE*.

To calculate absolute standard CIE $L^*a^*b^*$ values for X/X_n, Y/Y_n, or Z/Z_n greater than 0.01,

$$L^* = 116 \sqrt[3]{\left(\frac{Y}{Y_n}\right)} - 16$$

$$a^* = 500\left(\sqrt[3]{\frac{X}{X_n}} - \sqrt[3]{\frac{Y}{Y_n}}\right)$$

$$b^* = 200\left(\sqrt[3]{\frac{Y}{Y_n}} - \sqrt[3]{\frac{Z}{Z_n}}\right)$$

$$C^* = \sqrt{a^{*2} + b^{*2}}$$

Perform the same L^*, a^*, b^*, C^* calculations for the batch, then CIE $DL^*a^*b^*$, and DC^*:

$$DE^*_{ab} = \sqrt{DL^{*2} + Da^{*2} + Db^{*2}}$$

$$DH^*_{ab} = \sqrt{DE^{*2}_{ab} - DL^{*2} - DC^{*2}_{ab}}$$

Further, to calculate CMC DE,

$$DE_{\text{CMC}} = \sqrt{\left(\frac{DL}{2*_{sl}}\right)^2 + \left(\frac{DC}{sc}\right)^2 + \left(\frac{DH}{sh}\right)^2}$$

If $L \leq 16.0$, $sl = 0.0511$; otherwise

$$sl = \frac{0.040975 L}{1 + 0.01765 L}$$

$$sc = \frac{0.0638 C}{1 + 0.0131 C} + 0.0638 \quad \text{where } C = \sqrt{a^2 + b^2}$$

$$sh = (FT + 1 - F)sc \quad \text{where } F = \sqrt{\frac{C^4}{C^4 + 1900}}$$

Where sl, sc and sh are systematic corrective factors for light-dark, sl; chroma, sc; and hue, sh. Note also that $(2* sl)$ used in the DE_{CMC} equation is referred to as CMC(2:1), used for the estimation of acceptability for color difference evaluations.

If $H \geq 164°$ and $H \leq 345°$,

$$T = 0.56 + |0.2 \cos(168 + H)|$$

Otherwise

$$T = 0.36 + |0.4 \cos(35 + H)|$$

For formulas entered into a spreadsheet, if $A \tan(a, b) > 0$,

$$H = A \tan 2(a, b) \frac{180}{\pi}$$

Otherwise

$$H = A \tan 2(a, b) \frac{180}{\pi} + 360$$

Now you have the CMC DE and equally important DC and DH values, but what do I do with them? First, you must realize that not all color space was created equal. Here, CMC only provides a more level playing field. It takes the Rocky Mountain peaks found in Hunter and CIE color space and turns them into gently rolling plains. However, what if your present work environment still mandates single sample/testing?

Several years ago, when I approached sales and marketing and production management with my solution to the single sample–single test problem, they simply stared back in disbelief: "What do you mean you need 200% additional QC staffing because the sample/test variation contributes 30% of our lot-to-lot variation?" "Can't you just fix the test?"

A simple answer remains—NO! While minimizing test variability through cross-training and standard reliable methods is the same as any process, physical testing will continue to exhibit its own unique random variability. One can double, triple, or even quadruple the number of samples or tests performed per sample to improve precision. Then, minimizing sample/test variation yields

$$n_L = (3\sigma'_o/e)^2$$

Where

n_L = Number of lot samples required to estimate lot test error.
σ'_o = Estimated (process, lot, tesing) standard deviation.
e = Maximum allowable error between estimate made from the sample and result of measuring all the component lots.

Solving for e, error in lot testing (as a percentage, multiply by 100):

$$e = 100(3\sigma'_o/\sqrt{n_L})$$

For n_L = 2 samples, $e = 100 \times (3 \times 0.32/1.41) \simeq 21\%$
For n_L = 3 samples, $e = 100 \times (3 \times 0.32/1.71) \simeq 17\%$
For n_L = 4 samples, $e = 100 \times (3 \times 0.32/2.00) \simeq 15\%$

So, by quadrupling sampling and/or testing (and the number of warm bodies required to sample and test), you have in effect only cut the test variation in half. No wonder management just stared back in disbelief. Furthermore, what is at risk if a manufacturer stays at two or three samples per test per lot? Risk is divided into two categories:

1. the producer's risk (type 1), where the manufacturer may reject a lot based on one, two, or three tests when in fact the lot passed; and
2. the consumer's risk (type 11), where the manufacturer approved a lot when in fact it should have been rejected.

Both are important to understand, depending on whether you are approving incoming materials based on a COA or shipping a lot to your customer against their specifications. You now understand how to derive sample and test variability as well as how to better manage and reduce its contribution to perceived process variance, given the final cumulative variance. How does one go about reducing its effect on process control?

Returning to definitions, *process control* is the gathering of sample data from a process stream in order to establish a feedback loop and help prevent manufacturing nonconforming product. The SPC is simply a systematic measuring of variance in the manufacturing system.

Control charts display the data (and variance) in the order they occur with "statistically" determined upper and lower control limits. The SPC charts are used to monitor the process for maintenance of the "target to zero" and then to determine whether process changes have accomplished their "desired quality effect."

Much classical SPC charting is standardized around Walter Shewhart's [1] work from the 1930s, which combined the principles of quality control with statistics and probability. Shewhart's *plan–do–check–act* cycle became the W. Edwards Deming *plan–do–study–act* quality improvement cycle. Shewhart referred Deming to Homer Sarasohn to teach statistical methods to postwar Japan. Deming went on to develop his 14 obligations of management. Three keystones to these 14 points comprised of consistency of purpose, continual improvement, and profound knowledge.

W. Edwards Deming's profound knowledge involved four inseparable components: appreciation for a system, theory of variation, theory of knowledge, and psychology. Unlikely as it may seem, Deming proposed that a psychological urge to learn and create was an integral part of any continuous improvement methodology. Further, he made two specific points about variation: (1) that errors and inconsistencies are always present in any process and (2) that people take "wrong" corrective steps when dealing with these variations. Finally, Deming considered knowledge a "prediction that comes true"; that knowledge comes from theory, and without theory there can be no learning.

Remember that the goal of SPC is to make sense of the data and variation collected from direct observation of product quality or a process in a way that helps us to expose hidden "properties." Only then can we make necessary corrections and predict the outcome of those corrections. This whole process then brings us to an appreciation of a system, provides profound knowledge about the system, and fulfills our urge to learn and create.

Historically the tools used to learn and create centered primarily around $\overline{X}$ and R charts and, more recently for the process industries, on Cumsum, sums of the deviation from an actual process average or target value for that process and the exponentially weighted moving average (EWMA).

26.6. EXPONENTIALLY WEIGHTED MOVING AVERAGE

The EWMA control charting is a new alternative to conventional CUMSUM and individual, moving range statistical process control. However, unlike Shewhart control charting, an EWMA generates the next predicted process value and acts in an immediate adviser capacity for detecting lack of control in the process on that predicted value. Additionally, EWMA charting is best used where a process is assumed to be centered (around zero, as is the case with color control) and exhibits small step changes in processing over an extended period of time.

Starting with a hypothetical examination for a set of process DH color values, EWMA and control charting will be contrasted against conventional individual and modified moving range average charting techniques.

Table 26.3. CIE DH for Individual Batch Samples

0.36	−0.11	0.05	−0.10	0.04
0.19	−0.24	0.16	−0.26	0.02
0.34	0.01	0.01	0.06	−0.10
0.24	−0.10	0.07	0.08	0.04
−0.18	0.06	0.14	0.07	0.04
0.14	0.08	0.09	0.04	0.01
−0.27	−0.20	0.01	0.04	0.06
0.03	−0.26	−0.11	−0.13	0.09
0.09	0.06	−0.52	0.03	0.06
0.01	0.30	0.01	−0.19	0.04

26.6.1. EWMA Calculations and Charting

Using CIE LCH and given a series of observed DH color values for batch $1 = X_1$ to batch $t = X_t$:

$$X_1, X_2, X_3, \ldots, X_t$$

The EWMA, forecast (F_{t+1}) for time $t + 1$, is defined as

$$F_{t+1} = \alpha X_t + (1-\alpha) F_t$$

where α is the weighting factor applied at time t ($0 < \alpha < 1$) and F_1 would ordinarily be set at the historical process average from a sequence of the previous 30 DH values, $\{X_{t-30}, X_{t-29}, X_{t-28}, \ldots, X_{t-1}\}$:

$$F_t = \overline{X}$$

If a limited number of batches (say $t = 12$) are processed within a run, F_1 could also set the average DH value from the previous run $\{X_{t-12}, X_{t-11}, X_{t-10}, \ldots, X_{t-1}\}$.

26.6.2. Example Charting of Individual DH Batch Color Values

The data in Table 26.3 are CIE LCH DH values as measured against a reference standard for several consecutive production runs. The time order for these values is given by reading down each column, beginning with the column on the left. The DH average range $\overline{R}$ from previous runs of the same product is 0.158 units; when this value is divided by $d_2 = 1.128$, we get an estimated standard deviation of 0.14 DH units. The midpoint of the DH specification for this product is 0.0. If 0.0 is taken as the central line for a control chart for individual values, then this chart will have the following upper and lower control limits (LCL, UCL):

$$0.0 \pm 3(0.14) \quad \text{or} \quad \text{LCL} = -0.42 \quad \text{UCL} = +0.42$$

The individual control chart for this production color data is shown in Figure 26.2. To construct an EWMA for these data, let F_1 be the average of the last 12 DH values from the previous product run:

Table 26.4. EWMA Process CIE DH

0.28	−0.03	0.07	−0.11	−0.03
0.25	−0.11	0.11	−0.17	−0.01
0.28	−0.06	0.07	−0.08	−0.05
0.27	−0.08	0.07	−0.02	−0.01
0.09	−0.02	0.10	0.02	0.01
0.11	0.02	0.09	0.03	0.01
−0.04	−0.07	0.06	0.03	0.03
−0.01	−0.15	−0.01	−0.03	0.05
0.03	−0.06	−0.21	−0.01	0.06
0.02	0.08	−0.12	−0.08	0.05

$$F_1 = 0.23$$

Using $\alpha = 0.4$, and the first individual DH value from Table 26.3,

$$F_{t+1} = 0.4(0.36) + (1 - 0.4)0.23 = 0.28$$

The balance of EWMA F_{t+1} values in Table 26.4 are calculated in the same order as the original values in Table 26.3. Upper and lower control limits around the EWMA $\{F_{t-11}, F_{t-10}, F_{t-9}, \ldots, F_t\}$ at F_{t+1} are calculated as

$$\text{LCL}_{t+1}, \text{UCL}_{t+1} = F_{t+1} \pm \left[3\left(\frac{\overline{R}_{[i]}}{d_2} \sqrt{\frac{\alpha}{2-\alpha}} \right) \right]$$

where $[i]$ can represent the average number of batches per run and $d_2 = 1.128$. Note also that when an EWMA is used on a control chart, the limits are *constricted* relative to the mean by a factor of $\sqrt{\alpha/2-\alpha}$. Deviating from "classical statistics," suppose one wanted to compare the forecasted average (EWMA) against the process average, $\overline{X} = \{X_{t-11}, X_{t-10}, X_{t-9}, \ldots, X_t\}$, but apply the limits derived from the EWMA $\overline{R}$ to X_t. Remember, F_{t+1} is a dynamic predictor of the next forecasted value based on immediate, as well as historical, process information, while $\overline{X}$ is a static representation of past history. Upper and lower control limits for X_t could then be interpreted as

$$\text{LCL}_t, \text{UCL}_t = X_t \pm \left[3\left(\frac{\overline{R}_{[i]}}{d_2} \sqrt{\frac{\alpha}{2-\alpha}} \right) \right]$$

Because these are forecasted limits, the same exponential smoothing can be applied:

$$F_{t+1}(\text{LCL}, \text{UCL}) = \alpha(\text{LCL}, \text{UCL})_t + (1-\alpha)F_t$$

Plotting the individual DH, $\overline{X}$, EWMA, and upper and lower control limits against X_{t+1} yields Figure 26.3. This figure varies considerably from a conventional EWMA

396 Use of Statistics in the Coloring of Plastics

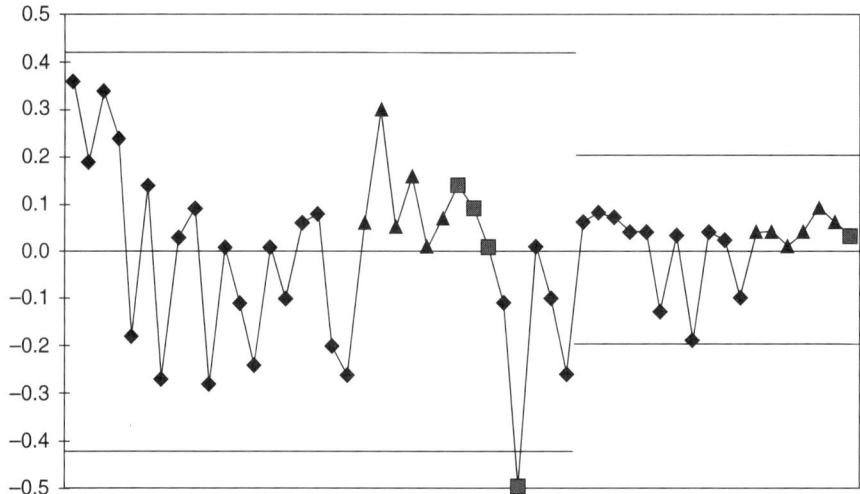

Figure 26.2. $\overline{X}$ chart for production batch CIE DH (historical LCL/UCL = ±0.42; improved LCL/UCL = ±0.20).

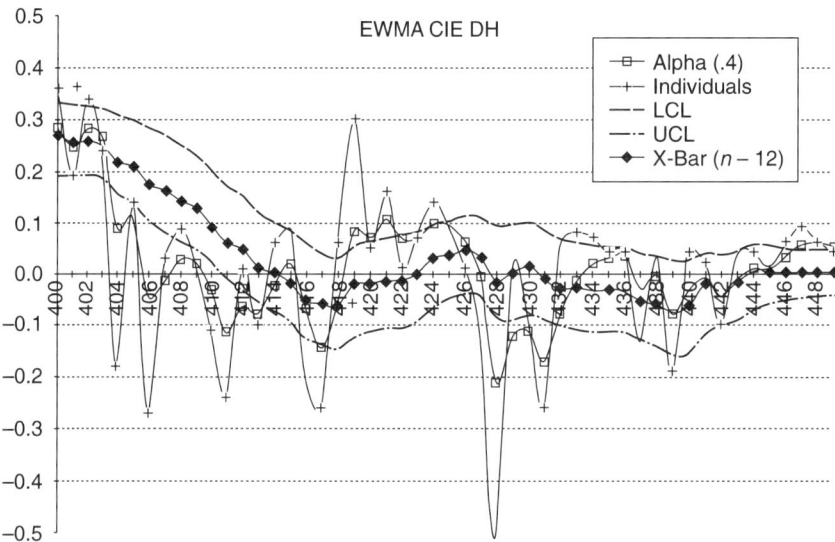

Figure 26.3. EWMA chart for production batch CIE DH.

chart. First note that the EWMA individual, moving range average is dynamic and "moves" with the process, as do the upper and lower control limits. Traditional MR_2 and EWMA charts depict a static center line and control lines calculated at some prior moment for $X_{t-?}$. Comparing Figures 26.2 and 26.3, historical means and limits from the Shewhart chart may or may not be relevant to present feedstocks and manufacturing conditions clearly depicted in the exponentially weighted individuals, moving range (EWIMA) chart.

This modified EWIMA chart at time t for the last $[i]$ number of batches reveals both the historical, moving average, $\{X_{t-11}, X_{t-10}, X_{t-9}, \ldots, X_t\}$ as well as a "weighted pulse" on the immediate average deviation from the target. Figure 26.3 clearly demonstrates the changing batch-to-batch variance around the process mean.

Note that the first four EWMA values range from 0.25 to 0.28. A process change was made at X_5 in an attempt to recenter DH around zero. Here, $\overline{X}_5 \cong 0.22$ represents the last 12 historical batches available for further manufacturing or to the customer for shipment. Also, $F_5 = 0.09$ predicts an immediate process shift and $\overline{X}_8 \cong 0.17$ while $F_8 = -0.01$ indicates the process is fully centered within four batches of the process change. At the start of the next run, X_{13}, the process begins with the same settings as the previous run.

Six batches into the run, X_{18} falls below the EWIMA lower control limit: $\overline{X}_{18} \cong -0.06$ and steadily dropping. Shewhart control methodology would not indicate a process change was needed. Using the EWIMA, a decision can be made to leave the process alone or make a minor adjustment. In this case such an adjustment was made. By the end of the run, $\overline{X}_{24} \cong 0.0$. At the start of the third run, at X_{27} and $\overline{X}_{27} \geq 0.0$, the process was again adjusted downward, this time in response to greater than seven individual DH values from the MR_2 chart above the mean.

The immediate response to the process adjustment is an out-of-control point well below zero. The individual moving range average, however, shows the DH recentered around zero by $\overline{X}_{29}$. No process change was made until X_{32} and $\overline{X}_{32} \leq 0.0$. Manufacturing started at X_{37}, where it left off from the previous run and continue through X_{48} with no process changes. By the end of the fourth run, $\overline{X}_{48} \cong 0.0$. This example shows the power of EWIMA as a tool to center a batch process around zero. Classical statistical charting methodology would consider this "overcontrolling" the process. However, given the noncontinuous nature of a batch process, short run lengths, and color manufacturing's extreme sensitivity to minor adjustments, EWIMA might be the preferred tool for "target-to-zero" and "zero-defect" manufacturing philosophies.

In conclusion, the EWIMA is best used as a directional tool to center a process around zero. However, EWIMA cannot foresee the unforeseen. When a process is operating at an acceptable level and in a reasonably stable mode, and one wants to sustain the same "physical state," the EWIMA process control illustrated in Figure 26.3 can be useful.

26.7. STATISTICAL PROBLEM SOLVING

If a process is not operating at permissible control levels or appears unstable, and when we need to improve the process output, then our first step should be to pay closer attention to the process inputs. We should begin by correlating each of the input variables relationship to each of the key output variables through statistical problem-solving techniques to answer the question: Is the process predictable, and therefore subject to being modeled satisfactorily, or is it unpredictable?

26.8. THE SPS PROCESS

Opportunities to model a process based on defined inputs and then predict an output generally begins with statistical problem solving (SPS) techniques detailed in Table 26.5. This simplified flow has shown up in numerous forms and publications. To elaborate further on SPS, here is a hypothetical solution path for one plastic's customer quality problem:

1. *Opportunity.* A three-color concentrate–two-component plastic: colorants A, B, and C and components D and E are mixed, then injection molded to form a dimensional product. The customer is pleased with dimensional characteristics but has noted some increased color variance to a more positive *DH* value in recent lots. While none of the lots has exceed the 0.4 maximum specification, all of the most recent shipment was at or near that tolerance.

    A team comprised of the sales, purchasing, quality control, process engineer, supervisor, and lead operator was assembled to review this color quality issue. Sales determined that this product line contributed greater than 20% of sales and 25% of overall gross profit. While the product was within specification, the team decided the customer's concern over an apparent shift in color warranted further investigation.

2. *Know the Process.* The management committee reviewed the quality improvement team's recommendations and agreed that formal investigation was needed. The team reassembled. Quality control provided SPC charts on CIE *DLCH* and CMC *DE* values for the last 10 shipments. The process engineer brought the process flow chart and sales provided the timing on lots shipped and any customer feedback on color quality. The supervisor and operator supplied production records, down time, maintenance, and run time for the last year of manufacturing, including any changes to equipment or the process. Reviewing all data and facts, the team found there was a mold change in midyear and mixer repair several months ago. All three color concentrates and plastic components were ordered from the same supplier. Quality control noted that the color of one of the three components varied more that the other two. The operator suggested that he noticed a slight change in flow characteristic for one of the plastics. Sales informed the group that the customer had recently received ISO-9002 certification and had also recently purchased a new spectrophotometer. Based on the combined knowledge of the customer's expectations, product quality and testing, historical process and component data, the process, and the manufacturing environment, the team developed a fishbone diagram (Fig. 26.4).

3. *Elucidate on Cause.* The investigative group went on to expand on each of the four groups of process inputs:

    Raw components:
    - Colorant concentration—adjustments made by QC will shift *DH*.
    - Colorant uniformity—supplier colorants shift in *DH* from lot to lot.
    - Plastic particle size—affect of flow characteristics on *DH* unknown.

    Equipment:
    - Scales—variations in weight would have a significant impact on *DH*.
    - Mixer—effects of mixer condition on *DH* unknown.
    - Injection molder—effects of change in molder condition on *DH* unknown.
    - Standard reliable methods—incomplete procedures could allow for changes in manufacturing "techniques."

Table 26.5. The SPS Process

SPS Steps	Detailed Implementation
Define the opportunity	To arrive at the level of control exhibited in Figure 17.3 requires fundamental questioning of the entire system, including, but not limited to, each raw material component and process step needed to make a final product. This process takes teamwork. The team is generally comprised of a group of experts who understand the process, the intermediate components that make up the finished product, and the testing, and often missed the customer's expectation's for quality, product market share, and profit contribution and margins.
Take time to Get to know the process	Once management has agreed to improve the process (or product quality) and all the key components are clearly defined and understood by the team members, the group can proceed by first clarifying the problem. Generally an agreed-upon process flow diagram is developed that defines all input and output variables. Then identify all critical measures, including a measure for success.
Elucidate on root causes	Brainstorming, the team generates a list of potential root causes, concentrating on the systems and processes, and not the people. These root causes may affect the process output mean or variation.
Collect and analyze process and quality data	As each component's chemical physical characteristics or process inputs are identified, simple scattergrams (XY plots) are developed using a spreadsheet. The best place to start gathering data is found in QC and production records. One key element often overlooked in many control plans is lot traceability. To succeed, production records must provide the investigation team with sufficient information to trace a finished batch back through the specific equipment, process setting under which it was produced, and each of the component lot numbers. Further, samples representing each of the series of components should be retained for additional analysis prior to, during, and after the team has completed its analysis.
Develop a corrective action plan	Identify all possible solutions. Using brainstorming, correlation analysis, polling, and consensus techniques, prioritize the list of solutions. Perform a barrier analysis, defining driving forces that will help the improvement process to succeed as well as restraining forces that may hold the team back.
Test the plan Optimize the process	Start by testing the top few suspected variables through observational analysis. Develop and update a regression model as each new variable is tested. Check the production model to verify any improvement in the process and quality that may have occurred. Now optimize the process using EWIMA, linear programming, iterative solving, and process simulation techniques.
Document	Incorporate all critical process control parameters into a process control plan. Develop a long-term measurement matrix to monitor success and improved capabilities to meet or exceed customer expectations. *Then reward the team!*

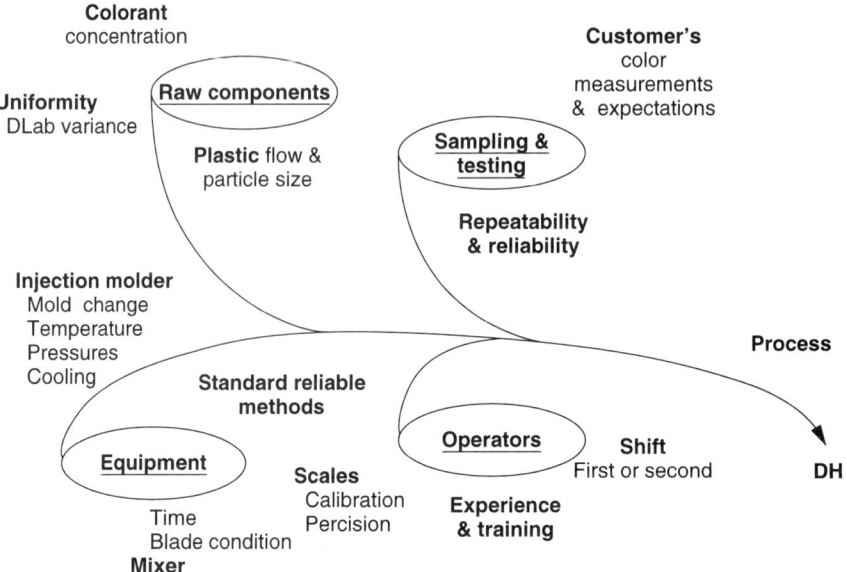

Figure 26.4. Example Fishbone Diagram.

Operators:
- Training and experience—changes in "technique" could occur between the more experienced senior and junior operators.
- Shift—most experienced operators are on first shift.

Sampling and testing:
- R&R—the team ruled out sampling and test precision and accuracy for the time being because the color had shown a consistent shift as opposed to an increase in variability.
- Customer's color measurements—the team did want to know if the customer could look at some of the most recent shipment on their old spectrophotometer and compare those values with the new unit.

Initially the team agreed to focus their attentions on "known" root causes, starting with the customer's R&R between spectrophotometers.

4. *Collect and Analyze Data.* After sales reluctantly approached the customer concerning the spectrophotometer R&R, results returned several weeks later revealed there was no significant difference in absolute or delta color values between the two units.

   The team reassembled and concluded their data investigation should begin with collecting color concentrate DH variability, formulation effects on DH, and scale R&R. Purchasing agreed to contact the color concentrate supplier for SPC charts on DH for the three components, starting with the component that exhibited increased variation. A purchase order was also cut for the scale service company to check the precision and accuracy of the scales used in manufacturing. Qualify control agreed to develop color concentrate loading versus averaged finished component Da scattergrams

(QC practice was to make all color adjustments using CIE $DLab$—it was felt that a shift in Da would best signal a shift in DH).

Before purchasing could obtain SPC charts from their concentrate supplier, QC called another meeting to discuss the findings illustrated in Figure 26.5. Reviewing the charts, two data points exhibited the lowest plastic component Da values and the highest Da were produced from concentrate A lots having the lowest and highest Da. Quality control confirmed that concentrate A was the colorant that exhibited the greatest hue variance. Purchasing had also received word that the concentrate supplier had in turn contacted its pigment manufacturer for color SPC charts for pigment used in preparing concentrate A. The scale service company had also inspected the scales and certified them to their original precision and accuracy tolerances. Additionally, the plant supervisor reviewed all procedures and the senior operator remained on the first part of the second shift to monitor the junior operator's work. Both reported that all standard operating procedures were complete, accurate, and up to date each shift. The group agreed to wait for SPC charts from the pigment manufacturer.

Once the Color SPC charts arrived for concentrate A, the team got back together to review that data against the concentrate manufacturer's charts and the original scattergram data. Not to anyone's surprise, while each of the lots sent to the concentrate company as well as color concentrate lots fell within their color tolerance, high- and low-Da pigment lots related to higher and lower Da concentrate lots and in turn related to even higher and lower component Da values. Much to the team's surprise, however, was how ineffectual QC color adjustments were to the concentrate A Da variance. In other words, even with extensive color matching, there was a point at which cross-blending color components to arrive at a final "match" still relied on the original pigment lot-to-lot color variance to control final component color variability.

5. *Corrective Action Plant*. From these findings, the group devised a two-step process for reducing final component variability:
   - Use incoming certificate-of-analysis data in possible cross-blend concentrate A lots within a mix to achieve final component $Da = 0$.
   - Specify tighter color tolerances on concentrate A (which would in turn demand tighter color tolerances on the pigment manufacturer or force the concentrate manufacture to cross-blend lots).

6. *Test the Plan*. Step 1 of the corrective action plan appeared to be the most difficult to achieve. Not all concentrate lots exhibited positive and negative Da values. The team quickly abandoned this approach, favoring the reduced concentrate A Da solution. The concentrate manufacture quickly contacted the pigment supplier requesting negative Da lots of color to correct the most recent positive Da material.

   Once the first lot of $Da < \pm 0.2$ material arrived, the team formulated this material into the plastic component mix. Again, not surprising, the team saw an immediate 0.15 drop in Da, while the customer saw an immediate shift in DH in the final plastic component.

7. *Optimize the Process*. Because concentrate A was used in many other formulations, the group decided to monitor the effects of reduced color vari-

ability in their other product lines. From this improvement, QC noted a 10–20% reduction in color variation for other concentrate A uses after a three-month study. This translated to a 10–20% improvement in color control and 50% reduction in scrap and rework for the same products, at no additional cost to the company.

8. *Document.* Documentation was simple: Agree to new color tolerances for concentrate A. The team took the savings from the first three months of reduced scrap and rework and purchased a plaque to record their achievements, displayed it proudly in the front entrance, then took themselves out to a dinner with the family to celebrate their success.

26.9. CONCLUSION

This SPS team was empowered to improve their system. Upper management simply said, "OK, see if you can fix it!" The team's solution used a simple path of least resistance. For a more complex solution set, they may have selected multivariate regression modeling or hired an outside statistical consultant. The main point is that they used a scientifically proven, systematic approach to solving a complex problem, through teamwork and simple observational and analytical techniques.

Their concentrate supplier's demand on the pigment manufacturer caused the pigment supplier to review sampling, testing, and process techniques. The pigment manufacturer in turn discovered that one raw material component was causing the hue shift. That raw material producer in turn reviewed its manufacturing capability and developed a corrective action plan.

The key to success is first understanding the customer's true expectations (not the specifications). Then each supplier must transform those process outputs (expectations) into its system inputs using simple (and sometimes more complex) statistical problem-solving techniques and methodology. The goal is always to impress the customer and make the stockholders happy—and of course to satisfy our psychological urge to learn and create. Statistics and SPC can only signal when something is "different." Combined with a little human ingenuity, SPS is just one of the many tools needed to help us arrive at the what, where, when, why, and how of a problem and eventually reach the goal.

Reference

1. W. A., Shewhart, Economic Control of Quality of Manufactured Product, Van Nostrand, New York, 1931.

Index

A
Absolute unit calculations, many-flux color matching, 54–55
Absorbance, Lambert/Beer laws, color matching theories, 52
Absorption, light objects, 10
Accelerated testing, color migration, 331
Acid dyes:
  development of, 177
  fluorescent dyes as, 250
Acid pasting, copper phthalocyanine blue, 115
Acid swelling, copper phthalocyanine blue, 115
Acrylonitrile-butadiene-styrene (ABS):
  carbon black pigment:
    dispersion, 169
    impact strength, 171
  color durability testing, heat stability, 327
  complex inorganic pigments (CIPS), 132, 141
  dry color concentrates, 278
  soluble dye manufacturing, 178
Active pixel/active column video technology, 31–32
Adams-Nickerson formula, color difference metrics, 36
Additives:
  carbon black pigment dispersion, 165
  color compounding, 270–271
  concentrate processing, 302–305
    dry concentrates, 303
    liquid concentrates, 305
    pellet products, 303–304
  fluorescent plateout reduction, 252–255
  non-formaldehyde-based fluorescent pigments, 253–255
  pearlescent pigments, 230–231
  plastics coloring:
    antioxidants, 346–347
    antistatics, 347
    coupling agents, 347
    flame retardants, 347–348

foaming agents, 348
heat stabilizers, 348
impact modifiers, 348–349
lubricants/mold releases, 349
polymer blends, 346
reinforcing agents, 349–350
research background, 340–344
translucent/opaque resins, 344–346
transparent resins, 344
UV stabilizers, 350
Additivity principle:
  color matching, 49–50, 62
  many-flux vs. Kubelka-Munk color matching, 55
Air pollution:
  concentrate production:
    air emissions, 311–312
    air quality standards, 315–316
    contaminant regulations and standards, 314–316
    workplace exposures and controls, 308–311
  EPA regulations, 362–367
Alizarin, soluble dyes from, 177
Alizarine, maroons, properties of, 93
Alston method, tolerance region derivation, 35
Aluminum flake:
  dry color concentrates, 286
  metallic pigment production, 204–206, 227–228
  pigment shapes, 209–211, 231–232
  visual color matching, 75
American Association of Textile Chemists and Colorists (AATCC), Test Method 173, 80
American Conference of Government Inudstrial Hygienists (ACGIH):
  defined, 377
  exposure limits of, 362
American National Standards Institute (ANSI), ISO 9000 and, 81
Analog simulation:
  spectrocolorimetry specifications, 31

visual colorimetry, 25–27
  schematic, 26
Analog-to-digital conversion, spectrocolorimetry specifications, 31
Anatase pigments, photochemistry and durability, 155–156
Anisotropic crystals, additive effects, 342–343
Anthracene, 177
Anthraquinone dyes:
  chemical properties, 181
  red dyes, structure and properties, 110
Antimony:
  lead chromate/molybdenate pigments, 135–136
  thermochromic materials, 198–200
Antioxidants, additive coloring effects, 346–347
Antistatics, additive coloring effects, 347
Appearance criteria, colorant selection process, 258
Application-specific integrated circuit (ASIC), spectrocolorimetry specifications, 31, 46
Application technology, colorant systems, 260–261
  color introduction, 263
  color matching checklist, 266
  color target, 262–263
  finishing systems, 264
  formula compounding, 266–267
  process conditions, 261
  process selection, 261
Appraiser variability, plastics coloring statistics, 385–387
Arsenic compounds, thermochromic materials, 198–200
Artificial intelligence, tolerance region derivation, 35
ASTM standards:
  D 1279, product color tolerances, 35
  D 1729-89, 7
  D 1979-89, 7
  D 3134:
    product color tolerances, 35
    tolerance region derivation, 35
  ISO equivalents, 80
  rheological testing, melt flow, 334–335
  strength properties testing, 332–334
Atomization, metallic pigment production, 204–206, 227–228

Automotive Industry Action Group (AIAG), plastics coloring statistics, 379
Auxochromes, defined, 101
Azamethine dyes, thermochromic properties, 199–200
Aziridines, photochromic behavior, 186–196
Azo compounds:
  chemical properties, 180–181
  metallized azo pigments:
    reds, 102–106, 123–126
      BON reds, 105
      lithol reds, 103–104
      lithol rubine red, 104–105
      permanent red 2B, 104
      pigment scarlet, 104
      red lake, 105–106
    yellows, organic structure and properties, 117
  nonmetallized azo reds, 106–107
  oranges, organic structure and properties, 121–123, 129
  photochromic behavior, 195–196, 209
  regulations concerning, 374

B
Banbury thermal processors, pellet concentrate processing, 304
Band gap:
  scattering properties of colorants, 11–12
  spectrocolorimetry, performance verification, 33–34
BASF company, soluble dye manufacturing, 177
Basic dyes, fluorescent dyes as, 250
Basic lead carbonate, pearlescent pigments, 236–237
Batch mixing, dry color concentrates, 281
Batch-to-batch variability, plastics coloring statistics, 383–384
Bayer Corporation, soluble dye manufacturing, 177
Beer's law:
  absorption, 10
  color matching, 52
  visual color matching, 74–75, 75
Benzene ring, rhodamine A structure, 247
Benzimidazolone compounds:
  reds, structure and properties, 110–111, 125

yellows/oranges/reds:
   organic pigments, 118, 119, 127
   properties of, 91
Benzo [c]-1, 8-naphthpyridine hydrochloride, photochromic materials, 185–196
Berns method, tolerance region derivation, 35
Best management practices (BMPs), 316
Biased errors, plastics coloring statistics, 380
Biconical instrumentation, 32
Binder selection, colorant application technology, 258
Biochemical applications, fluorescent dyes, 250
Bismuth oxychloride, pearlescent pigments, 237
Bismuth vanadate, inorganic pigment structure and properties, 136–137
Blackbodies:
   properties of, 7–8
   spectral power distribution, 8
Black light:
   fluorescent dyes and brighteners, 180
   fluorescent pigment lightfastness, 256–257
Blended polymers, additive coloring effects, 346
Blooming, color migration, 331
Blowing agents, color additives, 348
Blow molding, metallic pigment processing, 218
Blowpin plateout testing, fluorescent pigments, 255
Blue pigments:
   color match case study, 74
   organic blues, 111–115
      copperphthalocyanine blue, 111–115, 126
BON reds, organic structure and properties, 106
Brain function, psychological response of human observers, 13–15
Brighteners, fluorescent dyes and brighteners, 180
Brightness:
   color space, 17–18
   defined, 16
British Standards (BS):
   BS No. 6923, 80
   6923 (color difference metrics), 36–44
Brominated pyranthrone red, structure and properties, 110

Brunauer, Emmett and Teller Test method (BET), titanium dioxide pigment incorporation, 151–152, 161
Buss Kneader, metallic pigment masterbatching, 215
Byk-Gardner model, visual color matching, 62

C
Cabot Corporation testing, carbon black pigments, UV stability, 166–167, 178–179
Cadmium pigments:
   concentrate production, contaminant regulations and standards, 315
   inorganic structure and properties, 133–134
   regulations concerning, 374
   yellows/oranges/reds/maroons, 95–96
Calendering process, metallic pigment processing, 217–218
Calibration requirements, spectrocolorimeters, 31–33
Candela, defined, 32
Carbazole dioxazine violets:
   properties of, 93
   regulations concerning, 374
Carbazole violet, structure and properties, 115–116
Carbon black pigments:
   compound moisture adsorption (CMA), 173
   conductivity properties, 169–170, 180
   dispersion effects, 163–169, 177
   dry color concentrate carrier systems, 279
   impact strength, 171
   melt viscosity (processability), 173
   outdoor weathering, 168–169
   particle size predictor, 163
   performance evaluation, 163, 166, 173–174, 182
   physical properties, 170–171, 181
   plastics applications:
      dispersion effects, 165–166, 178
      furnace process schematic, 161, 176
      morphology, 160–162, 177
      research background, 159–160
   properties of, 94
   QUV-B Weatherometer testing, 167–169
   refractive index, additive effects, 342–344, 365

regulations concerning, 374
resin effects, 172–173, 181
selection criteria, 166–173
tensile strength, 171
UV stability, 166–169, 178–179
Carrier systems, dry color concentrates, 278–280
Categorical wastewater discharge (CWD), 312
"CC120" method, liquid color concentrate distribution, 290
Central blending, colorant application technology, 263
Central limit theorem, 31
CERAM Ceramic Colour Standards II, spectrocolorimetry performance verification, 34
Certificates of analysis (COAs):
    color compounding, 275
    plastics coloring statistics, 387
"Chalking" phenomenon, titanium dioxide pigments, 154–157, 162–164
Charpy impact strength, ISO test procedures, 80
Chemical foaming agent (CFA), additive coloring effects, 348
Chemical resistance:
    colorants, 87
    color durability testing, 332
Chemicals on Report Rules (CORR), 365
Chloride process, titanium dioxide pigment production, 147–148, 158
Chlorofluorocarbons, 363
Cholesteric liquid crystals, thermochromic materials, 198–200
Chroma:
    colorant properties, 86
    color difference metrics, 37–44
Chromaticity diagram (CIE), 18
Chromenes, photochromic behavior, 191–196, 207
Chrome oranges, properties of, 95
Chromium III pigments:
    inorganic structure and properties, 138
    regulations on, 375
Chromium oxide greens:
    classification, 95
    inorganic structure and properties, 139
Chromogen, rhodamine A structure, 247–250

Chromophore, defined, 101
CIEDE2000 tolerance equation, 44
CIELAB2000, color difference metrics, 44
CIELAB color metrics:
    C^*-L^* plane, 49–50
    L^*-C plane, 38–44
    plastics coloring, additive effects, 344
    properties of, 36
CIELAB coordinates, color difference metrics, 37–44
CIELUV color metrics, properties of, 36
CIE94 tolerance equation, color difference metrics, 43–44
Cis-trans isomerization, photochromic behavior, 195–196
Clarion red, organic structure and properties, 123
Clean Air Act Reauthorization (1990), 362–363
Clean Water Act, 363
CMC Delta E tolerance, 388–392
CMC 2:1 Lightness to Color ratio, 80
Coal tars, soluble dyes from, 176–177
Coating procedures, metallic pigment processing, 219
Cobalt pigments:
    blue complex inorganic pigments (CIPs), 132–133
    regulations for, 375
    thermochromic materials, 197–200
Coding of data, plastics coloring statistics, 380
Co-extrusion, metallic pigment processing, 217
Collaborative Testing Service (CTS), color standards, 82
Collection systems, plastic recycling, 352–353
Color:
    appearance, 20–22
    gloss, 21
    language of, 16–20
    observer, 21–22
    plastics coloring statistics, 380
    plastics compounding, 271–272
    psychological response to, 13–15
    rendition and metamerism, 21
    surround properties, 20–21
Colorants. *See also* Pigments
    absorption and degradation, 10
    application technology, 260–261

chemical resistance, 87
classification, 90
color approval criteria, 263
color compounding, 272–274
color matching checklist, 266
color match request form, 265
color target, 262–263
   matching process, 266
color values, 86
concentrate processing, 302–305
   dry concentrates, 303
   liquid colorants, 305
   pellet products, 303–304
defined, 85, 259, 335
dispersability, 87
end-use stability, 9
FDA-regulated colorants, 375–376
finished colorant system, 264
formula compounding, system requirements, 266–267
health and safety issues, 99
heat resistance, 87
inorganic pigment families, 94–97
introduction systems, 263–264
Kubelka-Munk color matching and, 51
Lambert/Beer laws, 542
lightfastness/weatherability, 88
mixing process, 68
organic pigment families:
   blues, 111–116, 126
   defined, 100–101
   dispersion mechanisms, 124–125
   greens, 123–124
   heat stability parameter, 125–126
   international nomenclature: CI system, 101–102
   oranges, 121–123
   properties, 90–94
   reds, 102–111
      high-performance reds, 107–111
      metallized azo reds, 102–106
      nonmetallized azo reds, 106–107
   selection criteria, 102
   yellows, 116–121
      benzimidazolone yellows, 118
      diarylide yellows, 117–118
      diazo condensation yellows, 119, 121
      heterocyclic yellows, 118–120
      metallized azo yellows, 117
      monoarylide yellows, 116–117

performance factors, 89–90
process conditions, 261
   central blending, 263
   metering systems, 264
processing stability, 89
process selection, 261
properties, 86
regulatory issues with "problem" colorants, 372–377
research background, 85–86
selection criteria, 264–266
   color matching, 263
   polymer selection, 259–260
   target matching, 262–263
solubility resistance, 88
soluble dyes, 98
special effect colorants, 286
specialty pigments, 97–98
strength and tint strength, 86–87
testing and evaluation, 98–100
toxicity/environmental properties, 88–89
visual color matching, 62–63
   light effects, 75
Color blindness, defined, 15
Color concentrate. *See also* Concentrate processing
   defined, 259
Color consistency, dry color concentrate quality testing, 284–285
Color deficiency, defined, 15
Color difference, colorimetric prediction, 36–44
Colored Pigment Manufacturers Association (CPMA), pigment definitions, 100–101
Colorimetry:
   color difference predictions, 36–44
   defined, 15–16, 24
   pearlescent pigments, 230–231
   product color tolerances, 35
   research issues, 44–45
   spectrocolorimetry, 28–36
      calibration and standardization, 31–33
      design and specifications, 29–31
      performance verification, 33–34
   tolerance region derivations, 35–36
   visual colorimetry, 24–25
      analog simulation, 25–27
      digital simulation, 27–28

Color matching:
  advanced techniques, 52–53
    many-flux theory, 53
    research issues, 59
    turbid-media theory, 52–53
  checklist for, 266
  color approval criteria, 263
  computerized matching, 48–49
  contemporary systems, 53–59
    applications, 59
    dynamic database, 57–58
    Kubelka-Munk vs. many-flux calculations, 54–56
    many-flux theory, 53–56
    spectral matching, 57
    tristimulus matching, 56
  full vs. limited sample sets, 58–59
  traditional techniques, 49–52
    additivity principle, 49–50
    Kubelka-Munk theory, 50–52
    Lambert-Beer theories, 52
    visual color evaluation, 49
  visual color matching:
    blue color match example, 74
    color standard, 64–65
    gray color match example, 72–73
    light source, 65–66
    maroon color match example, 73
    mixing process, 67
    plastic materials, 61–62, 69–72
    polymer system, 66
    red color match example, 73
    scientific principles, 67–68, 71–72
Color match request (CMR) form:
  sample form, 265
  visual color matching, 68–72, 75–76
Color migration, color durability testing, 331–332
Color release, dry color concentrate:
  defined, 276
  letdown ratios, 283–284
Color specifications, plastics coloring statistics, 380
Color standard, visual color matching, 64–65
Color temperature, light source, 7
Colour Index (CI):
  international pigment nomenclature, 101–102
  pigment dispersion testing, 324–325
Colour Measurement Committee (CMC):
  color difference metrics, 37–44

ISO testing procedures, 80
plastics coloring statistics, 388–392
Commercial factor, color difference metrics, 41–44
Commission Internationale de l'Eclairage (CIE):
  color categories, 17–19
  color difference metrics, 36–44
  illuminant light sources, 7
  plastics coloring statistics:
    exponentially weighted moving average (EWMA), 392–396
    process color control, 388–392
  spectrocolorimetry specifications, 30–31
    calibration and standardization, 33
    CIELAB unit color difference error, 33–34
  statistical problem solving (SPS) and, 396–401
  Technical Committee TC 1-28, 38–44
  Technical Committee TC 2-39, 34
  tristimulus values, 36, 45
  visual colorimetry:
    analog simulation, 26, 45
    digital simulation, 27–28
Compact Disc (CD) grade, dry color concentrate carrier systems, 279–280
Complex inorganic pigments (CIPs):
  defined, 96
  structure and properties, 130–133
Compliance guidelines, environmental regulations, 375–377
Compounding process. See also Mixing process
  color compounding:
    additives/components, 270–271
    basic principles, 269–270
    colorants, 272–274
    plastics, 271–272
    quality control, 274–275
    research background, 268–269
    safety, health, and environmental issues, 275
  dry color concentrates, equipment, 280–281
  metallic pigment processing, 215–216, 235
Compound moisture absorption (CMA), carbon black pigment on resins, 173
Computer-controlled pump systems, liquid color concentrates, 288
Computerized color matching, applications of, 48–49

Concentrate processing:
  carbon black pigment dispersion, 164–165
  colorant/additive production process, 302–305
    dry concentrates, 303
    liquid colorants and additives, 305
    pellet products, 303–305
  color compounding, 274
  distribution channels, 306–307
  dry color:
    carrier requirements, 278–280
    categories of concentrates, 277–278
    "color release" issues, 283–284
    compounding equipment, 280–281
    distribution vs. dispersion, 283
    loading levels, 284
    particle size issues, 282–283
    quality testing, 284–286
    special effect concentrates, 286
    stranding and dicing issues, 282
    terminology, 276–277
  employee education, 318
  environmental exposures, 311–313
    air emissions, 311–312
    solid waste processing, 312–313
    storm water, 312
    wastewater treatment, 312
  liquid color:
    advantages/disadvantages, 292, 294–295
    applications, 288
    chemical properties, 287
    comparison with other methods, 296–298, 314
    delivery systems, 291, 293–294
    dispersion and distribution, 290–291
    historical background, 287–288
    manufacturing process, 288–290, 311
    packaging techniques, 292
    physical properties, effect on, 298–300
    pump systems for, 291–294, 312–313
    rheology, 295–296
  regulatory controls and standards, 313–318
    Consumer Product Safety Commission (16 CFR 1000-1799), 313
    environmental protection regulations, 315–318
    Food and Drug regulations (21 CFR), 314
    Occupational Safety and Health Administration (OSHA) (29 CFR), 314–315
  research background, 301–302
  workplace exposures and controls, 307–311
    nonproduction areas, 308–309
    production areas, 309–311
Conductivity:
  carbon black pigments, 169–170, 180
  metallic pigments, 204
Cone plasticators, dry color concentrates, 281
Cones:
  human vision, 13
  visual color matching, 61–64
Conference of Northeastern Governors (CONEG):
  cadmium sulfide/sulfoselenide pigment regulations, 374
  concentrate processing regulations, 318
  current regulatory status, 365–366
  defined, 377
  fluorescent pigment standards, 256
  heavy metal colorants regulations, 373
Conservation of energy, fluorescence and, 244, 266
Consumer Product Safety Commission (16 CFR 1000-1799), concentrate processing, 313
Contamination:
  carbon black pigment on resins, 172
  color compounding, 273–274
Continuing coloration process, metallic pigment processing, 217–219
Contrast ratios, additive coloring effects, 344–346
Cool white fluorescence:
  spectral distribution, 8
  visual color matching, 65
Copper compounds:
  metallic pigments, 202–203
  thermochromic materials, 197–200
Copper-halogen systems, additive coloring effects, heat stabilization, 348
Copper phthalocyanine blue:
  metal-free pigment, 116
  structure and properties, 111–115, 126
Copper phthalocyanine green, organic structure and properties, 123–124
Corn flake shape, metallic pigments, 209, 231
Corrective action plan, statistical problem solving (SPS) and, 400–401
Cosmic radiation, 5

Coupling agents:
  additive coloring effects, 347
  metallized azo red pigments, 103
Crocking, soluble dye manufacturing, 178
Cryopreservation, thermochromic materials, 198–200
Crystallizable polyethylene terephthalate (CPET), physical properties testing, 334
Cube blends:
  defined, 277
  dry color concentrates, plastics coloration, 283
Cutting procedures, metallic pigment production, 209
Cyanine chromogens, rhodamine A structure, 248–249
Cycle times, colorant application technology, 261
Cyclic polyenes, rhodamine A structure, 249
Cyclopentadienone oxides, photochromic behavior, 186–196

D
Data:
  defined, 379
  statistical problem solving (SPS) and, 39–401
Database structure:
  computerized color matching, 48–492
  dynamic color matching, 57–58
Daylight, visual color matching, 65
Deflection temperature, ISO test procedures, 8
Delamination, dry color concentrate, defined, 276–277
Delivery systems, liquid color concentrate manufacturing, 288–290
Delta Lab (D Lab chart), plastics coloring statistics, 380, 388–392
De Ment's absorption, fluorescence, 243–244
Deming *plan-do-study-act* cycle, plastics coloring statistics, 392
Desensitizers, thermochromic materials, 196–200
Design of Experiments (DOEs), plastics coloring statistics, 380–381
Detectors:
  human observer:
    color deficiencies and color blindness, 15
    physiological response, 12–13
    psychological response, 13–15
  instrumental observer, 15–16
  spectrocolorimetry, 29–31
  as triad component, 12–16
Dianisidine oranges, properties of, 92
Diarylide pigments:
  organic yellow/orange, 117–118, 127
  properties of, 90–91
  regulation of, 374–375
Diaryl naphthopyrans, photochromic behavior, 192–196
Diasazo condensation reds, structure and properties, 111, 126
Diazotization, metallized azo red pigments, 103, 123
Dibromanthrone red, structure and properties, 109
Dichroism, soluble dye fastness, 182
Dicing problems, dry color concentrate manufacture, 282
Die gas emissions, 311
Differential shrinkage, 337. *See also* Warpage
Digital simulation, visual colorimetry, 27–28
spiro[1,8a]Dihydroindolizines (DHIs), photochromic behavior, 188–196
1,8a-Dihydronaphthalene, photochromism, 194–196
Diketo-pyrrolo-pyrrol reds/oranges, properties of, 92
Dilatant viscosity, liquid color concentrate rheology, 295–296
Dimensional stability testing, color products, 336–338
  shrinkage, 336–337
  warpage, 337–338
Dinitroaniline orange, organic pigment classification, 92
Dinitroaniline orange, organic structure and properties, 122
Direct sales systems, color concentrates, 306
Disazo condensation yellows/reds:
  organic pigment structure and properties, 119, 121
  properties of, 91–92
Discoloration, titania pigment photochemistry and durability, 156
Disperse dyes, properties, 179
Dispersion:
  carbon black pigments, 163–166
    UV stability *vs.*, 168

colorants, 87
color compounding, 269–270
defined, 321
dry color concentrate:
  defined, 276
  distribution vs., 283
fluorescent dyes, 250
liquid color concentrates:
  basic principles, 290–291
  manufacturing process, 290–291
metallic pigment incorporation, 221
organic pigments, 124–125
testing, product dispersion, 321–326
  methods, 321–322, 325–326, 351
  pigment dispersion, 322–325, 349–350
Dissolver systems, liquid color concentrate manufacturing, 288–290
Distinction of image (DOI), visual color matching, 66–67
Distribution:
  color compounding, 269–270
  concentrates, 306–307
  defined, 321
  dry color concentrate:
    defined, 276
    dispersion vs., 283
  liquid color concentrates, 290–291
Distributorships, color concentrates, 306
Document control, ISO 9000 standard concerning, 83
Donor acceptor groups, rhodamine A structure, 247–248
Drum plasticators, dry color concentrates, 281
Dry colorants:
  concentrate manufacture:
    carrier requirements, 278–280
    categories of concentrates, 277–278
    "color release" issues, 283–284
    compounding equipment, 280–281
    distribution vs. dispersion, 283
    loading levels, 284
    particle size issues, 282–283
    processing procedures, 303
    quality testing, 284–286
    special effect concentrates, 286
    stranding and dicing issues, 282
    terminology, 276–277
  defined, 276
  finished colorant systems, 264
  liquid color concentrate vs., 296–297

Dry milling process, metallic pigment production, 206
D65 source, 7
DuPont Corporation, soluble dye manufacturing, 178
Durability properties:
  color durability testing, 326–332
    chemical resistance, 332
    color migration, 331–332
    fade and weather resistance, 328–331, 352–353
    heat stability, 326–327
    titanium dioxide pigment photochemistry, 152–157, 162–164
Dust control, concentrate production, 313
Dwell time, metallic pigment incorporation, 220
Dyes. See also Soluble dyes; Vat dyes
  color compounding, 273–274
  defined, 321
  physical properties, 98
  strength properties testing, 334

E
Effect pigments, pearlescent pigments as, 203, 228–229
Electrical resistivity, carbon black pigments, 180
Electrocyclization, photochromism:
  1,3-electrocyclization, 186–196
  1,5-electrocyclization, 188–196
Electromagnetic shielding properties, metallic pigments, 04
Electromagnetic spectrum, visible light, 5
*Electromagnetic spectrum, visible light, 17
"Ellipselike" tolerance volume, color difference metrics, 38–41
EMA carrier systems, dry color color concentrates, 278
Emergency Planning and Community Right to Know regulations, concentrate production, 317
Employee education, concentrate processing hazards, 318
"Empress green," 177
Encapsulated concentrates:
  dry color concentrates, 278
  processing systems, 304–305
Encapsulation strategy, titanium dioxide pigments, 156
End-use stability, colorant properties, 89

Energy distribution, light sources, 7
Environmental management systems (EMS), ISO 14000 standard for, 369
Environmental Protection Agency (EPA):
  carbazole violet regulations, 374
  chromium pigments, 375
  current regulatory status, 362–367
  future direction, 368
  heavy metal colorants regulations, 373
  historical background, 358–362
  lead chromate/molybdenate pigment regulations, 373–374
  permit programs, 316
  regulatory strategies, 360–362
  33/50 initiative, 367–368
Environmental standards:
  colorant properties, 88–89
  color compounding, 275
  concentrate processing, environmental exposures, 311–313
    air emissions, 311–312
    regulations and standards, 315–318
    solid waste processing, 312–313
    storm water, 312
    wastewater treatment, 312
  goals and strategies, 357–358
  regulations:
    compliance guidelines, 371–372
    current status, 362–367
    glossary of terms, 377–378
    historical overview, 358–362
    ISO 14000 series for environmental management, 82–83
    liability issues, 372
    "problem" colorants, 372–377
    33/50 initiative, 367–368
    worker regulations, 358–362
Epoxy resins, glitter pigments, 211
Equipment variability, plastics coloring statistics, 385–387
Error sources:
  plastics coloring statistics, 380–383
  spectrocolorimeters, 33–34
Ethylene bis-stearamide (EBS) carrier system, dry color concentrates, 278–280
European Community (EC), ISO color standards, 79
EVA. *See* Polyethylene vinyl acetate (EVA)
Evaluation procedures, colorant evaluation, 98–99
Exponentially weighted individual moving average (EWIMA), plastics coloring statistics, 392–396
Exponentially weighted moving average (EWMA), plastics coloring statistics, 392–396, 417
Exposure limits. *See also* Permissible exposure limit (PEL); Threshold limit values (TLVs)
  OSHA regulations, 361–362
Extrusion systems:
  liquid color concentrates, 288
    delivery methods, 291–295
  metallic pigment co-extrusion, 217
  pellet concentrate processing, 304
  single-screw extrusion (SSE):
    color compounding, 269–270
    dry color concentrates, 280–281
  twin-screw extrusion (TSE):
    color compounding, 269–270
    dry color concentrates, 280–281
E-Z isomerization, photochromism, 195–196

F
Fade resistance, color durability testing, 328–331, 352–353
Farnsworth-Munsell 100-hue test:
  color blindness, 15
  visual color evaluation, 49
  visual color matching, 62
Fastness properties, soluble dyes, 181–182
Federal regulations, inorganic pigments, 139–145
Federal Water Pollution Control Act, concentrate production regulations, 317
Fiberglass compounds, additive coloring effects, 349–350
Fischer base, spiropyrans, photochromic behavior, 189–196
Fishbone diagrams, plastics coloring statistics, 399
Fish-eye defects, pigment dispersion testing, 324–325, 350
Flame-retardant materials, visual color matching, 66
Flame retardants, additive coloring effects, 347–348
Flash lamps, spectrocolorimetry, 29–31
Flavanthrone yellows, properties of, 91
Flexural properties, ISO test procedures, 80

Flop effect:
  pearlescent pigments, 235–236
  plastics coloring, additive effects, 342–344
  refractive index and transmission, 10
Flow characteristics, titania pigments, 156–157
Flow reduction, metallic pigment molding, 221–224
Fluorescein, properties of, 246–247
Fluorescence:
  decay phenomenon as, 246
  energy level diagram, 245
  luminescent pigments, 97–98
  pigments:
    additives, 254–255
    applications, 257
    blowpin plateout test procedures, 255
    formaldehyde-based pigments, 251–252
    lightfastness, 256–257
    metal ion additives, 253–254
    non-formaldehyde-based pigments, 252–253
    toxicity, 255–256
  soluble dyes and brighteners, 180
    applications, 250–251
    fluorescein, 246–247
    rhodamine A, 247–250
    rhodamine B, 247
  spectrocolorimetry, 29–31
    performance verification, 33–34
  theory and mechanics, 242–246
Fluorescent colorants, visual color matching, 75
Foaming agents, additive coloring effects, 34
Food and Drug Administration (FDA):
  cobalt pigment regulation, 375
  color migration testing, 332
  concentrate processing regulations, 314
  defined, 377
  heavy metal colorants regulations, 373
  inorganic pigment regulation, 139–145
  regulated colorants, 375–376
Food and Drug Cosmetic (FD&C) certified colorants, 176
Food packaging, color migration, 332
Formaldehyde-based pigments, fluorescent pigments, 251–252
Freeze-dried color concentrate, properties, 277
Friele, MacAclam Cheekering (FMCII) standard, plastics coloring statistics, 388–392

Fulgides, photochromic behavior, 191–196
Furnace black process, carbon black pigment, 161

G
Gardner impact testing, color product testing, 332–334
Gaylord packaging system, 305
Gear pumps, liquid color concentrate delivery, 291, 293
Genetics, physiological vision response, 13
Glass fibers, visual color matching, 66
Glitter, metallic pigment production, 211
Gloss, as color property, 21, 30
Gloss reduction, visual color matching, 64
Gold dust, origins of, 202–203
Granite effects, visual color matching, 75
Gravimetric metering, colorant application technology, 264
Gray color match, case study, 72–73
Green pigment:
  complex inorganic pigments (CIPs), 133
  organic structure and properties, 123–124
  soluble dyes, 177
Ground state, fluorescence and, 245–246

H
Hansa yellows, properties of, 91
"Hard settling," liquid color concentrates, 288
Hardy recording spectrophotometer, spectral bandwidth, 30–31
Hazard Communication Standard (HCS):
  concentrate production, 315
  historical background, 361
Hazardous waste management, concentrate production, regulations and standards, 316
Health, safety and environmental (HSE) function, 357
Health issues:
  colorant evaluation, 99
  color compounding, 275
  concentrate processing, workplace exposures and controls, 307–311
    nonproduction areas, 308–309
    production areas, 309–311
Heat deflection temperature (HDT):
  color product testing, 334
  liquid color concentrate, 300, 310
Heat resistance, colorants, 87

Heat stability parameter:
   color durability testing, 326–327
   fluorescent dyes, 250
   organic pigments, 125–126
Heat stabilizers, additive coloring effects, 348
Heavy metals:
   in colorants, regulation of, 372–373
   fluorescent pigments, 255–256
   inorganic pigments, federal regulations concerning, 139–145
Heterocyclic compounds:
   orange pigments, organic structure and properties, 123–124
   photochromic behavior, 186–196
   yellow pigments, organic structure and properties, 118–120
Heterolytic bond cleavage, photochromic behavior, 186–196
High aspect ratio:
   metallic pigments, 209
   pearlescent pigments, 232–233
High-density polyethylene (HDPE), recycling and collection, 353–354
High-performance red pigments, structure and properties, 107–111
Hindered amine light stabilizer, fluorescent pigment lightfastness, 257
Hoechst Chemical company, 177–178
Holographic gratings, 29
Hue:
   colorant properties, 86
   color difference metrics, 37–44
   color space, 17–18
   defined, 16
   soluble dye fastness, 181–182
Human observer:
   color deficienciees and color blindness, 15
   physiological response, 12–13
   psychological response, 13–15
Human vision:
   color deficiencies and color blindness, 15
   diagram, 14
   physiological response, 12–13
   psychological response, 13–15
   visual colorimetry and, 25–26
   visual color matching, 61–64
Humidity, metallic pigment incorporation, 220
Hunter values, plastics coloring statistics, process color control, 388–392
Hydrocarbon compounds, photochromic behavior, 188–196
Hydrogen transfer, photochromic behavior, 195–196
Hydroxyphenyltriazoles, photochromic behavior, 196

I

ICI company, soluble dye manufacturing, 178
Illuminants, standards, 7
Image analysis, pigment dispersion testing, 325–326, 351
Impact modifiers, additive coloring effects, 348–349
Impact strength, carbon black pigments, 171
Incandescent lamps:
   spectrocolorimetry, 29–31
   visual color matching, 65
Incorporation, metallic pigment processing, 219–221
Indanthrone blues:
   properties of, 93
   structure and properties, 115
Indenone oxides, photochromic behavior, 186–196
Indigo compounds, photochromic behavior, 195–196
Indigoid dyes, structure and properties, 181
Industrial Toxics Project (ITP), 367–368
Information sources, OSHA regulations, 370–372
Infrared radiation:
   defined, 5
   metallic pigments, 204
Injection molding:
   liquid color concentrates, 288
      delivery methods, 291–295
   metallic pigment processing, 217
Inorganic pigments:
   bismuth vanadate, 136–137
   cadmium pigments, 133–134
   chrome greens, 139
   chromium III pigments, 138
   color compounding, 272–274
   complex pigments, 130–133
   dry color concentrate carrier systems, 279
   FDA status, 139
   heavy metal usage, federal regulation, 140–141

iron blues, 138–139
iron oxides, 128–130
lead chromate/lead molybdate pigments, 135–136
major families, 94–97
metallic pigment surfaces, 212–213
performance criteria, 141–145
polymer suitability criteria, 141–142
rare earth pigments, 139
research background, 127–128
ultramarine pigments, 137–138
Insert molding, metallic pigment processing, 218
Instrumental observer, defined, 15–16
Intensity, defined, 16
Interference/, pearlescent pigments, properties, 97, 227
Intermediate media systems, color matching theory, 52–53
Intermixer, liquid color concentrate delivery, 291, 294, 313
Internal conversion, fluorescence, 246
International Agency for Research on Cancer (IARC):
  carbon black pigment regulations, 374
  defined, 377
  nickel-containing pigment regulations, 375
International Organization for Standardization (ISO):
  defined, 79–80
  document control standards, 83
  historical background, 79–80
  ISO 9000, 80–81, 84
  ISO 17025 laboratory standard, 81–82
  ISO 14000 series for environmental management, 82–83, 369
  preventive action guidelines, 83
  process audit, 84
  QS-9000 quality system requirements, 82
  quality system procedures, 83
  spectrocolorimetry calibration and standardization, 33
  strength properties testing, 332–334
  Test Procedure No. 105, 80
  test procedures, 80
Iodine, thermochromic materials, 197–200
Iron blues:
  classification, 96
  inorganic structure and properties, 138–139

Iron oxide pigments:
  natural iron oxides, 128–129
  pearlescent iron oxide-mica pigments, 235
  properties of, 94–95
  synthetic iron oxides, 129–130
Isoindolinone/isoindoline yellows/oranges/reds, properties of, 91
Izod impact strength:
  additive coloring effects, 350, 366
  carbon black pigments, 171
  color product testing, 332–334
  ISO test procedures, 80

J
Japanese Agency of Industrial Science and Technology, 198
Japanese Standards Institute (JSI), ISO 9000 and, 81
Jetness, carbon black pigments, 162–164

K
Kaolin, pearlescent pigments from, 240
K coefficient, 51–52, 57–58
  dry color concentrates, 285
Kelvin, defined, 7
Kneaders, dry color concentrates, 281
Kubelka-Munk theory:
  color matching, dynamic database, 58
  dry color concentrate quality testing, 285–286
  many-flux color matching vs., 54–55
  visual color matching, 50–52

L
$L*a*b*$ color space (CIE LAB), 17–20
  carbon black pigments, dispersion effects on polymers, 165–166
  color difference metrics, 37–44
  color durability testing, 330–331
  color target selection, 263
  computerized color matching, 48–49
  pearlescent pigment colorimetry, 230–231
  plastics coloring statistics, 388–392
  recycled plastics, 354, 372
  titanium dioxide pigments:
    durability, 155
    pearlescent pigment production, 233–235
  visual color matching, 62–63
Labeling systems, color concentrates, 306–307

Laboratory standards, ISO 17025 standard, 81–82
Lacing resistance, titanium dioxide pigments, 156
Lambert's law:
　absorption, 10
　color matching, 52
　visual color matching, 75
Laminating, metallic pigment processing, 218
Language of color, defined, 16–20
Laser marks, pearlescent pigments, 239–240
Laux process, synthetic iron oxides, 130
$l:c$ ratio, color difference metrics, 38–41
Leaching, color migration, 331–332
Lead carbonate. *See* Basic lead carbonate
Lead chromate pigments:
　government regulations, 373–374
　inorganic pigment structure and properties, 135–136
　yellow pigments, properties of, 95
Lead exposure:
　concentrate production, contaminant regulations and standards, 314
　heavy metal colorants regulations, 373
Lead molybdenate pigments:
　government regulations, 373–374
　inorganic pigment structure and properties, 135–136
　orange pigments, properties of, 95
Leistritz-type extrusion equipment, pellet concentrate processing, 304
Lenticular shape, metallic pigment production, 231
Lenticular (silver dollar) shape, metallic pigment production, 210
Letdown (L/D) ratios:
　dry color concentrate, 277
　　color release, 283–284
　　dispersion *vs.* distribution, 283
　　loading levels, 284
　　liquid color concentrate, 287–288, 308, 310
"Leuco" state, vat dyes, 179
Liability issues, environmental regulations, 372
Life cycle of color products, defined, 377
Light-emitting diodes (LEDs), analog colorimetry, 26
Lightfastness:
　defined, 88, 101–102

fluorescent dyes, 250–251
fluorescent pigments, 256–257
soluble dyes, 181–182
Light flux, spectrocolorimetry calibration and standardization, 32–33
Lightness:
　color difference metrics, 37–44
　color properties, 17–18
Light reflection. *See* Interference
Light source:
　blackbodies, 7–8
　spectrocolorimetry, 29–31
　as triad component, 5–9
　visual color matching, 61–62, 65–66
Linear Low Density Polyethylene (LLDPE) film, carbon black pigments, UV stability, 167
Liquid color:
　concentrates, 277
　　advantages/disadvantages, 292, 294–295
　　applications, 288
　　chemical properties, 287
　　comparison with other methods, 296–298, 314
　　delivery systems, 291, 293–294
　　dispersion and distribution, 290–291
　　historical background, 287–288
　　manufacturing process, 288–290, 311
　　mixing, 305
　　packaging techniques, 292, 305
　　performance evaluation, 298–300
　　physical properties, 298–300
　　pump systems for, 291–294, 312–313
　　rheology, 295–296
　　weight-out techniques, 305
　finished colorant systems, 264
Liquid crystal polymers (LCPs), additive coloring effects, 344–346
Liquid crystals, thermochromic materials, 198–200
Lithol reds, structure and properties, 103–104
Lithol rubine red, organic structure and properties, 104–105
Lithopones, cadmium pigments, 133–134
Loading levels, dry color concentrate, 284
Low-density polyethylene (LDPE):
　dry color concentrate carrier system, 278–280
　glitter production, 211
Lower control limit (LCL), 393–396
Low molecular weight, color migration, 331

Lubricants, additive coloring effects, 349
Luminescent pigments. *See also* Pearlescent pigments
　properties, 97–98
　Stoke's Law, 243–244
Luster effects, pearlescent pigments, 228–229, 235–236

M

Madder, soluble dyes from, 176–177
Manganese compounds, thermochromic materials, 198–200
Manifest/bill of lading systems, color concentrates, 306
Many-flux theory, color matching, 53–54
　computerized systems, 56
　Kubelka-Munk comparison with, 54–55
Maroon color match, case study, 73
Masstone, pearlescent pigments, 235
Masterbatch method:
　carbon black pigment dispersion, 165
　color compounding, 274–275
　dry color concentrates, 277–278
　finished colorant systems, powder masterbatching, 264
　fluorescent pigments, non-formaldehyde-based fluorescent pigments, 252–253
　metallic pigment production, 215
Material safety data sheets (MSDS):
　carbon black pigment regulations, 374
　color concentrates, 306
　concentrate production, Hazard Communication Standard, 315
　historical background, 361–362
Materials handling systems, concentrate processing, workplace exposures and controls, 308–311
Mauveing, soluble dyes from, 177
Measuring and test equipment (MT&E) standards, ISO 17025 standard, 81–82
Mechanical property retention, additive coloring effects, 349–350
Media mills, liquid color concentrate manufacturing, 289–290, 311
Melt flow:
　carbon black pigment viscosity on resins, 173
　rheological testing, 334–335, 354
Melt points, color migration, 331
Mercury arc light, spectral distribution, 9

Mercury-cadmium pigments:
　oranges/reds/maroons, properties of, 96
　structure and properties, 134
Mercury compounds, thermochromic materials, 197–200
Metal-free copper phthalocyanine blue, structure and properties, 116
Metal ions, fluorescent pigment additives and avoidance of, 253–254
Metallic pigments. *See also* Pearlescent pigments
　application problems, 219–225
　　incorporation process, 219–221
　　molding, 221–225
　dry color concentrates, 286
　glitter, 211
　nonoptical effects, 204
　optical effect, 203
　plastic applications, 224–225
　processing, 214–219
　　blow molding, 218
　　calendering, 217–218
　　coating and printing, 219
　　co-extrusion, 217
　　compounding process, 215–216
　　continuing coloration, 217–218
　　injection molding, 217
　　insert molding, 218
　　laminating, 218
　　masterbatches, 215
　　polyvinyl chloride (PVC) coloration, 216–217
　　pressing, 219
　　reactive resin systems, 219
　　vacuum thermoforming, 218
　production, 204–209
　　cutting procedure, 209
　　dry milling, 206
　　milling equipment, 204–206
　　pellet production, 207–209
　　raw materials, 204
　　wet milling, 206–207
　vs. "regular" pigments, 203
　research background, 202–204
　shapes:
　　irregular (corn flake) shape, 209
　　lenticular (silver dollar) shape, 210
　　spherical shape, 210–211
　special-effect pigments, 203
　supply form, 213–214
　surface treatment, 212–213

Metallized azo pigments:
  reds, organic structure and properties, 102–106
    BON reds, 105
    lithol reds, 103
    lithol rubine red, 104–105
    permanent red 2B, 104
    pigment scarlet, 105
    red lake, 105–106
  yellows, organic structure and properties, 117
Metal oxide-mica pigments:
  functional properties, 240
  plastics applications, 239–240
  structure and properties, 232–233
Metamerism:
  colorant selection criteria, 259
  color rendition, 21
  observer/detector, 21–22
  visual color matching, 27, 65–66
Metering systems, colorant application technology, 264
Methylsalicylates, photochromic behavior, 195–196
Metric hue, color difference metrics, 37–44
Metric hue angles, color difference metrics, 36–44
Mica, pearlescent pigment production, 232–233
  combination pigments, 235–236
  iron oxide-mica pigments, 235
  titanium dioxide mica pigments, 233–235
Mixed metal oxides (MMOs), defined, 96
Mixing process. *See also* Compounding process
  concentrate processing:
    dry color, 303
    liquid color, 305
    pellet products, 304
    workplace exposures and controls, 309–310
  dry color concentrate, 277
  liquid color concentrates, 288–290, 311
    distribution and dispersion, 290–291
    metallic pigment incorporation, 220
  visual color matching, 67
Molding techniques, metallic pigments, 221–225
Mold releases, additive coloring effects, 349
Monoarylide yellows, organic structure and properties, 116
Monochromatic, defined, 24

Montionless mixers, liquid color concentrate distribution, 290–291
Munsell nomenclature system, 17
Muscovite, pearlescent pigment production, 232–233

N
Nacreous pigments, defined, 227
Naphthol pigments:
  orange, 123
  reds, 106
    organic pigment classification, 92
National Bureau of Standards (NBS), ISO 17025 standard and, 81–82
National Institute of Standards (NIST), ISO 17025 standard and, 81–82
National Physical Laboratory (NPL), spectrocolorimetry, performance verification, 33–34
National Pollution Discharge Elimination System (NPDES) (EPA), concentrate production, 316
Natural pearl essence, structure and properties, 232
Near infrared region (NIR), spectrocolorimetry:
  performance verification, 33–34
  spectral analyzer design, 30–31
Nematic mesophase, 198
New Jersey Worker and Community Right to Know laws, concentrate processing regulations, 318
Newtonian viscosity, liquid color concentrate rheology, 295–296
Nickel-containing pigments:
  azo yellows, 91
  regulations for, 375
Non-formaldehyde-based pigments, fluorescent pigments, 252–253
Nonlinear relationships, many-flux *vs.* Kubelka-Munk color matching, 55
Nonmetallized azo reds, organic structure and properties, 106–107
Non-micaceous systems, pearlescent pigments, 240
Nonoptical effects, metallic pigments, 204
Nyquist bandwidth, spectrocolorimetry specifications, 31

O
Object:
  absorption, 10

gloss/surface structure, 21
refractive index and transmission, 10
scattering phenomenon, 11–12
as triad component, 9–13
Observer/detector:
human observer:
color deficiencies and color blindness, 15
physiological response, 12–13
psychological response, 13–15
instrumental observer, 15–16
metamerism, 21
as triad component, 12–16
Occupational Safety and Health Administration (OSHA):
cadmium sulfide/sulfoselenide pigment regulations, 374
concentrate processing:
regulations and standards, 314–315
workplace exposures and controls, 307–311
defined, 377
future direction, 370–374
historical background, 358–362
information sources, 370–372
Ochre pigments, structure and properties, 129
Oil absorption, dry color concentrate carrier systems, 279
Opacifiers, fluorescent pigment lightfastness, 257
Opacity factors, visual color matching, 66
Opaque resins, additive coloring effects, 344–346
Optical brighteners, visual color matching, 75
Optically thick media, color matching theory, 53
Optically thin media, color matching theory, 52
Optical microscopy, pigment dispersion testing, 325–326, 350–351
Optical properties:
metallic pigments, 203
pearlescent pigments, 228–232
Orange pigments:
fluorescent vs. nonfluorescent, 242
organic structure and properties, 121–123
Organic pigments:
blues, 111, 113–116
color compounding, 273–274
defined, 100–101

dispersion mechanisms, 124–125
dry color concentrate carrier systems, 279
greens, 123–124
heat stability parameter, 125–126
international nomenclature: CI system, 101–102
major families, 94–97
metallic pigment surfaces, 213
oranges, 121–123
properties, 90–94
reds, 102–111
high-performance reds, 107–111
metallized azo reds, 102–106
nonmetallized azo reds, 106–107
selection criteria, 102
titanium dioxide pigments, 147–148
warpage testing, 337–338
yellows, 116–121
benzimidazolone yellows, 118
diarylide yellows, 117–118
diazo condensation yellows, 119, 121
heterocyclic yellows, 118–120
metallized azo yellows, 117
monoarylide yellows, 116–117
Original Equipment Manufacturer, defined, 108
Orthonitroaniline orange, organic structure and properties, 122
Outdoor weather testing:
carbon black pigments, 168
color durability, 328–331, 352–353
Oxidation, titanium dioxide pigments, 154–157

P
Packaging systems:
dry color concentrates, 303
liquid color concentrates, 292, 305
pellet concentrate processing, 305
Pantone color standard, visual color matching, 64
Particle size:
carbon black pigments:
conductivity properties, 170–171
morphology, 162–163
outdoor weathering tests, 168
QUV-B weatherometer testing, 167
dry color concentrate manufacture, 282–283
liquid color concentrates, 295–296
metal oxide-mica pigments, 232–233
pearlescent pigments, 237–240

pigment dispersion testing, 324–325
plastics coloring, additive effects, 343–344, 365
titanium dioxide pigments, 150
Part-to-part variation (PV), 385–387
Pastes:
    finished colorant systems, 264
    metallic pigments, 213–214
Pauli exclusion principle, fluorescence and, 245–246
PBT compounds, additive coloring effects, 349–350
Pearlescent pigments. *See also* Luminescent pigments
    classification:
        bismuth oxychloride, 237
        combined mica pigments, 235–236
        iron oxide-mica pigments, 235
        lead carbonate, 236–237
        metal oxide-mica pigments, 232–233
        natural pearl essence, 232
        non-micaceous systems, 240
        plastics applications, 237–240
        titanium dioxidie-mica pigments, 233–235
    defined, 227
    iron oxide-mica pigments, 235
    metal oxide-mica pigments, 232–233
    non-micaceous systems, 240
    optical principles, 228–232
    properties and applications, 237–240
    research background, 226–232
    titanium dioxidie-mica pigments, 233–235
Pelletization:
    carbon black pigment dispersion, 164–165
    concentrate product processing, 303–305
        encapsulation, 304
        mixing, 304
        packaging, 305
        weigh-out, 303–304
        workplace exposures and controls, 311
    dry color concentrates, 278
        plastics coloration, 283
        stranding and dicing, 282
    finished colorant systems, 264
    liquid color concentrate *vs.,* 296–297, 314
    metallic pigment production, 207–208, 214
    supply forms, 234
Penniman-Zoph process, synthetic iron oxides, 130
Perception, visual color matching, 61–62

Performance evaluation:
    carbon black pigments, 173
    colorant evaluation, 98–99
    colorant properties, 89–90
    inorganic pigments, 140–145
    liquid color concentrates, 298–300
    spectrocolorimeters, 33–34
Peristaltic pump system, liquid color concentrate delivery, 291, 293, 312
Permanent red 2B, organic pigment classification, 93
Permanent red 2B, structure and properties, 104
Permissible exposure limit (PEL):
    defined, 377
    historical background, 362
Persian Gulf red oxide, structure and properties, 128–129
Personal liability, environmental regulations, 372
Personal protective equipment (PPE), concentrate processing:
    regulations and standards, 314
    workplace exposures and controls, 309–311
Perylene reds, structure and properties, 110
Phlogopite, pearlescent pigment production, 232
Phosphorescence, triplet state of electrons, 246
Phosphorescent colorants, visual color matching, 75
Photochemical properties, titanium dioxide pigments, 152–157
Photochromic materials, structure and properties, 185–196
Photometric scale, spectrocolorimetry calibration, 32–33
Photomultiplier tubes, spectrocolorimetry specifications, 31
Phthalocyanine pigments:
    blues:
        dry color concentrate carrier systems, 279
        properties of, 93–94
    greens, properties of, 94
Physical properties testing, colorants, 332–334
Physical properties testing, product testing, strength properties, 332–334
Physiological response, human observers, 12–13

Pigment loading systems:
  metallic pigment incorporation, 220
  plastics coloring, additive effects, 342–343
Pigments. *See also* Colorants
  color compounding, 272–274
  defined, 321
  dispersion testing, 322–326, 349–350
  fluorescent pigments:
    additives, 254–255
    applications, 257
    blowpin plateout testing, 255
    formaldehyde-based pigments, 251–252
    lightfastness, 256–257
    metal ion additives, 253–254
    non-formaldehyde-based pigments, 252–253
    plateout reduction, 254–255
    toxicity, 255–256
  inorganic pigments:
    bismuth vanadate, 136–137
    cadmium pigments, 133–134
    chrome greens, 139
    chromium III pigments, 138
    classification, 96–99
    complex pigments, 130–133
    FDA status, 139
    heavy metal usage, federal regulation, 140–141
    iron blues, 138–139
    iron oxides, 128–130
    lead chromate/lead molybdate pigments, 135–136
    major families, 94–97
    performance criteria, 141–145
    polymer suitability criteria, 141–142
    research background, 127–128
    ultramarine pigments, 137–138
  major specialty pigments, 97–98
  metallic pigments:
    application problems, 219–225
      incorporation process, 219–221
      molding, 221–225
    glitter, 211
    irregular shape, 209
    lenticular shape, 210
    nonoptical effects, 204
    optical effect, 203
    processing, 214–219
      blow molding, 218
      calendering, 217–218
      coating and printing, 219
      co-extrusion, 217
      compounding, 215–216
      continuing coloration, 217–219
      injection molding, 217
      insert molding, 218
      laminating, 218
      masterbatches, 215
      polyvinyl chloride (PVC) coloration, 216–217
      pressing, 219
      reactive resin systems, 219
      vacuum thermoforming, 218
    production, 204–209
      cutting procedure, 209
      dry milling, 206
      milling equipment, 204–206
      pellet production, 207–209
      raw materials, 204
      wet milling, 206–207
    vs. "regular" pigments, 203
    research background, 202–203
    shape
      irregular (cornflake) shape, 209
      lenticular (silver dollar) shape, 210
      spherical shape, 210–211
    special-effect pigments, 203
    spherical shape, 210–211
    supply form, 213–214
    surface treatment, 212–213
  organic pigments:
    blues, 111–115
    defined, 100–101
    dispersion mechanisms, 124–125
    greens, 123–124
    heat stability parameter, 125–126
    international nomenclature: CI system, 101–102
    major families, 90–94
    oranges, 121–123
    properties, 90–94
    reds, 102–111
      high-performance reds, 107–111
      metallized azo reds, 102–106
      nonmetallized azo reds, 106–107
    selection criteria, 102
    yellows, 116–121
      benzimidazolone yellows, 118
      diarylide yellows, 117–118
      diazo condensation yellows, 119, 121
      heterocyclic yellows, 118–120
      metallized azo yellows, 117
      monoarylide yellows, 116
  pearlescent pigments:

bismuth oxychloride, 237
combined mica pigments, 235–236
iron oxide-mica pigments, 235
lead carbonate, 236–237
metal oxide-mica pigments, 232–233
natural pearl essence, 232
non-micaceous systems, 240
optical principles, 228–232
plastics applications, 237–240
research background, 226–232
titanium dioxidie-mica pigments, 233–235
scattering properties, 11–12
visual color matching, 62–63
scientific theory, 68–71
warpage testing, 337–338
Pigment scarlet, properties of, 93, 105
Planck's constant, fluorescence and, 243–244
Plastic color standard, visual color matching, 68–72
Plasticizers, dry color concentrate carrier systems, 280
Plastics. *See* Polymers
Plateout effect, fluorescent pigments:
additives for reduction of, 254–255
blowpin test procedures, 255
Plateout reduction, non-formaldehyde-based fluorescent pigments, 252–255
Polishing process, metallic pigments, 212
Polyacrylonitrile/polyamides, soluble dyes, 177
Polyamides:
glitter production, 211
non-formaldehyde-based fluorescent pigments, 254–255
Polycarbonate (PC):
dry color concentrate carrier systems, 278–280
melt flow testing, 335
soluble dyes, 178
Polychromatic, defined, 8
Polycyclic oxiranes, photochromic behavior, 186–196
Polyenes:
photochromic behavior, 195–196
rhodamine A structure, 249
Polyetheretherketone (PEEK), soluble dye manufacturing, 178
Polyethersulfone (PES), color compounding, 270–271
Polyethylene, soluble dyes, 177

Polyethylene terephthalate (PET):
liquid color concentrate, 287
recycling and collection, 354–356
Polyethylene vinyl acetate (EVA):
dry color color concentrates, 277–278
liquid color concentrates, 297
Polymers. *See also* specific polymer compounds
absorption and degradation, 10
additive coloring effects:
antioxidants, 346–347
antistatics, 347
coupling agents, 347
flame retardants, 347–348
foaming agents, 348
heat stabilizers, 348
impact modifiers, 348–349
lubricants/mold releases, 349
polymer blends, 346
reinforcing agents, 349–350
research background, 340–344
translucent/opaque resins, 344–346
transparent resins, 344
UV stabilizers, 350
carbon black pigments:
dispersion effects, 165–166
morphology, 160–162
research background, 159–160
selection criteria, 162–173
colorant properties, 90
colorant selection and application criteria, 259–260
color compounding:
additives/components, 270–271
basic principles, 269–270
colorants, 272–274
plastics, 271–272
quality control, 274–275
research background, 268–269
safety, health, and environmental issues, 275
dry color concentrates, 283
fluorescent dye applications, 250
fluorescent pigment applications, 257
inorganic pigment suitability criteria, 139–145
liquid color concentrates:
applications, 298–300
physical properties, 297–298
pearlescent pigment applications in, 237–240
photochromic behavior, 189–196

soluble dyes for, 175–176
   categories and properties, 177–180
   coloring processes, 177–178
   fastness properties, 181–182
   thermoplastic requirements, 182–184
statistics on coloring:
   batch-to-batch variability, 383–384
   exponentially weighted moving average, 392–396
   Hunter, CIE, and CMC control processes, 388–392
   process measurement systems analysis, 384–387
   research background, 379–383
   statistical problem solving (SPS) applications, 396–401
   test variability, 387–388
thermochromic properties, 199–200
titanium dioxide pigment incorporation, 150–152
   durability and photochemistry, 152–157
visual color matching, 66
Polymethyl methacrylate (PMMA), metallic pigment processing, 216
Polyolefins:
   carbon black pigment dispersion, 165
   complex inorganic pigments (CIPS), 132
   metallic pigment processing, 216
   non-formaldehyde-based pigments, 252–253
   titanium dioxide pigment durability and photochemistry, 152–157
Polyphenylene sulfide (PPS):
   color compounding, 270–271
   dry color concentrates, 278
Polypropylenes, soluble dyes, 178
Polystyrenes:
   soluble dyes, 177
   titanium dioxide pigment durability and photochemistry, 152–157
Polysulfones, soluble dye manufacturing, 178
Polyvinyl acetate, soluble dyes, 177
Polyvinyl chloride (PVC):
   color compounding, 269–270
   color durability testing, 330–331
   dry color concentrates, compounding equipment, 281
   glitter production, 211
   liquid color concentrates, 297
   metallic pigment processing, 216–217
   soluble dyes, 177

Polyvinyls, titanium dioxide pigments:
   durability and photochemistry, 152–157
   incorporation, 150–152
Postconsumer recycle (PCR) products, applications, 353–354, 372
Powdered pigments, metallic pigments, 213
Powder masterbatching, finished colorant systems, 264
Powered air respiratory protection (PARP), concentrate processing, workplace exposures and controls, 309–311
Precolored compounds:
   color compounding, 274
   dry color concentrate, 277
   finished colorant systems, 264
   liquid color concentrate *vs.*, 296–297
Premanufacture notification (PMN) regulations, 364–365
Pressing procedures, metallic pigment processing, 219
Pressure rise test, pigment dispersion testing, 325–326, 351
Preventive procedures, ISO standards, 83
Primary pigment pastes, metallic pigments, 214
Printing procedures, metallic pigment processing, 219
Process audits, ISO guidelines, 84
Process color control, plastics coloring statistics, 388–392
Processing methods:
   colorant application technology, 259
   dry color concentrate, 276
   stability, colorant properties, 89
Process measurement systems analysis (PMSA), plastics coloring statistics, 384–387
Product color tolerances, colorimetric analysis, 35
Production part approval process (PPAP), color quality controls, 82
Product testing:
   dimensional stability, 336–338
      shrinkage, 336–337
      warpage, 337–338
   dispersion testing, 321–322
      methods, 321–322, 351
      pigment dispersion, 322–326, 349–350
   durability testing, 322
      chemical resistance, 332
      color migration, 331–332

fade and weather resistance, 328–331,
    352–353
  heat stability, 326–327
  physical properties testing, 322
    strength properties, 332–334
  research issues, 320–321
  rheological testing, 334–346
    melt flow, 334–336, 354
    spiral flow, 336
  terminology, 321
Progressive cavity pump, liquid color
    concentrate delivery, 291, 293, 312
Proposition 65 (California):
  concentrate processing regulations, 318
  current status, 371–372
Psychological response:
  human observers, 13–15
  visual color matching, 61–62
Publicly owned treatment works (POTW):
  defined, 377
  EPA regulations, 363–364
Pump systems, liquid color concentrates:
  delivery methods, 291–294
  pump categories, 291–294, 312–313
Purge resins, concentrate production, 313
Pyrazolone pigments:
  orange pigments, 122
  red pigments, 106

Q

QC software, color difference metrics,
    40–41
Quality control:
  color compounding, 274–275
  color durability testing, heat stability
    parameter, 326–327
  dry color concentrate manufacture,
    284–285
  ISO 9001 standard, 82
  melt flow testing, 335
  plastics coloring statistics, 384
  QS-9000 quality system requirements,
    82
Quanta, fluorescent laws and, 244
Quenching, fluorescent pigment
    lightfastness, 256–257
Quinacridone magentas/violets, properties
    of, 92, 107–108
Quinacridone reds, properties, 92, 107–109
QUV-B Weatherometer testing, carbon
    black pigments, 167, 178–179

R

Radiation, light source and, 5–7
Radio waves, 5
Rare earth pigments, structure and
    properties, 139
Raw materials, metallic pigments, 204
Reactive dyes, fluorescent dyes as, 250
Reactive resins, metallic pigment
    processing, 219
Recycling of plastics:
  applications for, 353–354, 371–372
  collection procedures, 352–353
  historical background, 352
  polyethylene terephthalate, 354–356
  reprocessing techniques, 353
Red Lake C, organic pigment classification,
    92
Red Lake C, structure and properties,
    105–106
Red pigments:
  color match case study, 73
  high-performance reds, 107–111
  metallized azo reds, 102–106
    BON reds, 105
    lithol reds, 103
    lithol rubine red, 104–105
    permanent red 2B, 104
    pigment scarlet, 105
    red lake, 105–106
  nonmetallized azo reds, 106–107
  organic structure and properties, 102–111
    high-performance reds, 107–111
    metallized azo reds, 102–106
    nonmetallized azo reds, 106–107
  synthetic iron oxides, 129–130
Reflectance:
  color matching and, 59
  Kubelka-Munk color matching and, 51
  spectrocolorimetry, 29–33
    calibration and standardization, 32–33
Refractive index:
  light object, 10
  material particles, 11
  pearlescent pigments, 228–232
  plastics coloring, additive effects, 341–344
  rutile titania, 147
  scattering properties and, 11–12
  titanium dioxide pigments, 146
Regulatory controls and standards:
  compliance guidelines, 375–377
  concentrate processing, 313–318

Consumer Product Safety Commission
   (16 CFR 1000-1799), 313
   environmental protection regulations,
      315–318
   Food and Drug regulations (21 CFR),
      314
   Occupational Safety and Health
      Administration (OSHA) (29 CFR),
      314–315
   state and local requirements, 317–318
environmental standards:
   compliance guidelines, 371–372
   current status, 362–367
   glossary of terms, 377–378
   historical overview, 358–362
   33/50 initiative, 367–368
   ISO 14000 series for environmental
      management, 369
   liability issues, 372
   "problem" colorants, 372–377
   worker safety regulations, 358–362
Reinforcing agents, additive coloring
   effects, 349–350, 366
Relative viscosity properties, carbon black
   pigment on resins, 173
Repeatability/reliability data sheet, plastics
   coloring statistics, 385–387
Reprocessing techniques, colored plastics,
   353
Resins:
   additive coloring effects, 340–346
      translucent/opaque resins, 344–346
      transparent resins, 344
   carbon black pigment effects, 171–172
   colorant selection and application
      criteria, 259–260
   color durability testing, 327
   color migration, 331–332
   dry color concentrates, 277–278
   formaldehyde-based fluorescent
      pigments, 251–252
   glitter pigments, 211
   metallic pigment processing, 219
   pearlescent pigments, 239–240
   purge resins, concentrate production,
      313
   recycled plastics, 354–356
   soluble dye manufacturing, 178
      thermoplastic properties, 182–184
   thermochromic materials, 196–200
Resource Conservation Recovery Act
   (RCRA), heavy metal colorants, 373

"Reverse photochromism," 190
Reversible thermochromic materials,
   structure and properties, 196–200
RGB values, visual analog colorimetry, 26
Rheology. *See also* Viscosity properties
   liquid color concentrates, 295–296
   product testing, 334–336
      melt flow, 334–335, 354
      spiral flow, 335
Rhodamine A, properties of, 247–250
Rhodamine B, properties of, 247
Rhodopsin, physiological vision response,
   13
Rigidity, rhodamine A structure, 249–250
Rods:
   human vision, 13
   visual color matching, 61–64
Rutile pigments:
   complex inorganic pigments (CIPs),
      130–133
   titania pigments:
      occurrence and sources, 147–148
      photochemistry, 154–157

S
Safety issues:
   colorant evaluation, 99
   color compounding, 275
   concentrate processing, workplace
      exposures and controls, 307–311
      nonproduction areas, 308–309
      production areas, 309–311
   environmental and worker safety
      regulations, 358–362
"Salt and pepper" blending:
   color compounding, 274
   dry color concentrate, 277
      plastics coloration, 283
   liquid color concentrate *vs.*, 296–297
Salt attrition, copper phthalocyanine blue,
   115
Saunderson correction:
   color matching theory, 51
   many-flux color matching *vs.* Kubelka-
      Munk, 55
Scattering properties:
   light objects, 11–12
   material particles, 11
   plastics coloring, additive effects, 341–344,
      365
S coefficient, 51–52, 57–58
   dry color concentrates, 285

SCORIM process, metallic pigment molding, 221–223
Screw slippage, liquid color concentrate distribution, 290
Secondary pigment pastes, metallic pigments, 214
Shaping process, metallic pigments, 224–225
Shear heating, color durability testing, 327
Shearing force, carbon black pigment dispersion, 165
Shear thinning, liquid color concentrate rheology, 295–296
Shewhart's *plan-do-check-act* cycle, plastics coloring statistics, 392
Shipping systems, color concentrates, 306–307
Shrinkage, color product testing, 336–337
Sienna pigments, structure and properties, 129
Significant New Use Rules (SNUR), 364–365
Silica:
  fluorescent pigments, plateout reduction, 254–255
  lead chromate/molybdenate pigments, 135–136
Silver, thermochromic materials, 197–200
Single-constant systems, Kubelka-Munk color matching theory, 50–51
Single-pigment dispersions (SPDs):
  additive coloring effects, 344–346
  dry color concentrate, 277
Single-screw extrusion (SSE):
  color compounding, 269–270
  dry color concentrates, 280–281
Singlet state excitation, fluorescence and, 244–245
Smectic mesophase, 198
Society of Automotive Engineers (SAE), SAE J1885 color durability standard, 328
Society of Dyers and Colourists (SDC), 101
Solid-state photodetector, spectrocolorimeters, 31–32
Solid waste treatment, concentrate production, 312–313
Solubility resistance:
  colorants, 88
  dry color concentrates, particle size issues, 282–283
Soluble dyes:
  categories, 178–179
  chemical constitution, 180–181
  defined, 175–176
  fastness properties, 181–182
  fluorescent dyes and brighteners, 180
    applications, 250–251
    fluorescein, 246–247
    rhodamine A, 247–250
    rhodamine B, 247
  historical background, 176–178
  plastics applications, 175–176
  plastics coloring, 178
  properties, 98
  thermochromic materials, 199–200
  thermoplastic applications, 182–184
Solvent dyes, properties, 178–179
Solvent-free salt attrition, copper phthalocyanine blue, 115
Sparkle pigments, properties, 97
Special effect colorants, dry color concentrates, 286
Specimen preparation and presentation, spectrocolorimetry performance verification, 34
Spectral analyzer, spectrocolorimetry, 29–31
Spectral bandwidth:
  color matching, *vs.* tristimulus matching, 56–57
  spectrocolorimetry specifications, 30–31
Spectral distribution:
  blackbodies, 8
  cool white fluorescence, 8
  light source, 6–8
  line sources, 9
Spectrocolorimetry:
  calibration and standardization, 31–33
  defined, 28–29
  design and specifications, 29–31
  limitations of, 44–45
  performance verification, 33–34
  product color tolerances, 35
  tolerance region derivation, 35–36
Spectrophotometry:
  color observation, 16
  computerized color matching, 48–49
  plastics coloring statistics, 380
Spherical shape, metallic pigment production, 210–211
Spinel compounds, complex inorganic pigments (CIPs), 130–133
Spiral flow test, color rheology, 335
Spiro[1,8*a*]dihydroindolizines (DHIs), photochromic behavior, 204–205

Spirooxazines (SOs), photochromic behavior, 191–196
Spiropyrans:
  photochromic behavior, 189–196
  thermochromic properties, 199–200
Stability properties, colorant properties, 89
Standard Industrial Classification (SIC) system, 316
Standardization:
  regulatory controls and standards, concentrate processing, 313–318
    Consumer Product Safety Commission (16 CFR 1000-1799), 313
    environmental protection regulations, 315–318
    Food and Drug regulations (21 CFR), 314
    Occupational Safety and Health Administration (OSHA) (29 CFR), 314–315
  spectrocolorimeters, 31–33
Statistical problem solving (SPS), plastics coloring statistics, 396–401
Statistical process control (SPC), 379, 383–384
Statistical quality control (SQC), 384
Statistics:
  defined, 379
  plastics coloring:
    batch-to-batch variability, 383–384
    exponentially weighted moving average, 392–396
    Hunter, CIE, and CMC control processes, 388–392
    process measurement systems analysis, 384–387
    research background, 379–383
    statistical problem solving (SPS) applications, 396–401
    test variability, 387–388
Stearates, fluorescent pigment additives, 253–254
Stilbene compounds, photochromic behavior, 195–196
Stobbe condensations, photochromic behavior, 193–196
Stoke's Law:
  fluorescence, 243–244
  liquid color concentrate rheology, 295–296
Storm water management, concentrate production, 312

Stranding problems, dry color concentrate manufacture, 282
Strength properties, color product testing, 332–334
Strontium compounds, thermochromic materials, 198–200
Styrene-butadiene rubber (SBR), soluble dye manufacturing, 177–178
Sulfate process, titanium dioxide pigment production, 147–148
Superfund Amendment and Reauthorization Act (SARA):
  concentrate production regulations, 317
  historical background, 361
Surface texture:
  color difference metrics, 38
  metallic pigments, 212–213
  titanium dioxide pigment:
    durability and photochemistry, 152–157
    modification process, 148–150
  visual color matching, polymer systems, 66
Surfactant demand, titanium dioxide pigment incorporation, 151–152
Surrounding, color in context of, 20–21
Synthetic iron oxides, structure and properties, 129–130

T
Talc, pearlescent pigments from, 240
Target selection:
  colorant application technology, 262–263
  color matching, 262–263
Tautomerism, photochromic behavior, 195–196
Temperatures:
  colorant application technology, 261
  metallic pigment incorporation, 220
10 degree observer, 4
Tensile flexural properties, color product testing, 332–334
Tensile properties, ISO test procedures, 80
Tensile strength:
  carbon black pigments, 171
  color product testing, 332–334
  liquid color concentrate, 298
Testing procedures:
  colorant evaluation, 98–99
  plastics coloring statistics, variability, 387–388
2,3,4,4-Tetrachloro-1-(4H)-naphthalenone, photochromic materials, 185–196

Textiles, fluorescent dyes in, 250
Thermal processing:
    color compounding, 269–270
    concentrate production, workplace exposures and controls, 310–311
    photochromic behavior, 185–196
Thermochromic materials:
    spectrocolorimetry performance verification, 34
    structure and properties, 196–200
Thermoplastics:
    color compounding, 270–271
    pearlescent pigments, 237–240
    shrinkage testing, 336–337
    soluble dyes, 182–184
    strength properties testing, 332–334
Thermosets, soluble dyes, 183–184
Thioindigoid reds/violets, properties of, 92
33/50 Initiative (EPA), 367–368
Three-dimensional color space:
    properties, 16–17
    visual color matching, 68–72
Three laws of fluorescence, 243–246
Three-roll mill, liquid color concentrate manufacturing, 288–290, 311
Threshold limit values (TLVs):
    defined, 378
    historical background, 362
Tint strength:
    colorants, 86–87
    complex inorganic pigments (CIPs), 133
    dry color concentrate quality testing, 284–285
Titanium dioxide pigments (titania pigments):
    fluorescent pigment lightfastness, 257
    future applications, 157
    particle size distribution and optics, 150
    pearlescent pigment production, 233–235
    photochemistry and durability, 152–157
    polymer incorporation, 150–152
    properties of, 94
    refractive index, additive effects, 342–344, 365
    research background, 146–147
    sources and occurrence, 147–148
    surface modifications, 148–150
Tolerance regions:
    color difference metrics, 36–44
    colorimetric derivation, 3536
Tolyl orange, organic structure and properties, 122–123

Toxic and hazardous substances standards:
    concentrate processing, 314–315
    Toxic Substances Control Act (TSCA), 317
Toxicity Characteristic Leaching Protocol (TCLP):
    cadmium sulfide/sulfoselenide pigment regulations, 374
    defined, 378
    heavy metal colorants, 373
    lead chromate/molybdenate pigment regulations, 373–374
Toxicity properties:
    colorants, 88–89
    fluorescent pigments, 255–256
Toxic Release Inventory (TRI) (EPA), 360, 366–367
Toxic Substances Control Act (TSCA):
    current status, 364–365
    defined, 378
    standards of, 331
Translucent resins, additive coloring effects, 344–346
Transmission electron micrographs, titanium dioxide pigments, 149
Transmission properties:
    light object, 10
    visual color matching, 74–76
Transmittance calculations:
    color matching and, 59
    Lambert/Beer color matching theories, 52
Transparent color matching, basic principles, 74–76
Transparent resins, additive coloring effects, 344
Triad components:
    color rendition and metamerism, 21
    defined, 5
    gloss/surface structure, 21
    light source, 5–7
    object, 9–13
    observer, 12–16
Triazole compounds, photochromic behavior, 195–196
Tribological properties, color compounding, 268–269
Trichromatic, defined, 17–20
Triplet state of electrons, fluorescence and, 245–246
Tristimulus values (CIE):
    color difference metrics, 36

color matching, *vs.* spectral techniques, 56–57
defined, 17–19
Turbid-media theory, color matching, 52–53
TV waves, 5
Twin-screw extrusion (TSE):
color compounding, 269–270
dry color concentrates, 280–281
Two-constant systems:
Kubelka-Munk color matching theory, 50–51
many-flux color matching *vs.*, 54
2 degree observer, 4
Two-dimensional color space, visual color matching, 68, 71–72
Two-tone pigments, pearlescent pigments, 235

U
Ultramarine pigments:
blues/violets/pinks, properties of, 96
inorganic structure and properties, 137–138
Ultraviolet (UV) radiation, 5
fluorescent pigment lightfastness, 256–257
metallic pigments, 204
photochromic behavior, 186–196
spectrocolorimetry, spectral analyzer design, 30–31
Ultraviolet (UV) stability:
additive coloring effects, 350
carbon black pigments, 166–167
dispersability *vs.*, 168
dry color concentrate carrier systems, 279
Umber pigments, structure and properties, 129
Universal concentrate, dry color blending, 277–278
Upper control limit (UCL), 393–396

V
Vacuum thermoforming, metallic pigment processing, 218
Value, colorant properties, 86
Vat dyes:
properties, 179
reds, structure and properties, 109–110
yellows/oranges, properties of, 91
VERSATINS, optical properties, 251
Vicat softening temperature, ISO test procedures, 80

Viewing angle, pearlescent pigments, 231–232, 250
Viscosity properties. *See also* Rheology
carbon black pigment on resins, 173
dry color concentrate carrier systems, 279
liquid color concentrates, 295–296
Visible spectrum, 5
Visual colorimetry:
analog simulation, 25–27
defined, 24–25
digital simulation, 27–28
product color tolerances, 35
schematic, 25
tolerance region derivation, 35
Visual color matching:
end-use applications, 66–67
evaluation protocols, 49
plastic materials:
basic principles, 62, 64, 68–72
blue color match example, 74
color standard, 64–65
gray color match example, 72–73
light source, 65–66
maroon color match example, 73
mixing process, 67
polymer system, 66
red color match example, 73
scientific principles, 67–68, 71–72
Volatile organic compounds (VOCs), concentrate processing, workplace exposures and controls, 307–311
Volumetric metering, colorant application technology, 264
Voluntary Protection Programs (VPP) (OSHA), 370–371

W
Warpage, color product testing, 337–338
Washout, pigment dispersion testing, 322–325, 349–350
Waste oil, concentrate production, 313
Waste products, concentrate production, 313
Wastewater treatment, concentrate production, 312
pretreatment residues, 313
Wavelength:
spectrocolorimetry calibration and standardization, 32–33
Stoke's Law, 243–244

Wax pigments:
  dry color concentrate carrier system, 278–280
  liquid color concentrate *vs.*, 296–297
Weather resistance:
  carbon black pigments, 167–168
  colorants, 88
  color durability testing, 328–331, 352–353
Weigh-out techniques:
  concentrate pellet products, 303–304
  concentrate processing, workplace exposures and controls, 309
  dry color concentrates, 303
  liquid color concentrates, 305
Wet milling process, metallic pigment production, 206–207
White pigments, titania pigments:
  future applications, 157
  particle size distribution and optics, 150
  photochemistry and durability, 152–157
  polymer incorporation, 150–152
  properties of, 96
  research background, 146–147
  sources and occurrence, 147–148
  surface modifications, 148–150
White vinyl test, color migration, 332
Woodward-Hoffman rules, photochromic behavior, 186–196
Worker safety regulations, historical background, 358–362
Workplace exposures and controls, concentrate processing, 311–313
  nonproduction areas, 308–309
  production areas, 309–311
Wuerz synthesis, metallic pigment incorporation, 220

X
Xenon arc lamps, spectrocolorimetry, 29–31
X-rays, 5
XYZ tristimulus values (CIE):
  analog colorimetry, 26
  defined, 17–19
  plastics coloring statistics, 388–392

Y
Yellow pigments:
  complex inorganic pigments (CIPs), 130–133
  organic structure and properties, 116–121
    benzimidazolone yellows, 118
    diarylide yellows, 117–118
    diazo condensation yellows, 119, 121
    heterocyclic yellows, 118–120
    metallized azo yellows, 117
    monoarylide yellows, 116
  synthetic iron oxides, 130
Ylide stability, photochromic behavior, 186–196
Y_{xy} color space, 17–18

Z
Zinc sulfide/oxide pigments, 97
  cadmium pigments, 133–134
Zircon, complex inorganic pigments (CIPs), 133